DYNAMICS OF CANCER

Mathematical Foundations of Oncology

Dominik Wodarz & Natalia L Komarova
University of California, Irvine, USA

World Scientific
NEW JERSEY • LONDON • SINGAPORE • BEIJING • SHANGHAI • HONG KONG • TAIPEI • CHENNAI

Published by

World Scientific Publishing Co. Pte. Ltd.
5 Toh Tuck Link, Singapore 596224
USA office: 27 Warren Street, Suite 401-402, Hackensack, NJ 07601
UK office: 57 Shelton Street, Covent Garden, London WC2H 9HE

Library of Congress Cataloging-in-Publication Data
Wodarz, Dominik, author.
Dynamics of cancer : mathematical foundations of oncology / by Dominik Wodarz (University of California, Irvine, USA) & Natalia L. Komarova (University of California, Irvine, USA).
pages cm
Includes bibliographical references and index.
ISBN 978-981-4566-36-0 (hardcover : alk. paper)
1. Oncology--Mathematics. 2. Cancer--Treatment--Data processing. 3. Computational biology.
I. Komarova, Natalia L., author. II. Title.
RC254.W66 2014
616.99'400285--dc23

2013033157

British Library Cataloguing-in-Publication Data
A catalogue record for this book is available from the British Library.

Printed in Singapore by Mainland Press Pte Ltd.

This book is dedicated to our parents and children.

Preface

Cancer is an age related disease that leads to the outgrowth of cell populations in the body, which can result in significant morbidity and mortality. Treatment approaches are based on our understanding of the biology of specific cancers, and a great amount of work has been and is being performed to improve this understanding. Significant progress is being made with respect to identifying cellular defects, genetic and epigenetic events, and molecular pathways towards malignant growth in the context of different cancers. Apart from focusing on cellular characteristics, the interactions of tumor cells with other cells in their vicinity, the microenvironment, is thought to be a crucial component in carcinogenesis. Experiments and clinical studies have elucidated many key mechanisms that underlie the development of cancer and that can be potentially exploited by treatments.

However, another approach to studying the development and progression of cancer has emerged and is becoming increasingly important: the field of mathematical oncology, which uses mathematical and computational techniques to study how cancer cells get generated, grow, and evolve. The concept of evolution plays a special role in this context. Cancers arise and progress by accumulating various mutations and epigenetic changes, which gives them a selective advantage over healthy cells, and lead to growth and pathogenesis. Mutation and selection, the hallmarks of evolution, are central components in the development of cancer. Mathematical work, in turn, has played a central role for elucidating evolutionary principles.

This book provides an introduction to the field of mathematical oncology, explaining some central mathematical and computational techniques that are at the core of this field, and highlighting how mathematical analysis has provided valuable biological insights into concepts and data. The book is not meant to be a comprehensive survey of the field, which has grown to be rather large. Instead it aims to explore a few topics in depth, with an emphasis on evolutionary aspects of cancer. The book is based on our own work. It provides the reader with a solid technical background to understand how mathematical results have been obtained, and shows how these mathematical results can by applied to advance biology.

The book is multi-faceted, and can be used in a variety of settings. Because it explains in detail the mathematical techniques, it can be used as a textbook for applied mathematics students learning aspects of mathematical biology, as well as as a reference for people that are interested in the mathematical techniques. As the book highlights biological insights and advances, it can also be used by students in population and evolutionary biology who are interested in biomedical applications. Similarly, it can serve as a summary for cancer biologists who are interested in learning about some of the biological lessons that have resulted from mathematical models.

We have tried to write the book in such a way that it is accessible and useful both for a mathematical and a biological readership by marking with gray background all the materials that are more suited for mathematicians and that might be difficult to understand for a biological readership without a mathematical background. It is the hope that future generations of scientists will be well-trained at the interface between mathematics and biology, such that there will be fewer communication issues between biologists and mathematicians. It is a field that is rising at a fast pace, and our book aims to contribute to the interdisciplinary education of the new generation of researchers.

We would like to thank a number of people, whose input and discussions have shaped the material that is discussed in this book: David Axelrod, Rick Boland, Hung Fan, Steve Frank, Ajay Goel, Andy Hofacre, Yoh Iwasa, Vincent Jancen, David Krakauer, John Lowengrub, Ignacio Rodriguez-Brenes, Alex Sadovski, Akira Sasaki, Fred Wan, and Ryan Zurakowski. We also thank members of our group, Alen Katouli, John Lau, Leili Shahriyari, Zhiying Sun, and Erin Urwin. We would like to especially thank Martin Nowak, whose guidance has shaped the work of both authors.

Dominik Wodarz and Natalia Komarova

Contents

Chapter 1

Teaching guide

This book can be used for teaching upper-division undergraduate of graduate classes in Cancer modeling. In this chapter we provide some hints on how to organize a course based on this book's materials.

1.1 How to use this book

The book is written with two audiences in mind. Those who have a solid mathematics background will find very detailed mathematical derivations of all the results. These are printed on a **gray background**.

In order to follow the "gray" parts of the book, we require some basic knowledge of applied mathematical techniques, such as solutions of linear systems of ODEs, linear stability analysis of ODEs, the method of characteristics for solving first order PDEs, and basic probability.

These parts of the book can be skipped by readers that have less of a mathematical background, without interrupting the logical flow of the exposition.

Each chapter contains a set of problems. There are three kinds of problems. (1) Basic problems are mathematical exercises aimed at a more detailed understanding of the mathematical derivations given in the text, or reviewing some concepts needed to understand the "gray" parts of the book. (2) "Numerical projects" are more advanced problems which require some computer coding. They require some knowledge of a programming language, such as Mathematica, Maple, Matlab, C++, Fortran, etc. These two types of problems should be offered in a course taught to an audience with some applied mathematics background. (3) Finally, the third type of problems termed "Research projects" do not require any knowledge of mathematics, and suggest topics of independent in-depth study of a biological topic or the history of a certain concept or discovery. These can be offered in a course taught to students with a biology background.

The book contains three parts. Part 1, "Basic growth dynamics and deterministic models", introduces concepts of growth (including single-species growth and two-species competition), and also shows how these concepts can be relevant for studying cancer. It further introduces the important issue of genetic instability and talks about angiogenesis, inhibitors, and promoters. Almost all the mathematical developments in this part of the book are deterministic. Important concepts that come up are exponential and logistic growth, two-species competition and stability analysis, axiomatic modeling, quasispecies equations, and optimization.

Part 2, "Evolutionary dynamics and stochastic models", introduces stochasticity in the description of cancer. With this powerful tool, it is possible to model cell population dynamics at low numbers, which is important when talking about cancer initiation (Chapters 9-14). In this context, we discuss oncogenes and tumor suppressor genes, sporadic and familial cancers, and stem cells. Stochasticity is also essential when talking about cancer treatment, in particular, the generation of resistance against drugs (Chapters 15-16).

Finally, Part 3, "Advanced topics", uses stochastic and deterministic methods, together with computational tools, to explore a variety of topics, such as epigenetic mechanisms, telomeres, gene therapy and oncolytic virus therapy, and immune responses.

A 30-hour course aimed at students with some mathematical background will normally only cover sections in Parts 1 and 2 of the book. A course taught in a biology department will be able to cover all three parts of the book.

1.2 A sample syllabus for a Mathematics course

Below we provide a sample syllabus for a 30-hour course aimed at Mathematics students:

(1) Chapter 2: Cancer and somatic evolution.
(2) Chapter 2, continued: Overview of cancer biology.
(3) Chapter 3: Mathematical modeling techniques.
(4) Chapter 4: Different growth laws for tumors.
(5) Chapter 5: Two-species growth.
(6) Chapter 5, continued: Axiomatic modeling.
(7) Chapter 6: Competition between genetically stable and unstable cells.
(8) Chapter 6, continued: Quasispecies equations.
(9) Chapter 7: Chromosomal instability.
(10) Chapter 7, continued: The optimization problem.
(11) Chapter 8: Angiogenesis, inhibitors, and promoters.
(12) Chapter 15: A birth-death process with mutations.
(13) Chapter 15, continued: Modeling cancer drug treatments.
(14) Chapter 15, continued: Example: the case of two drugs.
(15) Review.
(16) Midterm.
(17) Chapter 16: CML treatment with small molecule inhibitors.
(18) Chapter 16, continued: Tumor stem cells.
(19) Chapter 16, continued: Cyclic treatments vs combination treatments.
(20) Chapter 9: Oncogenes and gain-of-function mutations.
(21) Chapter 9, continued: Moran process and mutant fixation.
(22) Chapter 10: Tumor-suppressor genes and loss-of-function mutations.
(23) Chapter 10, continued: Two-step process and stochastic tunneling.
(24) Chapter 11: Chromosomal instability and colon cancer initiation.
(25) Chapter 11, continued: Sporadic and familial cancers.
(26) Chapter 12: Modeling hierarchical populations.
(27) Chapter 12, continued: symmetric and asymmetric divisions of stem cells.
(28) Chapter 14: Spatial tumor dynamics.
(29) Advanced topics: an overview.
(30) Review.

1.3 A sample syllabus for a Biology course

Below we provide a sample syllabus for a 30-hour course aimed at Biology students:

(1) Chapter 2: Cancer and somatic evolution.
(2) Chapter 2, continued: Overview of cancer biology.

two hits, although a single mutation might result in reduced function in the context of certain repair genes.

2.3 Multi-stage carcinogenesis and colon cancer

Cancer initiation and progression involves the sequential accumulation of mutations, most importantly in tumor suppressor genes and in oncogenes. A case study where this is understood in some detail so far is colorectal cancer [Goel and Boland (2010)]. The colon consists of a collection of so-called *crypts*.

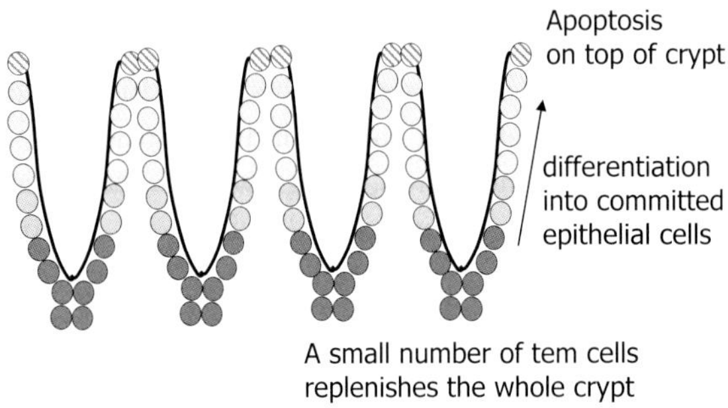

Fig. 2.2 Schematic diagram of crypts in the colon.

Crypts are involutions of the colonic epithelium (figure 2.2). Stem cells are thought to be located at the base of the crypts. These are undifferentiated cells which can keep dividing and which give rise to differentiated epithelial cells. The division patterns of tissue stem cells have been debated. In fruit flies, tissue stem cell division tends to be asymmetric, that is, stem cell division creates one new stem cell and one cell which embarks on a journey of differentiation [Wodarz (2005a); Wodarz and Huttner (2003); Hirth (2011)]. In this way, the stem cell population is maintained while also giving rise to differentiated cells. This mode of division is also thought to occur in humans, although it has been increasingly recognized that another pattern of cell division may be more important here [Simons and Clevers (2011a); Mascré *et al.* (2012); Snippert *et al.* (2010); Lopez-Garcia *et al.* (2010); Baumann (2010)]. It is thought that the type of daughter cells arising from a tissue stem cell division is determined probabilistically. For example, 50% of divisions lead to two daughter stem cells, while the other 50% of stem cell divisions give rise to two cells that are on the pathway of differentiation. On average, this also maintains a constant stem cell population while replenishing the pool of differentiated tumor cells. The differentiating cells travel up the crypt, perform their function, and die by apoptosis after about a week. Because the epithelial cells are relatively short lived, stem cell division has to give rise to new differentiated cells continuously in order to replenish the tissue. For this process to function in a healthy way, it is crucial that the differentiated cells die by apoptosis. If this cell death fails, we observe an accumulation of transformed cells around the crypts, and this gives rise to a mass of cells called a *dysplastic crypt* (figure 2.3).

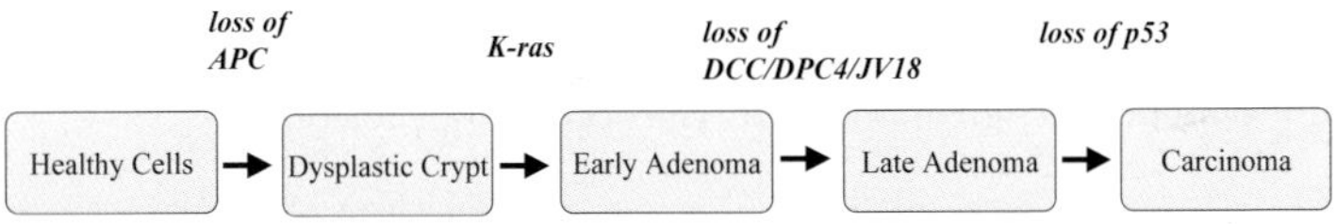

Fig. 2.3 Diagram describing the multi-stage progression of colon cancer. Drawn according to [Kinzler and Vogelstein (2002)].

This is the first stage of colon cancer [Vogelstein and Kinzler (2002)]. In molecular terms, the death of differentiated cells is induced by the APC gene. APC is a tumor suppressor gene. Data suggest that the majority of colon cancers are initiated through the inactivation of the APC gene (figure 2.3). A dysplastic crypt is also sometimes referred to as a polyp. As a subsequent step, many colon cancers activate the oncogene K-ras which allows the overgrowth of surrounding cells and an increase in the size of the tumor. This stage is called the early adenoma stage (figure 2.3). In more than 70% of the cases, this is followed by the loss of chromosome 18q which contains several tumor suppressor genes including DDC, DPC4, and JV18-1/MADR2. This results in the generation of late adenomas (figure 2.3). In the further transition from late adenoma to the carcinoma stage, p53 is typically lost in more than 80% of the cases (figure 2.3). Further mutations are assumed to occur which subsequently allow the colon cancer cells to enter the blood system and metastasize. Note that this sequence of event is not a hard fact, but rather a caricature. The exact details may vary from case to case, and new details emerge as more genetic and epigenetic research is performed (see below).

This is a good example of cells acquiring sequential mutations in a multi-step process while they proceed down the path of malignancy. This gives rise to an important question. The multi-step process requires many mutations. The inactivation of each tumor suppressor gene requires two mutations, and the activation of each oncogene requires one mutation. The physiological mutation rate has been estimated to be 10^{-7}per gene per cell division. Is this rate high enough to allow cells to proceed through multi-stage carcinogenesis during the life-time of a human? Some investigators argue that the process of clonal expansion involves a sufficient number of cell divisions in order to account for the accumulation of all the mutations [Tomlinson *et al.* (1996); Sieber *et al.* (2003)]. A competing argument says that the accumulation of the oncogenic mutations requires a loss of repair function and the generation of mutator phenotypes (i.e. genetically unstable cells) [Loeb (1991, 2011)]. Genetic instability is a defining characteristic of many cancers. It is reviewed in the following section.

2.4 Genetic instability

Many cancer cells show a large variety of genetic alterations which range from small scale mutations to large chromosomal aberrations [Vogelstein and Kinzler (2002); Hoeijmakers (2001)]. While this is an intriguing observation, this does not prove that the cells are genetically unstable. The alterations could come about through a variety of factors, such as the exposure to extensive damage at some point in time, or specific selective conditions. Genetic instability is defined by an increased *rate* at which cells acquire genetic abnormalities [Lengauer *et al.* (1998)]. That is, cells have a defect in specific repair genes which results in higher variability. Indeed, studies have shown that many cancer cells are characterized by an increased rate at which genetic alterations are accumulated and are truly genetically unstable [Lengauer *et al.* (1997)]. Different types of genetic instabilities can be distinguished [Lengauer *et al.* (1997); Negrini *et al.* (2010); Boland *et al.* (1998)]. They can be broadly divided into two categories. Small sequence instabilities and gross chromosomal instabilities (figure 2.4).

Small sequence instabilities involve subtle genetic changes which can dramatically speed up the process of cancer progression. Defects in mismatch repair mechanisms give rise to microsatellite instability or MSI [Popat *et al.* (2005); Boland and Goel (2010); Vilar and Gruber (2010)]. This involves copying errors in repeat sequences (figure 2.4). MSI is most

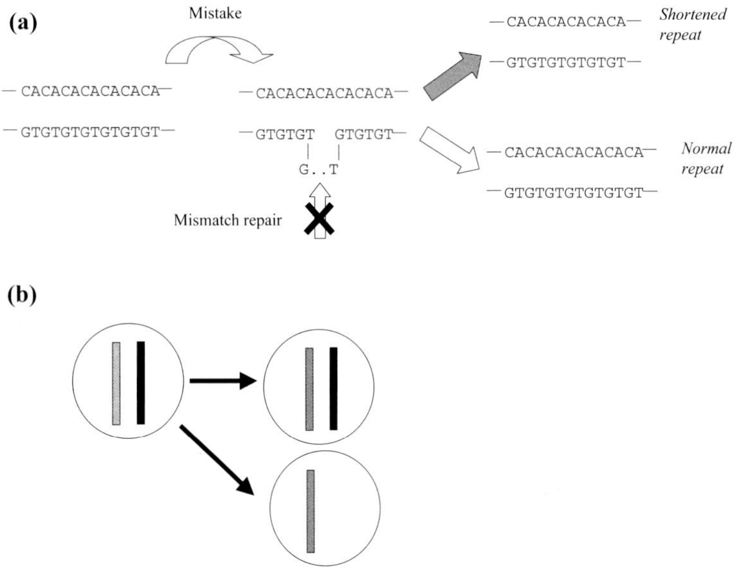

Fig. 2.4 Schematic diagram explaining the concept of genetic instability. (a) Small scale instabilities, such as MSI, involve subtle sequence changes. With MSI, mismatch repair genes are defect and this leads to copying mistakes in repeat sequences. (b) Chromosomal instability involves gross chromosomal changes, such as loss of chromosomes.

common in colon cancer. It is observed in about 13% of sporadic cases and is the mechanism of cancer initiation in the hereditary non-polyposis colorectal cancer (HNPCC). Another type of small scale instability comes about through defects in nucleotide excision repair genes. These are responsible for the repair of DNA damage caused by exogenous mutagens, most importantly ultraviolet light. It is thus most important in the development of skin cancers. A defect in such repair mechanisms has been found in a disease called *xeroderma pigmentosum*, which is characterized by the development of many skin tumors in sun exposed areas [Dworaczek and Xiao (2007); Cleaver (2005)].

Instabilities which involve gross chromosomal alterations are called chromosomal instability or CIN [Thompson *et al.* (2010); Pino and Chung (2010); Lengauer *et al.* (1998)] (figure 2.4). Cells which are characterized by CIN show a variety of chromosomal abnormalities. There can be alterations in chromosome numbers which involve losses and gains of whole chromosomes. This results in aneuploidy [Rajagopalan and Lengauer (2004)].

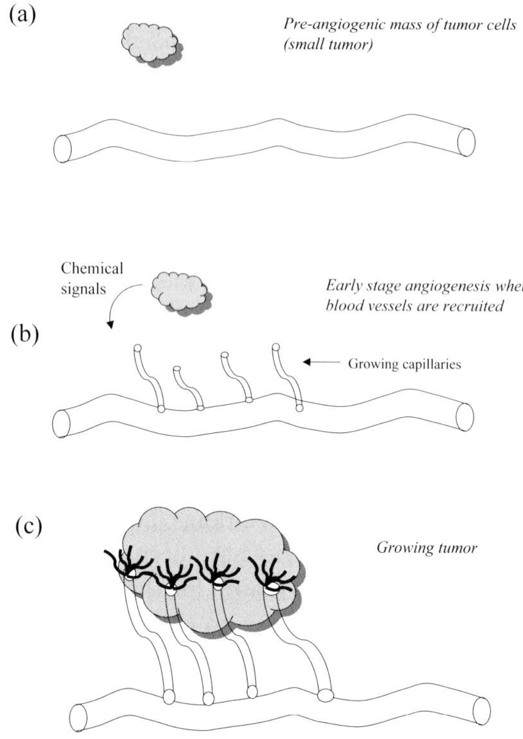

Fig. 2.5 Diagram explaining the concept of angiogenesis. (a) When a cancerous cell is created it can expand up to a small size without the need for blood supply. At this stage, the growth of an *avascular tumor* stops. (b) When angiogenic cell lines emerge, they send out chemical signals called promoters. This induces blood vessels to grow towards the tumor. (c) This process leads to the complete vascularization of the tumor, allowing it to grow to larger sizes.

ples of promoters are growth factors such as FGF, VEGF, IL-8, or PDGF. If the balance between inhibitors and promoters has been shifted sufficiently in favor of promotion, the cancer cells can grow to higher numbers and progress towards malignancy (figure 2.5). The mechanisms by which blood supply is recruited to the tumor, and the ways in which inhibitors and promoters affect cancer cells are still under investigation. New blood supply can be built from existing local endothelial cells. On the other hand, angiogenesis promoters may induce a population of circulating endothelial progenitor cells to be recruited to the local site where the blood supply needs to be built. Blood supply can affect cancer cells in different ways.

It can influence the rate of cell death. That is, in the absence of blood supply cells die more often by apoptosis as a result of hypoxia, and this is relaxed when sufficient blood supply is available. On the other hand, lack of blood supply can prevent cancer cells from dividing. In this case they remain *dormant*, That is, they do not divide and do not die.

2.6 Cellular hierarchies in cancer

When discussing the development of cancer we presented simplified scenarios, talking about cells that accumulate alterations and subsequently undergo clonal expansion. Of course, this is generally the correct picture, but it is more complex than this. Cancers are thought to be characterized by a hierarchical structure, similar to healthy tissue (figure 2.6). It is a wide-spread notion that cancers are maintained by so-called cancer stem cells, which make up a relatively small fraction of the total tumor cell population [Visvader and Lindeman (2008); Driessens *et al.* (2012); Chen *et al.* (2012); Schepers *et al.* (2012); Reya *et al.* (2001); Dalerba *et al.* (2007)]. The tumor stem cells give rise to cells that are more differentiated and that make up the bulk of the tumor and cannot maintain growth. This has important implications for a variety of aspects. Importantly, the size of the tumor is often a key variable when considering genetic diversity and responses to treatment. If a relatively large fraction of the tumor cell population cannot maintain growth, then the "effective population size" of the tumor could be a lot smaller. This in turn could imply that the tumor has a lower chance to have accumulated certain mutations that promote the disease or render the cells resistant to drugs. Population size is a crucial measure when investigating the evolutionary dynamics of cancer, and the uncertainty regarding the effective population size of the tumor cell population has to be kept in mind in the context of these considerations.

2.7 Genetic and epigenetic changes

So far, we have concentrated our discussion on genetic changes that are accumulated in cells and that allow the cells to escape tissue homeostasis and grow uncontrolled. However, epigenetic changes can also contribute to the generation of cancerous phenotypes, in particular changes in methylation patterns across the genome [Iacobuzio-Donahue (2009); Jones and Baylin (2002); Laird and Jaenisch (1996); Sharma *et al.* (2010); Esteller (2008);

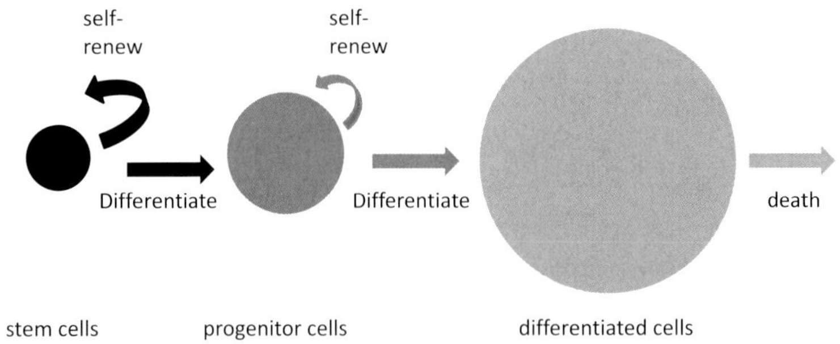

Fig. 2.6 The architecture found in healthy tissue is also thought to underlie the architecture of tumors. Healthy tissue is maintained by a small population of tissue stem cells that can self-renew. They differentiated to give rise to progenitor cells, which can also self-renew, although to a lesser extent. They eventually differentiate to give rise to terminally differentiate cells which do not have self-renewal capacity and die. A similar architecture is thought to apply to tumors. They are thought to be maintained by a relatively small population of cancer stem cells that can self-renew and seed new cancer growth. The bulk of the tumor is thought to be made up of more differentiated cells.

Sugimura and Ushijima (2000)]. Tumor cell genomes are often characterized by global hypomethylation. This has been suggested to contribute to the emergence of karyotypic instabilities as well as to the activation of oncogenes. On the other hand, CpG islands are susceptible to hypermethylation, and when it occurs in the promoter, it is associated with gene silencing and can induce the deactivation of tumor suppressor genes. Similar to the mutator phenotype concept, the "methylator phenotype" concept has emerged to account for the tendency to observe patterns of hypermethylation in certain cell lines. Tumor cell lines that are characterized by relatively high levels of CpG island methylation have been called CpG island methylator phenotypes (CIMP cells), which sets them apart from non-CIMP cells that are characterized by lower levels of CpG island methylation [Issa (2004); Nosho *et al.* (2008); Toyota *et al.* (1999); Toyota and Issa (1999)]. Both genetic and epigenetic changes are most likely involved in the processes that contribute to carcinogenesis, and they can be interconnected. An interesting example is colon cancer. As mentioned above, microsatellite instability (MSI) and chromosomal instability (CIN) can contribute the the development of colon cancer. Most cases of sporadic MSI colon cancers have been attributed to the epigenetic silencing of the mismatch repair genes *MLH1*, usually brought about by CIMP phenotypes [Toyota *et al.* (1999)]. Furthermore, a negative correlation has been found between the CIN and CIMP phenptypes in colon cancer [Goel *et al.* (2007); Cheng *et al.* (2008)], indicating that they are two distinct mechanisms for driving tumor initiation that do not overlap. This picture shows that while the genetic events that can contribute to the development of colon cancer are well documented, the disease process is a lot more complex than the simple accumulation of the various mutations outlined above in this chapter. Even in cancers that are thought to be relatively well-understood, we are only beginning to elucidate some of the molecular complexities that underlie and drive carcinogenesis.

2.8 Evolutionary theory and Darwinian selection

Theodosius Dobzhansky who, according to Stephen J. Gould, was the greatest evolutionary geneticist of our times, wrote that "nothing in biology makes sense except in the light of evolution". This also applies to our understanding of cancer. The process of carcinogenesis includes all the essential ingredients of evolutionary theory [Greaves (2002, 2007); Beerenwinkel *et al.* (2007); Wodarz (2005b); Frank and Nowak (2004); Merlo *et al.* (2006)]: reproduction, mutation, and selection (figure 2.7). Mathematical models play a crucial role for analyzing such evolutionary processes [Kimmel (2010)].

As outlined in detail above, the process of cancer initiation and progression is concerned with the accumulation of genetic and epigenetic changes which allow the cells to break out of normal regulatory mechanisms. Such cells will grow better than healthy cells and are advantageous. In evolutionary terms, they are said to have a higher *fitness*. The more oncogenic alternations the cells acquire, the better they are adapted to growing in their environments, and the higher their fitness. Cancer cells which grow best can be selected for and can exclude less fit types. Cancer cells can even adapt their "evolvability": genetically unstable cells may be able to evolve faster and adapt better than stable cells. This can be very important in the face of many selective barriers and changing environments. Barriers can include inhibitory effects which are exerted by the tissue microenvironment, or an adaptive immune system which can specifically recognize a variety of tumor proteins and mount new responses as the tumor evolves. The environment in our bodies can change over time and render different genotypes advantageous at different stages. An example is aging which involves the continuous rise in the rate of DNA damage as a result reactive oxygen species which are produced as a byproduct of metabolism [Benz and Yau (2008)].

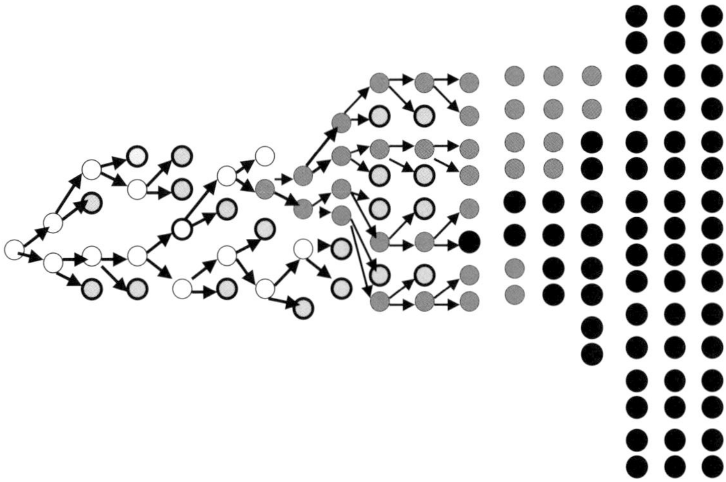

Fig. 2.7 Diagram explaining the concept of somatic evolution and cancer progression. Cancer originates with the generation of a mutant cell. This cell divides and the population grows. This is called clonal expansion. Further mutations can subsequently arise which have a higher fitness. They grow and expand further. Consecutive mutations and rounds of clonal expansion allow the cancer to grow to ever increasing sizes.

Chapter 3

Mathematical modeling of tumorigenesis

Broadly speaking, there are three major areas where theory has contributed the most to cancer research:

(i) **Modeling in the context of epidemiology and other statistical data.** One of the oldest and most successful methodologies in theoretical cancer research is using the available incident statistics and creating models to explain the observations. This field was originated by [Armitage and Doll (1954)], and then taken to the next level by Moolgavkar and colleagues [Moolgavkar and Knudson (1981)].

(ii) **Mechanistic modeling of tumor growth, including multiscale modeling.** An entirely different approach to cancer modeling is to look at the mechanistic aspects of tumor material and use physical properties of biological tissues to describe tumor growth, see [Preziosi (2003); Cristini and Lowengrub (2010)] for review.

(iii) **Modeling of cancer initiation and progression as somatic evolution.** In this area of research, methods of population dynamics and evolutionary game theory are applied to study cancer. First developed by ecologists and evolutionary biologists, these methods have been used to understand the collective behavior of a population of cancer cells, see [Gatenby and Gawlinski (2003); Gatenby and Vincent (2003b)].

In this chapter we review basic mathematical tools necessary to undertake different types of cancer modeling. These are: ordinary differential equations, partial differential equations, stochastic processes, cellular automata and agent based modeling.

and Li (2004)], latin hypercube sampling on large ranges of the parameters is employed, which is a method for systems with large uncertainties in parameters. This involves choosing parameters randomly from a range and solving the resulting system numerically, trying to identify the parameters to which the behavior is the most sensitive. In the paper by [Evans *et al.* (2004)], "structural identifiability analysis" is discussed, which determines whether model outputs can uniquely determine all of the unknown parameters. This is related to (but is not the same as) the confidence with which we view parameter estimation from experimental data. In general, questions of robustness and reliability are studied in *mathematical control theory*.

In this book, several chapters use ODEs as their main method of description. Most of Part 1 of the book ("Basic growth dynamics and deterministic models") uses ordinary differential equations as the method of choice. ODEs are also used in some of the advanced topics.

3.2 Extensions of ODE modeling

3.2.1 *Optimal control*

A particular branch of research related to ODEs that we would like to mention is optimal control theory. Mathematical theory of optimal control has been used in several areas of biosciences [Sontag (2004); Lenhart and Workman (2007)]. In the context of oncology, optimal control theory has been employed to design treatment strategies by methods of optimization [Swan (1990); Kirschner *et al.* (1997); Ledzewicz and Schättler (2007, 2008); Ledzewicz *et al.* (2012); Engelhart *et al.* (2011)].

One particular area of application concerns the interactions between cancer and the immune system [De Pillis and Radunskaya (2001, 2003); De Pillis *et al.* (2007)]. Typically, the problem is formulated as a system of ODEs for the number of cells in a colony obeying a birth-death dynamics. The growth and/or death parameters are affected by treatment, which is described as an unknown function of time whose values reflect the amount of drug administered. The goal is to find the treatment function which in some sense optimizes an "objective function", which for example could be equivalent to minimizing the number of cancer cells at a certain moment of time. The mathematical theory for finding the optimal strategy has been developed in the literature and allows to use analytical methods such as the Pontryagin Maximum Principle [Pontryagin *et al.* (1962)].

In this book, control theory is used in Chapter 7, but it is not applied to treatment design. Instead, we use control theory to argue about different strategies adopted by cancer in the context of chromosomal instability.

3.2.2 *ODEs and cancer epidemiology*

The idea is the following. A multi-stage model of carcinogenesis is formulated as a stochastic process, which includes a series of mutational events and clonal expansions. The mutation rates, the average rates of clonal expansions for each stage, and even the number of stages are variables of the model. Then, the probability of developing cancer by a certain age is calculated (usually, by means of numerical simulations), as a function of all the unknown parameters. The outcome of such calculations, for each set of parameters, is then compared with the existing data on cancer incidence, and the set of parameters which gives the best fit is identified. In their papers, [Luebeck and Moolgavkar (2002); Meza *et al.* (2008)] use the data on the incidence of colorectal cancers in the Surveillance, Epidemiology, and End Results (SEER) registry. They conclude that the statistics are most consistent with a model with two rare events followed by a high-frequency event in the conversion of a normal stem cell into an initiated cell that expands clonally, which is followed by one more rare event. The two rare events involved in the initiation are interpreted to represent the homozygous loss of the APC gene.

Several authors have analyzed age-incidence curves [Frank (2004, 2007); Hornsby *et al.* (2007)] and death statistics [Filoche and Schwartz (2004)]. In the latter paper, the statistics of fluctuations in cancer deaths per year lead to an intriguing discovery: there is a big difference between cancers of young ages and cancers after 40. The authors suggest that cancers attacking older people behave like "critical systems" in physics and can be considered as an avalanche of "malfunctions" in the entire organism.

3.3 Partial differential equations

The next method that we will mention here is partial differential equations (PDEs). In many cases, this is the tool of choice when studying tumor growth and invasion into surrounding tissue. In many models, tumor tissue is described as a mechanistic system, for instance, as a fluid (with a production term proportional to the concentration of nutrients) [Evans *et al.*

give several examples where various stochastic methods are used to describe tumorigenesis.

The process where the stochastic nature of events can be seen very clearly is the accumulation of mutations. This process is central to cancer progression, and therefore developing tools to describe it is of vital importance for modeling. In the simplest case, we can envisage cell division as a binary (or *branching*) process, where at regular instances of time, each cell divides into two identical cells with probability $1-u$, and it results in creating one mutant and one wild type cell with probability u. To complete the description of this simplified model, we assume that a mutant cell can only give rise to two mutant daughter cells. Let us start from one wild type cell and denote the number of mutants at time n as z_n. The random variable z_n can take nonnegative integer values; another way to say this is that the state space is $\{0\} \cup I$. This is a simple branching process, which is a discrete state space, discrete time process. We could ask the question: what is the probability distribution of the variable z_n? Possible modifications of this process can come from the existence of several consecutive mutations, a possibility of having one or both daughter cells mutate as a result of cell division, allowing for cell death, or from distinguishing among different kinds of mutations. As an example of a branching process type model, we will mention the paper by [Frank (2003)] which addressed the accumulation of somatic mutation during the embryonic (developmental) stage, where cells divide in a binary fashion, similar to the branching process. Two recessive mutations to the retinoblastoma locus are required to initiate tumors. In this paper, a mathematical framework is developed for somatic mosaicism in which two recessive mutations cause cancer. The following question is asked: given the observed frequency of cells with two mutations, what is the conditional frequency distribution of cells carrying one mutation (thus rendering them susceptible to transformation by a second mutation)? *Luria-Delbruck*-type analysis is used to calculate a conditional distribution of single somatic mutations.

Next, we consider another important process, the *birth and death process*. Suppose that we have a population of cells, whose number changes from time t to time $t + \Delta t$, where Δt is a short time interval, according to the following rules:

- With probability $L\Delta t$ a cell reproduces, creating an identical copy of itself,
- With probability $D\Delta t$ a cell dies.

All other events have a vanishingly small probability. The number of cells, $x(t)$, can take positive integer values, and it depends on the continuous time variable. That is, it can change at any time, and not just at prescribed intervals. Therefore, this is a continuous time, discrete state space process. One obvious modification to the above rules is to include mutations. Say, instead of $L\Delta t$, we could have the probability $L(1-u)\Delta t$ to reproduce faithfully, and probability $Lu\Delta t$ to create a mutant. Further, we could consider a chain of mutations, and describe the evolution of the number of cells of each type. This resembles Moolgavkar's description of multistage carcinogenesis [Moolgavkar and Knudson (1981)] which is discussed in Chapter 10.

In the birth-death type processes, the population of cells may become extinct, or it could grow indefinitely. Another type of process that is very common in tumor modeling corresponds to constant population size. An example is the Moran process. Whenever a cell reproduces (with the probability weighted with the cell's fitness), another cell is chosen to die to keep a constant population size. If we include a possibility of mutations (or sequences of mutations), which lead to a change of fitness in cells, we can model the emergence and invasion of malignant cells. Models of this kind are relevant for the description of cellular compartments [Komarova *et al.* (2003b)] or organs of adult organisms. In a series of stochastic models, [Frank and Nowak (2003)] discussed how the architecture of renewing epithelial tissues could affect the accumulation of mutations. They showed that a hierarchy of stem cells could reduce the accumulation of mutations by the mechanism that they term *stochastic flushing*. They assume that each compartment retains a pool of nearly quiescent proto-stem cells. The renewal of tissue happens in the usual way by stem cell divisions. If a stem cell dies, it is replaced from the pool of proto-stem cells.

Stochastic models of stem cell dynamics have been proposed by several authors, in particular, in the context of the hematopoietic system [Dingli *et al.* (2007)]. [Nowak *et al.* (2003)] employ a *linear process* of somatic evolution to mimic the dynamics of tissue renewal. There, cells in a constant population are thought to be put in a straight line. The first one is the symmetrically dividing stem cell, which places its offspring next to itself and moves the other cells by one position. The last cell is taken out of the system. This process has the property of canceling out selective differences among cells yet retaining the protective function of apoptosis. It is shown that this design can slow down the rate of somatic evolution and therefore delay the onset of cancer. A different constant population model is employed by [Calabrese *et al.* (2004); Kim *et al.* (2004)], where precancerous mutations in colon stem cell compartments (*niches*) are studied. Each niche contains multiple stem cells, and niche stem cells are lost at random with replacement. It is assumed that each stem cell can either divide asymmetrically, or give rise to two stem cells, or to two differentiated cells. This loss and replacement dynamics eventually leads to the loss of all stem cell lineages except one. The average time to cancer is calculated with this model, using five successive mutational steps. The results are compared with the existing age-incidence statistics. We will briefly mention statistical methods at the end of this chapter.

In this book, most chapters in Part 2 ("Evolutionary dynamics and stochastic models") involve stochastic modeling of the type described here (except Chapter 14 which uses cellular automata). All the models discussed so far describe stochastic effects, but do not include spatial structures of populations. The next type of models can do both.

3.5 Cellular automaton models

Traditionally, cellular automaton models are based on a spatial grid, where the dynamics are defined by some local rules of interaction among neighboring nodes. The interaction rules can be deterministic or stochastic (that is, dictated by some random processes, with probabilities imposed). Each grid point may represent an individual cell, or a cluster of cells; for simplicity we will refer to them as "cells". To begin, we present a very simple model of tumor growth which illustrates the method. We start from a rectangular, two-dimensional grid. Let us refer to a grid point as x_{ij}, where i and j are the horizontal and vertical coordinates of the point. Each node, x_{ij}, can be a healthy cell, a cancer cell, or a dead cell. We start from an initial distribution of tumor cells, healthy cells and dead cells. For each time-step, we update the grid values according to some local interaction rules. Let us denote the discrete time variable as $n = 1, 2, \ldots$. Here is an example of an update rule. At each time step, we update one site, in a random order; $x_{ij}(n + 1)$ is given by the following:

- If x_{ij} is surrounded by a layer of thickness δ of tumor cells, it dies.
- If x_{ij} is a tumor cell not surrounded by a δ-layer of tumor cells, it reproduces. This means that the site x_{ij} remains a tumor cell, and in addition, one of its neighbors (chosen at random) becomes a tumor cell. As a result, all the non-dead cells on that side of x_{ij} between x_{ij} and the nearest dead cell are shifted away from x_{ij} by one position.

- If x_{ij} is a dead cell, it remains a dead cell.
- If x_{ij} is a healthy cell, it remains a healthy cell.

Of course, this is only a toy model. Cellular automata models used to describe realistic situations are more complicated as they have to grasp many aspects of tumor biology. For instance, [Kansal *et al.* (2000)] used a very sophisticated three-dimensional cellular automaton model of brain tumor growth. It included both proliferative and non-proliferative cells, an isotropic lattice, and an adaptive grid lattice.

Cellular automata have been used to study a variety of questions. [Alarcón *et al.* (2003)] studied how inhomogeneous environments can affect tumor growth. They considered a network of normal healthy blood vessels and used an (engineering in spirit) approach to model the dynamics of blood flow through this fixed network. The outcome of this part of the model was the distribution of oxygen (red blood cells) throughout the network. Next, a cellular automaton model was run where, like in our toy model above, each element of the discrete spatial grid could take one of three values: "unoccupied", "has a normal cell", or "has a cancerous cell". The concentration of oxygen was fed into the local interaction rules.

In the paper by [Gatenby and Gawlinski (2003)], the acid-mediated tumor invasion hypothesis was studied. This hypothesis states that tumors are invasive because they perturb the environment in such a way that it is optimal for their proliferation, and toxic to the normal cells with which they compete for space and substrate. The authors considered a spatial tumor invasion model, using PDEs and cellular automata. The model was based on the competition of healthy and tumor cells, with elements of acid production by tumor cells, acid reabsorption, buffering and spatial diffusion of acid and cells. The authors proposed that the associated *glycolytic phenotype* represents a successful adaptation to environmental selection parameters because it confered the ability for the tumor to invade.

A cellular automaton model of tumor angiogenesis was designed by [Anderson and Chaplain (1998b)]. In their discrete model, the movement rules between states are based directly on a discretized form of the continuous model, which was considered in the previous section. The discretization is performed by using the Euler finite difference approximation to the PDEs. Then, numerical simulations allow for tracking the dynamics of individual endothelial cells, as they build blood vessels in response to TAFs. A qualitatively novel feature of this model is its ability to describe branching of new vessels by imposing some simple local rules. In particular, it is assumed that if (i) the density of TAFs is above critical, (ii) there is enough space for branching, and (iii) the current sprout is sufficiently "old", then there is a finite probability for the vessel to branch and form a new sprout. This behavior cannot be grasped by the continuous, PDE-based models.

In this book, cellular automata are used in Chapters 14, 17, 21, and 23.

The cellular automaton approach gives rise to a new class of behaviors which can hardly be seen in continuous, PDE-based models. It allows to track individual cells, and reproduce the dynamics of emerging structures such as tumor vasculature. A drawback of this approach is that it is almost universally numerical. It is difficult to perform any analysis of such models, which leaves the researcher without an ability to generalize the behavioral trends.

3.6 Hybrid and multiscale modeling

These modern approaches to tumor modeling [Alarcón *et al.* (2005)] are constantly being developed by several groups and combine elements of different methodologies described above.

As the first example, we mention the paper by [Anderson (2007)] which describes a hybrid discrete-continuous model of solid tumor growth and invasion. This model focuses on four variables: tumor cells, extracellular matrix, matrix-degradative enzymes, and oxygen. This model is considered to be hybrid because the latter 3 variables are concentrations and thus are continuous, while the tumor cells are discrete objects.

Hybrid models have been widely employed in modeling tumor vasculature. The need for hybrid models comes from the necessity to describe the process of sprouting together with several diffusion-type processes. A nice review of discrete, continuous, and hybrid models of angiogenesis is given in [Milde *et al.* (2008)]. The state of the art of angiogenesis modeling is presented in [Jackson (2012)], where a variety of hybrid models by different groups are presented.

We will also mention the multiscale modeling approach [Cristini and Lowengrub (2010); Deisboeck and Stamatakos (2011)] that incorporates modeling techniques on several spatio-temporal scales, starting from within-cell processes, all the way to the level of individual organs and even whole-body processes, see e.g. [Rejniak and Anderson (2011); Osborne *et al.* (2010); Stolarska *et al.* (2009)].

In this book, we will not cover these advanced methods. The reason is that the main focus of our studies is evolutionary cancer dynamics, and the philosophy of our approach is reductionist. We aim to reduce the very complex systems at hand to smaller, simpler problems which we then try to understand by analytical or relatively simple numerical methods. This approach should be considered complimentary to those working on very complex models described in this section.

PART 1

Basic growth dynamics and deterministic models

Chapter 4

Single species growth

As alluded to in Chapter 2, cancer is an incredibly complex disease, characterized by an array of cellular and micro-environmental defects that allow the cells to escape homeostatic control and to grow without bound. Molecular biology is aiming to unravel those defects, thus improving our understanding of the pathways that lead to the development of cancers and identifying possible drug targets for therapies. It is, however, equally important to understand the laws and principles according to which tumor cell populations grow. This fundamental aspect of cancer research is the focus of the current chapter. Different tumor growth patterns have been identified experimentally and clinically over the years. Mathematical models have been constructed to describe those observed patterns. The tumor growth models are partly rooted in ecological models that study the growth dynamics of single species populations. The simplest growth law is exponential growth, which results from unbounded reproduction of cells. More realistic models have introduced density dependence in a variety of ways, and included specific biological details to account for specific observations. This chapter reviews the main tumor growth models that have been described, and relates each growth model to experimental data. This is done in the context of a historic time-line [Rodriguez-Brenes *et al.* (2013)], describing the models in the chronological order in which they have been published. This timeline is summarized graphically in figure 4.1, and can be consulted for reference while reading the rest of this chapter.

4.1 Exponential growth

The exponential law is probably the simplest model used to describe tumor growth (figure 4.2(a)). In 1956 Collins [Collins (1956)] proposed that

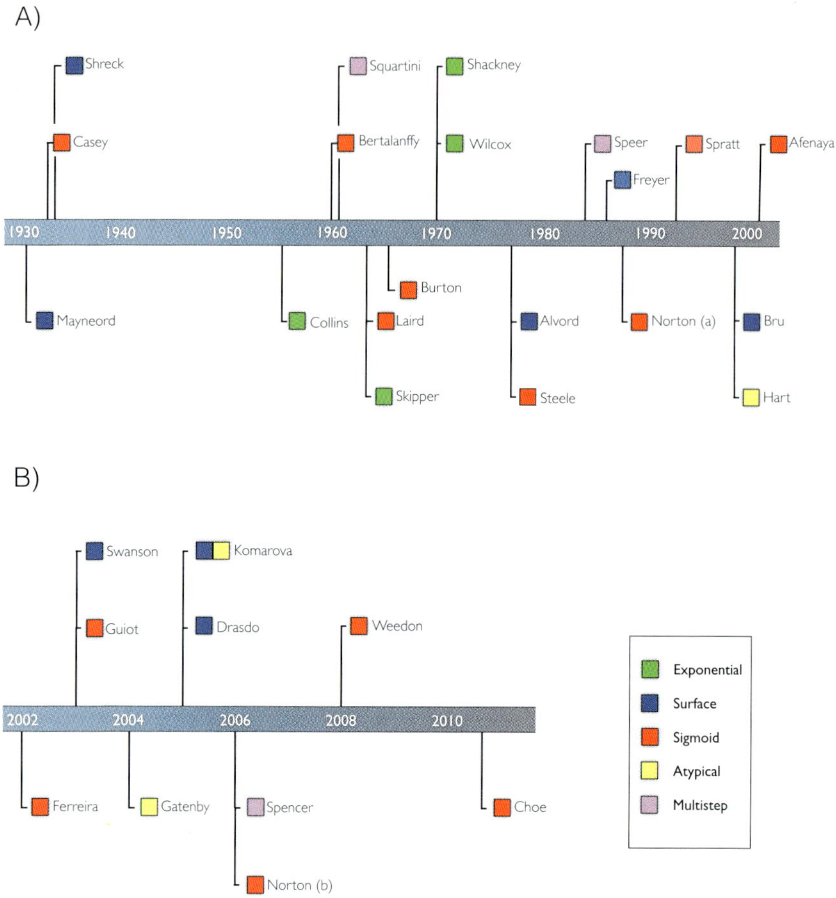

Fig. 4.1 Time line of representative models of tumor growth, organized historically. The models are depicted with different colors, representing the type of growth patterns predicted: exponential, surface, sigmoidal, atypical, and multistep. A) From 1930-2000. B) From 2000. Note: The time line does not include many valuable contributions. It only contains some representative examples. In particular it does not include influential 3D models, which do not focus explicitly on tumor growth patterns.

tumors grow exponentially with one cell giving rise to two cells, these two cells to four cells... and so on. This growth scheme leads to equation

$$\dot{W} = aW, \tag{4.1}$$

where W is the number of tumor cells at a given time and $\dot{W}$ is the derivative with respect to time.

Collins' work introduced the concept of doubling times $DT = \ln(2)/a$ to characterize tumor growth. Prior to this analysis, tumor growth was generally described in non-quantitative terms such as "slow" or "fast". Over the following decades, the doubling times of many tumors were computed (see e.g. [Spratt Jr and Spratt (1964); Steel (1977)]).

The exponential law was used to calculate the clinically observed portion of a tumor's existence, and it played a significant role in the field of chemotherapy. The discovery that a number of leukemias and lymphomas exhibited exponential growth –most notably the L1210 leukemia [Shackney (1970)]– lead Skipper, Schabel and Wilcox to develop the log-kill model [Skipper *et al.* (1964); Wilcox (1966)]. It proposed that a given chemotherapy drug dose would kill the same percentage, not the same number, of widely different-sized leukemic cell populations. Exponential growth has also been documented in solid tumors (for a review see [Friberg *et al.* (1997)]). However, it has also been suggested that this law might not be applicable over long time periods to most solid tumors, and instead might be only appropriate to describe the growth dynamics of certain non-solid cancer-types [Simon and Norton (2006)].

4.2 Surface growth

Other measurements of tumor growth suggest that a solid tumor's diameter grows primarily as a linear function of time [Mayneord (1932); Schrek (1935); Knighton *et al.* (1977)], resulting in a cubic growth law for the tumor's volume (figure 4.2(b)). In 1932 while studying Jensen's rat sarcoma, Mayneord [Mayneord (1932)] showed that this growth pattern could be explained in mathematical terms if the active growth of a solid tumor was limited to a thin surface layer of cells. Several recent models use the same basic idea in which most of the growth activity is concentrated at the tumor's boundary. [Komarova and Mironov (2005)] proposed that angiogenesis could explain cubic growth assuming that new blood vessels are formed near the surface of the existing tumor, making the cells near the surface divide more often than the core. [Brú *et al.* (2003)] observed a similar pattern in *in vitro* cell colonies, where most of the mitotic activity occurred near the tumor's boundary. This group developed a model that explains the growth kinetics as a consequence of the tumor's fractal structure [Brú *et al.* (1998, 2003)]. Other studies use contact inhibition to explain cubic growth dynamics, like the biophysical model by [Drasdo

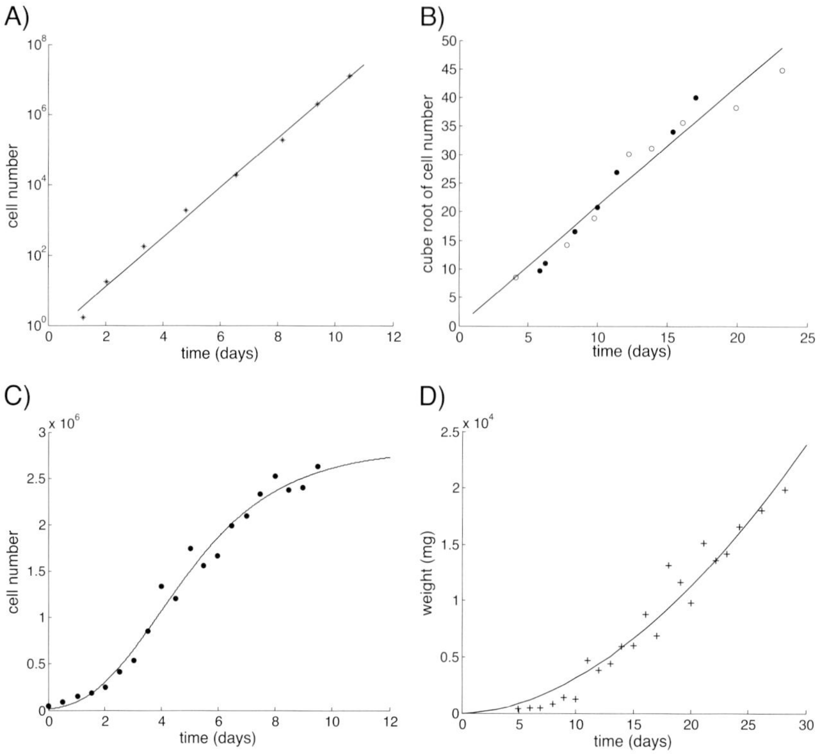

Fig. 4.2 Data samples depicting different types of tumor growth. A) Exponential growth. L1210 a mouse lymphocytic leukemia [Shackney (1970)]. B) Surface growth. Multicellular tumor spheroids of EMT6/Ro cells [Freyer and Sutherland (1986)], a mouse mammary tumor. C) sigmoidal growth. Jurkat T cell human leukemia [Reuss *et al.* (2004)]. The fit was produced using Gompertz law. D) Atypical growth, power law with exponent of size 1.84. Murine plasmacytoma [Simpson-Herren *et al.* (1970)].

and Höhme (2005)] describing the growth of multicellular tumor spheroids [Freyer and Sutherland (1985, 1986)]. This model starts with an exponential phase followed by cubic growth, as a result of a biomechanical form of contact inhibition. There is also evidence of surface growth from a study of 27 untreated low-grade gliomas [Mandonnet *et al.* (2003)], which suggests a linear relationship between time and a tumor's diameter. In 2003 Swanson et al. [Swanson *et al.* (2003)] developed a model that replicates this phenomenon as a product of cell proliferation and diffusion.

The basic model of surface growth can be presented as follows. In two dimensions, the surface scales with the square root of the volume, giving

$$\dot{W} = r\sqrt{W},$$

with the solution

$$W(t) = (W_0^{1/2} + rt/2)^2,$$

where we set the initial tumor load to be $W(0) = W_0$. The growth is quadrating in time. In three dimensions, the surface scales with the power 2/3 of the volume, and we have

$$\dot{W} = rW^{2/3} \Leftrightarrow W(t) = (W_0^{1/3} + rt/3)^3,$$

that is, the volume (and the cell number) grows as a cubic power of time. In these models, the growth is much slower than in the simple exponential model, but still as time advances, the volume increases to infinity. In the next section we review several types of self-limiting growth, where $W(t)$ increases and then reaches a horizontal asymptote.

4.3 Sigmoidal growth

Numerous models describe tumor growth as following a sigmoidal (S-shaped) curve (figure 4.2(c)). Here, we consider some noteworthy representative examples.

4.3.1 *Logistic growth*

The logistic equation is possibly the first used to model tumor dynamics [Robertson (1923)]:

$$\dot{W} = aW - bW^2. \tag{4.2}$$

Because of the widespread use of this equation in modeling, we will present its solution and discuss its long-term behavior. First we note that a conventional way to write down equation (4.2) is in terms of the carrying capacity,

$$\dot{W} = aW(1 - W/K), \tag{4.3}$$

where the carrying capacity constant is given by $K = a/b$.

To solve this nonlinear ODE, we note that

$$\int \frac{dW}{W(1-W/K)} = \int a\,dt.$$

The left hand side can be evaluated by partial fractions,

$$\int \left(\frac{1}{W} + \frac{1}{K(1-W/K)}\right) dW = at + c,$$

where c is the constant of integration. Integrating the left hand side, we obtain

$$\ln\frac{W}{K-W} = at + c \Leftrightarrow W = \frac{KCe^{at}}{1+Ce^{at}}, \quad C = e^c.$$

If the initial condition is given by

$$W(0) = W_0,$$

then the constant $C = \frac{W_0}{K+W_0}$, and the solution can be written as

$$W(t) = \frac{KW_0e^{at}}{K+W_0(e^{at}-1)}. \tag{4.4}$$

The logistic equation is very versatile: for example by making b small in (4.2), and thus the carrying capacity K large, the kinetics become almost exponential for small values of time. It is intuitively clear: far from the carrying capacity, the colony grows exponentially, but as it inceases, the restrictions start being felt more and more, slowing down the growth and eventually stopping it. We note that

$$\lim_{t\to\infty} W(t) = K,$$

that is, if $W_0 < K$, the tumor grows toward the carrying capacity limit. If $W_0 > K$, it will decrease to K.

An intuitive understanding of tumor growth without having to solve the ODE analytically is afforded by linear stability analysis. The fixed points of equation (4.3) are given by

$$W = 0 \quad \text{and} \quad W = K.$$

To investigate the stability of these, we note that the right hand side of equation (4.3) is a one-humped function of W (with roots at the fixed points), and it has a positive slope at $W = 0$ and a negative slope at

$W = K$. This means that the point $W = 0$ is unstable, and the point $W = K$ is stable.

The logistic model is mostly empirical as there is no clear physiological basis for the exponents' selection. It however fits several tumor data sets well and continues to be used (see e.g. [Choe *et al.* (2011); Weedon-Fekjær *et al.* (2008)]).

4.3.2 *Other sigmoidal laws*

All sigmoidal curves have a lower and upper asymptote (the upper asymptote is the maximum tumor size). In the logistic equation the curve approaches both asymptotes symmetrically. This differs from Gompertz curve (described below) where the maximum tumor size is approached more gradually than the lower valued asymptote. A proposed generalized version is [Weedon-Fekjær *et al.* (2008); Spratt *et al.* (1993a,b)]:

$$\dot{W} = aW - bW^{N+1}. \tag{4.5}$$

If the growth rate decays exponentially we get Gompertz law:

$$\dot{W} = aW - bW \ln(W). \tag{4.6}$$

Casey in 1934 first used it to fit tumor data. In 1964 [Laird (1964)] successfully fit the growth of 19 tumor lines (ten mice, eight rat and one rabbit) using the Gompertz equation. Since then it has been extensively used in this context (see e.g. [Steel (1977); Afenya and Calderón (2000); Norton (1988)]). Originally developed on an empirical basis, physiologically-based models have since been developed to explain it. For example, Burton in 1966 [Burton *et al.* (1966)] proposed that diffusion and the limitation of nutrient consumption to the tumor's surface could explain Gompertzian kinetics. More recently, in 2006, Norton and Massagué [Norton and Massagué (2006)] suggested that "self-seeding" of tumor cells would cause tumor proliferation to concentrate on the tumor's periphery, because of tumor cells in the bloodstream approaching the mass from the outside and by cells migrating outward from the tumor core.

Gompertz law also played an important role in chemotherapy, especially through Norton and Simon's work beginning in 1976 [Simon and Norton (2006)]. They argued that the log-kill hypothesis only applies to certain leukemias and that Gompertz growth is better suited for chemotherapy regimes dealing with solid tumors. Because chemotherapy is most effective when cancer cells are dividing rapidly, Gompertz growth kinetics implies that it is more efficient to attack cancer cells at their earliest stage of growth. Hence, they proposed reducing the intervals between treatment cycles (a scheduling plan known as *dose-dense*) giving tumors less time to regrow between treatments [Simon and Norton (2006)].

Other sigmoidal growth models are based on allometric principles. In 1960 Bertalanffy [von Bertalanffy (1960)] proposed that the tumor growth and degradation rates are proportional to a power of the tumor's size. The model assumes that (i) every tumor cell might die, but (ii) only cells near the boundary have access to nutrients and are capable of division (the addition of a death rate distinguishes it from surface growth). If a tumor is modeled as a sphere, then:

$$\dot{W} = aW^{2/3} - bW. \tag{4.7}$$

In 2003 Guiot et al. [Guiot *et al.* (2003)] proposed a growth law that is very similar in structure:

$$\dot{W} = aW^{3/4} - bW. \tag{4.8}$$

In 2001 West et al. [West *et al.* (2001)] proposed this law as a model for ontogenetic growth, deriving (4.8) from physical principles like energy conservation and allometric scaling. In particular, the 3/4 exponent was derived from the fractal-like structure of the energy distribution network [West *et al.* (2001)]. This structure ultimately determines the scaling law for the total number of terminal supply units (capillaries) in the network [West *et al.* (2001)]. With the exception of logistic growth, all the sigmoidal models discussed are based on the same principle: proliferation occurs in a tumor's region that has a lower fractal dimension than the region where cell death occurs.

Several authors have compared the fits produced by competing sigmoid models (e.g. [Skehan (1986); Marušić *et al.* (1994)]), but there is no consensus on a single optimal model to fit all data. This may be partially attributed to the versatility and the functional similarities found in these models. Marušić [Marušić *et al.* (1994)] provides a good comparison from a mathematical point of view. If we consider the Gompertz equation as a special limiting case of the generalized logistic, the sigmoidal models considered here are described by the "generalized two parameter equation"

[Marušić *et al.* (1994)]:

$$\dot{W} = aW^{\alpha} - bW^{\beta}. \tag{4.9}$$

Finally, sigmoidal curves follow three distinct phases: the initial exponential phase, the linear phase and the plateau. This makes it possible to fit data that show no decrease in the growth rate with sigmoidal curves by arguing that saturation occurs at some point in the future not observed in experiments. This practice often results in data fits where the curves' theoretical upper limit is several times larger than the maximum tumor size observed. Given that simpler models (i.e., with fewer parameters that predict no saturation) may also fit this type of data well, the prevalence of sigmoidal curves in tumor data fitting could be partially the result of over-parametrization.

4.4 Atypical growth

Other data sets of solid and non-solid tumors do not conform to the growth laws described so far. These data sets show sub-cubic growth for solid tumors and sub-exponential growth for non-solid tumors. Moreover they show no real retardation in their growth rate, indicating no reason to assume a sigmoidal pattern (figure 4.2(d)). Hart et al. [Hart *et al.* (1998)] reported this growth type in 1998, where data from primary breast cancer is shown to be inconsistent with the exponential, logistic and Gompertz laws, but supports instead a power law with exponent size equal to 0.5. Other examples of this type of growth pattern have been reported in several types of cancer, including the human ovarian carcinoma cell line A2780 [Simeoni *et al.* (2004)], murine leukemia [Simpson-Herren *et al.* (1970)], and Ehrlich's ascites tumor [Klein and Révész (1953)].

Gatenby and Friden in 2004 [Gatenby and Frieden (2004)] proposed a model that predicts sub-cubic growth in solid tumors, using information theory and Monte Carlo methods to study the role of information in tumorigenesis. They predicted a power growth law with an exponent of 1.62, which agreed very well with data from six studies demonstrating power law growth in small human breast cancers with an exponent of 1.72 ± 0.24. [Komarova and Mironov (2005)] developed a model based on the dynamics of endothelial cells in neovascularization that predicts the same type of pattern. They argued that if vasculogenesis is the dominant process by which tumors form new vasculature, tumor mass will be characterized by a linear growth in time.

4.5 Multistep growth

Other authors argue in favor of irregular growth patterns. The theory of multistage carcinogenesis also played a role in the development of tumor growth laws. According to this, cancer is primarily a genetic disease that requires cells to sequentially accumulate several random mutations and epigenetic changes [Moolgavkar and Knudson (1981); Meza *et al.* (2008)]. In 1984 Speer et al. [Speer *et al.* (1984)] introduced a model characterized by periods of tumor growth alternated by periods of dormancy (for a tumor dormancy review see [Aguirre-Ghiso (2007)]). The Speer-Retsky model predicts the occurrence of spontaneous changes in the growth rate, causing the overall growth pattern to proceed in a stepwise fashion. It uses published clinical data that show irregular growth kinetics [Squartini (1961); Retsky *et al.* (1990)] and argues that traditional growth laws (e.g. exponential, Gompertz) are better suited to describe tumor growth averages, but are not valid for all individual tumors. According to this model, adjuvant chemotherapy should be applied at low doses over a prolonged period of time as to coincide with the growth spurts.

Several authors used the multistage carcinogenesis theory to develop models that predict a multi-step pattern, such that sharp increases in the tumor growth rate correspond to the acquisition of new mutations. A recent example [Spencer *et al.* (2004)] describes the trajectory of tumor growth in terms of ordinary differential equations that explore the balance between angiogenesis, cell death rate and genetic instability. Ashkenazi et al. [Ashkenazi *et al.* (2008)] (2008) offer another such model that simulates the progressive acquisition of mutations that lead to an increase in cancer cells fitness. Spencer et al. [Spencer *et al.* (2006)] give a stochastic model predicting a multistep growth pattern.

4.6 Conclusions

This chapter summarized different tumor growth patterns that have been found in the literature, and mathematical models that describe them. This is the most fundamental information from a population dynamic perspective. Most models in the field of mathematical oncology, no matter how complex, are based on specific assumptions about the laws of tumor growth, and these assumptions obviously can influence the predictions of the model. This is important to kept in mind when considering the robustness of results. With this information in mind, the following chapters will build increasingly complex models of tumor dynamics that are all based on the simple types of growth dynamics outlined here.

Problems

Problem 4.1. *Generalize the equation for surface growth to d dimensions, and write down its solution.*

Problem 4.2. *Study the equation describing logistic growth. (a) Find fixed points of equation (4.3) and investigate their stability. (b) Find the limit as* $t \to \infty$ *of solution (4.4) of the logistic equation. Discuss in the context of the solution of part (a) of the problem.*

Problem 4.3. Numerical project. *By using numerical simulations, compare the properties of logistic and Gompertzian growth laws.*

Problem 4.4. Research project. *Find out about different reasons for the growth "plateau" often observed in cancer growth patterns.*

Chapter 5

Two-species competition dynamics

In Chapter 4 we presented a number of models describing growth of one species of (cancer) cell. Because of the heterogeneous nature of cancers, it is important to develop tools to study the growth and interactions of more than one species of cells. While many types of interactions are possible among cancer cells, competition for resources or space is of central importance. Such dynamics can determine whether different cell clones can coexist or whether only one cell clone persists and the other one goes extinct. These dynamics are very related to the topic of interspecific competition in ecology, and the models explored here will be similar in nature. We start by exploring a specific, simple model that describes the competition for space among two species or clones of cancer cells. We subsequently generalize this analysis by considering axiomatic models of competition among two species of cells.

5.1 Logistic growth of two species and the basic dynamics of competition

Here, we describe the simplest model for the logistic growth of two species that compete with each other. We make the simplifying assumption that the two cell clones share the same space, defined by the carrying capacity K, and that they only differ in their rates of cell division. This can describe a variety of scenarios where the total population size cannot exceed a certain threshold, determined by e.g. limitation of blood supply or space. Later on in the chapter, we consider more general models where the two species of cells can differ in other parameters as well, such as in their respective carrying capacities.

Thus, equation (4.2) or (4.3), describing the logistic growth of one species, can be generalized to two species in the following way. Let us denote the abundance of two species of cells as $x(t)$ and $y(t)$, and describe their joint dynamics as

$$\dot{x} = r_x x \left(1 - \frac{x + \alpha_{yx} y}{K}\right), \tag{5.1}$$

$$\dot{y} = r_y y \left(1 - \frac{y + \alpha_{xy} x}{K}\right). \tag{5.2}$$

This type of model is often discussed in ecological texts, see e.g. [Begon *et al.* (2009)], and the properties are summarized as follows. If $y(t) = 0$, the variable $x(t)$ satisfies equation (4.3) with $a = r_x$, a logistic growth law. Similarly, if $x(t) = 0$, the variable $y(t)$ satisfies logistic equation (4.3) with $a = r_y$. The constants α_{xy} and α_{yx} contain information on the extent to which crowding affects the growth of the two species of cells. If $\alpha_{xy} \neq \alpha_{yx}$, this means that the species affect each other in an asymmetric way. Larger values of α_{xy} or α_{yx} correspond to a higher degree of suppression that one species exercises toward the other.

While an analytical solution of a one-species logistic equation was possible (Section 4.3.1), we cannot solve system (5.1-5.2) directly. Instead, we will study the long-term behavior of its solutions by performing a linear stability analysis of its fixed points.

Let us suppose that $\alpha_{xy} \neq 1$ and $\alpha_{yx} \neq 1$. To find the fixed points, we solve equations $\dot{x} = \dot{y} = 0$, which yields four different solutions:

$$S_0 = (0, 0), \tag{5.3}$$

$$S_x = (K, 0), \quad S_y = (0, K), \tag{5.4}$$

$$S_{xy} = \left(\frac{K(\alpha_{yx} - 1)}{\alpha_{xy}\alpha_{xy} - 1}, \frac{K(\alpha_{yx} - 1)}{\alpha_{xy}\alpha_{yx} - 1}\right). \tag{5.5}$$

The first fixed point corresponds to the extinction of both species. The second and the third point correspond to the extinction of one of the species, while the second species resides at the carrying capacity; such type of a dynamic outcome is called "competitive exclusion". Finally, the fourth point corresponds to the coexistence of both species.

We would like to investigate stability properties of these points. For a comprehensive treatment of stability analyses we refer the reader to standard textbooks such as [Arrowsmith and Place (1990)]). The Jacobian of system (5.1-5.2) is given by

$$J = \begin{pmatrix} r_x\left(1-\frac{x+\alpha_{yx}y}{K}\right)-\frac{r_x}{K}x & -\frac{r_x\alpha_{yx}}{K}x \\ -\frac{r_y\alpha_{xy}}{K}y & r_y\left(1-\frac{y+\alpha_{xy}x}{K}\right)-\frac{r_y}{K}y \end{pmatrix}.$$

If both eigenvalues of the Jacobian evaluated at a fixed point have a negative real part, the corresponding fixed point is stable. It is unstable if at least one eigenvalue has a positive real part. Neutral stability corresponds to both eigenvalues having a zero real part.

Starting from the extinction equilibrium, S_0 (equation (5.3)), we can see that the Jacobian has eigenvalues r_x and r_y, which are positive numbers. Therefore, the extinction equilibrium is always unstable.

Next, let us consider the first of the competitive exclusion equilibria, S_x (equation (5.4)). The Jacobian becomes

$$J = \begin{pmatrix} -r_x & -\alpha_{xy}r_x \\ 0 & (1-\alpha_{yx})r_y, \end{pmatrix},$$

and the eigenvalues are given by $-r_x$ (which is always negative) and $(1-\alpha_{xy})r_y$, which is negative as long as $\alpha_{xy} > 1$. We conclude that the first species can dominate if it exhibits more suppression on the second species than the second species on itself. In other words, competitive exclusion occurs if the degree of inter-specific competition is stronger than the degree of intra-specific competition. Similarly, we can show that species 2 can dominate (equilibrium S_y stable) if $\alpha_{yx} > 1$.

Finally, we consider the coexistence equilibrium S_{xy}, equation (5.5). The Jacobian is given by

$$J = \begin{pmatrix} -\frac{r_x(1-\alpha_{yx})}{1-\alpha_{xy}\alpha_{yx}} & -\frac{\alpha_{yx}r_x(1-\alpha_{yx})}{1-\alpha_{xy}\alpha_{yx}} \\ -\frac{\alpha_{xy}r_y(1-\alpha_{xy})}{1-\alpha_{xy}\alpha_{yx}} & -\frac{r_y(1-\alpha_{xy})}{1-\alpha_{xy}\alpha_{yx}} \end{pmatrix}.$$

The characteristic polynomial of this matrix is given by

$$P(\lambda) = \lambda^2 + b\lambda + c, \tag{5.6}$$

with

$$b = \frac{(1-\alpha_{xy})r_x + (1-\alpha_{yx})r_y}{1-\alpha_{xy}\alpha_{yx}}, \quad c = \frac{(1-\alpha_{xy})r_x(1-\alpha_{yx})r_y}{1-\alpha_{xy}\alpha_{yx}}.$$

In order for both roots of the polynomial $P(\lambda)$ to have negative real parts, we need to require simultaneously,

$$b > 0, \quad c > 0. \tag{5.7}$$

These conditions are satisfied if and only if,

$$\alpha_{xy} < 0, \quad \alpha_{yx} < 0.$$

That is, in order to have a stable coexistence equilibrium, we need to require for both species to exhibit less suppression on the other species than on themselves. In other words, the degree of intra-specific competition needs to be greater than the degree of inter-specific competition.

All these cases can be represented graphically by drawing phase portraits, which show the direction of trajectories in the (x, y) space at each point (x, y), see figure 5.1. There are four logical cases. If both α_{xy} and α_{yx} are greater than one, then the long-term behavior of the system is determined by the initial conditions. Either species one or species two will dominate, and the other one will go extinct, see figure 5.1(a). This situation when two equilibria are simultaneously stable is called bistability. If one of the coefficients α_{xy}, α_{yx} is greater than 1 and the other one is smaller than 1, then in the long run, only one of the species will survive, and the other one will go extinct, resulting in competitive exclusion, figure 5.1(b,c). Finally, if both coefficients α_{xy} and α_{yx} are smaller than 1, we have a stable coexistence of an internal equilibrium, S_{xy}, figure 5.1(d).

5.2 Two-species dynamics: the axiomatic approach

System (5.1-5.2) presents a particular model of the growth rate of the two species. There, the growth rate of species one is given by $r_x[1 - (x + \alpha_{yx}y)/K]$, and the growth rate of species two is given by $r_y[1 - (y + \alpha_{xy}x)/K]$. When $x \ll K$ and $\alpha_{yx}y \ll K$, species one grows nearly exponentially with a constant rate r_x (a similar statement can be made about the growth of species two). For larger values of x and y, the growth slows down because of the crowding effect. Usually, while the verbal description holds true, it is not precisely known in what way the growth slows down. In other words, the functional form describing the growth rate of species is unknown. Can we still say anything about the dynamics, without specifying the growth terms?

This question brings us to an axiomatic modeling approach, where instead of studying specific equations, we will spell out the biological assumptions and try to derive the information about the system's behavior without using specific, often arbitrary, functional forms.

Let us assume that we want to describe a growth of two interacting species, which inhibit each other's (and their own) growth as their abun-

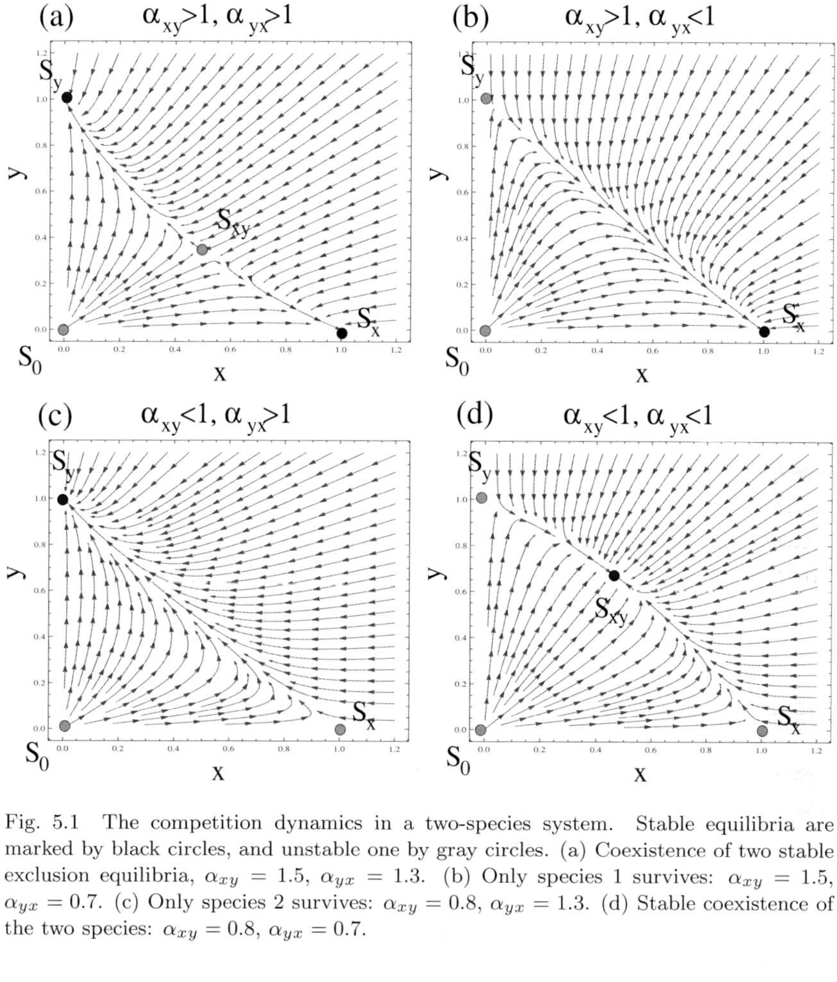

Fig. 5.1 The competition dynamics in a two-species system. Stable equilibria are marked by black circles, and unstable one by gray circles. (a) Coexistence of two stable exclusion equilibria, $\alpha_{xy} = 1.5$, $\alpha_{yx} = 1.3$. (b) Only species 1 survives: $\alpha_{xy} = 1.5$, $\alpha_{yx} = 0.7$. (c) Only species 2 survives: $\alpha_{xy} = 0.8$, $\alpha_{yx} = 1.3$. (d) Stable coexistence of the two species: $\alpha_{xy} = 0.8$, $\alpha_{yx} = 0.7$.

dances increase. We will ignore spatial effects and stochasticity, and describe the system as two ODEs:

$$\dot{x} = r_x x F(x, y), \tag{5.8}$$

$$\dot{y} = r_y y G(x, y), \tag{5.9}$$

where

$$F(0, 0) = G(0, 0) = 1,$$

such that r_x and r_y describe the growth rate of the species at very low levels. To incorporate the phenomenon of crowding, we must assume that

$$\frac{\partial F}{\partial x} \leq 0, \quad \frac{\partial F}{\partial y} \leq 0, \quad \frac{\partial G}{\partial x} \leq 0, \quad \frac{\partial G}{\partial y} \leq 0. \tag{5.10}$$

The functions $F(x, y)$ and $G(x, y)$ represent the net growth rate of the two species. They can be thought of as a balance between divisions and deaths. If the division rate is higher than the death rate, the corresponding net growth function is positive. If the death rate is higher than the division rate, the corresponding function is negative. Our assumption is that when the space becomes too crowded, the death rate increases, and/or division rate decreases, eventually tipping the balance such that the net growth rate becomes negative. This means that when x and y are large enough, the functions $F(x, y)$ and $G(x, y)$ become negative.

We will formulate this mathematically and graphically in the following way, see figure 5.2. The quadrant $x \geq 0$, $y \geq 0$ consists of two connected regions: R_F^+ and R_F^-, such that for all $(x, y) \in R_F^+$, $F(x, y) > 0$, and for all $(x, y) \in R_F^-$, $F(x, y) < 0$. We have $(0, 0) \in R_F^+$. We will call the curve separating these regions C_F: if $(x, y) \in C_F$, then $F(x, y) = 0$. In figure 5.2 the regions of positivity of function $F(x, y)$ are denoted by right-slanted, dashed lines.

Similarly, for the function $G(x, y)$, the two regions are R_G^+ and R_G^-. For all $(x, y) \in R_G^+$, $G(x, y) > 0$, and for all $(x, y) \in R_G^-$, $G(x, y) < 0$. We have $(0, 0) \in R_G^+$. We will call the curve separating these regions C_G: if $(x, y) \in C_G$, then $G(x, y) = 0$. In figure 5.2 the regions of positivity of function $G(x, y)$ are denoted by left-slanted, thick lines.

Let us assume that the two curves intersect at (at most) one point, $(x^*, y^*) \in C_F$, $(x^*, y^*) \in C_G$. In figure 5.2, only the curves in panels (a) and (d) intersect. Let us denote by $\bar{x}$ the point of intersection of the curve C_F with the x-axis, that is, $(\bar{x}, 0) \in C_F$. Similarly, $(0, \bar{y}) \in C_G$. Note that the values $\bar{x}$ and $\bar{y}$ have the meaning of the carrying capacity of each of the species in the absence of the other one.

The possible arrangements of intersecting and non-intersecting lines C_F and C_G are shown in figure 5.2. For example, in panels (a) and (b), point $\bar{x}$ belongs to the region of positivity of function G, R_G^+, because it is inside the region marked by left-slanted, thick lines. In panels (c) and (d), $\bar{x}$ is outside R_G^+. Similarly, in panels (a) and (c), point $\bar{y}$ is inside region R_F^+ (right-slanted, dashed lines), and in panels (b) and (d) it is outside that region.

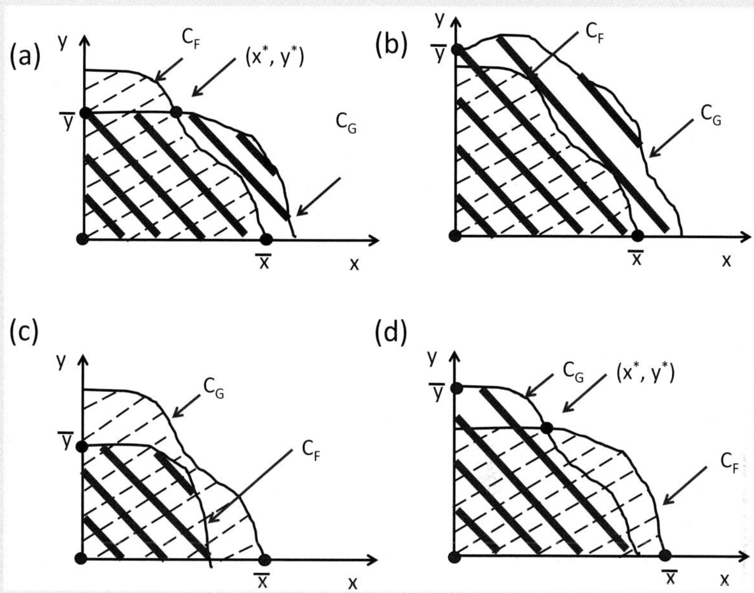

Fig. 5.2 The competition dynamics in an axiomatic two-species system, equations (5.8-5.9). Stable equilibria are marked by black circles, and unstable one by gray circles. The lines C_F and C_G outline the regions where $F(x, y)$ and $G(x, y)$ are positive, respectively. The regions of positivity of F are marked by right-slanted, dashed lines. The regions of positivity of G are marked by left-slanted, thick lines. (a) Coexistence of two stable exclusion equilibria. (b) Only species 1 survives. (c) Only species 2 survives. (d) Stable coexistence of the two species.

System (5.8-5.9) has four equilibria:

$$S_0 = (0, 0), \tag{5.11}$$

$$S_x = (\bar{x}, 0), \quad S_y = (0, \bar{y}), \tag{5.12}$$

$$S_{xy} = (x^*, y^*), \quad \text{if exists.} \tag{5.13}$$

As in Section 5.1, the first fixed point corresponds to the extinction of both species, the second and the third point - to the competitive exclusion outcome, and the fourth point to the coexistence of both species.

Next we demonstrate how stability analysis can be performed for these equilibria. The Jacobian of system (5.8-5.9) can be written:

$$J = \begin{pmatrix} r_x \left(F + x\frac{\partial F}{\partial x}\right) & r_x x \frac{\partial F}{\partial y} \\ r_y y \frac{\partial G}{\partial x} & r_y \left(G + y\frac{\partial G}{\partial y}\right) \end{pmatrix}.$$

For point S_0, equation (5.11), we have

$$J = \begin{pmatrix} r_x & 0 \\ 0 & r_y \end{pmatrix},$$

and thus the extinction equilibrium is always unstable.

For point S_x, equation (5.12), we have

$$J = \begin{pmatrix} r_x\bar{x}\frac{\partial F}{\partial x}|_{(\bar{x},0)} & r_x\bar{x}\frac{\partial F}{\partial y}|_{(\bar{x},0)} \\ 0 & r_y G(\bar{x},0) \end{pmatrix}.$$

The first eigenvalue, $r_x\bar{x}\partial F/\partial x|_{(\bar{x},0)} \leq 0$ because of condition (5.10). The second eigenvalue has the same sign as $G(\bar{x},0)$. As shown in figure 5.2, there are two ways in which we can arrange intersecting C_F and C_G with respect to their intersections with the x-axis: either $(\bar{x},0) \in R_G^+$ or $(\bar{x},0) \in R_G^-$. The equilibrium S_x is stable in the latter case, panels (a,b), and it is unstable in panels (c,d). We can see that the first species can drive the second one extinct in the long run if $G(\bar{x},0) < 0$. This latter condition implies that there is a region in the space (x,y) where y is low, and

$$G(x,y) < F(x,y),$$

that is, species x exhibits a larger crowding effect on the y species than on itself.

A similar argument leads to the conclusion that point S_y is stable as long as $F(0,\bar{y}) < 0$. This point is stable in panels (a,c) and unstable in panels (b,d).

Finally, we consider the stability of the internal equilibrium $S_{xy} = (x^*, y^*)$, equation (5.13). The Jacobian is given by

$$J = \begin{pmatrix} r_x x\frac{\partial F}{\partial x} & r_x x\frac{\partial F}{\partial y} \\ r_y y\frac{\partial G}{\partial x} & r_y y\frac{\partial G}{\partial y} \end{pmatrix}.$$

The characteristic polynomial is given by equation (5.6) with

$$b = -(r_x x^* F_x + r_y y^* G_y), \quad c = r_x x^* F_x r_y y^* G_y - r_x x^* F_y r_y y^* G_x,$$

where we denoted by F_x, F_y, etc. the partial derivatives of the functions F and G evaluated at equilibrium S_{xy}. Conditions (5.7) must be satisfied for stability of the coexistence equilibrium. Both conditions are equivalent to

$$F_x G_y > F_y G_x.$$

The latter inequality has a transparent biological meaning. The two multipliers on the left characterize the effect each species exhibits on itself. The multipliers on the right measure the effect of each species on the other species. Thus the condition for stability of the coexistence equilibrium states that each of the species must suppress itself more than it suppresses the other species. The internal equilibrium is stable in figure 5.2(d) and it is unstable in figure 5.2(a).

5.3 Summary

The simple models examined here demonstrated several things. First of all, in the presence of competition, the long-term behavior can follow different patterns depending on system parameters. For instance, one species can outcompete the other driving it extinct (competitive exclusion). Under some circumstances both species may in principle be capable of outcompeting the other, in which case the long-term outcome depends on the initial conditions. Alternatively, one species can be "weaker", such that for any initial conditions, it will be dominated by the other species and go extinct. Finally, both species may be able to coexist at a certain balance.

Apart from these interesting scenarios associated with a seemingly simple competition model, we also saw how universal these findings are. Instead of considering a specific (logistic) model of interactions, we formulated a system based on several very general biological assumptions. We call this approach axiomatic modeling. It was possible to show that consistent with these assumptions, and independent of the particular shape of the interaction functions, the same outcomes can be expected. The conditions defining which outcome is expected are formulated in terms of inequalities involving the (unknown) interaction functions. These conditions have an intuitive interpretation. Competitive exclusion occurs if the degree of inter-specific competitions is greater than the degree of intra-specific competition. In the opposite case, coexistence is possible. Graphic analysis of the growth rates of the two species (see figure 5.2) allows to identify stability properties of the various equilibria.

The simple two-species system considered here is a stepping stone toward more complex and more realistic systems describing aspects of cancer. In Chapter 6 we use this type of description to study the competition between stable and unstable phenotypes in cancer. More details will be added to take account of further complexities, such as the heterogeneity within each phenotype.

Problems

Problem 5.1. *Suppose that* $\alpha_{xy} = 1$ *and* $\alpha_{yx} \neq 1$ *in system (5.1-5.2). Perform the linear stability analysis.*

Problem 5.2. *Suppose that* $\alpha_{xy} = \alpha_{yx} = 1$ *in system (5.1-5.2). What kind of interaction between the two species does such a system describe? Perform the linear stability analysis. Learn about the phenomenon of neutral stability.*

Problem 5.3. *Perform the general analysis described in Section 5.2 by using the specific example of system (5.1-5.2). Sketch figure 5.2 for this example. Rederive stability conditions of the various equilibria.*

Chapter 6

Competition between genetically stable and unstable cells

The last two chapters covered basic tumor growth dynamics. Chapter 4 described different tumor growth laws, and Chapter 5 was concerned with competition between two different tumor cell variants. This is the simplest way to capture a very important aspect of tumors: heterogeneity. In a growing tumor, mutations give rise to different cell variants, and these variants can often compete for limiting resources, such as space, growth factors, and nutrients. The current chapter studies tumor heterogeneity and complexity in more detail. In particular, we investigate the growth and competition dynamics of different cell types that vary in their ability to accumulate mutations. On the one hand, we consider tumor cells that are characterized by a mutation rate similar to that in healthy cells. In such cells, repair mechanisms are still intact. We call such cells "stable cells". On the other hand, we consider "mutator phenotypes" [Loeb (1991, 2001, 2011); Loeb *et al.* (2003); Fox *et al.* (2013); Kolodner *et al.* (2011); Stratton *et al.* (2009)]. They are characterized by the lack of appropriate repair mechanisms and can hence accumulate mutations with a faster rate than healthy cells. As explained in Chapter 2, this phenomenon is called genetic instability, and we term the affected cells "mutator cells" or "unstable cells". We will examine the competition dynamics between stable and unstable cells and investigate the conditions under which genetically unstable cells enjoy a selective advantage and thus grow to become the dominant population. While different types of genetic instability can be distinguished [Cahill *et al.* (1999); Vogelstein and Kinzler (2002)], we do not delve on these differences in this chapter; see Chapters 7, 11 for more details. Here, we concentrate on the following trade-off.

Mutations occur when DNA becomes altered or damaged. How does

DNA damage influence the growth process of the cell populations? High amounts of DNA damage have the following consequences for **stable cells**. On the one hand, the intact repair system allows these cells to maintain relatively stable genomes, because environmentally induced DNA damage gets repaired. On the other hand, repair takes time and is usually manifested in cell cycle arrest or in stalling of the replication process. Repair is therefore costly because it slows down the overall growth of the cell population. **Unstable cells** are influenced by high levels of DNA damage in the following way. They avoid repair and therefore do not enter cell cycle arrest. On the other hand, they pay an alternative cost. Many mutants are created, and a large proportion of the mutants are likely to be non-viable.

With this trade-off in mind, under what conditions are stable cells advantageous, and when are unstable cells selected for? This depends on the level of DNA damage that is occurring, and this in turn can be determined by environmental factors. DNA damage can come from a variety of sources. Carcinogens contained in food we eat or in the air we breathe can damage DNA. UV radiation can break DNA. Chemotherapeutic agents can lead to various forms of DNA damage. Most importantly perhaps, aging leads to an increased amount of DNA damage. This is because metabolic activities produce reactive oxygen species which are toxic for our genome [Hasty *et al.* (2003); Campisi and Vijg (2009); Lim and Campisi (2001); Campisi and Warner (2001); Jackson *et al.* (1998)].

This chapter presents a mathematical model to investigate whether and how DNA damage can influence the growth processes of stable and unstable cells. This is done by examining the competition dynamics between stable and unstable cells. Which cell type wins? Can an increase in the level of DNA damage reverse the outcome of competition?

6.1 Competition dynamics

We start by exploring the competition dynamics between a stable and a mutator cell population [Komarova and Wodarz (2003)]. Let us denote the abundance of stable and mutator cells as S and M, respectively. The competition dynamics are given by the following pair of differential equations, which describe the development of the cell populations over time,

$$\dot{S} = r_s S(1 - u + \beta\epsilon_s u) + \alpha u r_s S(1 - \epsilon_s) - \phi S, \tag{6.1}$$

$$\dot{M} = r_m M(1 - u + \beta\epsilon_m u) + \alpha u r_m M(1 - \epsilon_m) - \phi M. \tag{6.2}$$

The model is explained graphically in figure 6.1. The cells replicate at a rate r_s or r_m. These parameters reflect how often cells reproduce and die; we will call this the *intrinsic replication rate* of the cells. The two cell populations compete for a shared resource. Competition is captured in the expressions ϕS and ϕM, where ϕ is defined as follows:

$$\phi = Sr_s\left[1-u\left(1-\beta\epsilon_s-\alpha(1-\epsilon_s)\right)\right] + Mr_m\left[1-u\left(1-\beta\epsilon_m-\alpha(1-\epsilon_m)\right)\right]. \tag{6.3}$$

During replication a genetic alteration can occur with a probability u. We call this the DNA hit rate. DNA damage can occur both spontaneously (most likely at low levels), or it can be induced by DNA damaging agents which corresponds to a high value of u. If a genetic alteration has occurred, it gets repaired with a probability ϵ_s or ϵ_m. The probability of repair is assumed to be higher for stable cells than it is for mutators:

$$\epsilon_s > \epsilon_m.$$

During repair, there is cell cycle arrest, and this is captured in the parameter β. The value of β can lie between zero and one and thus reduces the rate of cell division (given by βr). If $\beta = 0$, the repairing cells never replicate and this is the maximal cost. If $\beta = 1$, there is no cell cycle arrest and no cost associated with repair. With a probability ($1 - \epsilon_s$ or $1 - \epsilon_m$) the genetic alteration does not get repaired, and a mutant is generated. A mutation is therefore the result of the occurrence of DNA damage combined with the absence of repair. The mutant is viable with a probability α, while it is non-viable with a probability $1-\alpha$. Therefore, the model captures both the costs and benefits of repair: Efficient repair avoids deleterious mutations but is associated with cell cycle arrest. Absence of efficient repair can result in the generation of deleterious mutants, but avoids cell cycle arrest.

Note that in this first model, we assume that mutants that are created are either non-viable (and thus do not participate in the competition dynamics) or neutral (and thus have the same intrinsic reproductive rate as the wild type). We will include the possibility of advantageous and disadvantageous (but viable) mutants later.

Let us explore how the competition dynamics depends on the rate at which cells acquire genetic alterations (DNA hit rate, u). In general, if

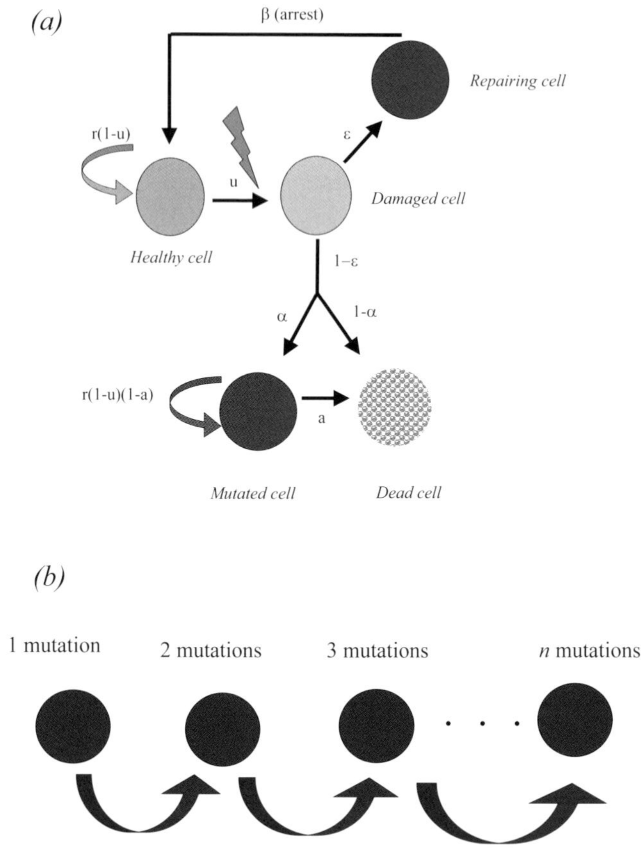

Fig. 6.1 Schematic diagram of the model. (a) Process of cell reproduction, DNA damage, repair, cell cycle arrest, mutation, and death. (b) When DNA damage is not repaired, the cells can accumulate mutations. In the model cancer progression corresponds to the successive accumulation of mutations, also referred to as the mutation cascade.

two cell populations compete, the cells with the higher fitness wins. The fitness of the cells in this model is given by $r_{s,m}[1 - u[1 - \alpha + \epsilon_{s,m}(\alpha - \beta)]]$. Note that the quantity $1 - \alpha$ has the meaning of the cost of production of

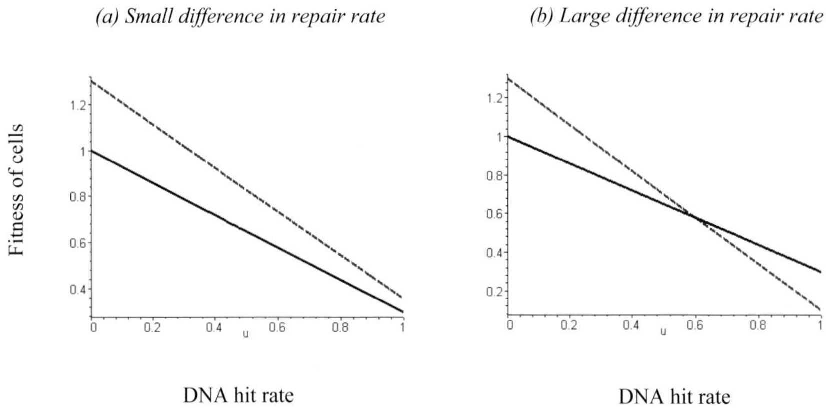

Fig. 6.2 Effect of the DNA hit rate, u, on the fitness of two cell populations. At low DNA hit rates, the population with the higher intrinsic replication rate wins. An increase in the DNA hit rate decreases the fitness of both cell populations. However, the degree of fitness reduction of the population characterized by the higher intrinsic replication rate is stronger than that of the slower population of cells. If there is a sufficient difference in the repair rates (degrees of genetic stability) between the two cell populations (b), an increase in the DNA hit rate can result in a reversal of the relative fitnesses, and thus in a reversal of the outcome of competition. If the difference in repair rates between the two cell populations is not sufficient, (a), we do not observe such a reversal. Parameter values were chosen as follows: $r_s = 1$; $r_m = 1.3$; $\alpha = 0.05$; $\beta = 0.3$; $\epsilon_s = 0.99$. For (a) $\epsilon_m = 0.1$. For (b) $\epsilon_m = 0.9$.

deleterious mutants; we will refer to it as

$$C_{del} = 1 - \alpha.$$

Similarly, the quantity $1 - \beta$ is the cost of cell cycle arrest,

$$C_{arr} = 1 - \beta.$$

In these notations, we can rewrite the expression for the fitness in a more intuitive way,

$$r_{s,m} - ur_{s,m}[C_{del} + \epsilon_{s,m}(C_{arr} - C_{del})]. \tag{6.4}$$

If the DNA hit rate is low (low value of u), the fitness of the cells is approximately given by their intrinsic rate of replication (r_s and r_m). Thus, the cell population with the higher intrinsic replication rate has a higher fitness than the cell population with the lower intrinsic replication rate.

On the other hand, when the DNA hit rate, u, is increased, the fitness depends more strongly on other parameters. In particular, the fitness of both populations can depend on the DNA hit rate, u. Notably, an increase in the value of u may result in a stronger decline in fitness of the cell population with the faster intrinsic rate of replication relative to the slower cell population (figure 6.2). Therefore, if the DNA hit rate crosses a critical threshold, $u > u_c$, the outcome of competition can be reversed. The value of u_c is given by

$$u_c = \frac{r_s - r_m}{(r_s - r_m)C_{del} + (r_s\epsilon_s - r_m\epsilon_m)(C_{arr} - C_{del})}. \tag{6.5}$$

We are interested to find out, under what circumstances reversal of competition can occur. One condition required for the reversal of competition is that the stable and mutator cells are characterized by a sufficient difference in the repair rate (figure 6.2) which is defined as

$$\Delta\epsilon \geq \frac{|r_s - r_m|[(1 - C_{del})(1 - \epsilon_s) + \epsilon_s(1 - C_{arr})]}{|C_{arr} - C_{del}|r_m}. \tag{6.6}$$

Further, we need to distinguish between two scenarios.

(1) In the first case we assume that the stable cells have a faster intrinsic rate of replication than the mutator cells (i.e., $r_s > r_m$). Therefore, at low DNA hit rates, the stable cells win. An increased DNA hit rate, u, can shift the competition dynamics in favor of the unstable cells. In other words, unstable cells gain a selective advantage as the DNA hit rate becomes large. This is because the population of stable cells frequently enters cell cycle arrest when repairing genetic damage and this slows down the overall growth rate. For this outcome to be possible, the following condition has to be fulfilled: The cost of cell cycle arrest, C_{arr}, must be greater than the cost of producing non-viable mutants, C_{del}. If this condition is not fulfilled, reversal of competition at high DNA hit rates is not observed.

(2) In the second case we assume that the stable cells have a slower intrinsic replication rate than the mutator cells (i.e., $r_s < r_m$). Therefore, at low DNA hit rates, the unstable cells win. An increased DNA hit rate, u, can shift the competition dynamics in favor of the stable cells. In other words, a high DNA hit rate selects against genetic instability. This is because the unstable cells produce more non-viable mutants and this reduces the effective growth rate significantly. In contrast to the previous scenario, this requires that the cost of producing non-viable mutants, C_{del}, must be higher than the cost of cell cycle arrest, C_{arr}.

If this condition is not fulfilled, reversal of competition at high DNA hit rates is not observed.

Table 6.1 Summary of the basic competition dynamics. If the mutators (M) have a lower intrinsic replication rate than the stable cells (S), a high DNA hit rate can select in favor of M. If the intrinsic replication rate of M is higher than that of S, then a high DNA hit rate can select for S.

	M slower than S	**M faster than S**
Low DNA hit rate	S win	M win
High DNA hit rate	M win if $C_{arr} > C_{del}$	S win if $C_{arr} < C_{del}$

To summarize, this analysis gives rise to the following results (Table 6.1). A high DNA hit rate, u, can reverse the outcome of competition in favor of the cell population characterized by a slower intrinsic growth rate if the competing populations are characterized by a sufficient difference in their repair rates. The higher the difference in the intrinsic replication rate of the two cell populations, the higher the difference in repair rates required to reverse the outcome of competition. If the intrinsic replication rate of the genetically unstable cell is slower, a high DNA hit rate can select in favor of genetic instability. On the other hand, if the intrinsic growth rate of the genetically unstable cell is faster, a high DNA hit rate can select against genetic instability.

6.2 Competition dynamics and cancer evolution

6.2.1 *A quasispecies model*

In the previous section, we considered the competition dynamics between stable and unstable populations of cells, assuming that they are characterized by different and fixed rates of replication. We further assumed that mutations are either non-viable or neutral. However, mutations are unlikely to be neutral, and will change the replication rate of the cells. In other words, cells may evolve to grow either at a faster or a slower rate, depending on the mutations generated. Here, we extend the above model to take into account such evolutionary dynamics.

The competition problem. As before, we consider two competing cell populations: a genetically stable population, S, and a mutator population, M, see figure 6.1(b). We start with unaltered cells which have not accumulated any mutations. They are denoted by S_0 and M_0, respectively.

Both are assumed to replicate at the same rate, r_0. When the cells become damaged and this damage is not repaired, mutants are generated. If the mutants are viable, they can continue to replicate. During these replication events, further mutations can be accumulated if genetic alterations are not repaired. We call the process of accumulation of mutations the *mutational cascade*. Cells which have accumulated i mutations are denoted by S_i and M_i, respectively, where $i = 1, \ldots, n$. They are assumed to replicate at a rate r_i. Stable and unstable cells differ in the rate at which they proceed down the mutational cascade. In addition to the basic dynamics of cell replication described in the previous section, we assume that during cell division, mutated cells can undergo apoptosis, since oncogenic mutations can induce apoptotic checkpoints [Seoane *et al.* (2002); Vogelstein *et al.* (2000)]. Thus, the intrinsic replication rate of mutated cells is given by $r_i(1 - a)$, where a denotes the probability to undergo apoptosis upon cell division. These processes can be summarized in the following equations:

$$\dot{S}_0 = R_0 S_0 (1-u_s) - \phi S_0, \tag{6.7}$$

$$\dot{S}_i = \alpha u R_{i-1} S_{i-1} (1-\epsilon_s) + R_i S_i (1-u_s) - \phi S_i, \quad 1 \le i \le n-1, \tag{6.8}$$

$$\dot{S}_n = \alpha u R_{n-1} S_{n-1} (1-\epsilon_s) + R_n S_n [1-u_s+\alpha u(1-\epsilon_s)] - \phi S_n, \tag{6.9}$$

$$\dot{M}_0 = R_0 M_0 (1-u_m) - \phi M_0, \tag{6.10}$$

$$\dot{M}_i = \alpha u R_{i-1} M_{i-1} (1-\epsilon_m) + R_i M_i (1-u_m) - \phi M_i, \quad 2 \le i \le n-1, \tag{6.11}$$

$$\dot{M}_n = \alpha u R_{n-1} M_{n-1} (1-\epsilon_m) + R_n M_n [1-u_m+\alpha u(1-\epsilon_m)] - \phi S_n, \tag{6.12}$$

$$\dot{w} = (1-\alpha) u \left[(1-\epsilon_s) \sum_{i=1}^{n} R_i S_i + (1-\epsilon_m) \sum_{i=1}^{n} R_i M_i \right] - \phi w, \tag{6.13}$$

where we introduced the following short hand notations: R_i is the effective intrinsic reproductive rate, $R_i = r_i(1 - a)$ for $1 \le i \le n$ and $R_0 = r_0$, and $u_{s,m}$ are the two effective mutation rates, $u_{s,m} = u(1-\beta\epsilon_{s,m})$. The variable w denotes the non-viable mutants produced by the cells. The equations are coupled through the function ϕ, the average fitness, which is given by

$$\phi = (1 - u_s) \sum_{j=0}^{n} R_j S_j + (1 - u_m) \sum_{j=1}^{n} R_j M_j.$$

Solving quasispecies equations. Equations (6.7-6.13) are an example of a *quasispecies*-type system, which is a well-known population dynamical model in evolutionary biology. Quasispecies equations were first derived for molecular evolution by [Eigen and Schuster (1979)], and since then have found applications in many areas of research, including biochemistry, evolution, and game theory.

In order to analyze system (6.7-6.13), we would like to review some of the techniques for solving quasispecies equations. Let the variable $\mathbf{x} = (x_0, x_1, \ldots, x_{n+1})$ satisfy the system

$$\dot{x}_0 = a_0 x_0 - \phi x_0, \tag{6.14}$$

$$\dot{x}_i = b_i x_{i-1} + a_i x_i - \phi x_i, \qquad 1 \le i \le n, \tag{6.15}$$

$$\dot{x}_{n+1} = \sum_{i=0}^{n} c_i x_i - \phi x_{n+1}, \tag{6.16}$$

where

$$\phi = (a_0 + c_0) x_0 + \sum_{k=1}^{n} [(a_k + c_k) x_k + b_k x_{k-1}].$$

We have $\sum_{k=0}^{n+1} x_k = 1$. System (6.14-6.16) is nonlinear. However, the nonlinearity can be removed by the following trick. Let us consider the variable $\mathbf{z} = (z_0, z_1, \ldots, z_{n+1})$ which satisfies the following system:

$$\dot{z}_0 = a_0 z_0, \tag{6.17}$$

$$\dot{z}_i = b_i z_{i-1} + a_i z_i, \qquad 1 \le i \le n, \tag{6.18}$$

$$\dot{z}_{n+1} = \sum_{i=0}^{n} c_i z_i. \tag{6.19}$$

If we set

$$x_i = z_i / \sum_{k=0}^{n+1} z_k, \quad 0 \le i \le n+1, \tag{6.20}$$

then the variable $\mathbf{x}$ satisfies system (6.14-6.16). The general solution of system (6.17-6.19) is given by

$$\mathbf{z}(t) = \sum_{j=0}^{n} \alpha_j \mathbf{v}^{(j)} e^{a_j t} + \alpha_{n+1} \mathbf{v}^{(n+1)}, \tag{6.21}$$

where α_j are constants determined from the initial condition, and $\mathbf{v}^{(j)}$ are eigenvectors of the appropriate triangular matrix corresponding to the eigenvalues a_j. The eigenvector $\mathbf{v}^{(n+1)} = (0, 0, \ldots, 0, 1)^T$ corresponds to the zero eigenvalue, and for the rest of the eigenvectors we have,

$$v_i^{(j)} = \begin{cases} 0, & i < j, \\ 1, & i = j, \\ (-1)^{i-j} \prod_{k=1}^{i-j} \frac{b_{j+k}}{(a_{j+k} - a_j)}, & j+1 \le i \le n. \end{cases} \tag{6.22}$$

From solution (6.21) and transformation (6.20) we can see that as time goes to infinity, the solution $\mathbf{x}(t)$ tends to the normalized eigenvector corresponding to the largest of the eigenvalues $a_0, \ldots, a_n$.

The exact solution corresponding to the initial condition $\mathbf{z}(0) = (1, 0, \ldots, 0)^T$ can be found. The appropriate coefficients in equation (6.21) are

$$\alpha_j = \prod_{m=1}^{j} \frac{b_m}{(a_j - a_{m-1})}, \qquad 0 \le i \le n.$$

For the ith component, we obtain,

$$z_i(t) = \left(\prod_{k=1}^{i} b_k \right) \sum_{j=0}^{i} e^{a_j t} \prod_{m=0,\, m \neq j}^{i} \frac{1}{a_j - a_m}, \qquad 1 \le i \le n. \tag{6.23}$$

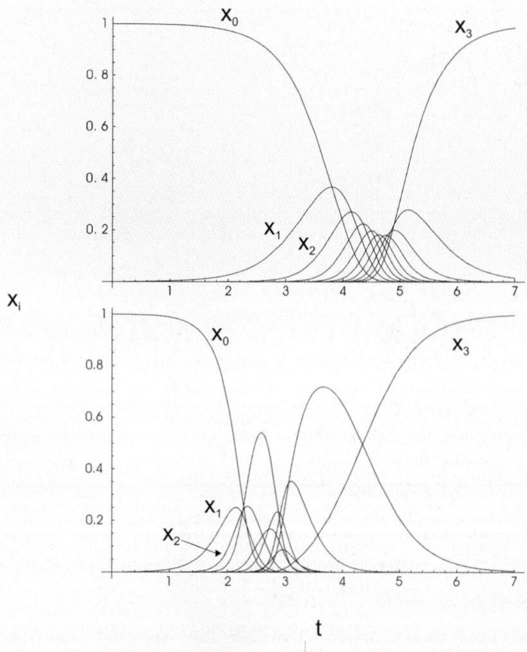

Fig. 6.3 Simulation of mutation cascades. In the picture above, we have $a_i = a_{i-1} + 2$, for $1 \le i \le n$ and $b_i = \epsilon$. In the picture below, we have $a_i = a_{i-1} + 2(1+\xi_i)$ and $b_i = \epsilon(1+\zeta_i)$, where ξ_i and ζ_i are some random numbers drawn from a uniform distribution between zero and one. For both pictures, $n = 9$, $c_i = 0$ for all i, $\epsilon = 0.001$ and $a_0 = 1$.

The short-time behavior of this quantity is given by

$$z_i(t) = \left(\prod_{k=1}^{i} b_k\right) t^i, \quad 1 \le i \le n, \quad a_j t \ll 1 \quad \forall j.$$

The expression for z_{n+1} can also be obtained but is slightly more cumbersome.

Mutation cascades. Let us assume that $a_0 < a_n$, and in addition we have $a_i - a_{i-1} \sim a_i \sim 1$. Then the system exhibits the following behavior (figure 6.3).

Starting from the "all x_0" state, the fraction of type x_0 goes down steadily, and the population acquires some amount of x_1. Upon reaching a maximum, the fraction of x_1 decreases and the fraction of x_2 experiences a "hump", to be in turn replaced by x_3, etc. The characteristic time at which each type experiences its maximum abundance can be estimated if we replace the expressions for $z_j(t)$ in (6.23) by the leading term, i.e., the term which has the largest exponent, so that

$$z_i(t) \approx \left(\prod_{m=1}^{i} \frac{b_m}{a_i - a_{m-1}}\right) e^{a_i t}.$$

Then type i is at its maximum near

$$t \approx t_i = \frac{1}{a_{i-1} - a_i}\left(\log \frac{b_i}{a_i - a_{i-1}} + \sum_{m=0}^{i-2} \log \frac{a_{i-1} - a_m}{a_i - a_m}\right). \tag{6.24}$$

In particular, after time t_n, the type x_n will dominate.

Multidimensional competition dynamics. Equations (6.7-6.13) represent two parallel mutation cascades, that is, two sets of quasispecies equations, coupled via the common fitness term, ϕ. In order to use the techniques developed above, let us write the equations for the mutational cascade in a simpler form,

$$\dot{z}_0 = a_0 z_0, \quad \dot{z}_i = b_i z_{i-1} + a_i z_i, \qquad 1 \le i \le n, \tag{6.25}$$

$$\dot{z}'_0 = a'_0 z'_0, \quad \dot{z}'_i = b'_i z'_{i-1} + a'_i z'_i, \qquad 1 \le i \le n, \tag{6.26}$$

$$\dot{z}_{n+1} = \sum_{k=0}^{n} (c_k z_k + c'_k z'_k), \tag{6.27}$$

by introducing the following obvious notations:

$$z_i \to S_i, \quad z'_i \to M_i, \quad 0 \le i \le n, \quad z_{n+1} \to w, \tag{6.28}$$

$$a_i = R_i(1-u_s), \quad a'_i = R_i(1-u_m), \quad 0 \le i \le n-1, \tag{6.29}$$

$$a_n = R_n[1-u_s+\alpha u(1-\epsilon_s)], \quad a'_n = R_n[1-u_m+\alpha u(1-\epsilon_m)], \tag{6.30}$$

$$b_i = \alpha u R_{i-1}(1-\epsilon_s), \quad b'_i = \alpha u R_{i-1}(1-\epsilon_m), \quad 1 \le i \le n, \tag{6.31}$$

$$c_i = (1-\alpha)R_i u(1-\epsilon_s), \quad c'_i = (1-\alpha)R_i u(1-\epsilon_m), \quad 0 \le i \le n. \tag{6.32}$$

In a matrix notation, equations (6.7-6.13) read:

$$\dot{\mathbf{z}} = \hat{f}_s \mathbf{z}, \quad \dot{\mathbf{z}}' = \hat{f}_m \mathbf{z}', \tag{6.33}$$

where the fitness matrices $\hat{f}_{s,m}$ are found from (6.25-6.27). The solution of the nonlinear system can be found by re-normalizing the solution of system (6.25-6.27), as before.

Fitness landscape. In order to analyze the dynamics of system (6.7-6.13), we have to make assumptions on the fitness landscape for the consecutive mutants (figure 6.4).

Since we are interested in cancer progression, we assume that the intrinsic rate of cell division of the consecutive mutants, r_i, increases ($r_{i+1} > r_i$). Such mutations could correspond to alterations in oncogenes or tumor suppressor genes. Because an accumulation of mutations cannot result in an infinite increase in the division rate of cells, we assume that the division rate plateaus. Once the cells have accumulated n mutations, we assume that further viable mutants are neutral because the division rate cannot be increased further. (This end stage of the mutational cascade is thus mathematically identical to the simple model discussed in the previous section.) While we assume that the consecutive mutants can divide faster, they can also carry a disadvantage: the mutations can be recognized by the appropriate checkpoints which induce apoptosis. With this in mind, we will consider two basic types of fitness landscapes. If $r_0 > r_n(1-a)$, the intrinsic growth rate of the mutated cells, S_i and M_i, will be less than that of the unaltered cells, S_0 and M_0 (figure 6.4). While the mutations allow the cells to divide more often, the mutated cells are killed at a fast rate by apoptosis upon cell division. This scenario corresponds to the presence of efficient apoptotic mechanisms in cells. On the other hand, if $r_0 < r_n(1-a)$, the accumulation of mutations will eventually result in an intrinsic growth rate which is larger than that of unaltered cells (figure 6.4). While mutated

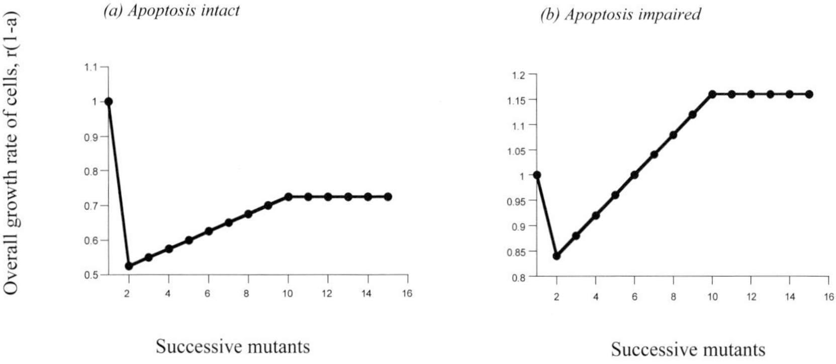

Fig. 6.4 Fitness landscape as a result of the successive accumulation of mutations by cells. We distinguish two scenarios. (*a*) If apoptosis is intact, accumulation of mutations results in a lower fitness compared to unaltered cells. Even if the mutations result in an increased rate of cell division, the induction of apoptosis in mutated cells prevents them from attaining a higher fitness than the unaltered cells. (*b*) If apoptosis is impaired, the accumulation of successive mutations will eventually result in a higher fitness compared to unaltered cells. The exact shapes of the curves are not essential. What is important is whether the mutants will eventually have a lower (a) or higher (b) intrinsic reproductive rate.

cells can still undergo apoptosis upon cell division, apoptosis is not strong enough to prevent an increase in the intrinsic growth rate. Hence, this scenario corresponds to impaired apoptosis in cells. In the following sections, we study the competition dynamics between stable and mutator cells in an evolutionary setting, assuming the presence of relatively strong and weak apoptotic responses.

Time scale separation. In what follows we will assume that the dynamics of the two cell populations happen on two different time scales. In other words, we require that the stable population is still in the state S_0 while the unstable population has already produced all mutants and reached a quasistationary state.

The typical time, t_1^s, of change for the type S_0 is found from equation (6.24). Similarly, we can find the time, t_n^m, it takes to reach the state M_n. It is given by the same equation (6.24) except the coefficients in the nth equation must be replaced by the corresponding coefficients with primes. The mapping to the biological parameters is found from (6.28-6.32). Note that using formula (6.24) has its restrictions, and in the case where it is not applicable, one can directly calculate t_n^m by estimating the time it takes for z_n to reach its maximum (see formula (6.23)). The inequality

$$t_1^s \ll t_n^m \tag{6.34}$$

guarantees that by the time the unstable population has traveled down the mutation cascade to approach its quasistationary distribution, the stable population of cells is still dominated by S_0.

The conditions for the reversal of competition. In the multidimensional competition problem, equations (6.33), the outcome is determined by the largest eigenvalue of the fitness matrices, $\hat{f}_s$ and $\hat{f}_m$. As time goes by, the unstable cell population will approach its stationary distribution (defined by the eigenvector corresponding to the principal eigenvalue), and its fitness is given by the eigenvalue. Because of the time-scale separation, we will assume that during this time, the stable population remains largely at the state S_0. Thus the "winner" of the competition is defined by comparing the two eigenvalues, a_0 and a'_n, see equations (6.29-6.30).

Let us define the value of u, u_c, so that for $u = u_c$, we have $a_0 = a'_n$. As the hit rate passes through u_c, the result of the competition reverses. We have

$$u_c = \frac{R_0 - R_n}{R_0 - R_n - \beta(R_0\epsilon_s - R_n\epsilon_m) + R_n\alpha(1-\epsilon_m)}.$$

In order to determine whether competition reversal takes place for each scenario (see below), we need to make sure that the following condition is satisfied:

$$0 \le u_c \le 1.$$

In the next sections we will examine different parameter regimes and conclude that competition reversal may or may not take place; we derive the exact conditions for this. In what follows, we will use several definitions. Let us set

$$\Delta\epsilon = \epsilon_s - \epsilon_m,$$

and denote by $\bar{\epsilon}_s$ the following threshold value of ϵ_s,

$$\bar{\epsilon}_s = \frac{\alpha R_n}{\beta R_0 + R_n(\alpha - \beta)}. \tag{6.35}$$

This quantity is defined from setting $u_c = 1$ and $\Delta\epsilon = 0$. Finally, we define the critical *gap*, $\Delta^*\epsilon$, between the two values of ϵ, by setting $u_c = 1$:

$$\Delta^*\epsilon = \epsilon_s \left[1 + \frac{\beta R_0}{(\alpha - \beta)R_n}\right] - \frac{\alpha}{\alpha - \beta}.$$

Now, let us go back to the two types of fitness landscape, figure 6.4, and examine the scenarios of strong and weak apoptosis separately.

6.2.2 *Strong apoptosis*

Here we assume that the apoptotic mechanisms in cells are strong. That is, $r_0 > r_n(1 - a)$ (figure 6.4(a)). This means that although the successive mutations will allow the cell to divide faster, the induction of apoptosis ensures that the intrinsic growth rate of the mutants is lower compared to unaltered cells. Note that it is not necessary to assume that oncogenic mutations allow cells to divide faster. Indeed, some cancer cells may progress more slowly through the cell cycle than healthy cells. The important assumption is that accumulation of mutations lowers the intrinsic growth rate of the cells.

In this scenario, the intrinsic growth rate of the stable cells, S, is higher than that of the unstable cells, M. The reason is as follows. The population of stable cells, S, has efficient repair mechanisms. Thus, most cells will remain at the unaltered stage, S_0. Because population M is unstable, a higher fraction of this cell population will contain mutations. Since these mutations impair reproduction (e.g. because of induction of apoptosis), the intrinsic growth rate of the unstable cells, M, is lower than that of the stable population, S.

At low DNA hit rates, the cells with the faster intrinsic growth rate win the competition. Thus, at low DNA hit rates (low value of u), the stable phenotype, S, wins (figure 6.5(a)). On the other hand, at higher DNA hit rates (high value of u), the outcome of competition can be reversed because frequent cell cycle arrest significantly reduces growth. That is, the genetically unstable cells, M, may win and take over the population. As in the simple model discussed above, it requires that the cost of cell cycle arrest is higher than the cost associated with the generation of deleterious mutants (i.e., $C_{arr} > C_{del}$). Furthermore, reversal of competition may require that the repair rate of stable cells (ϵ_s) lies below a threshold, and that there is a sufficient difference in the repair rate between stable and unstable cells.

As the population of unstable cells wins, they accumulate mutations. Even if the sequential mutants are disadvantageous because of the induction of apoptosis, the high mutation rate pushes the population down the mutational cascade. While all variants, M_i, persist, the distribution of the variants becomes skewed toward M_n as the DNA hit rate is increased.

These results can be obtained by a very simple analysis of the relative values of the relevant eigenvectors, see (6.29-6.30), and by finding conditions under which reversal can occur. In mathematical terms, strong apoptosis corresponds to the situation where

$$R_0 > R_n.$$

From definition (6.35), $0 \leq \bar{\epsilon}_s \leq 1$. Also, we will use the fact that $\Delta^*\epsilon$ grows with ϵ_s, so that $\Delta^*\epsilon > 0$ for $\epsilon_s > \bar{\epsilon}_s$ and $\Delta^*\epsilon < 0$ for $\epsilon_s < \bar{\epsilon}_s$. We can distinguish the following two cases:

- If $\beta > \alpha$ (which is the same as $C_{arr} < C_{del}$), we have $a_n > a'_n$, which means that M_n *never* corresponds to the largest eigenvalue. This means that the stable cells always win and the competition reversal does not happen. (Technically speaking, the reversal happens between a_0 and a_n rather than a_0 and a'_n.)
- If $\beta < \alpha$ (which is the same as $C_{arr} > C_{del}$), then competition reversal will happen if the following condition is satisfied: $\Delta\epsilon > \Delta^*\epsilon$ (this is because the function u_c decays with $\Delta\epsilon$). We also observe that in this case, $\Delta^*\epsilon$ is a growing function of ϵ_s, which reaches zero at $\epsilon_s = \bar{\epsilon}$, with $0 \leq \bar{\epsilon} \leq 1$. We have two subcases:

  (a) For $\epsilon_s < \bar{\epsilon}_s$, we have $\Delta^*\epsilon < 0$, and the reversal happens for *any* difference between ϵ_s and ϵ_m.

  (b) For $\epsilon_s > \bar{\epsilon}_s$, $\Delta^*\epsilon > 0$, and we need a finite gap between ϵ_s and ϵ_m, $\Delta\epsilon > \Delta^*\epsilon$. We also have to make sure that $\Delta^*\epsilon < \epsilon_s$, which gives the condition

$$\epsilon_s < \frac{\alpha R_n}{\beta R_0}.$$

The biological interpretation of these conditions was given in the beginning of this section.

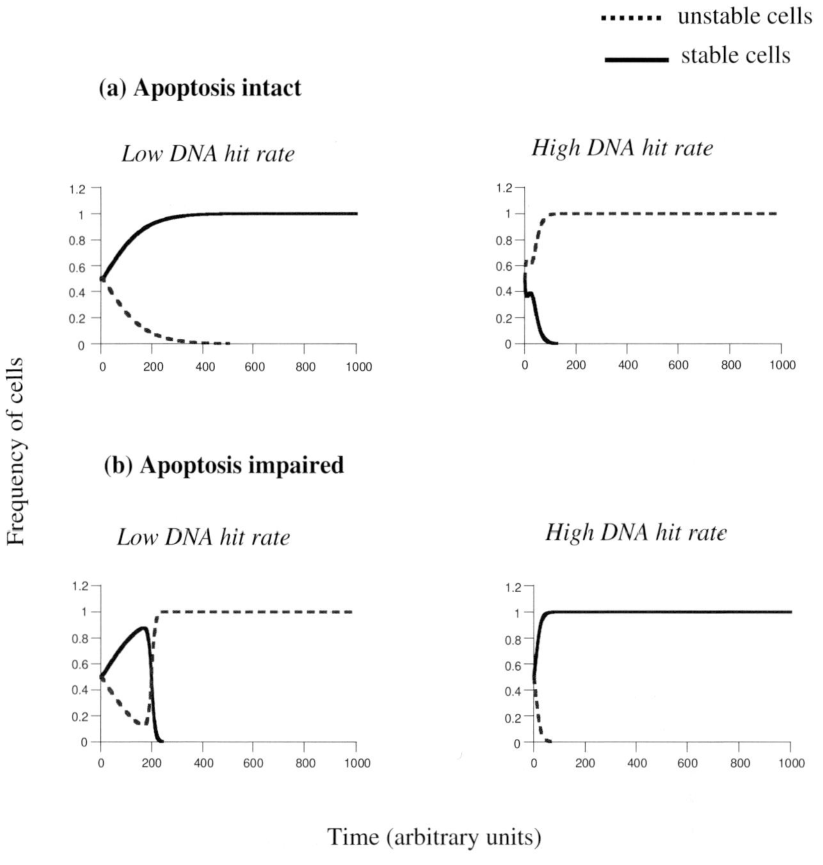

Fig. 6.5 DNA damage and the selection of genetic instability. (a) Cells have intact apoptotic responses. At low DNA hit rates stability wins. At high DNA hit rates instability wins. (b) Cells have impaired apoptotic responses. At low DNA hit rates, instability wins. At high DNA hit rates stability wins. Parameter values were chosen as follows: $\epsilon_s = 0.99$; $\epsilon_m = 0.1$; $\beta = 0.2$. For (a) $\alpha = 0.6l$ $a = 0.5$. For (b) $\alpha = 0.1$; $a = 0.2$. Low DNA hit rate corresponds to $u = 0.07$, and high DNA hit rate corresponds to $u = 0.7$. Fitness landscapes for successive mutants are given in figure 6.4.

The model tells us that in the presence of intact apoptotic mechanisms, a high DNA hit rate selects in favor of genetic instability, while the tissue remains stable and unaltered if the DNA hit rate is low. These results have important practical implications. A high DNA hit rate can be brought about both by the presence of carcinogens, or by chemotherapy. Therefore, if healthy tissue is exposed to carcinogens, we expect genetic instability to rapidly emerge and this can give rise to cancer progression. In the same way, chemotherapy can select for genetic instability in otherwise healthy tissue and thus induce new tumors as a side effect.

6.2.3 Weak apoptosis

Now we assume that the apoptotic mechanisms in cells are impaired. That is, $r_0 < r_N(1-a)$, figure 6.4(b). This means that accumulation of mutations will eventually result in the generation of variants which have a faster intrinsic growth rate compared to unaltered cells. Thus, in principle, both populations are expected to eventually evolve toward the accumulation of mutations and progression to cancer. Hence, both stable and unstable cancers can be observed. However, as we noted before, we assume that these processes occur over different time scales for the two populations of cells, condition (6.34).

If the stable and unstable populations compete, the unstable population will have a higher intrinsic growth rate than the stable population (because the induction of apoptosis in response to mutation is inefficient). Therefore, at low DNA hit rates, u, the mutator phenotype, M, wins the competition (figure 6.5). If the DNA hit rate is increased, the competition can be reversed in favor of the stable cell population, S. This requires that the cost of generating deleterious mutants be greater than the cost of cell cycle arrest (i.e., $C_{del} > C_{arr}$). Furthermore, a sufficient difference in the repair rate of stable and unstable cells is required to reverse the outcome of competition.

Here is the reasoning behind these conclusions. For weak apoptosis, we have

$$R_0 < R_n.$$

In this case, if $(R_n - R_0)/R_n < \alpha/\beta$, then $\bar{\epsilon}_s > 1$. If on the other hand $(R_n - R_0)/R_n > \alpha/\beta$, then $\bar{\epsilon}_s < 0$. We have the following two cases:

- If $\beta > \alpha$, then the function u_c decays with $\Delta\epsilon$, so for reversal to occur we need to have $\Delta\epsilon > \Delta^*\epsilon$.

  (a) For $(R_n - R_0)/R_n < \alpha/\beta$, the function $\Delta^*\epsilon$ decays with ϵ_s and crosses zero at $\epsilon_s = \bar{\epsilon} > 1$. This means that for all ϵ_s, $\Delta^*\epsilon > 0$. We need to require that $\Delta^*\epsilon < \epsilon_s$, which gives the condition

$$\epsilon_s > \frac{\alpha R_n}{\beta R_0}.$$

If this condition holds then the reversal occurs, as long as $\Delta\epsilon > \Delta^*\epsilon$; that is, the difference in repair rates must be larger than the critical value.

(b) For $(R_n - R_0)/R_n > \alpha/\beta$, the function $\Delta^*\epsilon$ grows with ϵ_s and crosses zero at $\epsilon_s = \bar{\epsilon} < 0$. This means that for all ϵ_s, $\Delta^*\epsilon > 0$. We need to require that $\Delta^*\epsilon < \epsilon_s$, which gives again the condition

$$\epsilon_s > \frac{\alpha R_n}{\beta R_0}.$$

If this condition holds then the reversal occurs as long as $\Delta\epsilon > \Delta^*\epsilon$.

- If $\beta < \alpha$, then the function u_c grows with $\Delta\epsilon$, so we need to have $\Delta\epsilon < \Delta^*\epsilon$.

(a) Condition $(R_n - R_0)/R_n > \alpha/\beta$ is impossible to satisfy, so reversal does not happen in this case.

(b) If $(R_n - R_0)/R_n < \alpha/\beta$ then $\Delta^*\epsilon$ is a growing function of ϵ_s which crosses zero at $\epsilon_s = \bar{\epsilon} > 1$. This means that for all ϵ_s, $\Delta^*\epsilon < 0$, and reversal is again impossible.

Our results have practical implications. If cells develop a mutation resulting in impaired apoptotic responses, then genetic instability has a selective advantage if the DNA hit rate is low. Therefore, even if there is no exposure to carcinogens, a chance loss of apoptosis can result in the outgrowth of genetic instability and thus progression of cancer. On the other hand, if there is a growing cancer with impaired apoptotic responses, our results suggest that an elevation of the DNA hit rate by chemotherapeutic agents can reverse the relative fitness in favor of stable cells, and this can result in cancer reduction or slower progression.

A note of clarification: in the above arguments we assumed for simplicity that apoptosis is inefficient in both the unstable and the stable cells. The arguments about chemotherapy, however, remain robust even if we assume that only the mutator phenotypes have impaired apoptosis, while the stable and healthy population of cells has intact apoptotic responses. The reason is that over the time frame considered, the population of stable cells remains genetically unaltered (i.e., at stage S_0). Since the cells are unaltered, the presence or absence of apoptosis does not change the dynamics.

6.3 Overview of the insights obtained so far

The equations have examined the competition dynamics between genetically stable and unstable populations of cells. They identified under which circumstances genetic instability is selected for or against in the context of cancer progression. In particular, they examined the role of the rate at which DNA is damaged.

Table 6.2 Summary of the results gained from the model which takes into account evolution and mutation cascades. If apoptosis is intact, mutators (M) have a lower intrinsic growth rate than stable cells (S). Hence, a high DNA hit rate can select for M. If apoptosis is impaired, M have a higher overall intrinsic growth rate than S. Thus, a high DNA hit rate can select in favor of S.

	Apoptosis intact	**Apoptosis impaired**
Low DNA hit rate	S win	M win
High DNA hit rate	M win if $C_{arr} > C_{del}$	S win if $C_{arr} < C_{del}$

A change in the DNA hit rate can reverse the outcome of competition. In the simplest setting, an increase in the DNA hit rate can switch the outcome of competition in favor of cells characterized by a slower intrinsic growth rate. This requires a sufficient difference in the repair rate between the stable and mutator cells, and a condition on the relative values of costs associated with cell cycle arrest and creation of deleterious mutants. The conditions under which genetic instability is selected for depends on the efficacy of apoptosis. In terms of cancer evolution and progression, this gave rise to the following insights (Table 6.2).

- If apoptosis is strong, accumulation of mutations by unstable cells slows down the intrinsic growth rate because of the frequent induction of cell death. Thus, stable cells have a higher intrinsic growth rate than mutators. Consequently, at low DNA hit rates, the stable cells win. The presence of high DNA hit rates can, however, result in the selection and emergence of the genetically unstable cells. This occurs if the cost of cell cycle arrest upon repair is higher than the cost of creating deleterious mutations. These dynamics can have implications for conditions, in which inflammatory processes increase the amount of DNA damage and lead to the prevalence of genetically unstable cells in otherwise healthy tissue. This in turn can predispose the tissue to the development of cancer. An example of this is ulcerative colitis

[Rabinovitch *et al.* (1999)]. These notions are further supported by data demonstrating that increased inflammation and DNA damage can provide an advantage for unstable cells [Breivik (2001); Bardelli *et al.* (2001)].

- On the other hand, if apoptotic responses are impaired, accumulation of mutations by unstable cells will not result in frequent cell death upon division. Therefore, the intrinsic growth rate of unstable cells can be higher than that of stable cells if adaptive mutations are acquired. In this case, genetic instability is expected to emerge at low DNA hit rates. At high DNA hit rates, however, genetic instability can be selected against and mutators can go extinct. This occurs if the cost of creating deleterious mutations is higher than the cost of cell cycle arrest. These results have implications for chemotherapies, which is explored in the next section.

6.4 Can competition be reversed by chemotherapy?

The results derived in this chapter have implications for the use of chemotherapy (figure 6.6). Chemotherapy essentially increases the degree of DNA damage. Therefore, it can be used to reverse the relative fitness of stable and unstable cells such that unstable cells are excluded (figure 6.6). This can drive progressing cancer cells extinct and result in the persistence of stable cells. These may either be healthy cells or less aggressive and slowly progressing tumor cells. In order to achieve this outcome, there needs to be a sufficient difference in the repair rate between stable and unstable cells. If this is not the case, therapy can merely slow down cancer progression.

Since in this scenario, chemotherapy acts by modulating the competition between stable and unstable cells, it is not a requirement that every last cell is killed by the drugs. Selection and competition will make sure that the unstable cancer cells are driven extinct. This argument, however, requires that there is an element of competition between unstable and stable cells. Whether and under which circumstances this is the case remains to be determined.

This is a different mechanism of action compared to the traditional view which assumes that chemotherapy only acts by killing dividing cells. For chemotherapy to reverse the fitness of stable and unstable cells, two conditions are required. (i) There needs to be a sufficient difference in the repair

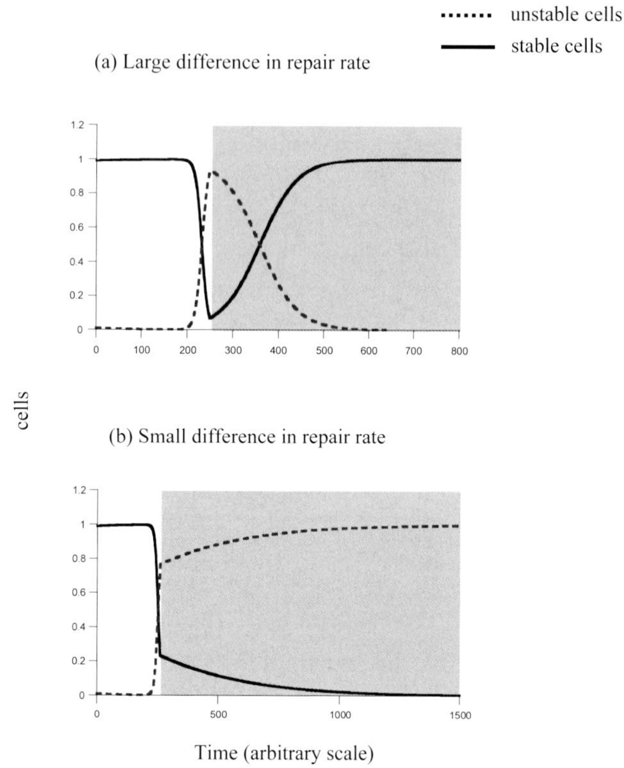

Fig. 6.6 Simulation of chemotherapy, modeled by an increase in the DNA hit rate, u. We start with a situation where cells which are unstable and have impaired apoptosis spontaneously give rise to cancer growth and progression. (a) If there is a large difference in repair rate between stable and unstable cells, therapy can exclude the unstable cells. (b) If there is a smaller difference between stable and unstable cells, then therapy fails to exclude instability. Parameters were chosen as follows: $\epsilon_s = 0.99$; $\beta = 0.2$; $\alpha = 0.1$. We assume that the degree of apoptosis differs between stable and unstable cells. Stable cells have intact apoptosis ($a = 0.5$), while unstable cells have impaired apoptosis ($a = 0.2$). For (a) $\epsilon_m = 0.1$. For (b) $\epsilon_m = 0.4$. Low DNA hit rate corresponds to $u = 0.07$, and high DNA hit rate corresponds to $u = 0.8$. Fitness landscapes for successive mutants are given in figure 6.4.

rate between stable and unstable cells. The higher the replication rate of unstable cells relative to stable cells, the higher this difference in the repair rate required to achieve success. Therefore, contrary to traditional views, a faster rate of cell division of cancerous unstable cells renders successful treatment more difficult in this scenario. (ii) The cost of generating lethal mutants in unstable cells must be higher than the cost of cell cycle arrest in stable cells. If this is not the case, it does not pay to retain repair mechanisms, and the fitness of unstable cells can never be reversed. In this case, treatment has a higher negative impact on stable than on unstable cells, and the mutators are resistant.

6.5 Summary

This chapter explored how the environment, and in particular the mutation rate, can influence the relative fitness of genetically stable and unstable cells. One of the core results was that an increase in the mutation rate can lead to a reversal of the competition outcome. Depending on the exact conditions and circumstances, an increased hit rate can either lead to a selective advantage or disadvantage of genetically unstable cells. As mentioned in the previous section, this can be utilized by chemotherapy, where a treatment-induced increase in the rate at which genomes get altered can lead to the selection against unstable cells that might be responsible for driving pathogenesis. This highlights the clinical importance of understanding *in vivo* ecological processes that are inherent in the population dynamics of cancer cells. Rather than directly killing pathogenic cell lines, it might be more effective to create environmental conditions that select against those cell lines and thus indirectly lead to their disappearance. This concept goes beyond the competition between genetically stable and unstable cells. Environmental conditions are likely crucial with respect to selection of several different cellular characteristics that drive progression, and therapy can be aimed at changing these environmental conditions such that those pathogenic cell lines become disadvantageous.

Problems

Problem 6.1. Research project. *Learn about quasispecies equations, the first derivation by M. Eigen and P. Schuster [Eigen and Schuster (1979)], and different ways in which they are used in various branches of quantitative biology.*

Problem 6.2. Research project. *Learn about the concept of "mutator phenotype" and the work of L. Loeb who introduced this concept (see references in this chapter).*

Problem 6.3. *Show that equations (6.1-6.2) add up to zero under the condition* $S + M = 1$. Hint: *use expressions (6.3) for the average fitness,* ϕ.

Problem 6.4. *Derive formula (6.6) by first finding the expression* u_c *for which the values for cell fitness defined in (6.4) are equal to each other (this is given in equation (6.5)), and then requiring that requiring that* $0 \leq u_c \leq 1$.

Problem 6.5. *Show that equations (6.14-6.16) add up to zero under the condition* $\sum_{k=0}^{n+1} x_k = 1$.

Problem 6.6. *Show that quantity (6.20) satisfies system (6.14-6.16).*

Problem 6.7. Numerical project. *Consider the following version of quasispecies equations:*

$$\dot{x}_0 = a_0(1-u)x_0 - \phi x_0, \tag{6.36}$$

$$\dot{x}_i = a_{i-1}ux_{i-1} + a_i x_i(1-u) - \phi x_i, \qquad 1 \leq i \leq n-1, \tag{6.37}$$

$$\dot{x}_n = a_{n-1}ux_{n-1} + a_n x_n - \phi x_n, \tag{6.38}$$

where

$$\phi = \sum_{i=0}^{n} a_i x_i.$$

(a) Show that the ODEs add up to zero under the condition $\sum_{i=0}^{n} x_i = 1$. *(b) Suppose that all values* a_i *are distinct, and* a_k *with* $k \neq 0$ *is the largest of them:* $a_k = \max_i\{a_i\}$. *Write a code to simulate the behavior of system (6.36-6.38) starting with the initial condition* $x_0 = 1$, $x_i = 0$ *for* $1 \leq i \leq n$. *What is the long-term behavior? (c) Use the transformation to the variable* z *as in equation (6.20) to solve the system analytically and confirm the numerical results. (d) What happens if there are two values,* $a_k = a_m$ *with* $m \neq k$ *such that* $a_k = \max_i\{a_i\}$ *(that is, two phenotypes have a maximum fitness)?*

Chapter 7

Chromosomal instability and tumor growth

Chapter 6 considered competition dynamics between genetically stable and unstable cells, and examined the conditions that can lead to the selection of genetically unstable cell populations *in vivo*. Unstable cells, in turn, can accumulate selfish, advantageous mutations, that allow them to undergo uncontrolled clonal expansion, thus forming a tumor. This chapter investigates the effect of genetic instability on tumor formation in more detail, taking into account both the required mutant generation and the clonal expansion of the mutated cell population.

As reviewed in Chapter 2, genomic instability can be divided into two broad categories: small scale mutations (such as MSI) and gross chromosomal abnormalities (such as CIN). After a mutant cell has been generated, it needs to give rise to clonal expansion for the cancer to be established. If the instability induces the generation of subtle sequence changes, the process of clonal expansion is not likely to be influenced to a significant degree. If the instability induces destructive genomic changes, such as imbalances in genes and chromosome numbers and loss-of-heterozygocity (LOH) events, then clonal expansion can be compromised significantly. Although the cells can undergo uncontrolled growth, genome destruction can result in frequent cell death and this could counteract the establishment of a cancer. Therefore, while CIN can speed up the generation of a mutated cell with an inactivated tumor suppressor gene, it might impair the growth of these cells and slow down clonal expansion. In the light of this tradeoff, what is the overall effect of CIN on the establishment and progression of a cancer? This is the subject of the current chapter.

7.1 The effect of chromosome loss on the generation of cancer

We study the role of chromosomal instability in the context of the inactivation of tumor suppressor genes (TSG). We will concentrate on a specific event, namely, the LOH event [Thiagalingam *et al.* (2001); Sotillo *et al.* (2009); Janssen and Medema (2013); Burrell *et al.* (2013)]. Other features of CIN such as mitotic recombination or chromosome duplication, may contribute to an activation of oncogenes or gene dosage effects [Luo *et al.* (2000); Tischfield and Shao (2003); Wijnhoven *et al.* (2001)], but such events cannot turn off a TSG. Thus, focusing on cancers with a TSG allows us to isolate one feature of CIN, that is, an increased frequency of LOH, and identify its role in cancer progression.

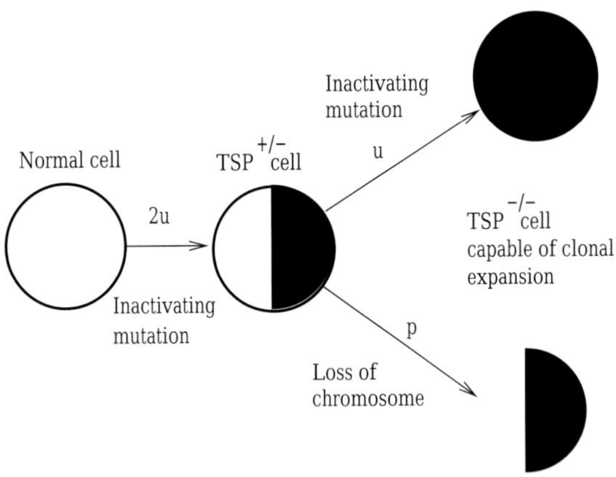

Fig. 7.1 Two mechanisms of a TSG inactivation. First, one allele of the TSG must be inactivated by a small-scale event, e.g. a point mutation. Then, there are two possibilities. Either the second allele experiences another small-scale hit (the phenotype with two inactivated copies of the gene is represented by a black circle). Or, the whole chromosome containing the second, functional copy of the TSG could be lost (this phenotype is represented by a black semicircle).

Let us start our quantitative study by identifying exactly how LOH may influence the inactivation of a TSG, see figure 7.1 [Komarova and Wodarz (2004)], see also Chapter 11. In a normal cell (an empty circle), both maternal and paternal chromosomes are present, and both alleles of the TSG are intact. An inactivating mutation can occur which turns off one of the alleles of the TSG (this is represented by a half of the circle turning black). The corresponding phenotype is denoted by TSG$^{+/-}$. For "classical" TSG's there is no noticeable change in function of such cells; both alleles must be inactivated before a phenotypic change is observed. This second event, the inactivation of the remaining allele of the TSG, can happen in two ways. First of all, another inactivating small-scale event could occur (both halves of the circle become black). Alternatively, the second allele can be lost by a loss-of-chromosome event (this is depicted by means of a "missing" half of a circle). This will unmask the mutated copy of the TSG and lead to a phenotypic change in the cell.

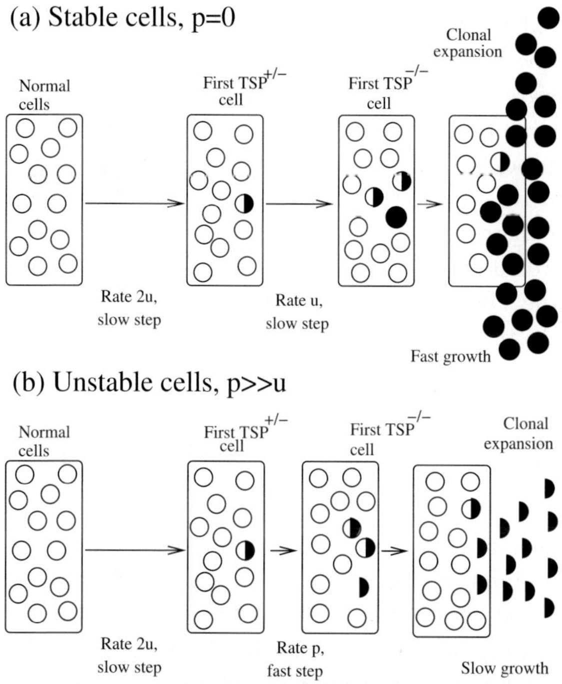

Fig. 7.2 TSG inactivation and clonal expansion. (a) In the case where chromosome losses occur rarely ($p = 0$), we have the following sequence of events: first, a mutant appears which has one copy of the TSG inactivated. After a while, a second mutation may occur producing a cell with two inactivated copies of the TSG. This leads to clonal expansion. (b) If losses of chromosomes are possible, then one of the progeny of the first TSG$^{+/-}$ cell may lose the chromosome containing the functional copy of the TSG, thus giving rise to cells with one inactivated TSG copy and one "missing" TSG copy. Such cells enter a phase of clonal expansion, but this happens at a slower rate compared to (a) because of frequent chromosome loss events resulting in dead or non-reproductive cells.

Let us denote the rate at which small-scale genetic events happen by u (per cell division per gene), and the rate of LOH by p (per cell division per chromosome). The basic rate at which such mutation events occur in stable cells has been estimated to be approximately $u = 10^{-7}$ per cell division per gene. The inactivation of the first allele of the TSG will happen with the rate $2u$, because there are two alleles. The inactivation of the second allele can happen with the rate u by a mutation, and with the rate p by loss of chromosome, see figure 7.2. Let us first suppose that the rate of chromosome loss is zero, $p = 0$, figure 7.2(a); a TSG can only be inactivated by two consecutive, independent (small scale) genetic events. This is possible, but the probability of such a double mutation is very low. Next, let us consider the opposite extreme, where the rate of LOH is very high, such that $p \gg u$, figure 7.2(b). Now, the second inactivation event happens with probability p, that is, it is greatly accelerated compared to the case $p = 0$. However, the price that the cell lineage has to pay is a very high rate at which non-viable mutants are produced. This will considerably slow down the expansion of the TSG-negative phenotype.

Therefore, there must be an intermediate, optimal (for cancer!) value of the rate of chromosome loss, for which wild-type cells have a high chance of inactivating the TSG, without having to pay too high a price in non-viable or non-reproductive mutants.

7.2 Calculating the optimal rate of chromosome loss

The model set-up. We model epithelial tissue organized into compartments. In the simplest case, there is one stem cell per compartment. For example, in colon this would correspond to crypts with a stem cell situated at the base of each crypt. Stem cells divide asymmetrically producing one (immortal) stem cell and one differentiated cell. Here we concentrate on the dynamics of the stem cells. Each division event is equivalent to a replacement of the old stem cell with a copy of itself. Upon division of a stem cell, the immortal daughter cell might (i) acquire a silencing mutation in one of its alleles of the APC gene with probability u per cell division, or (ii) lose one of its chromosomes, with probability p per cell per cell division per chromosome. Once both copies of the TSG have been inactivated, the cell will be able to escape homeostatic control and create a growing clone. We will describe the clonal expansion by a deterministic model.

Uncertainties still exist about the exact cellular origins of cancer, see Chapter 12 for a discussion. According to the stem cell theory, it is the stem cells which are at risk. The de-differentiation theory suggests that partially differentiated cells could be targets for cancerous mutations. If we do not want to restrict ourselves to one or the other theory, we can solve the problem of optimization for a number of different assumptions. According to one scenario, cancer is initiated in adult stem cells. There are one or several stem cells per compartment, such as the crypt of the colon. Alternatively, we can assume that a healthy compartment contains a population of partially differentiated cells, which are subject to a constant turnover, but still maintain a constant size of the compartment. Depending on the number of cells, this can be described either with a stochastic or a deterministic model. Again, inactivation of a TSG results in clonal expansion. It turns out that the results remain very similar in the context of the different assumptions.

Optimal rate of chromosome loss. Suppose that a stem cell loses a chromosome with probability p per chromosome per cell division. First we calculate the probability to inactivate the TSG gene by time t. The sequence of events can be expressed by the following simple diagram,

$$\begin{array}{ccccc} Y_0 & \xrightarrow{\quad 2u \quad} & Y_1 & \xrightarrow{\quad p \quad} & Y_2 \\ \Big\downarrow{\scriptstyle d_0(k)} & & \Big\downarrow{\scriptstyle d_1(k)} & & \\ \text{lose } 2k & & \text{lose } 2k-1 & & \\ \text{chromosomes} & & \text{chromosomes} & & \end{array} \qquad (7.1)$$

Here Y_i is the probability for the stem cell to have i inactivated copies of the TSG. The first event of inactivation happens by a fine-scale genetic event (probability u times two for two alleles), and the second event is a loss of the chromosome with the remaining copy of the TSG gene (probability p). The parameter k is related to the cost of chromosome loss, as explained below.

A very important issue here is the exact cost of LOH events for the cell and its reproductive potential. At one end of the spectrum, (a), there is no reduction in fitness due to the loss of any other chromosomes: the only chromosome that "counts" is the one containing the TSG. At stage Y_0, a loss of either copy damages the cell, and at stage Y_1, a loss of the chromosome with the *mutated* copy of the TSG is harmful, and a loss of the other copy leads to a clonal expansion. An alternative interpretation of this extreme case is that while loss of a single chromosome copy would reduce fitness, this is buffered by duplication events. At the other end of the spectrum, (b), a loss of any chromosome results in cell death, unless it leads to a TSG inactivation. It is safe to say that the reality is somewhere between these extreme scenarios.

For scenario (b), we set $d_0(k) = 1-(1-p)^{2k}$ and $d_1(k) = 1-(1-p)^{2k-1}$, where $k = 23$ is the number of chromosomes. For scenario (a), the death rates can be expressed by the same formulas with $k = 1$.

Diagram 7.1 allows us to write down the Kolmogorov forward equations for all the probabilities, see e.g. [Karlin and Taylor (1975)]. For example, in an infenitesimally-small time-interval, Δt, a wild-type stem cell can die with probability $d_0(k)\Delta t$, or become mutated with probability $(1 - d_0(k))2u\Delta t$. Similarly, a stem cell containing one inactivated copy of a TSG can die with probability $d_1(k)\Delta t$, or become a double-hit mutant with probability $(1 - d_1(k))p\Delta t$. Skipping the argument k of d_0 and d_1, we have the following system:

$$\dot{Y}_0 = [(1 - d_0)(1 - 2u) - 1]\, Y_0, \tag{7.2}$$

$$\dot{Y}_1 = (1 - d_0)2uy_0 + [(1 - d_1)(1 - p) - 1]\, Y_1, \tag{7.3}$$

$$\dot{Y}_2 = (1 - d_1)pY_1, \tag{7.4}$$

with the initial condition $Y_0(0) = 1$. Note that while these equations are valid for a single stem cell, more sophisticated methods exist for describing stochastic dynamics of populations of cells, see Chapter 10. We need to calculate the probability distribution of creating a TSG$^{-/-}$ mutant as a function of time, which is given by $\dot{Y}_2$. We have,

$$\dot{Y}_2(t) = \frac{up(e^{-ut} - e^{-(p+d_1)t})}{p + d_1 - u}, \tag{7.5}$$

where we assumed that $ut \ll 1$. Once a TSG$^{-/-}$ cell has been produced, it starts dividing according to some law which is (at least, initially) close to exponential. Starting from one cell at time $t = 0$, by time t we will have $Z_y(t)$ cells, with

$$Z_y(t) = e^{a[1-d_1(k)]t}, \tag{7.6}$$

where the parameter a is the growth rate of the initiated cells. The factor $[1 - d_1(k)]$ comes from the probability for a CIN cell to produce a non-viable mutant, which for scenario (a) only happens if only one particular chromosome is lost, and for scenario (b) - if any chromosome is lost.

If we now include the mutation stage, we will need to evaluate the convolution,

$$Z_y(t) = \int_0^t \dot{y}_2(t')e^{a(1-p)^{2k-1}(t-t')}\,dt'.$$

The integral yields the following law of growth:

$$Z_y(t) = \frac{upe^{a(1-d_1)t}}{a(1-d_1)},$$

where we assumed that for relevant times, $at > 1$.

Let us ask the following question: *how long does it take, on average, for a TSG$^{-/-}$ clone to reach a certain size?* The answer will depend on all the parameters of the system, and in particular, on the rate of chromosome loss, p. For the reasons explained above, the waiting time will be relatively large both for $p = 0$ and for very high values of p. Indeed, for very small p the mutations that lead to a TSG inactivation will take too long, and for very large values of p, the clonal expansion will be too slow because of the amount of non-viable or non-reproductive cells produced. The waiting time will have a minimum for an intermediate value of $p = p_*$, which we call the *optimal* (for cancer) value of the rate of chromosome loss. With this value of p, a cancer will appear and grow at the fastest rate. This approach is equivalent to the "minimum-time-to-target" method in optimization theory.

To find an optimal value of p that maximizes the growth, we solve $Z_y(t) = M$ for t, which gives,

$$T = \frac{1}{a(1-d_1)} \log\left[\frac{aM}{u}\frac{1-d_1}{p}\right],$$

and then we minimize this as a function of p. This can be done easily if we assume that $p \ll 1/(2k)$ (it will turn out that the result for p_* satisfies this assumption). Expanding the expression dT/dp in terms of p, we obtain the equation for p,

$$\frac{1}{p} = (2k-1)\log\frac{aM}{up},$$

where we formally have $k = 1$ for scenario (a), and $k = 23$ for scenario (b).

Parameter dependence of the result. The result for p_* turns out to be amazingly robust, see Table 7.1. We can see that p_* depends logarithmically (that is, weakly), on the combination $\kappa = aM/u$. As we vary these parameters over many decades, so that κ changes from 10^5 to 10^{20}, the result for the optimal value of p varies only slightly. Interestingly, it also does not significantly depend on the overall fitness cost for the cell brought about by chromosome loss. The results for scenarios (a) and (b) are presented in Table 7.1. The optimal value of p is lower in scenario (b), which is not surprising because this case assumes a higher penalty for chromosome loss events. The remarkable fact is that the values of p_* for the two scenarios are so close to each other, and that they depend so little on the assumptions of the model.

Table 7.1 Calculated values for the optimal rate of chromosome loss, p_*, for different values of the parameter, κ, and for each of the two scenarios, (a) and (b).

Scenario	$\kappa = 10^{20}$	$\kappa = 10^5$
(a)	2×10^{-2}	8×10^{-2}
(b)	5×10^{-4}	2×10^{-3}

Large initial number of cells. In order to handle the scenario where a large number of cells are competing in a compartment, we can model the dynamics as a deterministic quasispecies set of equations, see [Eigen and Schuster (1979)] and also Chapter 6. Suppose that the abundance of wild-type cells is y_0, the abundance of cells with one copy of the TSG inactivated is y_1, and the abundance of double-hit mutants is y_2. We have,

$$\dot{y}_0 = (1 - 2u - d_0)y_0 - \phi y_0, \tag{7.7}$$

$$\dot{y}_1 = 2uy_0 + (1 - u - p - d_1)y_1 - \phi y_1, \tag{7.8}$$

$$\dot{y}_2 = (p + u)y_1 + (1 - d_1)y_2, \tag{7.9}$$

where

$$\phi = \frac{(1 - d_0)y_0 + (1 - d_1)y_1}{N}.$$

Cells reproduce and die, and the rate of renewal is taken 1 for types y_0 and y_1. In the absence of double mutants, y_2, the total number of cells y_0 and y_1 stays constant (that is, the sum of equations (7.7-7.9) with $y_2 = 0$ is zero). The optimal rate of chromosome loss in this case can be calculated numerically, and is close to the rate p_* found from the stochastic model.

Experimental measurements of the rate of LOH. We can compare these results with the value of the rate of chromosomal loss obtained by [Lengauer *et al.* (1997)] in vitro for several CIN colon cancer cell lines. In their paper, Lengauer et al. allowed cell colonies to grow from a single cell for 25 generations, after which FISH analysis was performed on a subset of the progeny. This allowed to count the number of individual chromosomes in cells. The average number of chromosomal copies was calculated for each cell line, for each chromosome, and this was compared with the mode number, equivalent to the number of chromosome copies in the original cell. This was the first (and only) experiment which allowed to calculate the *rate* of chromosome loss and gain, as opposed to the estimates of the *frequency* of various chromosomal aberrations in a given lesion/cell colony. Two types of cancer cells have been used: some known to possess mismatch repair instability, and some characterized by CIN. In the cell lines with microsatellite instability, the rate of chromosome loss was the same as control (and indistinguishable from the background). In the chromosomally unstable cell lines, the rate of chromosome copy change was highly elevated. The value that emerges from experiments of Lengauer et al. is $p = 10^{-2}$ per chromosome per cell division, which is almost exactly in the middle of the range that we obtained theoretically.

7.3 The optimal rate of LOH: a time-dependent problem

So far in this chapter we asked: What is the optimal level of instability which makes the cancer progress in the fastest way? The mathematical problem is finding the most efficient (from the point of view of cancer) rate at which genetic changes occur in cells. It was shown that "too much" instability is detrimental for the cells because of an increased death rate. "Too little" instability also slows down the progress because the basic rate at which cancerous mutations are acquired is low. An optimal level of genetic instability was identified which maximizes the rate of progression. This was quantified in terms of the probability of chromosomal loss per cell division.

Next, we take this model a step further and study the temporal change of the level of instability. As cancer progresses, the microevolutionary pressures inevitably change. What might have been a good strategy at the beginning of the growth may be detrimental for the colony later on. A highly elevated rate of LOH calculated in this chapter may indeed be optimal during early and intermediate stages of cancer progression as long as they involve TSG's. However, it is well-known that a lesion cannot grow above a certain small size (about 2 mm) without extra blood supply (angiogenesis). A larger or metastasizing tumor is hard to maintain, and the price of losing chromosomes becomes too high to be balanced by an elevated variability. Therefore, it is possible that at later stages, the optimal rate of LOH will decrease to nearly zero [Komarova (2004); Komarova *et al.* (2008a)]. This is consistent with observations that late stage cancers are sometimes (surprisingly) stable [Albertson *et al.* (2003)].

Similarly, if the activation of oncogenes (rather than the inactivation of tumor suppressor genes) plays a major role in the progression of cancers, then chromosomal instability is likely to be detrimental to the cancer. Indeed, to turn on an oncogene, a small scale mutation is often needed rather than a chromosome loss event or another crude chromosomal change. Moreover, a chromosome loss event may lead to the inactivation of a functioning oncogene which will revert the process of oncogenesis.

In [Chin *et al.* (2004)] it is argued that the level of genetic instability in breast cancers first increases, reaches a peak and then decreases as the cancer passes through telomere crisis. A paper by [Rudolph *et al.* (2001)] reports data on intestinal carcinoma in mice and humans which is consistent with a similar model: telomere dysfunction promotes chromosomal instability which drives carcinogenesis at early stages, and telomerase activation restores stability to allow further tumor progression. The mechanism of telomerase activation and subsequent prevention of chromosomal instability is also described in the papers by [Samper *et al.* (2001)] and [Artandi and DePinho (2000)]. It is shown that short telomeres can make mice resistant to skin cancer because of an increased cell death rate [Gonzalez-Suarez *et al.* (2000)] which also suggests that telomerase activation and reduction in the level of chromosomal instability may be a necessary step for cancer to develop.

The idea that instability may be beneficial for cancer at an early stage and can become a liability later on is developed mathematically in the rest of this chapter. We formulate the time-dependent optimization problem to investigate the way to maximize cancerous growth. To model growth and mutations, we employ ODEs similar to equations (7.7-7.9). Using methods of optimal control theory, we find strategies most advantageous for the tumor's growth. The degree of instability (the rate of mutations) appears as an unknown function of time, sought to maximize the growth of the mutants.

7.3.1 *Formulation of the time-dependent problem*

We will consider two types of processes: a simpler one-step process (see diagram (7.10) below) and a two-step process (see diagram (7.11)):

$$\begin{array}{ccccc} x_1 & \xrightarrow{u+p} & x_2 & \longrightarrow & \text{growth} \\ {\scriptstyle d(p)}\big\downarrow & & \big\downarrow{\scriptstyle d(p)} & & \\ \text{death} & & \text{death} & & \end{array} \tag{7.10}$$

$$\begin{array}{ccccccc} x_0 & \xrightarrow{2u} & x_1 & \xrightarrow{u+p} & x_2 & \longrightarrow & \text{growth} \\ {\scriptstyle d(p)}\big\downarrow & & \big\downarrow{\scriptstyle d(p)} & & \big\downarrow{\scriptstyle d(p)} & & \\ \text{death} & & \text{death} & & \text{death} & & \end{array} \tag{7.11}$$

In the first scenario, we assume that a colony of cells (of abundance x_1) is currently at a constant population size, near a selection barrier. The growth is stalled, and the cellular population remains near the "carrying capacity", which is defined by the available space, nutrients, and the cells' ability to divide and die. This barrier can be overcome by the offspring of a mutant whose properties are different (the abundance of these mutants is denoted by x_2). Here we assume that the transformation proceeds according to a gain-of-function scenario (Chapter 9). For instance, the mutant cells could have an activated oncogene and show an increased division rate, or a decreased death rate. We assume that such transformed cells are created by means of one molecular event, genetic or epigenetic.

In the second scenario, the growth advantage is acquired by means of a loss-of-function mutation, such as an inactivation of a TSG (Chapter 10). Here, x_i with $i = 0, 1, 2$ stands for the abundance of cells with i copies of the TSG inactivated. There is a certain asymmetry between the first and the second inactivation events. The inactivation of the first copy of the TSG happens at rate $2u$, where u is the inactivation rate per copy. The inactivation of the remaining copy of the TSG happens at rate u. In addition, it can also be inactivated by an LOH event at rate p. In principle, the first allele can also be inactivated by an LOH event. However, the fitness of a cell with a missing chromosome and one active copy of the TSG is very low. Such cells will quickly die out and will make no difference for the present analysis. On the other hand, a cell with one chromosome missing and both copies of the TSG inactivated, has a selective advantage, because we assume that the inactivation of the TSG leads to an increase in the cell's growth rate. Such cells are produced by the sequence of events depicted in diagram 7.11.

The systems of quasispecies–like ODEs describing these processes can be formulated as follows. For the one-step process (diagram (7.10)) we have

$$\dot{x}_1 = (1 - u - p - d(p))x_1 - \phi x_1, \tag{7.12}$$
$$\dot{x}_2 = (p + u)x_1 + a(1 - d(p))x_2 - \phi x_2, \tag{7.13}$$

where $\phi = (1 - d(p))x_1/N$, and $x_1(0) = N$, $x_2(0) = 0$. The term ϕ is similar to logistic growth in the absence of mutants (see also Chapter 4), and accounts for the homeostatic control present in a system of x_1 cells. x_2-cells break out of regulation and enter a phase of exponential growth. The term with x_1 in the equation for x_2 is added to represent a partial, non-symmetric, homeostatic control that may play some role at the beginning of the growth of x_2 cells. Later on that term is simply a correction to the growth rate of the x_2 cells.

For the two-step process (diagram (7.11)) we have

$$\dot{x}_0 = (1 - 2u - d(p))x_0 - \phi x_0, \tag{7.14}$$
$$\dot{x}_1 = 2ux_0 + (1 - u - p - d(p))x_1 - \phi x_1, \tag{7.15}$$
$$\dot{x}_2 = (p + u)x_1 + a(1 - d(p))x_2, \tag{7.16}$$

where

$$\phi = \frac{(1 - d(p))(x_0 + x_1)}{N}.$$

Cells reproduce and die, and the rate of renewal is taken 1 for types x_0 and x_1. In the absence of dangerous mutants, x_2, the total number of cells x_0 and x_1 stays constant (that is, the sum of equations (7.14-7.16) with $x_2 = 0$ is zero). The mutants x_2 expand at rate $a > 1$.

This way of modeling the dynamics is not unique, and a detailed discussion of the robustness of the model is presented in [Komarova *et al.* (2008a)].

In diagrams (7.10,7.11), the death rate, $d(p)$, is a function of the mutation rate, p. If p is small, then chromosome losses do not happen, and if p is large, a cell often loses chromosomes which results in an increased death rate. Therefore, in general, the function $d(p)$ will be a monotonically increasing function of p. Here we consider an example of a parameterization of the death rate as a function of p. It is convenient to introduce a normalized rate of chromosome loss, $\tilde{p}$, and express the death rate in terms of this parameter:

$$d(\tilde{p}) = d_m \left(1 - (1-\tilde{p})^\alpha\right), \quad \tilde{p} = \frac{p - p_{min}}{p_{max} - p_{min}}, \quad p_{min} \leq p \leq p_{max}, \quad \alpha > 0, \tag{7.17}$$

see figure 7.3. The case $\alpha > 1$ corresponds to a concave function, and the case $\alpha < 1$ - to a convex function. The motivation for this particular dependency is as follows. Let us suppose that a cell dies if it loses one of α essential chromosome copies (out of the total of 2×23 copies in a human). Then, if we set $p_{min} = 0$ and $p_{max} = 1$, the death rate (7.17) can be written in a form

$$d(\tilde{p}) = d_m \times [\text{Probability of cell death by chromosome loss}].$$

The constant d_m defines the magnitude of the death rate, and the values p_{min} and p_{max} define a biologically relevant range of the LOH rate, p. We allow α, the exponent in equation (7.17), to be a real positive number. In particular, we investigate the influence of the concavity of this function on the optimal solution (cases $\alpha < 1$ and $\alpha > 1$). The special case $\alpha = 1$ yields a control problem where the controls enter linearly, a case much studied in the optimal control literature and rich with analytical results.

In this chapter we adopt the theoretical framework where it is possible to set the rate of genetic instability to an arbitrary (but meaningful) value at each moment of time. Mathematically, this means specifying the LOH rate, $p(t)$, as a function of time. Every choice of such a function determines a growth process of the tumor. We will seek the choice of $p(t)$ that allow the cancerous population to reach a given size, M, in the shortest possible time. In the terminology of optimization and control theory, the population size M is called the "target", the possible values of LOH rate p are called "admissible controls", and each choice of the function $p(t)$ is called a "strategy". A strategy steering the system to the target faster than any other strategy is said to be an "optimal strategy", or "optimal control". In this terminology, we seek a strategy for controlling the system to reach the target as soon as possible. A meaningful qualitative comparison between two strategies is now possible: the "better", or "more advantageous", strategy is the one allowing the system to reach the target sooner. Thus, an "advantageous strategy" is advantageous for the cancer in the sense of Darwinian microevolution in an individual organism.

To find the optimal strategy, we consider the quantity, T, which is the solution of the equation

$$x_2(T) = M,$$

where x_2 is the solution of system (7.12-7.13) or (7.14-7.16). The growth time, T, depends on all the parameters of the system, including the time-dependent mutation rate, $p(t)$. The optimal strategy is the one that minimizes the value of T.

7.3.2 *Mathematical apparatus*

In this section we develop a mathematical framework for the one-step process. Similar calculations lead to the corresponding formulation for the two-step problem, see [Komarova *et al.* (2008a)].

7.3.2.1 *Statement of the one-step problem*

Equations of State: Let us define the following parameter combinations: $\sigma = M/N$, $\mu = u + p_{min}$, $p_m = p_{max} - p_{min}$. Introducing the scaled quantities $x_1^* = x_1/N$, $x_2^* = x_1/M$, and dropping the asterisks for simplicity, we can rewrite system (7.12-7.13) as

$$\dot{x}_1 = -(\mu + p_m\tilde{p})x_1 + [1 - d(\tilde{p})](1 - x_1)x_1 \equiv g_1(x_1, x_2, \tilde{p}), \quad (7.18)$$

$$\dot{x}_2 = \frac{1}{\sigma}(\mu + p_m\tilde{p})x_1 + [1 - d(\tilde{p})](a - x_1)x_2 \equiv g_2(x_1, x_2, \tilde{p}). \quad (7.19)$$

The death rate, $d(\tilde{p})$, is given by equation (7.17). Recall that $\tilde{p}$ is the normalized LOH rate with $0 \leq \tilde{p} \leq 1$. As we show below, the three cases $\alpha > 1$, $\alpha = 1$ and $\alpha < 1$ may have to be treated separately in some cases.

Boundary Conditions: The two ODEs (7.18-7.19) are subject to the following three auxiliary conditions:

$$x_1(0) = 1, \qquad x_2(0) = 0, \qquad x_2(T) = 1, \qquad (7.20)$$

where T is the time when the dangerous mutant cell population reaches the target size.

Problem: Choose the control function $\tilde{p}(t)$ to minimize the time T needed to reach the target population size of the dangerous mutants subject to the inequality constraint,

$$0 \leq \tilde{p} \leq 1, \qquad (7.21)$$

on the control $\tilde{p}(t)$ and non-negative constraints on the cell populations,

$$x_1 \geq 0, \quad x_2 \geq 0. \qquad (7.22)$$

Re-statement of the Optimal Control Problem: To apply the Hamiltonian system approach, we re-cast the problem described above by choosing the control function $\tilde{p}(t)$ to minimize the performance index:

$$J = \int_0^T 1\,dt, \tag{7.23}$$

subject to the equations of state (7.18) and (7.19), the boundary conditions (7.20) and the inequality constraints (7.21) and (7.22) with $T > 0$ as a part of the solution.

7.3.2.2 *The Maximum Principle*

The Hamiltonian and Adjoint Variables. Optimal control problems can be effectively analyzed through the Pontryagin's Maximum Principle and its associated Hamiltonian formalism [Bryson and Ho (1969); Wan (1995)]. In this section we develop components of the Hamiltonian formalism for our system. The *Hamiltonian* for our problem is

$$H = 1 + \lambda_1(t)g_1 + \lambda_2(t)g_2, \tag{7.24}$$

where λ_1 and λ_2 are the two continuous and piecewise differentiable *adjoint* (or co-state) *variables* for the problem chosen to satisfy two *adjoint ODE*,

$$\lambda_1' = -\left(\lambda_1 \frac{\partial g_1}{\partial x_1} + \lambda_2 \frac{\partial g_2}{\partial x_1}\right), \tag{7.25}$$

$$\lambda_2' = -\left(\lambda_1 \frac{\partial g_1}{\partial x_2} + \lambda_2 \frac{\partial g_2}{\partial x_2}\right), \tag{7.26}$$

and (for the given auxiliary conditions on the state variable x_1 and x_2) one *transversality condition*

$$\lambda_1(T) = 0. \tag{7.27}$$

Note that (7.18), (7.19), (7.25), and (7.26) form a Hamiltonian system for the Hamiltonian given in (7.24).

Formulation of the Maximum Principle. The optimal solution of our minimum terminal time problem requires the optimal control function $\bar{p}(t)$ to satisfy the following necessary conditions, known as the Maximum Principle [Gelfand and Fomin (1963); Pontryagin *et al.* (1962); Wan (1995)]:

(1) Four continuous and piecewise differentiable functions $\{\bar{x}_1, \bar{x}_2, \bar{\lambda}_1, \bar{\lambda}_2\}$ exist and satisfy the four differential equations (7.18), (7.19), (7.25), (7.26), and four auxiliary conditions in (7.20) and (7.27) for the admissible control $\bar{p}(t)$.

(2) The minimum terminal time T obtained with $u = \bar{p}(t)$, satisfies a free end condition

$$[H]_{t=T} = \left[1 + \bar{\lambda}_2 \bar{g}_2\right]\Big|_{t=T} = 0, \tag{7.28}$$

with

$$\bar{g}_2 = g_2(\bar{x}_1, \bar{x}_2, \bar{\lambda}_1, \bar{\lambda}_2, \bar{p}(t)),$$

where the adjoint boundary condition (7.27) has been used to simplify the expression for H.

(3) For all t in $[0, T]$, the Hamiltonian achieves its minimum for $\tilde{p} = \bar{p}(t)$, i.e.,

$$H(\bar{x}_1(t), \bar{x}_2(t), \bar{\lambda}_1(t), \bar{\lambda}_2(t), \bar{p}(t)) = \inf_v \left[H(\bar{x}_1(t), \bar{x}_2(t), \bar{\lambda}_1(t), \bar{\lambda}_2(t), v)\right]. \tag{7.29}$$

for all v in the set of admissible controls restricted by (7.21).

(4) If there should be a change in the control $\bar{u}$ at the instance T_s that involves a finite jump discontinuity in the value of the control, optimality requires that the Hamiltonian be continuous at T_s [Bryson and Ho (1969); Wan (1995)]:

$$[H]_{t=T_s-}^{t=T_s+} = 0. \tag{7.30}$$

Given the admissible controls as specified by (7.21), the optimal control $\bar{p}(t)$ can have finite jump discontinuities in $(0, T)$. It follows from the way $\tilde{p}(t)$ appears in the state and adjoint ODE that the state and adjoint variables are continuous in $(0, T)$, including instances of a control jump discontinuity.

The Interior Control. Let us for simplicity take $d_m = 1$, and denote by d' the derivative of $d(\tilde{p})$ with respect to $\tilde{p}$:

$$d' = \alpha(1 - \tilde{p})^{\alpha - 1}. \tag{7.31}$$

Suppose the optimal control strategy $\bar{p}(t)$ satisfies $0 \le \bar{p}(t) \le 1$ and the equation

$$\left[\frac{\partial H}{\partial \tilde{p}}\right]_{\tilde{p} = \bar{p}} = 0 \tag{7.32}$$

with

$$\frac{\partial H}{\partial \tilde{p}} = \lambda_1\{-[p_m + d']x_1 + d' x_1^2\} + \frac{\lambda_2}{\sigma}\{p_m x_1 - \sigma d' x_2(a - x_1)\}$$
$$= p_m x_1 \left(\frac{\lambda_2}{\sigma} - \lambda_1\right) - \{\lambda_1 x_1(1 - x_1) + \lambda_2 x_2(a - x_1)\}d'. \quad (7.33)$$

Then we call $(x_1, x_2, \lambda_1, \lambda_2, \bar{p})$ an *interior solution* for which the optimal control $\bar{p} = p_{int}(t)$ is an extremum of the Hamiltonian. The *stationary* condition (7.32) can be written as

$$\{\lambda_1 x_1(1 - x_1) + \lambda_2 x_2(a - x_1)\}d' = \frac{p_m}{\sigma} x_1\{\lambda_2 - \lambda_1\sigma\}. \quad (7.34)$$

Using the expression for the death rate (7.17) with expression (7.31), we obtain from condition (7.34) the following formula for the interior solution,

$$\tilde{p}(t) = \tilde{p}_{int}(t) = 1 - \left(\frac{p_m x_1(\lambda_2 - \sigma\lambda_1)}{\alpha\sigma\{\lambda_1 x_1(1 - x_1) + \lambda_2 x_2(a - x_1)\}}\right)^{\frac{1}{\alpha-1}}. \quad (7.35)$$

It can be shown [Wan *et al.* (2010)] that the interior solution above violates the inequality constraints (7.21) in some part of the solution domain for some range of system parameter values. Consequently, some combination of the "*upper corner solution*" ($\tilde{p}(t) = 1$), the "*lower corner solution*" ($\tilde{p}(t) = 0$) and the interior solution have to be considered for the optimal solution. Whenever a corner control is applicable, the adjoint variables (and the corresponding adjoint ODE and auxiliary conditions) may or may not play a role in the solution process since the control variable $\tilde{p}(t)$ is completely specified (and *not* determined by the stationary condition (7.32)).

A Vanishing Hamiltonian, $H(t) = 0$. For an autonomous control problem, the Hamiltonian is constant for the optimal solution $\bar{p}(t)$ [Wan (1995); Wan *et al.* (2010)]. The free end condition (7.28) and the continuity condition (7.30) then require $H(t) = 0$. This result (together with the formula for the interior solution, equation (7.35)) will be used to find an optimal control. It can also be used to check how far a candidate control function is from the actual optimal control.

Given the approach sketched above, a combination of analytical and computational methods was used in [Komarova *et al.* (2008a)] to characterize the optimal control function, $\bar{p}(t)$. The results are summarized below.

7.3.3 The optimal strategy for cancer

We used methods of control theory to find the shape of the time-dependent LOH rate, $p(t)$, which maximizes the growth of cancer [Komarova *et al.* (2008a)]. The results are qualitatively similar for the two-step and the three-step models. In both cases, it was found that large mutation rates at first, and lower mutation rates later on, constitute the optimal strategy. The exact shape of the optimal mutation strategy depends on the concavity of the function $d(\tilde{p})$, the instability-dependent death rate. There are two cases. For non-convex death rates, figure 7.3(a), the best performance is achieved if the rate of LOH, $p(t)$, jumps (in a discontinuous, abrupt fashion) from a maximum to a minimum (what is referred to as a "bang-bang" control in the literature). For convex death rates (figure 7.3(b)), the decay of the function $p(t)$ is gradual. In both cases, having the highest possible mutation rate is advantageous at first; later on, it pays off to switch to a lower mutation rate. Figure 7.4 shows some numerically found shapes of the function $\tilde{p}(t)$, the optimal rate of LOH, as a function of time, for the one-step and the two-step models.

Genetic instability is "blind", that is, it does not necessarily "hit" the exact genes necessary for cancer progression. It may cause defects in other genes thus creating deleterious cells. The question is whether the gain in progression speed due to the increased mutation rate would outweigh the losses suffered by the cells as a result of spurious, deadly mutations created by the instability. In order to model this, we introduce two parameters, p_m, the maximum rate of LOH, and d_m, the magnitude of the death rate. Small values of p_m mean a small gain in creating cancerous mutations. Large values of d_m mean a large penalty paid by the colony as a result of many LOH-related deaths. While the function form of the time-dependent rate of LOH, $p(t)$, is always as shown in figure 7.3, the "switch" (abrupt or gradual) from a high degree of instability to lower degree of instability happens at a different time. In some cases, the function $p(t)$ drops down almost immediately, which simply means that instability does not pay off in such cases. In other cases, $p(t)$ remains very high for most of the time-course. And again, there are regimes where the switch happens at intermediate times. Which strategy is optimal largely depends on the parameters p_m and d_m.

Depending on these parameters, there are three qualitatively different forms of optimal strategies. In figure 7.5, black shading means "no instability", white – "instability all the way", and shades of gray correspond to

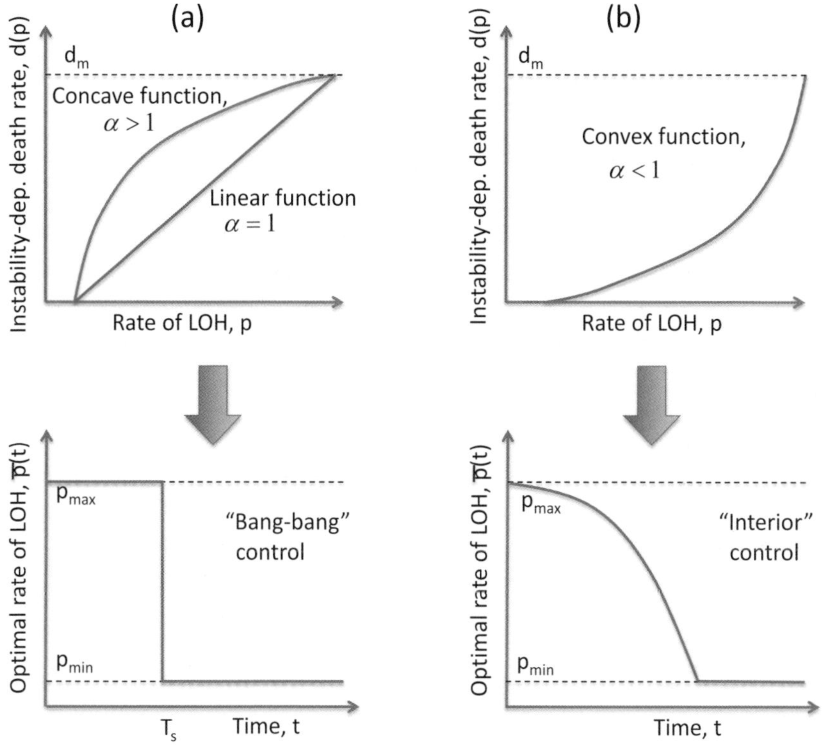

Fig. 7.3 The instability-dependent death rate and the shape of the optimal control, $\bar{p}(t)$. (a) The instability-dependent death rate, $d(p)$, is a linear or concave function, $\alpha \geq 1$. In this case the optimal control is a "bang-bang" control, with one switch at time T_s. (b) If the instability-dependent death rate is a convex function, $\alpha < 1$, the we have an interior optimal control. In both cases, the optimal rate of LOH starts off at its maximum value and then drops down as cancer progresses.

an intermediate switch from high to low levels of instability. In one scenario, the instability makes a minimal contribution to creating carcinogenic mutations (small p_m), but significantly increases the death rate (large d_m). Consequently, it does not pay to be unstable at any stage of the growth in this case. At the other extreme (large p_m and small d_m), instability is useful and it "comes cheap", because the death toll paid by the affected cells is small. Therefore, the colony is better off being unstable at all times. Finally, between these two extremes, optimal controls start at a large rate of LOH and then drop to the lowest possible value at some time during the

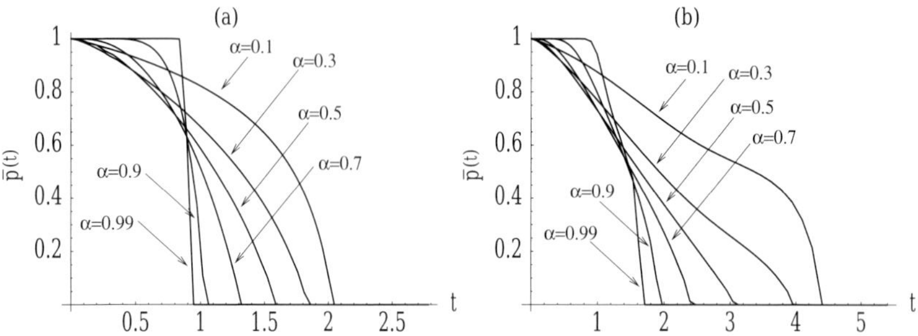

Fig. 7.4 The optimal control found numerically, for different values of $\alpha < 1$, for (a) a one-step process and (b) a two-step process. The other parameters are $a = 2$, $\sigma = 10$, $p_m = 1$, $u = 10^{-1}$.

dynamics. This corresponds to genetic instability being advantageous at the beginning and becoming a liability later. This explains the growing experimental literature suggesting that tumors switch from genetic instability to stability some time in the course of cancer progression.

7.4 The bigger picture

Mathematical theory of optimal control has been used in many areas of biosciences [Sontag (2004); Lenhart and Workman (2007)]. In biomedical applications, control theory has usually been employed to design treatment strategies by methods of optimization, see e.g. [Swan (1990); Kirschner *et al.* (1997); De Pillis *et al.* (2007)]. In this chapter, we applied optimal control theory to studying cancer in a very different way. We solved an optimization problem for the dynamics of cancerous growth in order to understand why cancer behaves the way it does. This approach is similar in spirit to the work of [Iwasa and Levin (1995)], who analyzed the optimal timing of life-strategies of breeding and migrating organisms. In a sense, we study the "ecology" of cancer, based on our current knowledge of carcinogenesis, and hypothesize that the observed behavior of tumors is essentially a consequence of the process of optimization.

We studied optimal strategies for a cancerous colony with respect to the magnitude of the rate of LOH, as the colony acquires carcinogenic mutations and enters a phase of a clonal expansion. Biologically, a large LOH rate corresponds to genetic instability, and a small LOH rate to stability of cancerous cells. The two types of carcinogenic mutations that we considered

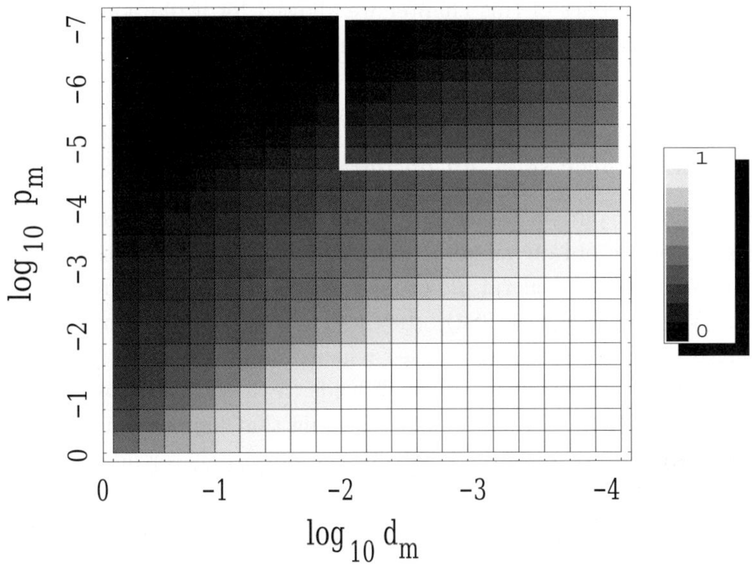

Fig. 7.5 The relative switching time, T_s/T, for the optimal strategy, depending on the parameters p_m and d_m. Black shading corresponds to $T_s/T = 0$ (an immediate switching) and white shading to $T_s/T = 1$ (no switching). The parameters are $a = 2$, $\alpha \geq 1$, $\sigma = 10$, $u = 10^{-4}$.

are activation of an oncogene (the one-step model) or inactivation of a tumor suppressor gene (the two-step model).

In order for a cancer to progress, a cell colony first has to generate carcinogenic mutants, and then to grow in size. Genetic instability may expedite the first of these processes and slow down the second. Therefore, this process can be examined as an optimization problem.

The first part of this chapter assumed that the rate of LOH has a constant value. We calculated the optimal rate of LOH assuming that cancer is initiated by the inactivation of a TSG followed by a clonal expansion. The resulting rate, $p_* \approx 10^{-2}$ per cell division per chromosome, is similar to that obtained experimentally by [Lengauer *et al.* (1997)]. This is a thought-provoking result. A hypothesis consistent with our finding is that the rate at which cancerous cells lose chromosomal material is under selection pressure, and as a result, the optimal rate, p_*, is the one that survives the competition. In other words, out of many possibilities, we will mostly see the cancers that have the rate of LOH close to optimal, because these are cancers that are initiated and progress at the fastest rate.

In the second part of this chapter we allowed for a temporal variation of the rate of LOH, and asked what exact function of time, $p(t)$, would optimize cancer growth.

7.4.1 *Does cancer solve an optimization problem?*

Not literally, of course. However, by solving the optimization problem described, one can obtain valuable information about the growth of cancer. This is similar to the general philosophy of the evolutionary game theory, see e.g. [Maynard Smith (1982)]. In the latter paper, different strategies are played against each other to see which one wins. In principle, one could design an "ideal" strategy which leads to the maximal payoff in the game. The game can be a model of something that happens in nature, for instance the behavior of animals in different situations, or adaptations of cells in various environments. The "ideal" (optimal) strategy may not even be realistic (there are many constraints in nature which escape modeling, but can make a strategy impossible). What occurs in reality, however, tends to approximate an optimal strategy. Finding the "evolutionary stable strategy" or the "Nash equilibrium" in the system helps us understand the general experimentally observed trend. A plausible explanation for the survival of those animals is that they have won the evolutionary game against other animals who used an inferior strategy.

In this chapter, we solved the optimization problem for cancerous growth and found optimal strategies. Does cancer always use an optimal strategy? Probably not. One obstacle to optimality is that cancer is unable to adjust its level of instability instantaneously throughout the entire population to optimize the growth. However, a cancer which follows the general trend, that is, a strategy close to an optimal one, will grow faster. These are the colonies that "succeed" causing a disease. Other pre-cancerous colonies of cells might exist in any organ of an organism, but if they do not use a strategy sufficiently close to optimal, they do not succeed with further growth and are not observed.

7.4.2 *Summary*

Throughout this chapter we formulated the problem of cancer growth from the point of view of cancerous cells, in order to find the most optimal mutation strategy. Selection in this problem takes place on two levels. One level is the level of individual cells, where normal cells are competing for space and nutrients (this is one of the mechanisms of homeostatic control), and cancerous cells escape this control to enter a phase of exponential growth. The forces of selection are manifested in the nonlinearities of the basic ODEs. The second level of selection is the level of cell colonies. Different colonies are characterized by different functions $p(t)$. They are not assumed to be in direct interaction with one another. Winning this competition simply means that a given colony will reach a cancerous state quickly and become observable. The mathematical problem we solve is to find the optimal strategy $p(t)$, such that the colony with this strategy will be the first one to "make it". This work highlights the costs and benefits of chromosomal instability, which is important for understanding the potential role of chromosomal instability for the generation and progression of cancers. It also helps us interpret data on patterns of chromosomal instabilities found at different time points in specific cancers.

Problems

Problem 7.1. Research project. *Learn about different large-scale aberrations observed in different cancers with CIN.* Hint: *Read papers by Mitelman and colleagues, e.g. [Mitelman (2000); Höglund* et al. *(2001, 2002b,a); Frigyesi* et al. *(2003)], and visit Mitelman Database of Chromosome Aberrations and Gene Fusions in Cancer,* http://cgap.nci.nih.gov/Chromosomes/Mitelman

Problem 7.2. Research project. *Find out about the connection between CIN and telomeres.*

Problem 7.3. *It is mentioned that systems (7.12-7.13) and (7.14-7.16) are not the only way to model the processes depicted in diagrams (7.10) and (7.11) respectively. What are some other quasispecies–type models compatible with these diagrams?* Note: *[Komarova* et al. *(2008a)] provides an analysis of robustness of the results with respect to different modeling choices.*

Problem 7.4. Research project. *Mathematical theory of optimal control has been used by many authors to study cancer treatment, see e.g. [Swan (1990); Kirschner* et al. *(1997); De Pillis and Radunskaya (2001); De Pillis* et al. *(2007); Ledzewicz and Schättler (2008)]. Use some of these references to learn about different ways in which control theory has been applied in oncology.*

Problem 7.5. Numerical project. *Suppose that* $p_m = d_m = 1$. *In [Komarova* et al. *(2008a)], the following algorithm was described which allowed to approximate numerically the optimal control function,* $\bar{p}(t)$, *in the case of the one-step process with* $\alpha < 1$:

(1) Start from any control, $0 \leq \tilde{p}_0(t) \leq 1$, *e.g.* $\tilde{p}_0(t) = 1$.
(2) Solve the initial value problem (7.18-7.20) for $0 \leq t \leq T_1$ *such that* $x_2(T_1) > 1$.
(3) Find the solution, $t = T_0$, *of the equation* $x_2(t) = 1$. *This is the zeroth approximation to the best terminal time.*
(4) Solve the boundary value problem (7.25-7.28) on $0 \leq t \leq T_0$ *with the functions* $x_1(t)$, $x_2(t)$ *known from the previous step.*
(5) Use the obtained state and adjoint variables to calculate the function $\tilde{p}(t)$ *by formula (7.35). Call this function* $\bar{p}_1(t)$.
(6) Take $\tilde{p}_1(t) = \bar{p}_1(t)\theta(\bar{p}_1(t))$ *(*θ *is the Heaviside step-function). That is, replace all negative values of the control by zero. This gives us the next approximation to the optimal control,* $\tilde{p}_1(t)$.
(7) Go to step 2, *and repeat all the operations.*

Implement this algorithm to find and plot the optimal controls for several different values of α.

Chapter 8

Angiogenesis, inhibitors, promoters, and spatial growth

In previous chapters, a variety of tumor growth models were considered that differed in their complexity. The simplest growth model is exponential growth, while more realistic models include some form of saturation or density dependence, which leads to a reduced growth rate and eventually to cessation of growth when the cell population reaches a certain level. The cell populations can break out of the growth plateaus by the emergence of mutants that overcome the relevant selective barriers. The barriers can be determined by a variety of factors, including spatial, nutrient, or metabolic limitations.

This chapter considers a more complex set of models, which assume that tumor cells can produce tumor-promoting and inhibiting factors, and where the relative balance between promoters and inhibitors determines whether tumor growth plateaus at a relatively small size, or whether more aggressive growth is observed [Folkman *et al.* (1992)]. Therefore, the exact tumor growth pattern emerges from these dynamics that are intrinsic to the tumor rather than being modeled more phenomenologically as in the previous chapters. The dynamics of tumor promotion and inhibition are especially important in the context of the formation of a solid blood supply for the tumor, a process that is called angiogenesis [Folkman (2006)]. These considerations are part of the notion that tumor growth is not only determined by the properties of the tumor cells themselves, but also by their surroundings, i.e., their microenvironment [Nyberg *et al.* (2008); Mbeunkui and Johann Jr (2009); Dvorak *et al.* (2011)]

Let us consider in some more detail the process of angiogenesis, i.e., the formation of blood supply which provides cancer cells with oxygen, the necessary nutrients, and factors required for replication and survival. A given tissue or organ must have a sufficient blood supply in order to function. No extra blood supply is available though, which will hinder any potential abnormal growth. Cancer cells have to induce the generation of new blood supply in order to sustain their growth, i.e., they have to induce angiogenesiss. Research on the role of angiogenesis for cancer progression has been pioneered by Judah Folkman in the 1960s and 70s [Folkman (1971)], and work from his laboratory has been very influential in the literature, e.g. [Folkman (1995, 2002, 2006)]. In early experiments, Folkman and colleagues placed a small number of rabbit melanoma cells on the surface of the rabbit thyroid gland. They observed that the tumor cells initially grew but subsequently stopped growing once they reached a relatively small size comparable to that of a pea. The reason is that the tumor cells run out of blood supply.

It is now clear that growth to larger sizes requires the emergence of so-called *angiogenic tumor cells*. The ability of the cancer to grow depends on the balance between so-called angiogenesis inhibitors, and angiogenesis promoters. Examples of inhibitors are thrombospondin, tumstatin, canstatin, endostatin, angiostatin and interferons. Examples of promoters are growth factors such as FGF, VEGF, IL-8, and PDGF. Normal tissue produces mostly angiogenesis inhibitors. So do cancer cells. This serves as a preventative measure against abnormal growth. Angiogenic cancer cells, on the other hand, have mutations which allow the balance between inhibitors and promoters to be shifted away from inhibition, and towards promotion. This is done by activating the production of angiogenesis promoters, or by inactivating genes which encode inhibitors. Once such angiogenic cells have evolved, it is possible for the cancer to recruit new blood vessels and hence to grow to larger sizes. Folkman's research has also given rise to exciting new avenues of therapies [Hahnfeldt *et al.* (1999); Weis and Cheresh (2011); Gastl *et al.* (2009); Abdelrahim *et al.* (2010)]: Administration of angiogenesis inhibitors can destroy blood supply and result in remission of cancers. While encouraging results have been obtained in laboratory animals, our understanding is far less complete in the context of human pathologies.

This chapter reviews ordinary differential equation models which have examined the dynamics of angiogenesis-dependent tumor growth, in particular the dependence of growth on the relative balance of promoters and inhibitors. This is just one example of mathematical work that has been done in this area, and for more details on this modeling topic, the reader is referred to the literature (e.g. [Mantzaris *et al.* (2004); Chaplain (1996); Chaplain *et al.* (2006); Anderson and Chaplain (1998a); Anderson *et al.*

(2012); Milde *et al.* (2008); Ledzewicz and Schättler (2008)]). Conditions will be discussed that lead to self-limited tumor growth up to a relatively small plateau size, and those that allow more advanced and uncontrolled tumor growth. Moreover, we will examine partial differential equation models that show how differences in the ranges over which promoters and inhibitors act can lead to specific spatial patterns that have been observed in patients.

8.1 Model 1: Angiogenesis inhibition induces cell death

We describe and analyze a model for the evolution of angiogenic tumor cell lines [Wodarz and Krakauer (2001)]. The model consists of three basic variables (figure 8.1).

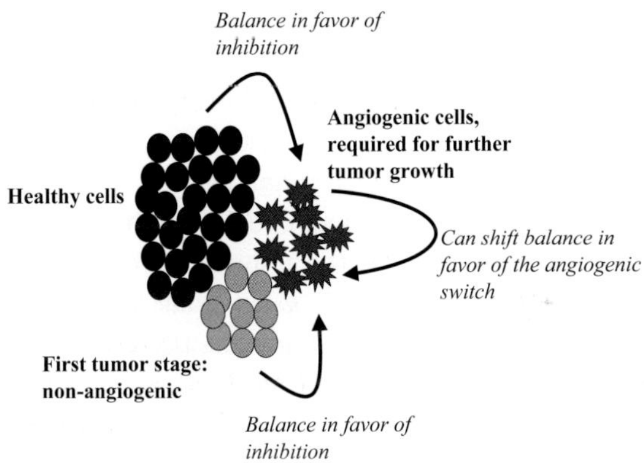

Fig. 8.1 Schematic diagram illustrating the central assumptions underlying the mathematical model.

Healthy host tissue, x_0; a first transformed cell line, x_1, which is non-angiogenic and cannot grow above a given threshold size; an angiogenic tumor cell line which has the potential to progress, x_2. It is thought that the formation of new blood vessels depends on a balance of angiogenesis inhibitors and promoters. If the balance is in favor of the inhibitors, new blood vessels are not formed. On the other hand, if it is in favor of the promoters, angiogenesis can proceed. Hence, the model assumes that healthy tissue, x_0, and stage one tumor cells, x_1, produce a ratio of inhibitors and promoters that is in favor of angiogenesis inhibition. On the other hand, it is assumed that angiogenic tumor cell lines have the ability to shift the balance in favor of angiogenesis promotion. We first consider progression from the wildtype cells to a first transformed cell line. The basic model is given by the following pair of differential equations,

$$\dot{x}_0 = r_0 x_0 \left(1 - \frac{x_0}{k_0}\right)(1 - \mu_0) - d_0 x_0, \tag{8.1}$$

$$\dot{x}_1 = \mu_0 r_0 x_0 \left(1 - \frac{x_0}{k_0}\right) + r_1 x_1 \left(1 - \frac{x_1}{k_1}\right)(1 - \mu_1) - d_1 x_1. \tag{8.2}$$

Healthy cells are assumed to replicate at a density dependent rate $r_0 x_0(1 - x_0/k_0)$. The value of k_0 represents the maximum size this population of cells can achieve, or the carrying capacity. The cells die at a rate $d_0 x_0$. We assume that the rate of mutation is proportional to the rate of replication of the cells, and is thus given by $\mu_w r_0 x_0(1 - x_0/k_0)$. The mutations give rise to the first stage of tumor progression, x_1, i.e., to a tumor cell line that is not angiogenic. This cell line will depend on the blood supply of the healthy tissue and will not be able to grow beyond a small size. These cells replicate at a density dependent rate $r_1 x_1(1 - x_1/k_1)$, where the carrying capacity k_1 is assumed to be relatively small ($k_1 << k_0$). They die at a rate $d_1 x_1$, and mutate to give rise to an angiogenic tumor cell line, x_2, at a rate $\mu_x r_1 x_1(1 - x_1/k_1)$. In the model, the population of healthy cells attains a homeostatic setpoint given by $x_0^* = k_0(r_0 - d_0)/r_0$. The mutation rate μ_0 can be assumed to be very small, since healthy tissue has intact repair mechanisms that ensure faithful replication of the genome. Once mutation gives rise to the first tumor cell line, it will grow to its small homeostatic set point level defined by $x_1^* = k_1(r_1 - d_1)/r_1$.

The wildtype cell population and the small population of first stage tumors are assumed to reach constant levels in a relatively short time. In other words, they reach an equilibrium abundance. Further tumor growth requires the emergence of the angiogenic cell line, x_2. In the following we investigate the conditions required for angiogenic tumor cell lines to evolve assuming a constant background abundance of x_0 and x_1.

The angiogenic cell line replicates at a density dependent rate $r_2 x_2(1 - x_2/k_2)$. As these cells can potentially influence the balance of inhibitors and promoters in favor of promoters, we have to take these dynamics into account. The death rate of these cells is determined by two components. The angiogenic tumor cells are characterized by a composite background death rate $d_2 x_2$, as with x_0 and x_1. In addition, the model assumes that the death rate can be increased if the balance between angiogenesis inhibition and promotion is in favor of inhibition. Hence, this death rate is expressed as $(p_0x_0 + p_1x_1 + p_2x_2)/(qx_2 + 1)$. Thus all three cell types lead to the inhibition of angiogenesis, whereas inhibition of angiogenesis can only be overcome by cell line x_2.

As we have assumed that x_0 and x_1 are at equilibrium, we start our analysis by ignoring mutation and simply looking at the dynamics of the angiogenic cell line, x_2. These dynamics are described by the equation,

$$\dot{x}_2 = r_2x_2\left(1 - \frac{x_2}{k_2}\right) - d_2x_2 - \frac{x_2\left(p_0x_0^* + p_1x_1^* + p_2x_2\right)}{qx_2 + 1},$$

where x_0^* and x_1^* are defined above. Two outcomes are possible. (*i*) The cell line x_2 cannot invade, resulting in equilibrium E0 where $x_2^{(0)} = 0$. (ii) The cell line x_2 can invade and converges to equilibrium E1 described by

$$x_2^{(1)} = \frac{-Q + \sqrt{Q^2 - 4r_2q_2\left[k\left(d_2 - r_2\right) + kp_Ix_I\right]}}{2r_2q_2},$$

where $Q = kq_2\left(d_2 - r_2\right) + r_2 + kp_2$ and subscript I refers to the inhibitory cell lines: $p_Ix_I = p_0x_0^* + p_1x_1^*$.

In the following we examine the stability properties of these two equilibria which are summarized in figure 8.2. If $p_Ix_I < r_2 - d_2$, then the equilibrium describing the extinction of the angiogenic tumor, $x_2^{(0)}$, is not stable. The equilibrium describing the invasion of the angiogenic tumor cell line, $x_2^{(1)}$, is stable. In other words, if the above condition is fulfilled, then the degree of angiogenesis inhibition is too weak, and the angiogenic tumor cell line can emerge, marking progression of the disease.

On the other hand, if $p_Ix_I > r_2 - d_2$, the degree of inhibition is stronger and the situation is more complicated (figure 8.2). The equilibrium describing the extinction of the angiogenic cell population, $x_2^{(0)}$, becomes stable. However, equilibrium $x_2^{(1)}$, describing the emergence of angiogenic tumor cells, may or may not be stable (figure 8.2).

(1) If the degree of angiogenesis inhibition lies above a certain threshold, equilibrium $x_2^{(1)}$ is unstable and the angiogenic cell line cannot invade. It was not possible to define this threshold in a meaningful way.
(2) If the degree of angiogenesis inhibition lies below this threshold, equilibrium $x_2^{(1)}$ remains stable. Now, both the extinction and the emergence equilibria are stable (figure 8.2). This means that two outcomes are possible and that the outcome depends on the initial conditions. Either the angiogenic cell line fails to emerge, or the angiogenic cell line

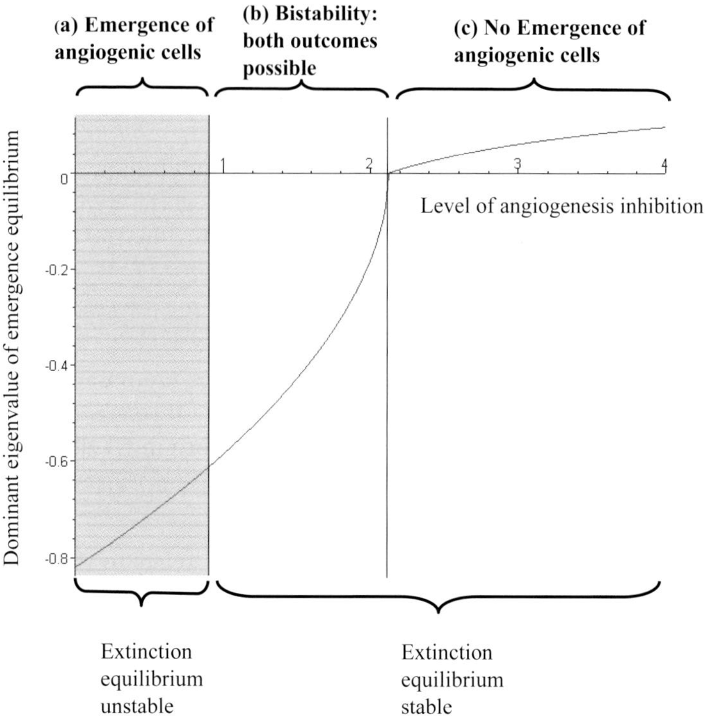

Fig. 8.2 Graph showing the stability properties and the outcome of the model. Parameters were chosen as follows: $r_2 = 1; k_2 = 1; d_2 = 0.1; p_2 = 1; q = 10$.

does emerge, resulting in tumor progression. As shown in figure 8.3, a low initial abundance of angiogenic tumor cells results in failure of growth. On the other hand, a high initial number of angiogenic tumor cells results in growth of the tumor and progression (figure 8.3).

To summarize, the model shows the existence of three parameter regions (figure 8.2). If the degree of angiogenesis inhibition by healthy tissue and stage one tumor cells lies below a threshold, angiogenic tumor cell lines always invade resulting in progression of the disease. If the degree of inhibition lies above a threshold, the angiogenic cell lines can never emerge and pathology is prevented. Between these two thresholds, both outcomes are possible depending on the initial conditions. A high initial number of

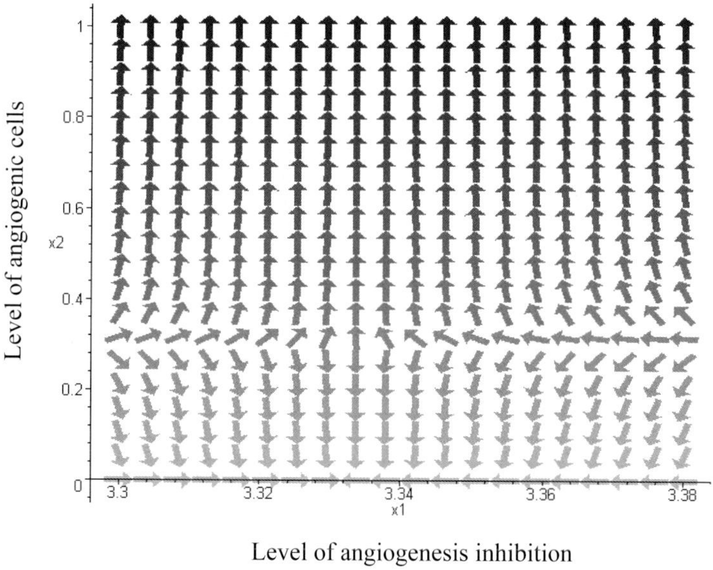

Fig. 8.3 Direction field plot showing how the outcome of the model can depend on the initial conditions. Parameters were chosen as follows: $r_2 = 1; k_2 = 100; d_2 = 0.1; p_2 = 1; q = 10$. For the purpose of simplicity the populations of non-angiogenic cells were summarized in a single variable and assumed to converge towards a stable setpoint (characterized by the parameters *r=0.15; d=0.1; k=10*).

angiogenic tumor cells results in growth of this cell line and progression of the disease.

What does the initial number of angiogenic cells mean in biological terms? The dependence of growth on the initial number of angiogenic tumor cells presents an effective barrier against pathologic tumor growth. Given that a small number of non-angiogenic tumor cells exists, it will be difficult to create a sufficiently large number of angiogenic mutants to overcome the blood supply barrier. This difficulty could explain why, upon autopsy, people tend to show small tumors which have failed to grow to larger sizes. The initial number of the angiogenic cells could be determined by the mutation rate μ_1, which gives rise to the angiogenic cells. If the mutation rate is high, the initial number of angiogenic cells will be high. On the other hand, if the mutation rate is low, the initial number of the angiogenic cells will be low. Hence, in the parameter region where the outcome of the dynamics depends on the initial conditions, a high mutation rate promotes the emergence and growth of angiogenic tumor cells (figure 8.4). If a high mutation rate by tumor cells defines genetic instability, then it is possible that genetic instability might be required for the invasion of angiogenic tumor cells.

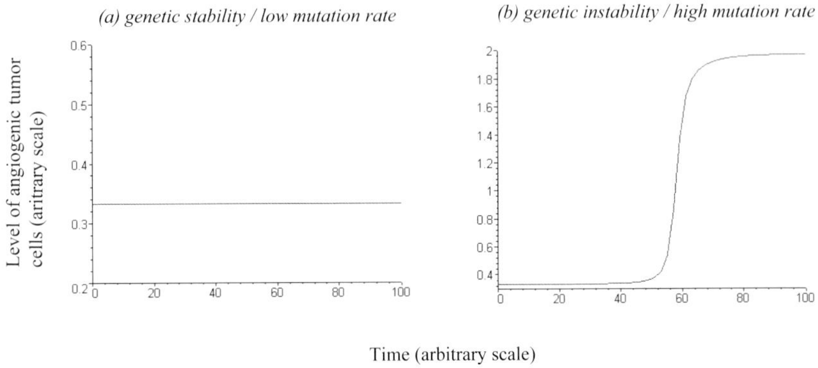

Fig. 8.4 Genetic instability and the emergence of angiogenic cell lines. (a) If the mutation rate is low (genetic stability), the initial number of angiogenic cells created is low. Consequently they cannot emerge. (b) On the other hand, if the mutation rate is high (genetic instability), a higher initial number of angiogenic cells is created. Hence, they emerge and become established. Parameters were chosen as follows: $r_0 = 0.11; k_0 = 10; \mu_0 = 0.001; d_0 = 0.1; r_1 = 0.12; k_1 = 2; d_2 = 0.1; r_2 = 2.5; k_2 = 2; d_2 = 0.1; p_0 = 2; p_1 = 2; p_2 = 2; q = 10$; for (a) $\mu_1 = 0.001$; For (b) $\mu_1 = 0.01; \mu_2 = \mu_1$.

8.2 Model 2: Angiogenesis inhibition prevents tumor cell division

We consider a basic mathematical model which describes the growth of a cancer cell population, assuming that the amount of blood supply does not influence cell death, but the rate of cell division [Wodarz *et al.* (2004)]. This model will also be used to consider the effect of diffusion of cells and soluble molecules across space; this is done in the next sections. Therefore, the model will take into account explicitly the dynamics of promoters and inhibitors. This is in contrast to the last section where for the purpose of simplicity inhibitors and promoters were assumed to be proportional to the number of cells which secrete them. The new model includes three variables: the population of cancer cells, C; promoters, P; and inhibitors, I. It is assumed that both promoters and inhibitors can be produced by cancer cells. In addition, inhibitors may be produced by healthy tissue. The model is given by the following set of differential equations which describe cancer growth as a function of time,

$$\dot{C} = \left(\frac{rC}{\epsilon C + 1}\right)\left(\frac{P}{I+1}\right) - \delta C, \tag{8.3}$$

$$\dot{P} = a_P C - b_P P, \tag{8.4}$$

$$\dot{I} = \xi + a_I C - b_I I. \tag{8.5}$$

The population of cancer cells grows with a rate r. Growth is assumed to be density dependent and saturates if the population of cancer cells becomes large (expressed in the parameter ϵ). In addition, the growth rate of the cancer cells depends on the balance between promoters and inhibitors, expressed as $P/(I+1)$. The higher the level of promoters relative to inhibitors, the faster the growth rate of the cancer cell population. If the level of promoters is zero, or the balance between promoters and inhibitors is heavily in favor of inhibitors, the cancer cells cannot grow and remain dormant [O'Reilly *et al.* (1996, 1997); Ramanujan *et al.* (2000)]. Cancer cells are assumed to die at a rate δ. Promoters are produced by cancer cells at a rate a_p and decay at a rate b_p. Inhibitors are produced by cancer cells at a rate a_I and decay at a rate b_I. In addition, the model allows for production of inhibitors by normal tissue at a rate ξ.

8.2.1 *Linear stability analysis of the ODEs*

Let us simplify system (8.3-8.5) by using a quasistationary approach, that is, we will assume that the level of promoters adjusts instantaneously to its steady-state value ($P = Ca_P/b_P$). It is convenient to denote

$$W = \frac{ra_P}{\delta b_P}, \quad \gamma = \frac{a_I}{b_I}.$$

Now we have a two-dimensional system,

$$\dot{C} = \delta C\left(\frac{WC}{(1+\epsilon C)(1+I)} - 1\right), \tag{8.6}$$

$$\dot{I} = b_I(\gamma C - I). \tag{8.7}$$

There can be up to three fixed points in this system,

$$(C, I) = (0, 0) \quad \text{and} \quad (C, I) = (\hat{C}_{\pm}, \hat{I}_{\pm}),$$

where $\hat{I}_{\pm} = \gamma \hat{C}_{\pm}$, and

$$\hat{C}_{\pm} = \frac{-(\gamma + \epsilon - W) \pm \sqrt{(\gamma + \epsilon - W)^2 - 4\epsilon\gamma}}{2\epsilon\gamma}.$$

It is obvious that if $\gamma + \epsilon - W < 0$ and $(\gamma + \epsilon - W)^2 - 4\epsilon\gamma > 0$, then there are exactly three positive equilibria in the system. If either of these conditions is violated, the $(0, 0)$ solution is the only (biologically meaningful) stable point.

Stability analysis can be performed by the usual methods. It shows that for the $(0, 0)$ equilibrium, the Jacobian is

$$\begin{pmatrix} -\delta & 0 \\ b_I\gamma & -b_I \end{pmatrix},$$

that is, this equilibrium is always stable. For the points $(C_{\pm}, I_{\pm})$, we get the following Jacobian,

$$\begin{pmatrix} \frac{-\delta(\epsilon-\gamma-W\pm\Gamma)}{2W} & \frac{\delta(\epsilon-\gamma-W-\Gamma)}{2\gamma W} \\ b_I\gamma & -b_I \end{pmatrix},$$

where we denote for convenience, $\Gamma \equiv \sqrt{(\epsilon + \gamma - W)^2 - 4\epsilon\gamma}$. It is easy to show that the eigenvalues of this matrix for the solution $(\hat{C}_-, \hat{I}_-)$ are given by

$$\frac{1}{4W}\left(-Y_- \pm \sqrt{Y_-^2 + 16 b_I \delta W \Gamma}\right),$$

and for the solution $(\hat{C}_+, \hat{I}_+)$ we have eigenvalues

$$\frac{1}{4W}\left(-Y_+ \pm \sqrt{Y_+^2 - 16 b_I \delta W \Gamma}\right),$$

where $Y\pm \equiv 2b_I W + \delta(\epsilon - \gamma - W \pm \Gamma)$. We can see that solution $(\hat{C}_-, \hat{I}_-)$ is always unstable and we will not consider it any longer. Solution $(\hat{C}_+, \hat{I}_+)$, which we call for simplicity $(\hat{C}, \hat{I})$ from now on, is stable as long as

$$Y_+ > 0. \tag{8.8}$$

8.2.2 *Conclusions from the linear analysis*

As we can see this model has very similar properties compared to the last one, and they are summarized as follows. There are two outcomes. *(i)* The cancer cells cannot grow and consequently go extinct. That is, $C = 0$, $P = 0$ and $I = \xi/b_I$. The cancer goes extinct in the model because we only consider cells which require the presence of promoters for division. If the level of promoters is not sufficient, the rate of cell death is larger than the rate of cell division. In reality, however, it is possible that a small population of non-angiogenic tumor cells survives. This was modeled in more detail in the previous section. Here, we omit this for simplicity. *(ii)* The population of cancer cells grows to significant levels, that is, $C = \hat{C}$.

How do the parameter values influence the outcome of cancer growth? The cancer extinction outcome is always stable. The reason is as follows. The cancer cells require promoters to grow. The promoters, however, are produced by the cancer cells themselves. If we start with a relatively low initial number of cancer cells, this small population cannot produce enough promoters to overcome the presence of inhibitors. Consequently, the cancer fails to grow and goes extinct. This outcome is always a possibility, regardless of the parameter values. Significant cancer growth can be observed if the intrinsic growth rate, r, lies above a threshold relative to the death rate of the cells, δ, and degree of tumor cell inhibition (a_p and b_p relative to a_I and b_I). The exact condition is given by (8.8). In this case, the outcome is either failure of cancer growth, or successful growth to large numbers. Which outcome is achieved depends on the initial conditions. Successful growth is only observed if the initial number of cancer cells lies above a threshold. Then, enough promoters are initially produced to overcome inhibition. This is the same result as presented in the previous section; in biological terms this may mean that mutant cells which produce promoters must be generated frequently (e.g. by mutator phenotypes) in order to initiate tumor growth to higher levels [Wodarz and Krakauer (2001)].

8.3 Spread of tumors across space

In this section, we introduce space into the above described model. We consider a one-dimensional space along which tumor cells can migrate. The model is formulated as a set of partial differential equations and is written as follows,

$$\frac{\partial C}{\partial T} = \left(\frac{rC}{\epsilon C + 1}\right)\left(\frac{P}{I+1}\right) - \delta C + D_c \frac{\partial C^2}{\partial x^2}, \tag{8.9}$$

$$\frac{\partial P}{\partial T} = a_P C - b_P P, \tag{8.10}$$

$$\frac{\partial I}{\partial T} = a_I C - b_I I + D_I \frac{\partial I^2}{\partial x^2}, \quad 0 \leq x \leq L. \tag{8.11}$$

The model assumes that tumor cells can migrate, and this is described by the diffusion coefficient D_c. Inhibitors can also diffuse across space, and this is described by the diffusion coefficient D_I. It is generally thought that inhibitors act over a longer range, while promoters act locally [Folkman (2002); Ramanujan *et al.* (2000)]. Therefore, we make the extreme assumption that promoters do not diffuse. For simplicity we assume that inhibitors are only produced by cancer cells and ignore the production by normal tissue (that is, $\xi = 0$). This simplification is justified because this model concentrates on the tumor dynamics, and numerical simulations show that the results considered here are not altered by this simplification. As mentioned above, the model considers tumor spread across space. It is important to point out that we do not consider long-range metastatic spread. Instead, we consider local spread of a tumor within a tissue, such as the breast, liver, brain, or esophagus.

These equations must be equipped with appropriate initial and boundary conditions. In the simulations we used the following (Neumann) boundary conditions:

$$\left.\frac{\partial C}{\partial x}\right|_{x=0} = \left.\frac{\partial I}{\partial x}\right|_{x=0} = \left.\frac{\partial P}{\partial x}\right|_{x=0} = \left.\frac{\partial C}{\partial x}\right|_{x=L} = \left.\frac{\partial I}{\partial x}\right|_{x=L} = \left.\frac{\partial P}{\partial x}\right|_{x=L} = 0.$$

The dependence of the results on the initial conditions is discussed above.

Here we investigate the process of tumor growth and progression in relation to the degree of inhibition and promotion. First we will present a mathematical analysis and then biological insights and results of simulations.

8.3.1 *Turing stability analysis*

Again, we are going to assume that promoters adjust instantaneously to their equilibrium level. By replacing P with C defined by $P = \frac{a_P}{b_p}C$, we can rewrite equation (8.9) as

$$\frac{\partial C}{\partial t} = \left(\frac{Cra_p}{b_p(1+\epsilon C)(1+I)} - \delta \right) C + D_y \frac{\partial^2 C}{\partial x^2}. \tag{8.12}$$

This equation together with equation (8.11) gives a Turing model.

Let us go back to the system of ODEs, (8.6-8.7), and assume that solution $(\hat{C}, \hat{I})$ is a stable equilibrium. Of course, this solution also satisfies the system of PDEs, (8.12, 8.11). Let us consider a wave-like deviation from this spatially uniform solution:

$$C(x,t) = \hat{C} + A\cos(\omega x)e^{\lambda t},$$

$$I(x,t) = \hat{I} + B\cos(\omega x)e^{\lambda t}.$$

Here, the amplitudes of the perturbation, A and B, are small compared to the amplitude of the spatially uniform solution, and we assume an infinitely large space. The equation for the new eigenvalue, λ, is

$$\det \begin{pmatrix} \alpha - D_c\omega^2 - \lambda & -\beta \\ a_I & -b_I - D_I\omega^2 - \lambda \end{pmatrix} = 0, \tag{8.13}$$

where we define

$$\alpha = \frac{\hat{C}ra_p}{b_p(1+\epsilon\hat{C})^2(1+\hat{I})} > 0, \quad \beta = \frac{\hat{C}^2 ra_p}{b_p(1+\epsilon\hat{C})(1+\hat{I})^2} > 0.$$

Equation (8.13) can be written as

$$\lambda^2 + \lambda(b_I - \alpha + (D_C + D_I)\omega^2) + a_I\beta - (b_I + D_I\omega^2)(\alpha - D_C\omega^2) = 0. \tag{8.14}$$

This is the dispersion relation which connects the growth-rate, λ, with the spatial frequency of the perturbation, ω. The stability conditions are now given by

$$b_I - \alpha + (D_C + D_I)\omega^2 > 0, \tag{8.15}$$

$$a_I\beta - (b_I + D_I\omega^2)(\alpha - D_C\omega^2) > 0. \tag{8.16}$$

Note that the stability conditions for solution $(\hat{C}, \hat{I})$ of the system of ODEs, (8.6-8.7), are obtained automatically from the conditions above by setting $\omega = 0$:

$$b_I - \alpha > 0, \tag{8.17}$$

$$a_I\beta - b_I\alpha > 0. \tag{8.18}$$

Inequality (8.15) is always satisfied because of inequality (8.17). Let us derive conditions under which the spatially uniform solution is unstable.

This requires that condition (8.16) be reversed. This can be expressed as follows:

$$F(\omega) \equiv D_I D_C \omega^4 - \omega^2 \gamma_1 + \gamma_2 < 0, \tag{8.19}$$

where we denoted for simplicity,

$$\gamma_1 = \alpha D_I - b_I D_C, \quad \gamma_2 = a_I \beta - \alpha b_I > 0.$$

This is a fourth order polynomial, symmetrical with respect to the line $\omega = 0$, with a positive leading term. The points, $\pm|\omega|$, satisfying

$$\omega^2 = \frac{\gamma_1}{2 D_I D_C}, \tag{8.20}$$

correspond to the two minima of the left hand side of inequality (8.19). Let us call these values of ω, $\pm\omega_c$. The condition $F(\omega_c) < 0$ defines that the uniform solution $(\hat{C}, \hat{I})$ is unstable.

[th]

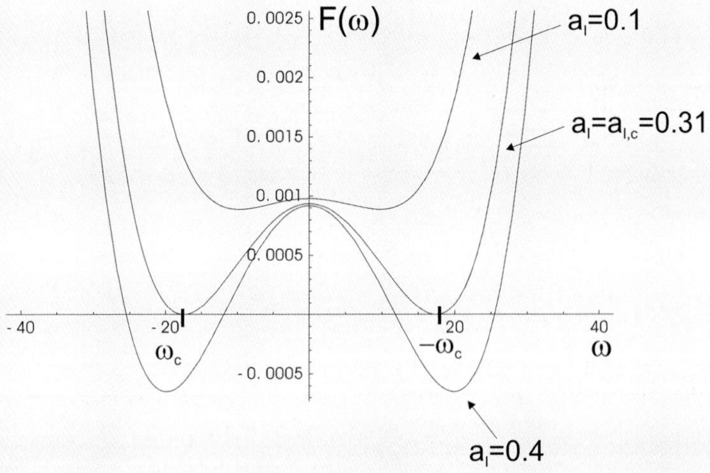

Fig. 8.5 Emergence of Turing instability. As a_I increases and through its critical value, the function $F(\omega)$ (equation (8.19)) crosses zero. Negative regions of $F(\omega)$ correspond to unstable wave-numbers. The wave-number which becomes unstable first is denoted by ω_c. The parameters are as follows: $r = 1; \delta = 0.1; a_P = 5; b_P = 0.1; b_I = 0.01; D_C = 0.00001; D_I = 0.001$.

Let us plot the function $F(\omega)$ for different values of a_I, see figure 8.5. For small values of a_I, $F(\omega)$ is strictly positive, and the spatially uniform solution is stable. As a_I increases, the function $F(\omega)$ crosses the line $F = 0$. The critical value of a_I, $a_{I,c}$, for which $F(\omega_c) = 0$, is determined from

$$(\alpha D_I - b_I D_C)^2 = 4 D_I D_C (a_I \beta - \alpha b_I),$$

where α and β both depend on a_I. We solved this equation numerically to find the critical value of $a_{I,c}$, see figure 8.5.

The applicability of the above analysis depends on the parameters of the system. First of all, we need conditions (8.17-8.18) to be satisfied. They mean that without diffusion, a positive, spatially uniform solution is stable. Next, we need to be in a *weakly nonlinear regime*, where the function $F(\omega)$ has only very narrow regions of ω corresponding to negative values. More precisely, $\Delta\omega \sim L^{-1}$, where L is the spatial dimension of the system. In terms of parameter a_I, we require that it is sufficiently close to $a_{I,c}$. Then, we can calculate the "most unstable" wave-number, that is, ω_c defined by equation (8.20), with $\omega_{c,I}$. This value will determine the spatial period of the solution,

$$\text{Period} = \frac{2\pi}{\omega_c}. \tag{8.21}$$

8.3.2 *Stationary periodic solutions*

Let us start from the value a_I below the critical, $a_I < a_{I,c}$. The system exhibits bistability. If we start in the vicinity of a $(0, 0)$ solution, then cancer will not grow and decay to zero. If we start from a point (C, I) in the domain of attraction of the solution $(\hat{C}, \hat{I})$, then the system will develop towards this positive spatially homogeneous stationary solution.

Next, let us suppose we have $a_I > a_{I,c}$, but make sure that it is sufficiently close to $a_{I,c}$ (the exact meaning of "close" is specified in the analysis above). Again, if the initial conditions are close to the zero solution, then the zero state will be the state that the system will attain. However, if we start in the vicinity of the $(\hat{C}, \hat{I})$ state, we will observe interesting behavior. Solution $(\hat{C}, \hat{I})$ is now unstable, and we will see "ripples" developing on top of this solution. This is Turing instability. The spatial period of the ripple was calculated in the previous section. Long-time evolution of this state is of course not in the realm of linear stability analysis, but we can predict that the spatial scale of the resulting solution will be given by (8.21).

Finally, let us assume that a_I is much higher than critical. Now, solution $(\hat{C}, \hat{I})$ is unstable even in the system of ODEs. However, a periodic solution will develop, unless the initial condition is in the domain of attraction of the zero solution. The spatial scale of the periodic solution is determined intrinsically by the parameters of the system, and it grows with a_I. Intuitively this is easy to understand, because higher values of a_I correspond to higher levels of inhibition, so the distance between regions of large C will become larger. Note that the exact period of the periodic solution is adjusted to fit the boundary conditions of the system. For instance, with the Neumann boundary conditions, the boundary points are forced to be troughs of the wave-like pattern. In other words, the period of the solution must be an integer fraction of L.

8.3.3 *Biological implications and numerical simulations*

We start with a scenario where the degree of inhibition is much larger than the degree of promotion ($a_I/b_I >> a_p/b_p$). This corresponds to the early stages when the tumor is generated. We then investigate how tumor growth changes as the degree of inhibition is reduced relative to the level of promotion (i.e., the value of a_I/b_I is reduced). We consider the following parameter regions (figure 8.6).

(1) If the degree of inhibition is strong and lies above a threshold, growth of the cancer cells to higher levels does not occur (not shown). Only a small number of cells which do not require promotion for survival would remain.

(2) If the degree of inhibition is weaker, the cancer cells can grow. The spread across space is, however, self-limited (figure 8.6(a)). The cancer cells migrate across space. The inhibitors produced by the cancer cells also spread across space, while the promoters do not. Therefore, as the cancer cells migrate, they enter regions of the tissue where the balance of inhibitors to promoters is heavily in favor of inhibitors. Consequently, these cells cannot grow within the space. They remain dormant and may eventually die. In biological terms, this corresponds to a single coherent but self-limited lesion (*uni-focal*). Note that this does not mean that it is in principle impossible to generate more lesions. It means that the space between lesions is bigger than the space provided for cancer growth within the tissue.

(3) As the production of inhibitors is further reduced, we enter another

parameter region. Now fewer inhibitors diffuse across space. We observe that multiple lesions or foci are formed (figure 8.6(b)). They are separated by tissue space which does not contain any tumor cells. The separate lesions produce some inhibitors, and they diffuse across space. This explains the absence of tumor cells between lesions. Because the production of inhibitors is weakened, however, tumor growth is only inhibited in a certain area around the lesion, and not across the whole space. How many lesions are found within a tissue depends on the parameters in the model, in particular on the relative strength of inhibition and promotion (figures 8.6(b) and (c)). The stronger the degree of inhibition, the larger the space between lesions, and the fewer lesions we expect. The weaker the degree of inhibition, the smaller the space between lesions, and the larger the expected number of lesions. In biological terms, the occurrence of multiple lesions within a tissue which arise from a single tumor is often referred to as *multi-focal* cancers. Spatially distinct lesions have also been studied mathematically in other contexts, e.g. [Marciniak-Czochra and Kimmel (2008)].

(4) If the degree of inhibition is further reduced and lies below a threshold, spread of inhibitors is sufficiently diminished such that the tumor cells can invade the entire space and tissue (figure 8.6(d)). In biological terms, this corresponds to the most extensive tumor growth possible within a tissue.

In summary, as the relative degree of inhibition is reduced, the patterns of tumor growth change from absence of significant growth, to a single self-limited tumor, to the occurrence of multiple foci, and to the maximal invasion of the tissue by tumor cells. Multi-focal cancers may arise through the dynamical interplay between long range inhibition and local promotion. The following section will examine this in the light of somatic evolution.

8.4 Somatic cancer evolution and progression

The previous sections have shown how the pattern of cancer growth can depend on the relative balance of promoters and inhibitors. Here we consider these results in the context of somatic evolution. Initially, the balance between inhibitors and promoters is in favor of inhibition. Inhibitors are likely to be produced by healthy cells (e.g. in the context of angiogenesis), which are more abundant than an initiating population of transformed

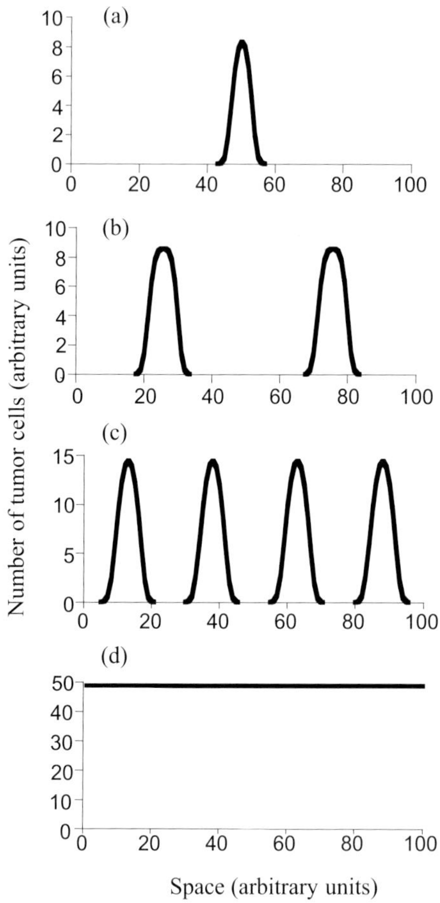

Fig. 8.6 Outcome of the spatial model depending on the relative balance of promoters and inhibitors, captured in the variable a_i. Parameters were chosen as follows: $r = 1; \delta = 0.1; a_P = 5; b_P = 0.1; b_I = 0.01; D_C = 0.00001; D_I = 0.001; L = 2$ For (a) $a_I = 3$, (b) $a_I = 2$, (c), $a_I = 1$, (d) $a_I = 0.1$.

cells. In the context of angiogenesis, specific mutations have been shown to result in the enhanced production of promoters or reduced production of inhibitors in cancer cells. Our model has shown that such mutants have to be produced at a relatively high frequency, so that a sufficient number of promoting cells are present in order to ensure that enough promoters are produced to overcome the effect of inhibition.

Once the promoting cells have succeeded to expand, cancer progression can occur in a variety of ways according to the model. How the cancer progresses depends on how much the balance between promotion and inhibition has been shifted in favor of promotion. We distinguish between three possibilities (figures 8.7, 8.8 and 8.9).

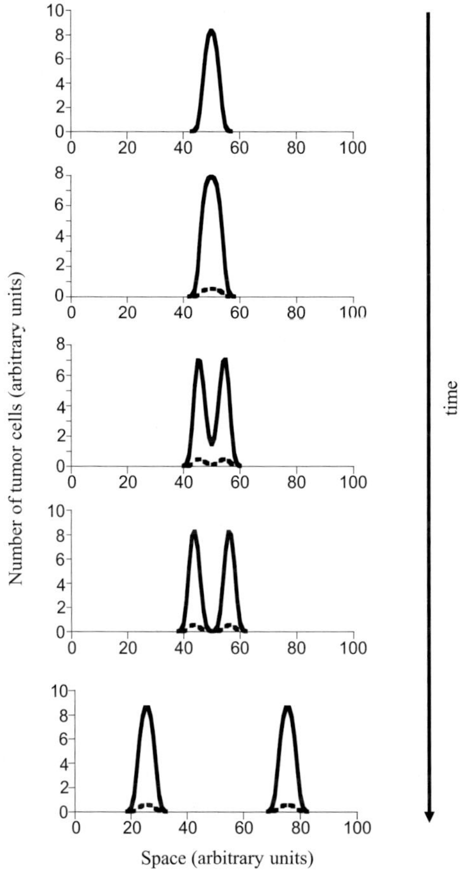

Fig. 8.7 Tumor progression if the initial mutant cell line has only shifted the balance between promoters and inhibitors slightly in favor of promotion. This cell line can only give rise to self limited growth. Further tumor growth requires the generation of further mutants. The new mutant in the simulation is depicted by the dashed line. Parameters were chosen as follows: $r = 1; \delta = 0.1; a_P = 5; b_P = 0.1; a_I = 3; b_I = 0.01; D_c = 0.00001; D_I = 0.001, L = 2$. For mutant: $a_I = 0.5; a_P = 20$.

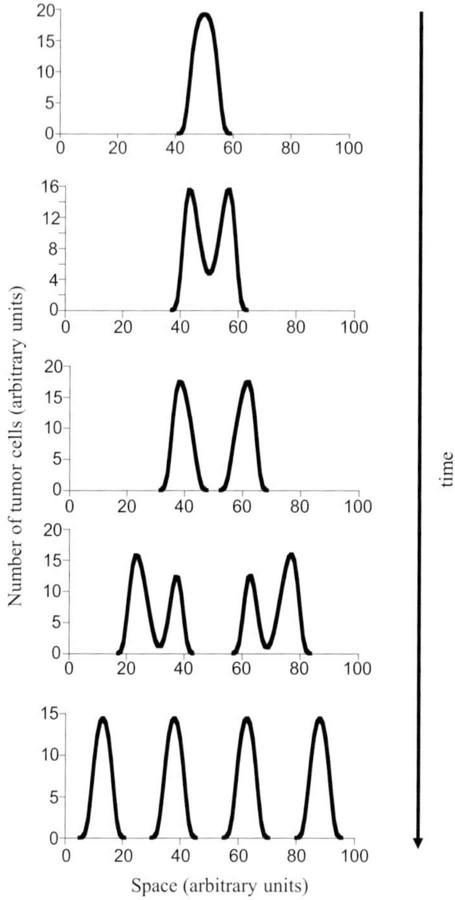

Fig. 8.8 Tumor progression if the initial mutant cell line has shifted the balance between promoters and inhibitors more substantially towards promotion. Now, multiple foci can develop without the need for further mutations. The multiple foci develop, however, by first generating a single lesion which subsequently splits to give rise to two lesions during the natural growth process. Parameters were chosen as follows: $r = 1; \delta = 0.1; a_P = 5; b_P = 0.1; a_I = 1; b_I = 0.01; D_C = 0.00001; D_I = 0.001, L = 2$.

(i) The balance between inhibition and promotion has been shifted only slightly in favor of promotion, such that self-limited growth of the cancer is observed (figure 8.7). That is, we observe a single lesion which can grow to a certain size but which is limited in the spread through the tissue. In order to progress further towards the occurrence of multiple lesions or towards more extensive invasion of the tissue, further mutants have to be generated which are characterized by enhanced production of promoters or by reduced production of inhibitors. This introduces a new problem: such a mutant will not have a selective advantage, but is selectively neutral relative to the other cells. This is because the promoters and inhibitors secreted from one cell affect the whole population of cells. If the mutant produces more promoters, not only the mutant, but the entire population of tumor cells benefits. This means that a mutant characterized by enhanced production of promoters will not invade the tumor cell population. Instead, we observe genetic drift which is stochastic and not described by the equations considered here. The model does, however, suggest the following (figure 8.7): if the population of mutant cells remains below a given threshold relative to the rest of the tumor cells, it will not alter the growth pattern. If the population of mutant cells grows beyond a threshold relative to the rest of the tumor cells, it can change the pattern of cancer growth, even if the mutants do not become fixed in the population (figure 8.7). The change can either be the generation of multiple lesions, or invasion of the whole tissue, depending on the amount by which the level of promotion has been enhanced by the mutant cell population. The chances that the mutant cell population drifts to levels high enough to cause such a change in tumor growth depend on the population size of the lesion. The larger the number of tumor cells, the lower the chance that the relative population size of the mutants can cross this threshold. If this cannot occur, further cancer progression not only requires the generation of a mutation which enhances the level of promotion, but an additional mutation which gives the promoter mutant a selective advantage over the rest of the cell population. That is, in addition to the mutation which shifts the balance in favor of promotion, a mutation is required either in an oncogene or a tumor suppressor gene so that the mutant can grow to sufficiently high numbers or fixation.

(ii) The first mutation shifts the balance between promoters and inhibitors to a lager extent which is sufficient to result in the generation of multiple lesions (figure 8.8). The multiple lesions do not, however, occur immediately. First, the tumor grows as a single and self limited lesion (figure 8.8). Over time, this lesion bifurcates to give rise to two lesions, or further lesions if the degree of promotion is large enough relative to the degree of inhibition (figure 8.8). The temporal sequence from a single and self-controlled lesion to the occurrence of multiple lesions is the same as in the previous case. But in contrast to the previous case, no further mutations are required. This is because multiple foci arise from the split and

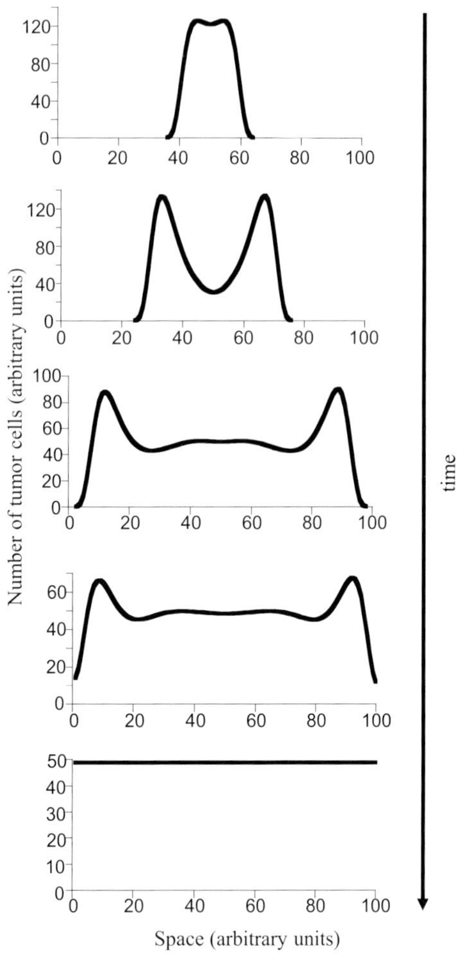

Fig. 8.9 Tumor progression if the initial cell line has largely escaped inhibition, and promotion is the dominant force. Now the tumor grows in space as a single lesion until the whole tissue is invaded. Parameters were chosen as follows; $r = 1; \delta = 0.1; a_P = 5; b_P = 0.1; a_I = 0.1; b_I = 0.01; D_C = 0.00001; D_I = 0.001, L = 2$.

migration of a single lesion. The number of foci that form depends on the exact degree of promotion which was achieved by the initial mutation. The higher the degree of promotion, the larger the number of lesions. Growth beyond this number of lesions (which will eventually result in maximal invasion) then requires higher levels of promotion. This is in turn achieved by further mutational events according to the same principles as described in the previous section.

(iii) Finally, assume that the initial mutation shifts the balance so much in favor of promotion that maximal invasion of the tissue is possible (figure 8.9). Now we observe cancer progression without the generation of multiple foci. Instead, a relatively small single lesion expands in space until all the tissue has been invaded.

In summary, the model predicts different modes of cancer progression in relation to the evolution away from tumor inhibition and towards promotion. A single cancer lesion may spread across the tissue without the occurrence of multiple lesions. Alternatively, the cancer can first grow as a single, self-contained lesion. This can then bifurcate to give rise to multiple foci, either as a result of additional mutations, or as a result of the natural pathway by which multiple foci are generated, depending on the degree of tumor promotion conferred by the initial mutation. Further evolutionary events can then induce the multiple foci to become a single, maximally invasive mass. The occurrence of multiple foci therefore represents an intermediate stage in tumor progression towards malignancy.

8.5 Summary and clinical implications

The occurrence of multiple lesions is observed in a variety of cancers. That is, not one, but several lesions are observed within a given tissue [Weissenbacher *et al.* (2010); Boyages *et al.* (2010), Wang *et al.* (2009); Andreoiu and Cheng (2010)]. Multiple lesions can occur by two basic mechanisms [Wilkens *et al.* (2000); Tsuda and Hirohashi (1995); Ruijter *et al.* (1999); Hartmann *et al.* (2000); Hafner *et al.* (2002)]. Either they originate independently by separate carcinogenic events, or they are generated by a single transformation event (monoclonal origin). Sometimes, the term "multicentric cancers" is used to describe the occurrence of clonally unrelated lesions, while the term "multi-focal" refers to a monoclonal origin [Teixeira *et al.* (2003)]. Clinically, it is important to determine the nature of multiple lesions. The occurrence of multiple lesions can be indicative of a familial cancer, especially if they occur at a relatively young age. Examples are familial adenomatous polyposis (FAP) in the colon, and familial retinoblastoma [Marsh and Zori (2002)]. The genetic predisposition of such individuals renders multiple independent carcinogenic events likely. Alternatively, multiple independent lesions can arise because a large area of tissue has been altered and is prone to the development of cancer, such as Barrett's esophagus [Van Dekken *et al.* (1999)], or by other mechanisms which are not yet understood. On the other hand, genetic analysis has indicated that multiple lesions in several cases have a monoclonal origin [Antonescu *et al.* (2000); Holland (2000); Junker *et al.* (2002); Kupryjanczyk *et al.* (1996); Louhelainen *et al.* (2000); Middleton *et al.* (2002); Miyake *et al.* (1998); Noguchi *et al.* (1994); Rosenthal *et al.* (2002); Simon *et al.* (2001); Van Dekken *et al.* (1999)]. Examples are mammary carcinoma, gliomas, renal cell carcinoma, hepatocellular carcinoma, and esophageal adenocarcinoma.

The models discussed here show that multiple foci with a monoclonal origin can develop through a dynamical interplay between tumor promoters and inhibitors. The cancer can only grow to high loads as a single mass if it has largely escaped all inhibitory effects. Otherwise, the cancer is likely to grow via the generation of a relatively small and self limited tumor which then bifurcates into multiple foci until it finally invades the entire tissue. The occurrence of multiple foci is therefore an intermediate stage in cancer progression. The higher the number of foci, the further advanced the stage of cancer progression.

A clinically important step in carcinogenesis is the process of metastasis. That is, the spread of tumor cells to the lymph node, entry into the blood supply, and the spread to other tissues. Various studies have investigated the metastatic potential of multi-focal compared to uni-focal cancers [Andea *et al.* (2002); Junker *et al.* (1997, 1999)]. In uni-focal cancers, tumor size has been found to be a predictor of metastatic potential. For staging multi-focal breast carcinomas, it has been suggested to use the diameter of the largest tumor only [Andea *et al.* (2002)]. This, however, assumes that the other foci do not significantly contribute to tumor progression. According to our arguments, this would under-stage the cancer. According to the model, the number of foci correlates with the stage of the disease. This has also been concluded in clinical studies, and is supported by data which show reduced patient survival with multi-focal compared to uni-focal cancers [Andea *et al.* (2002)]. Moreover, because our model suggests that multi-focality can occur as a result of reduced tumor cell inhibition, successful metastatic growth might be easier to achieve. Although under debate, some data suggest that inhibitors produced by the primary tumor can prevent metastatic cells from growing [Ramanujan *et al.* (2000)]. If multi-focality correlates with reduced inhibition, then it could also correlate with an increased chance that metastatic cells grow and do not remain dormant.

Further, it is important to note that studies which aim to assess the correlation between multi-focality and metastatic potential should not only concentrate on the number of foci, but also on the size of the foci. As we have shown with the model, cancer progression might start with a small single lesion which can be considered uni-focal. It can then bifurcate to give rise to multiple foci, and finally spread through the entire tissue. When such spread occurs, the multiple foci turn into a big and single mass, and this would again be considered uni-focal. Hence, the cumulative size or volume of the tumor is likely to be the best predictor of malignant progression.

Problems

Problem 8.1. *In the first model describing the invasion of angiogenic cell lines (model 8.1), how does the mutation rate threshold for* μ_1 *change as the carrying capacity* k_1 *is lowered?*

Problem 8.2. Research project. *Read more about multi-focal cancers. How do the number of foci relate to prognosis in clinical data?*

PART 2

Evolutionary dynamics and stochastic models

Chapter 9

Evolutionary dynamics of tumor initiation through oncogenes: the gain-of-function model

9.1 Introduction

The question about the origins of cancer is among the most important in our understanding of the disease. There is no universal answer to this question, as different cancers are initiated by different mechanisms. There are however certain patterns that can be recognized. Among the most prominent ones is cancer initiation via (1) an activation of oncogenes and (2) an inactivation of tumor suppressor genes. The first mechanism is studied in this chapter. The second one is addressed in the following chapter.

Some mutations directly lead to the generation of advantageous mutants. This is characteristic of the *gain-of-function* mutations (see e.g. [Strachan and Read (1996)]) which activate oncogenes. An oncogene is a modified gene that promotes tumor growth. The term "oncogene" was coined by [Huebner and Todaro (1969)] who proposed that viral genes could be integrated into the cellular genome where they might remain dormant or become activated, thus causing cancer.

The modern paradigm is that eukaryotic cells contain certain genes (called proto-oncogenes) involved in the control of cellular growth and differentiation. Changes in expression of these genes can give rise to malignant transformation. Following the hypothesis of Huebner and Torado, by using a variety of molecular techniques, scientists have subsequently identified a number of proto-oncogenes, which after activation to oncogenes become causative factors for a wide variety of human cancers. For example, the SRC oncogene was discovered by [Stehelin *et al.* (1976)]. Bishop and Varmus received the Noble Prize in 1989 for their discovery of cellular origins of retroviral oncogenes. Other important oncogenes are the RAS oncogene found in colon cancer [Vogelstein and Kinzler

(2002)], and the Brc-Abl fusion gene found in chronic myeloid leukemia [Heisterkamp *et al.* (1985)].

Three dominant genetic mechanisms of oncogene activation have been identified [Pierotti *et al.* (2003)]:

(i) Mutations lead to the uncontrolled, continuous activity of the mutated protein. For example, point mutations are frequently detected in the RAS family of proto-oncogenes (K-RAS, H-RAS, and N-RAS). It has been estimated that as many as 15-20% of human tumors may contain a RAS mutation. Mutations in K-RAS predominate in carcinomas. Studies have found K-RAS mutations in about 30% of lung adenocarcinomas, 50% of colon carcinomas, and 90% of carcinomas of the pancreas.

(ii) Another mechanism of oncogene activation is gene amplification, whereby an expansion in copy number of a gene is observed. The process of gene amplification occurs through redundant replication of genomic DNA, often giving rise to karyotypic abnormalities. About 20-30% of breast and ovarian cancers show c-myc amplification, and an approximately equal frequency of c-myc amplification is found in some types of squamous cell carcinomas.

(iii) Chromosomal rearrangements are often detected in hematologic malignancies as well as in some solid tumors. These rearrangements consist mainly of chromosomal translocations. The so-called "fusion genes" can be created by chromosomal rearrangements when the chromosomal breakpoints fall within the loci of two different genes. The resultant juxtaposition of segments from two different genes gives rise to a composite structure consisting of the head of one gene and the tail of another. For example, the Philadelphia chromosome is a translocation, in which parts of two chromosomes, 9 and 22, swap places. The result is that a fusion gene is created by juxtapositioning the Abl1 gene on chromosome 9 to a part of the BCR gene on chromosome 22, thus creating the Bcr-Abl oncogene responsible for chronic myeloid leukemia (CML).

Apart from these mechanisms, oncogenes can also be activated by viruses, whereby viral DNA is integrated into the cellular genome, which can cause the onset of cancer by activating an oncogene [Lodish *et al.* (2000)]. Moreover, the gene sequences may be picked up by certain viruses (in which they occur as viral oncogenes) and transduced to other cells. Viruses which have acquired cellular sequences of this kind are called oncogenic retroviruses

[Paul (1984)]. It is estimated that 15% of all human tumors worldwide are caused by viruses [Butel (2000)].

In this chapter we will set up a stochastic framework to discuss the generation of gain-of-function mutations. We will use an abstraction of space-free stochastic processes with a constant population. This is called the Moran process. We will discuss the dynamics of mutants in a sea of wild-type cells. What is the probability that starting from a single mutant, the whole colony of cells will eventually be replaced by progeny of this mutant? What is the expected time of an advantageous mutant fixation? These and other questions are addressed below.

9.2 Mutation-selection diagrams and the stochastic Moran process

In order to model a cellular turnover in tissues of constant size, we will employ the Moran process, whereby each cell division is coupled with a cell death, thus keeping the population constant. We will study the generation of mutants and also trace the fate of their clones (a clone is a set of all progeny of a given cell).

Let us assume that there could be two types of cells in a population, which we will call type "A" and type "B", and the total population size equals N. The two cell types differ by their fitness parameter. We let the cells of type "A" have fitness 1 and the cells of type "B" - fitness r. At each time step, one cell reproduces, and one cell dies, figure 9.1. We assume that all cells have an equal chance to die (this is equal to $1/N$). On the other hand, reproduction happens differentially depending on the type, and the relative probability of being chosen for reproduction is given by 1 and r for the cells of types "A" and "B" respectively. Obviously, in this setting the total number of cells is preserved.

We further include the process of mutations. Mutations happen with a small probability upon cell divisions, whereby one of the daughter cells differs from the original cell. In the simple process described here, only one type of mutations is considered. Namely, we assume that the probability that a cell of type "A" reproduces faithfully is $1-u$, and with probability u it will mutate to type "B". Cells of type "B" are assumed to reproduce faithfully. We will use the following convenient short-hand representation of the processes of divisions and mutations:

$$A_{(1)} \xrightarrow{\quad u \quad} B_{(r)} \tag{9.1}$$

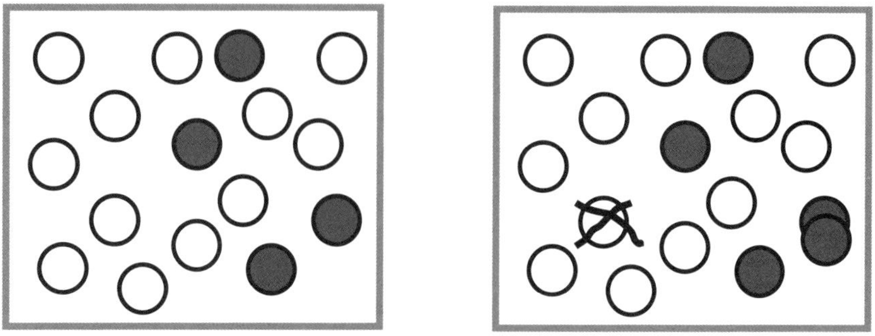

Fig. 9.1 A schematic of one update of the Moran process. Empty circles represent wild-type cells (type A cells), and filled circles represent mutants (type B cells). One cell is chosen for death and another cell is chosen for reproduction. Spatial locations of cells are not taken into account.

Here the fitness parameter of each type is given in brackets and the mutation rate is marked above the arrow. We will refer to such diagrams as *mutation-selection networks.* Mutation-selection networks can be very complex; network (9.1) represents one of the simplest examples.

The relative fitness parameter r tells us how likely a mutant cell is to divide compared to a wild-type cell. In the context of oncogene activation, the mutants' fitness is larger than that of wild-type cells. That is, $r > 1$. A mutation-selection network (9.1) can be relevant not only for the description of oncogene activation, but also for studying cancer initiation in patients with familial disorders, where the first allele is mutated in the germ line, and the inactivation of the second allele leads to a fitness advantage of the cell, see Chapter 11. In these cases, we can also assume $r > 1$. In the more general case, we can view the one-hit model as the process of any one genetic alteration, resulting in an advantageous ($r > 1$), disadvantageous ($r < 1$) or a neutral ($r = 1$) mutant.

Let us denote the number of cells of type "A" as a, and the number of cells of type "B" as b, so that $a+b = N$. The probability that a cell of type "A" reproduces is proportional to its frequency and relative fitness (equal to 1), and is given by $a/(a+rb)$. Similarly, the probability that a cell of type "B" reproduces is $rb/(a+rb)$. Thus the probability that the new cell is of type "A" or type "B" is given respectively by

$$P_{+A} = (1-u)\frac{a}{a+rb}, \quad P_{+B} = u\frac{a}{a+rb} + \frac{rb}{a+rb}.$$

Cells of both types have a probability to die proportional to their abundance, i.e., the probability that a cell of type "A" (or "B") dies is given respectively by

$$P_{-A} = \frac{a}{N}, \quad P_{-B} = \frac{b}{N}.$$

We will refer to an event consisting of one replication and one cell death by an *elementary event.*

The resulting population dynamics is a Markov process with states $b = 0, 1, \ldots, N$. We set the length of each time step to be $1/N$, so that during a unit time interval, N cells are chosen for reproduction and N cells die. The choice of this scaling (the "generation" time-scale) is dictated by the fact that during a physical time-unit, the number of cell divisions should scale with the system size, N.

9.3 Analysis

The probability that an elementary event results in an increase of the number of cells of type "B", is equal to $P_{+B}P_{-A}$, and the probability that the number of cells of type "B" decreases is equal to $P_{-B}P_{+A}$. If P_{ij} is the probability to go to state $b = j$ from state $b = i$, then the transition matrix is given by

$$P_{ij} = \begin{cases} \frac{u(N-i)+ri}{\mathcal{N}_i}\frac{N-i}{N}, & j = i+1, \\ \frac{(1-u)(N-i)}{\mathcal{N}_i}\frac{i}{N}, & j = i-1, \\ 1 - P_{i,i+1} - P_{i,i-1} & j = i, \\ 0 & \text{otherwise}, \end{cases} \tag{9.2}$$

where $0 \le i, j \le N$, and we introduced the notation

$$\mathcal{N}_i = N - i(1-r). \tag{9.3}$$

The corresponding Markov process is a biased random walk with one absorbing state, $b = N$. Let us set the initial condition to be $b = 0$ (all cells are of type "A") and study the dynamics of absorption into the state $b = N$.

An equation for the absorption time for the Markov process can be written down in a straightforward way, [Karlin and Taylor (1975)]. If we denote the number of elementary events until absorption starting from state i as t_i, we have

$$t_i = N + \sum_{m=0}^{N-1} P_{im}t_m, \quad 0 \le i \le N-1. \tag{9.4}$$

The absorption time is then given by $T \equiv t_0$. The factor N appears in the equation above because of our definition of a unit time: $T = 1$ corresponds to N elementary events. Solving system (9.4) directly is however cumbersome, so we will use some approximations.

9.3.1 *The method of differential equations*

Let us denote the probability to be in state $a = i$ at time t as $\varphi_i(t)$. Using the transition matrix for two types, (9.2), we can write down the Kolmogorov forward equation for φ_i:

$$\begin{aligned}\frac{\partial \varphi_i}{\partial t} &= (1-u)\left[\varphi_{i-1}\frac{(i-1)[N-(i-1)]}{rN-(i-1)(r-1)} - \varphi_i\frac{i(N-i)}{rN-i(r-1)}\right] \\ &+ \varphi_{i+1}\frac{[r(N-(i+1))+(i+1)u](i+1)}{rN-(i+1)(r-1)} \\ &- \varphi_i\frac{[r(N-i)+iu](N-i)}{rN-i(r-1)}. \qquad (9.5)\end{aligned}$$

It is convenient to introduce the variable $\eta = i/N$. Taking the continuous limit and expanding into the Taylor series up to the second order, we obtain the following partial differential equation for $\varphi(\eta, t)$:

$$\frac{\partial \varphi}{\partial t} = \frac{\partial}{\partial \eta}(M\varphi) + \frac{1}{N}\frac{\partial^2}{\partial \eta^2}(V\varphi), \qquad (9.6)$$

where

$$M = \frac{\eta(1-r)(1-\eta)-u}{\eta(r-1)-r}, \quad V = -\frac{1}{2}\frac{\eta[(1-\eta)(1+r)-u(1-2\eta)]}{\eta(r-1)-r}.$$

When $|1-r| \ll 1$, we have the following equation:

$$N\frac{\partial \varphi}{\partial t} = N(r-1)\frac{\partial}{\partial \eta}(\eta(1-\eta)\varphi) + \frac{\partial^2}{\partial \eta^2}(\eta(1-\eta)\varphi).$$

This equation is studied in [Kimura (1995)].

Neutral mutants. In the case $|1-r| \ll 1/N$ the principal term in the expression for $\varphi(\eta, t)$ is proportional to $e^{-\mu_0 t}$, where

$$\mu_0 = \frac{1}{N}(1 + O((N(r-1))^2)).$$

This sets the typical time-scale of the process.

Positively and negatively selected mutants. We can also study the case $1/N \ll |1-r| \ll 1$. In that limit, for $r > 1$ (advantageous mutants), the region of interest is $\eta \ll 1$ (remember that $\eta = 0$ corresponds to the all "B" state). Thus the equation simplifies to

$$N\frac{\partial \varphi}{\partial t} = N(r-1)\frac{\partial}{\partial \eta}(\eta\varphi) + \frac{\partial^2}{\partial \eta^2}(\eta\varphi).$$

This equation could be solved in terms of Laguerre polynomials, in general:

$$\varphi(\eta, t) = e^{-N(r-1)\eta} \sum_{n=0}^{\infty} c_n L_n^1(N(r-1)\eta)e^{-(r-1)(1+n)t}.$$

The Laguerre polynomials, $L_n^\alpha(x)$ satisfy the differential equation

$$\left(x\frac{d^2}{dx^2} + (\alpha + 1 - x)\frac{d}{dx} + n\right) L_n^\alpha(x) = 0.$$

Note that the leading transient gives

$$\mu_0 = |1 - r|$$

in this limit. One could similarly treat the case of $r < 1$ (disadvantageous mutants). In general, $\mu_0 = \frac{1}{N} f(N(r-1))$ where $f(s) = 1 + O(s^2)$ for small s but $f(s) \approx |s|$ for large s. The quantity μ_0 will be used in Sections 9.4 and 9.4.1 to find an approximation to the absorption time and to justify the coarse-grained description of system (9.5).

9.3.2 *The probability of absorption*

Equation (9.5) in principle provides all the information about the stochastic behavior of the cell compartment, but it is difficult to solve. Instead of solving the full system and obtaining values for all the probabilities, we can concentrate on some important aspects of the problem and obtain solutions in the limiting cases of interest.

First, let us consider a simpler problem, where cells do not mutate. Let us suppose that initially, cells of both types are present in the system. With a constant cellular turnover in the Moran process, and in the absence of mutations, as time $t \to \infty$, there are exactly two outcomes that we can envisage: (1) all cells are uniformly wild-type, and the mutants have gone extinct, or (2) all the cells are mutant, and the wild-type cells have gone extinct. Which outcome is more likely depends on the fitness of the mutants, and on the initial number of mutant cells. Of biological interest is the probability for a mutant to reach fixation (that is, to reach the state where all cells in the system are replaced by mutants). Below is a mathematical description of this problem.

For $u = 0$, the system has two absorbing states, $b = 0$ and $b = N$. Let us denote the probability to get absorbed in $b = N$ starting from the state $b = i$ as π_i. These quantities are given by the system:

$$\pi_i = P_{iN} + \sum_{m=1}^{N-1} P_{im}\pi_m; \qquad (9.7)$$

note that we set $u = 0$ in the expression for P. System (9.7) can be rewritten as

$$-\pi_{i-1} + (r+1)\pi_i - r\pi_{i+1} = 0, \qquad 1 < i < N-1,$$

where we canceled the common multiplier in the terms of the matrix $I - P$ in the same row. The boundary conditions are

$$(r+1)\pi_1 - r\pi_2 = 0,$$
$$-\pi_{N-2} + (r+1)\pi_{N-1} = r.$$

We can look for a solution in the form $\pi_i = \alpha^i$. The quadratic equation for α gives the roots $\alpha = 1/r$ and $\alpha = 1$. Substituting $\pi_i = Ar^{-i} + B$ into the boundary conditions we obtain the solution,

$$\pi_i = \frac{r^{N-i}(1-r^i)}{1-r^N} = \frac{1-(1/r)^i}{1-(1/r)^N}, \quad r \neq 1. \qquad (9.8)$$

For $r = 1$ we obtain in the limit,

$$\pi_i = \frac{i}{N}, \quad r = 1. \qquad (9.9)$$

The latter result can be derived (i) as a limit of expression (9.8) when $r \to 1$, and (ii) from symmetry considerations: if all cells have equal fitness, they all have an equal chance to eventually dominate the population.

9.4 Probability and timing of mutant fixation

There are two processes that go on in the system: mutation and selection. If the characteristic time scales of the two processes are vastly different, we can treat them separately, and obtain some very useful insights. Let us assume that the mutation rate u is very small, so that once a mutant of type "B" is produced, it typically has time to get fixated or die out before a new mutation occurs. In other words, once a mutant is produced, it is safe to assume that during its life-time no other mutations occur.

Below we will discuss the behavior of mutants whose fitness can be smaller than, equal to, or larger than that of the wild type, which corresponds to negatively selected, neutral, or positively selected mutants. In the context of oncogene activation, the assumption $r > 1$ (positively selected mutants) is usually made. We however will consider the whole range of possibilities. In particular, neutral and negatively selected mutants can be important in the context of tumor suppressor gene inactivation, which is studied in detail in Chapter 10.

Let us study the fate of individual mutant lineages. If the fitness of mutants is equal to or smaller than that for the wild-type cells, then most of the time the lineages will die out. Rarely will they expand to large numbers or even reach fixation, that is, replace the wild type cells and come to completely dominate the cell compartment. What is the probability of this happening? Starting from i mutants, the probability of fixation is given by

$$\pi_i = \begin{cases} \frac{1-(1/r)^i}{1-(1/r)^N}, & r \neq 1, \\ \frac{i}{N}, & r = 1. \end{cases} \tag{9.10}$$

The quantities π_i are probabilities of fixation of a mutant of relative fitness r starting from i cells, in a population of $N-i$ wild-type cells. Figure 9.2(a) shows the graphs of probability π_i as a function of the initial number of mutants for advantageous, neutral, and disadvantageous mutants.

Of particular importance is quantity π_1, the probability of fixation starting from exactly one mutant. Let us reserve the notation ρ for the probability π_1:

$$\rho \equiv \pi_1 = \begin{cases} \frac{1-1/r}{1-1/r^N}, & r \neq 1, \\ \frac{1}{N}, & r = 1. \end{cases} \tag{9.11}$$

Expressions given in (9.11) play an important role in the dynamics of one-hit mutants, spreading through a population of wild-type cells. A log-log plot of the probability ρ as a function of the fitness parameter r is presented in figure 9.2(b). Below $r = 1$, it quickly drops down to very low numbers. This means that it is unlikely for disadvantageous mutants to reach fixation in a population. For $r > 1$, the fixation probability approaches 1 as r increases.

Next, let us study the timing of fixation. Starting from the all-"A" population of cells, how long does it typically take for mutants "B" to be produced and to dominate the compartment? In the case of rare mutations,

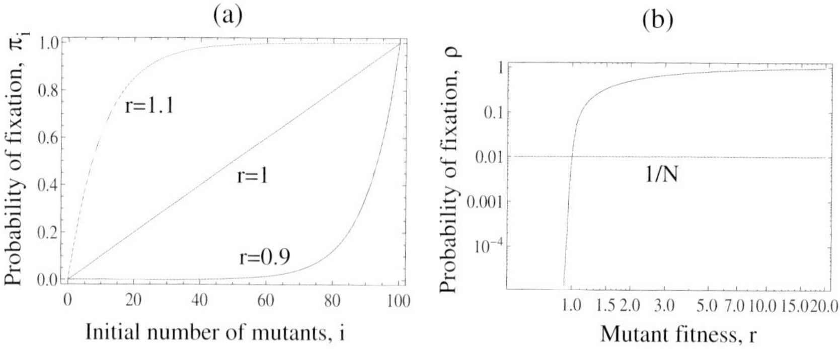

Fig. 9.2 The probability of mutant fixation. (a) The probability of fixation, π_i, as a function of the initial number of mutants, i, plotted for three different fitness values. (b) The probability of fixation starting from one cell, plotted as a function of the fitness parameter, r. The horizontal line represents that probability of fixation of neutral mutants, equal to $1/N$. We used the value $N = 100$.

the inverse time to fixation is roughly u times the probability for a given mutant to get fixated:

$$T = \frac{1}{Nu\rho}. \tag{9.12}$$

In order for approximation (9.12) to be valid, we need to make sure that the time-scale related to mutation ($(Nu)^{-1}$) is much longer than the time-scale of the fixation/extinction processes. Only the fraction ρ of all mutants will successfully reach fixation, whereas the rest will be quickly driven to extinction. In order for each mutant lineage to be treated independently, we need to require that the time it takes to produce a *successful* mutant, $(\rho N u)^{-1}$, is much larger than the typical time-scale of fixation, μ_0^{-1} (Section 9.3.1). In the case of neutral mutations, $\rho = 1/N$, $\mu_0 = 1/N$ and we arrive at the condition,

$$uN \ll 1, \qquad \text{if } |1-r| \ll \frac{1}{N}. \tag{9.13}$$

In the case where the mutation is positively or negatively selected, we have

$$uN \ll r^{-(N-1)}, \qquad \text{if } r < 1, \qquad \frac{1}{N} \ll |1-r| \ll 1, \tag{9.14}$$

$$uN \ll r, \qquad \text{if } r > 1, \qquad \frac{1}{N} \ll |1-r| \ll 1. \tag{9.15}$$

If these conditions hold, we can say that mutations are rare, and each mutant and its clone typically have a chance to either invade or die off before another mutant is generated. Furthermore, under these conditions, approximation (9.12) is valid. Finally, condition (9.13) (or (9.14), (9.15), depending on the mutant fitness) allows us to describe the system in a very simple and intuitive way, as explained next.

9.4.1 *The approximation of "almost absorbing" states and the growth of mutants*

We will call a state of the system *homogeneous*, or pure, if all the N cells are of the same type. States containing more than one type of cells ($1 < b < N$) will be referred to as *heterogeneous*, or mixed states.

If conditions (9.13-9.15) hold, then the mutation rate is very low relative to the time-scale of mutant lineage dynamics, and the probability of finding the system in a heterogeneous state is very low. The system spends most of the time in the pure states, $b = 0$ and $b = N$. This allows us to make a further approximation of "almost absorbing" states.

Let us use the capital letters A and B for the probability to find the system in the state $b = 0$ and $b = N$ respectively. Strictly speaking, the state $b = 0$ is not absorbing, but it is long-lived. We have approximately, $A + B \approx 1$. Let us define the following "coarse-grained", continuous time stochastic process: the system jumps between two states, $A = 0$ and $A = 1$, with the following probabilities:

$$P(A = 0, t + \Delta t | A = 1, t) = u\rho\Delta t, \quad P(A = 0, t + \Delta t | A = 0, t) = 1,$$
$$P(A = 1, t + \Delta t | A = 1, t) = 1 - u\rho\Delta t, \quad P(A = 1, t + \Delta t | A = 0, t) = 0.$$

The Kolmogorov forward equations for this simple system can be written down, which describe the dynamics of the two-species model, (9.1):

$$\dot{A} = -uN\rho A \qquad A(0) = 1, \tag{9.16}$$
$$\dot{B} = uN\rho A, \qquad B(0) = 0, \tag{9.17}$$

where A is the probability to find the entire system in state "A", B is the probability to find the entire system in state "B", and ρ is given by equation (9.8). Equations (9.16-9.17) lead to the solution $A(t) = \exp(-uN\rho t)$ and

$$B = 1 - \exp(-uN\rho t), \tag{9.18}$$

the probability of fixation of mutants of type "B".

A short-hand notation for coarse-grained differential equations (9.16-9.17) is as follows:

$$A \xrightarrow{R_{A\to B}} B \tag{9.19}$$

where the transition rate, $R_{A\to B} = uN\rho$, is a product of the total populations size, the mutatin rate, and the probability of mutant fixation.

Equations (9.16-9.17) give us a very powerful and simple method of studying the stochastic system of dividing, dying and mutating cells. Instead of keeping track of each cell and calculating all the probabilities, we can simply think of the cell compartment as a two-state system: all-"A" and all-"B", with a simple transition rate between the states. This type of dynamics is schematically illustrated in figure 9.3(a).

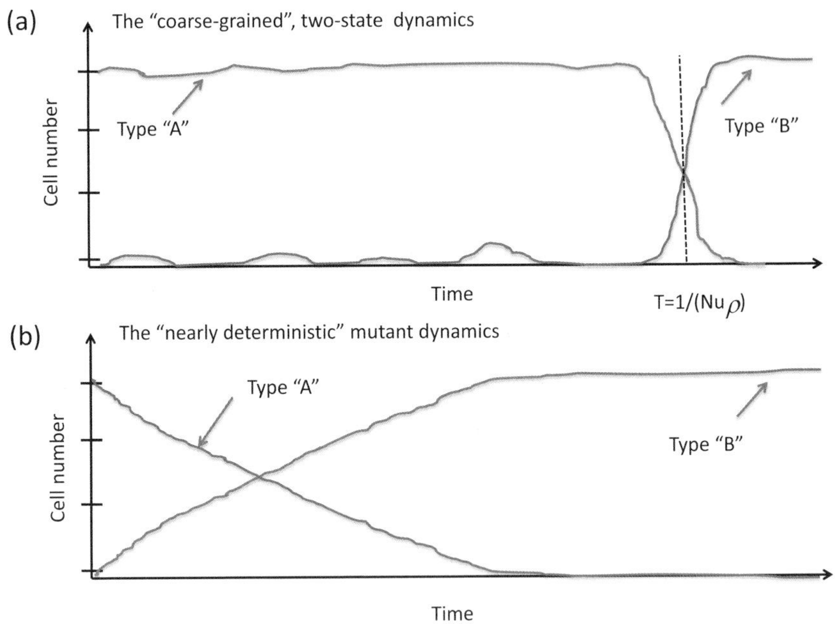

Fig. 9.3 A schematic illustration of the two types of behavior, which we expect to see depending on whether or not conditions (9.13-9.15) hold. (a) A jump between the all-"A" and the all-"B" states, described by the coarse-grained equations (9.16-9.17), when conditions (9.13-9.15) hold. (b) The nearly-deterministic rise of mutants in the case of neutral or advantageous mutants. For disadvantageous mutants such that condition (9.25) holds, the mutants do not reach fixation but are maintained at level (9.26).

9.4.2 *Nearly-deterministic regime*

Finally, we need to see what happens if the opposite of conditions (9.13-9.15) hold. In this case, we can think of the system as very large, and/or having a high mutation rate, such that new mutants are produced frequently, and accumulate in a nearly-deterministic fashion. In this regime, neutral mutants' dynamics will resemble the behavior of advantageous mutants, because in the absence of back-mutations, they will constantly grow in abundance, see figure 9.3(b).

In this case, the dynamics of the mutants can be described by using a deterministic set of quasispecies equations (Chapter 6). If x_0 is the abundance of type "A" cells, and x_1 the abundance of type "B" cells, we have

$$\dot{x}_0 = (1-u)x_0 - \phi x_0, \tag{9.20}$$

$$\dot{x}_1 = ux_0 + rx_1 - \phi x_1, \tag{9.21}$$

$$x_0(0) = N, \quad x_1(0) = 0, \tag{9.22}$$

where the average fitness is given by

$$\phi = \frac{x_0 + rx_1}{N}.$$

In these equations, we have $x_0 + x_1 = N$, and the solution for the number of type "B" mutants reads:

$$x_1(t) = N\left(1 - \frac{e^t(r+u-1)}{ue^{t(r+u)} + (r-1)e^t}\right). \tag{9.23}$$

In particular, if $r = 1$, this expression simplifies to

$$x_1(t) = N(1 - e^{-ut}). \tag{9.24}$$

If $r > 1 - u$, the abundance of one-hit mutants will rise and reach fixation. For disadvantageous mutants with

$$r < 1 - u, \tag{9.25}$$

the type-"B" mutants will be maintained at a selection-mutation balance,

$$x_1(t) = Nu/(1-r). \tag{9.26}$$

9.5 Summary

In this chapter we studied the (mathematically) simplest pattern of cancer initiation, which is described by a gain-of-function model. We discussed the probability of mutant fixation, the time to fixation, and a coarse-grained description of the process. A detailed understanding of these processes is required to gain insights about the development of tumors that are driven by the activation of an oncogene, such as chronic myeloid leukemia (CML). If the appropriate parameters are measured, it is possible to use the calculations developed here to predict the timing of cancer emergence and thus to interpret age-incidence patterns. Next, we look at loss-of-function mutations, which proceed in a sequence of two inactivation mutations.

Problems

Problem 9.1. Research project. *Learn about the connection between viruses and oncogene inactivation.*

Problem 9.2. *Derive equation (9.24) by taking the limit $r \to 1$ in expression (9.23).*

Problem 9.3. *Derive equation (9.26) by assuming that $r < 1 - u$ and taking the limit $t \to \infty$ in expression (9.23). What happens as $t \to \infty$ if $r > 1 - u$?*

Problem 9.4. *Derive equation (9.9) by taking the limit $r \to 1$ in expression (9.8).*

Problem 9.5. Numerical project. *Absorption time in a Markov chain can be calculated directly by using equation (9.4). Solve algebraic system (9.4) numerically for particular values of parameters u, r, and N. Show that result (9.12) holds approximately if condition (9.13) (or (9.14, 9.15) is satisfied. Note: Formula (9.12) can also be obtained analytically if one solves system (9.4) explicitly and then takes the first term in the Taylor expansion of T_0 in u.*

Chapter 10

Evolutionary dynamics of tumor initiation through tumor-suppressor genes: the loss-of-function model and stochastic tunneling

10.1 Introduction

In the previous chapter we concentrated on a mathematical description of oncogenes, that are activated by a gain-of-function mutation. Here we introduce another very common pattern in cancer initiation and progression, which is described by a loss-of-function model. A loss-of-function mutation results in a gene product having less or no function. Two independent loss-of-function mutations are necessary to inactivate a gene, because after the first mutation, the second copy of the gene is still active. In the context of cancer, the loss-of-function mechanism is involved in the inactivation of tumor suppressor genes.

The concept of tumor-suppressor genes has evolved during the last 30 years. A defining landmark was the discovery of the Rb gene. Retinoblastoma is a rare and deadly cancer of the eye that afflicts children. It comes in two versions. One affects newborn infants and is characterized by multiple tumors. The other hits children when they are older and is usually characterized by only a single tumor. In 1971, Alfred Knudson proposed an explanation, which became known as the famous Knudson's "two-hit hypothesis" [Knudson (1971)]. According to his theory, in the early-onset version of retinoblastoma, children inherit a defective gene from one parent. These children are halfway to getting the disease the moment they are born. Then, an error in DNA replication in a single eye cell, causing a defect in the normal gene that was inherited from the other parent, would send that cell on its way to becoming a tumor. In contrast, children who develop retinoblastoma later in childhood are probably born with two good copies of the gene but acquire two hits in both copies of the gene in a cell. This would take longer, causing the cancer to show up at a later age.

Knudson proposed the tumor suppressor gene hypothesis of oncogenesis after detecting a partial deletion of chromosome 13 in a child with retinoblastoma. This was a revolutionary concept, that is, cancer was not caused by the *presence* of an oncogene, but rather the *absence* of an "anti-oncogene". He concluded that the retinoblastoma tumor suppressor gene would be found at band 13q14. It wasn't until the late 1980s when scientists eventually cloned the gene Rb which mapped exactly where predicted by Knudson.

Other genes with similar properties were discovered, including p53, WT1, BRCA1, BRCA2 and APC. The generic definition of a *tumor suppressor gene* comprises the idea of a *loss of function*. Only when both alleles of the gene are inactivated, does the cell acquire a phenotypic change. Many of tumor suppressor genes are involved in familial cancers. The mechanism is similar to the one described by Knudson in retinoblastoma. If a defective allele is present in the germline, the affected individuals will have a higher chance of developing a cancer as only one remaining allele must be inactivated to initiate an early stage lesion.

In collaboration with Suresh Moolgavkar, Knudson went on to develop mathematical models for this hypothesis, which were the first to coalesce clinical-epidemiological observations with putative mutation rates and molecular genetics [Moolgavkar and Knudson (1981)]. In their publications, Moolgavkar and colleagues have created a rigorous methodology of studying two-stage carcinogenesis [Moolgavkar (1978); Moolgavkar *et al.* (1980, 1988)]. In this chapter we will review some of the main ideas of the two-hit models in a general context. We will also introduce the concept of "stochastic tunneling", which is an important computational tool to study the process of loss-of-function mutations.

10.2 Process description and the mutation-selection diagram

The process we are studying is characterized by the possibility of three types of cells: the wild-type cells, the intermediate mutants, or one-hit mutants, and the double-mutants. According to this, we suppose that there are three types of cells: type "A", type "B" and type "C", and the mutation-selection network that governs the dynamics is as follows:

$$\mathrm{A}_{(1)} \xrightarrow{\;u_1\;} \mathrm{B}_{(r)} \xrightarrow{\;u_2\;} \mathrm{C}_{(r_1)} \tag{10.1}$$

The fitness parameters are denoted respectively by 1, r and r_1. As before, the fitness parameters must be interpreted as *relative* probabilities to be chosen for reproduction, rather than parameters defining the time-scale. We assume that type "A" can mutate into type "B" with probability u_1, and type "B" can mutate to type "C" with probability u_2. There are no other mutation processes in the system.

This model describes several biologically relevant situations. For instance, it may be directly applied for the two-hit hypothesis, that is, the process of the inactivation of a tumor suppressor gene. In the simplest case, the inactivation of the first allele of a tumor suppressor gene (TSP) does not lead to a phenotypic change, which would correspond to the value $r = 1$. This rigid definition can be relaxed to allow for certain *gene dosage effects*. For instance, the loss of one copy can lead to a certain change in the phenotype, and the loss of both copies will increase this effect. In this case, we could have $1 < r < r_1$. Finally, the case $r < 1$, $r_1 > 1$ means that the intermediate cell has a disadvantage compared to wild type cells. For example, this may correspond to the situation where the inactivation of the first allele is achieved by a large scale genomic alteration, such as a loss-of-heterozigocity event where many genes have been lost. This would lead to the intermediate product having a disadvantage compared to the wild type cells. Losing the remaining allele of the tumor suppressor gene will give the cell a growth advantage which may override the fitness loss of the previous event, resulting in $r_1 > 1$.

In general, the two-hit model described above refers to any two consecutive mutations, such that the first one may be positively or negatively selected (or neutral), and the second one confers a significant selective advantage to the cell.

Let us specify the states of the system by the variables a, b and c, the number of cells of species "A","B" and "C", respectively. In the framework of the Moran process, they satisfy the constraint $a + b + c = N$. We can characterize a state as a vector (b, c). In this notation, the state we start with is $(0, 0)$, which is all "A". The final state, which is the state of interest, is $(0, N)$, or all "C". The question we will study is again, the time of absorption in the state $c = N$.

10.3 Three regimes: a two-step process, stochastic tunneling, and a nearly-deterministic regime

Let us start from the all "A" state. If we are in the regime of homogeneous states, conditions (9.13-9.15), we can consider the lineages of each mutant of type "B" separately. Once a mutant of type "B" is created, its lineage can either go extinct, or become fixated. A mutant of type "C" can be created before or after type "B" reaches fixation. This gives rise to two possible scenarios, figure 10.1(a,b).

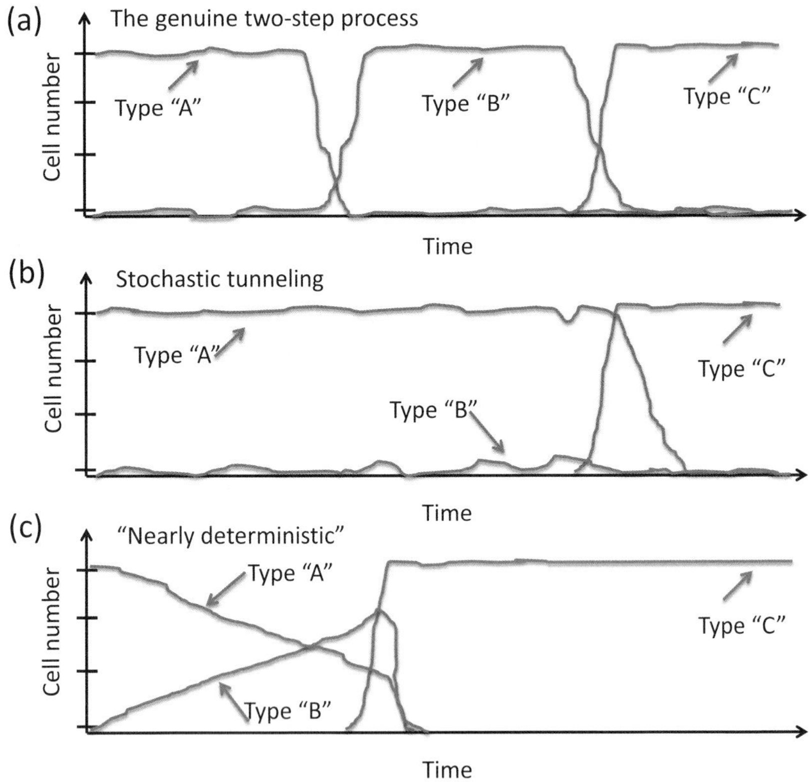

Fig. 10.1 A schematic depicting two possible processes of two-hit mutant creation. (a) The genuine two-step process. (b) Stochastic tunneling. (c) Nearly-deterministic process.

We will call a *genuine two-step process* a sequence of steps where starting from $(0,0)$, after some time the system finds itself in the state $(N,0)$ and then gets absorbed in the state $(0, N)$. In other words, starting from the all "A" state, the system gets to the state where the entire population consists of cells of type "B", and finally reaches fixation in the all "C" state, see figure 10.1(a).

We will use the term *stochastic tunneling* for processes where the system goes from $(0, 0)$ to $(0, N)$ without ever visiting state $(N, 0)$. This means that from the all "A" state the system gets absorbed in the all "C" state, skipping the intermediate fixation of type "B", see figure 10.1(b). The concept of stochastic tunneling was introduced by [Nowak *et al.* (2002); Komarova *et al.* (2003b)] when studying the first step in colon cancer initiation, the inactivation of the tumor suppressor gene APC. The concept has later been investigated by several groups in the context of cancer initiation, escape dynamics [Iwasa *et al.* (2004b)], and more broadly as a means of crossing a fitness valley by an evolving species [Weissman *et al.* (2009)]. The basic Moran process in a homogeneous tissue has been used as the underlying mathematical model. A generalization to a model of renewing epithelial tissue will be described in Chapter 12, and a spatial generalization for the tunneling rate will be considered in Chapter 13.

Finally, there is a third distinct regime of the double-hit mutant generation, whereby one-hit mutants grow in a nearly-deterministic fashion, and eventually a double-hit mutant is produced in their midst. This regime is schematically shown in figure 10.1(c).

It turns out that the computation of the waiting times for a mutant of type "C" to appear will be different in the three regimes.

10.4 The transition matrix

We are interested in the case where the type "C" has a large selective advantage, i.e., $r_1 \gg (1, r)$, so that once there is one cell of type "C", this type will invade instantaneously with probability one. Under this assumption we can use a trick which allows us to view the dynamics as a one-dimensional process. Namely, let us consider the following reduced Markov process with the independent stochastic variable b: the states $b = i$ with $0 \leq i \leq N$ correspond to $a = N - i$, $b = i$, $c = 0$, and the additional state E contains all states with $c > 1$. The state E is the only absorbing state of the system, because we assume that once a mutant of type "C" appears, then cells "C" invade, so the system cannot go back to a state with $c = 0$. Using notation 9.3,

$$\mathcal{N}_i = ri + N - i,$$

the transition probabilities are given by

$$P_{ij} = \begin{cases} \frac{u_1(N-i)+(1-u_2)ri}{\mathcal{N}_i}\frac{N-i}{N}, & j = i+1, \\ \frac{(1-u_1)(N-i)}{\mathcal{N}_i}\frac{i}{N}, & j = i-1, \\ \frac{u_2 ri}{\mathcal{N}_i}, & j = E, \\ 1 - P_{i,i+1} - P_{i,i-1} - P_{i,E}, & j = i, \\ 0 & \text{otherwise}, \end{cases} \quad (10.2)$$

for $0 \le i \le N$, $P_{E,E} = 1$ and $P_{E,j} = 0$ for all $0 \le j \le N$. In some special cases, the absorption time can be found from equation (9.4). A direct solution is not possible in the general case, and we will use some approximations.

10.5 Mathematical theory

10.5.1 *The Kolmogorov forward equation in the absence of intermediate mutant fixation*

Let us denote by i the number of one-hit mutants in a population of N cells. We start the process with all wild-type cells, and stop when the first double-hit mutant appears. We also assume that the number of intermediate, one-hit mutants does not become large, that is,

$$i \ll N. \quad (10.3)$$

The conditions when this approach holds true are discussed later in this chapter. The following transitions are possible, starting from state i:

- The number of mutants decreases by 1 with probability
$$P_{i\to i-1} = \frac{i}{N}\frac{N-i}{\mathcal{N}}(1-u_2) \approx \frac{1}{N}i(1-u_2), \quad (10.4)$$
where the first multiplier corresponds to the probability of death of a one-hit mutant, and the second multiplier is a faithful reproduction of one of the wild-type cells. Here and below we neglected i/N compared to 1 in the approximation.
- The number of one-hit mutants increases by 1 with probability
$$P_{i\to i+1} = \frac{N-i}{N}\left(\frac{ri}{\mathcal{N}}(1-u_2) + \frac{N-i}{\mathcal{N}}u_2\right) \approx \frac{1}{N}\left(ri(1-u_2) + Nu_2\right), \quad (10.5)$$

where the first multiplier corresponds to the probability of death of a wild-type cell, and the second multiplier contains two processes whereby a one-hit mutant is created: a faithful division of a one-hit mutant and a mutation of a wild-type cell.

- A two-hit mutant is created with probability

$$P_{i\to E} = \frac{ri}{N}u_2 \approx \frac{1}{N}riNu_2, \tag{10.6}$$

which corresponds to death of any cell (probability 1) and a division of a one-hit mutant with a mutation.

Let us denote by $\varphi_i(t)$ the probability to find the system in state i at time t. We have

$$\sum_{i=0}^{N}\varphi_i(t) + \varphi_E(t) = 1. \tag{10.7}$$

The function $\varphi_i(t)$ satisfies the following Kolmogorov forward equation:

$$\dot{\varphi}_i = \varphi_{i-1}P_{i-1\to 1} + \varphi_{i+1}P_{i+1\to i} - \varphi_i(P_{i\to i+1} + P_{i\to i-1} + P_{i\to E}). \tag{10.8}$$

Let us scale the time variable by a factor of N to adopt the generation time scale (see Section 9.2). This eliminates the factor $1/N$ in front of all the probabilities. We have:

$$\begin{aligned}\dot{\varphi}_i = {} & \varphi_{i-1}[r(i-1)(1-u_2) + Nu_2] + \varphi_{i+1}(i+1)(1-u_2) \\ & - \varphi_i[i(1-u_2) + ri + Nu_2]. \end{aligned} \tag{10.9}$$

10.5.2 *The probability generating function*

System (10.9) is a coupled infinite system of ODEs, which we do not intend to solve exactly. Instead, we will extract useful information from this system which does not require the detailed knowledge of all the probabilities $\varphi_i(t)$. We will use the method of the probability generating function:

$$\Psi(x,t) = \sum_{i=0}^{N}\varphi_i(t)x^i, \tag{10.10}$$

see e.g. [Karlin and Taylor (1975)]. The transition from $\{\varphi_i(t)\}$ to $\Psi(x,t)$ is a variable transformation that allows us to reduce an infinite system of ODEs, equations (10.9), to a single equation, a PDE for $\Psi(x,t)$.

Before we derive this PDE, we note that the quantity $\Psi(1,t)$ has the meaning of the probability to find the system in one of the states $0, 1, \ldots, N$. Therefore, by equation (10.7), the quantity

$$P_2(t) = 1 - \Psi(1,t)$$

is the probability to have produced a double-hit mutant by time t. This is the quantity we would like to calculate.

Multiplying equation (10.9) by x^i and summing over all i, we obtain the following PDE for the probability generating function:

$$\frac{\partial \Psi}{\partial t} = \frac{\partial \Psi}{\partial x}[r(1-u_2)x^2 + (1-u_2) - (1-u_2+r)x] + \Psi N u_2(x-1). \quad (10.11)$$

10.5.3 *The method of characteristics and the Riccati equation*

We can solve the first order linear PDE, equation (10.11), by the standard method of characteristics. We have

$$P_2(t) = 1 - exp\left(-\int_0^t N u_1(1 - x(t'))\, dt'\right), \quad (10.12)$$

where $x(t)$ is a solution of the initial value problem,

$$\dot{x} = r(1-u_2)x^2 + (1-u_2) - (1-u_2+r)x, \quad (10.13)$$

$$x(0) = 1. \quad (10.14)$$

Equation (10.12) is sometimes called the doubly-stochastic approximation, and can be derived in the theory of branching processes, see e.g. [Parzen (1962)]. The quantity

$$P_1(t) \equiv 1 - x(t) \quad (10.15)$$

is the probability to generate a double-hit mutant by time t, starting from exactly one cell of type "B".

The equation that $x(t)$ satisfies is a Riccati equation and can be solved exactly. Let us consider problem (10.13-10.14) with general (constant) coefficients:

$$\dot{x} = ax^2 - bx + c, \quad (10.16)$$

$$x(0) = 1, \quad (10.17)$$

where in our particular case,

$$a = r(1-u_2), \quad b = 1 + r - u_2, \quad c = 1 - u_2.$$

The solution of this equation is given by

$$x(t) = -\frac{A\beta_1 e^{(\beta_1-\beta_2)t} + \beta_2}{a(Ae^{(\beta_1-\beta_2)t} + 1)}, \quad A = -\frac{a+\beta_2}{a+\beta_1}, \tag{10.18}$$

where $\beta_1 < \beta_2$ are the two roots of the quadratic equation $\beta^2 + b\beta + ac = 0$.

Equation (10.16) has a globally-stable fixed point. We will describe the trajectory approaching this fixed point from initial condition (10.17),

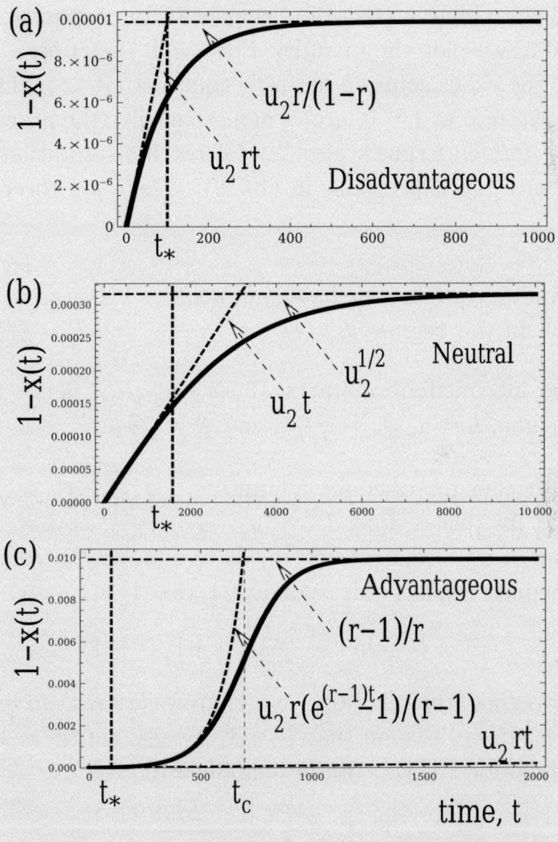

Fig. 10.2 The quantity $P_1(t) = 1 - x(t)$ from system (10.13-10.14) is plotted as a function time for (a) disadvantageous intermediate mutant ($r = 0.99$), (b) neutral intermediate mutants ($r = 1$), and (c) advantageous intermediate mutants ($r = 1.01$). The approximations given by formula (10.19) are shown by dashed lines. The mutation rate is $u_2 = 10^{-7}$.

in terms of function $P_1(t)$, equation (10.15), instead of the function $x(t)$. This function grows linearly with slope L for small values of $t \ll t_*$, and eventually saturates to a constant R for large values of $t \gg t_*$. We have

$$L = b - a - c, \quad t_* = \frac{1}{\beta_2 - \beta_1} = \frac{1}{\sqrt{b^2 - 4ac}}, \quad R = 1 + \frac{\beta_2}{a}. \quad (10.19)$$

10.5.4 *Tunneling for disadvantageous, neutral, and advantageous intermediate mutants*

We would like to characterize the behavior of the exponent in (10.12). In figure 10.2, we plot the quantity $P_1(t) = 1 - x(t)$ as a function of time, using the exact solution for $x(t)$, equation (10.18). The function $P_1(t)$ starts at zero at $t = 0$, grows monotonically (the growth is linear with slope L for short time scales), and it reaches saturation at level R. Let us evaluate the expressions in (10.19). There are three important limits:

(i) Disadvantageous intermediate mutants, $r < 1$, $|1-r| \gg \sqrt{u_2}$, figure 10.2(a). In this regime, $L = ru_2$, $t_* = \frac{1}{1-r}$, and $R = \frac{ru_2}{1-r}$.

(ii) Neutral intermediate mutants, $|1 - r| \ll \sqrt{u_2}$, figure 10.2(b). In this regime, $L = u_2$, $t_* = \frac{1}{2\sqrt{u_2}}$, and $R = \sqrt{u_2}$.

(iii) Advantageous intermediate mutants, $r > 1$, $|1 - r| \gg \sqrt{u_2}$, figure 10.2(c). In this regime, $L = ru_2$, $t_* = \frac{1}{r-1}$, and $R = \frac{r-1}{r}$. In the case of advantageous intermediate mutants, there is a distinct intermediate regime, corresponding $1/(r-1) < t \ll t_c$, where

$$t_c = \ln\left(\frac{(r-1)^2}{r^2 u_2} + 1\right) / (r-1). \quad (10.20)$$

In this regime, the function $1 - x(t)$ grows faster than linear, figure 10.2(c). If we assume that $|r - 1| \gg u_2$, and $r > 1$, then the expression for $1 - x(t)$ can be simplified to give

$$P_1(t) = \frac{u_2 r \left(e^{(r-1)t} - 1\right)}{r - 1}. \quad (10.21)$$

In the limit of small t, the above function becomes linear with the slope ru_2. Therefore, we can say that there are two regimes in the growth of function $P_1(t)$: for $t \ll t_c$ it can be approximated by formula (10.21), and for $t \gg t_c$, we have $P_1(t) \approx (r-1)/r$.

10.5.5 *Genuine two-step process vs tunneling*

The method used so far assumes the independence of the lineages of the intermediate mutant. Thus for the method to work, the probability of fixation of intermediate mutants must be small compared to the probability of "tunneling". The former probability is given by equation (9.18), and the latter - by equation (10.12). Therefore, the inequality,

$$\rho t \gg \int_0^t P_1(t')\,dt', \tag{10.22}$$

guarantees that fixation of an intermediate mutant typically happens before a double-hit mutant is produced. The quantity ρ is the probability that a one-hit mutant becomes fixated and is given by equation (9.11). The condition simplifies depending on the exact regime (growth, saturation) and the types of intermediate mutant.

If condition (10.22) holds, then the description of Section 10.5.4 breaks down, and we have a consecutive fixation first of type "B" mutants and then type "C" mutants. This is the genuine two-step process, figure 10.1(a).

10.5.6 *Time-scales of the process*

As mentioned above, the quantity $P_1(t) = 1 - x(t)$ has the meaning of the probability that starting from exactly one intermediate mutant, a double-mutant is created at time t. This quantity defines a certain time scale (t_* for neutral and disadvantageous mutants, and t_c for advantageous mutants). These time-scales separate the regime of growth and the regime of saturation of probability $P_1(t)$.

Another time-scale is defined by the left hand side of equation (10.22), and corresponds to the typical time to fixation of an intermediate mutant.

Finally, there is another time-scale coming from the factor Nu_1 (intermediate mutant production) in expression (10.12). This time-scale, which we call t_1, is the time when the expression in the exponent of (10.12) becomes of the order one,

$$u_1 N \int_0^{t_1} P_1(t')\,dt' = 1. \tag{10.23}$$

For $t \gg t_1$, we have $P_2(t) \approx 1$. Therefore, it makes sense to restrict the values of the time-variable to the domain of interest, $t \in [0, t_1]$.

The interplay between the three time scales is key for calculating the probability of double-mutant creation in the Moran process. This is shown next, in the context of the three distinct types of intermediate mutants.

10.5.7 *Neutral intermediate mutants*

Neutral mutants in this context are defined by the condition $|1-r| \ll \sqrt{u_2}$. We will separate two limiting cases: the case where $t_* \ll t_1$ and the case where $t_* \gg t_c$, see figure 10.3.

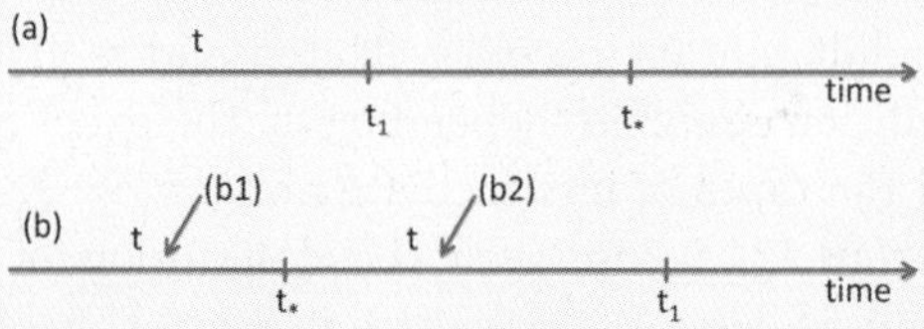

Fig. 10.3 Two configurations of the time-scales t_* (for neutral of disadvantageous mutants) or t_c (for advantageous mutants) and t_1. (a) $t_1 \ll t_*$. (b) $t_* \ll t_1$.

(a) Assume that $t_* \gg t_1$, figure 10.3(a). This means that for all relevant values of time, we have $P_1(t) = u_2 t$, a linear growth regime. This allows us to calculate $t_1 = \sqrt{\frac{2}{u_1 u_2 N}}$ from equation (10.23), and given that $t_* = 1/(2\sqrt{u_2})$, condition $t_* \gg t_1$ is equivalent to requiring

$$u_1 N \gg 1. \tag{10.24}$$

To decide if a two-step process takes place, we use inequality (10.22). A quick calculation shows that the result depends on time. It turns out that if $t \ll 2/(Nu_2)$, the fixation of a one-hit mutant typically has a chance to happen before a two-hit mutant is created. The two-step process is described in Section 10.6.1. Here we just mention that for $t \ll 2/(Nu_2)$, it is easy to show that condition (10.37) holds, and we have

$$P_2(t) \approx N^2 u_1 u_2 t^2/2. \tag{10.25}$$

When the opposite condition holds, that is, when $t \gg 2/(Nu_2)$, we can use formula (10.12) under the linear approximation of $P_1(t)$, to get

$$P_2(t) = 1 - exp\left(-N^2 r u_1 u_2 t^2/2\right), \tag{10.26}$$

which is approximately equal to the expression in (10.25), the result obtained by the two-step process.

Finally, we notice that condition (10.24) is the condition of being in a nearly-deterministic regime, see inequality (9.13) for neutral mutants. This regime is characterized by a steady growth of intermediate mutants, eventually producing a double-hit mutant. The formula for $P_2(t)$ calculated for the nearly-deterministic regime for neutral intermediate mutants coincides with equation (10.26).

(b) Assume that $t_* \ll t_1$, figure 10.3(b). To determine the value of t_1, we note that the greatest contribution to the integral in (10.12) comes from the saturated regime in $P_1(t)$. Therefore, we replace $P_1(t)$ by $R = \sqrt{u_2}$, and obtain from (10.23) that $t_1 = (u_1\sqrt{u_2}N)^{-1}$. Condition $t_* \gg t_1$ is equivalent to requiring

$$u_1 N \ll 1, \tag{10.27}$$

the opposite of condition (10.24). Now, depending on the time-scale of interest, there could be two separate cases: $t \ll t_*$ and $t_* \ll t \ll t_1$.

(b1) Suppose that $t \ll t_*$. This case is similar to case (a) in that we have $P_1(t) \approx u_2 t$. If $t \ll 2/(Nu_2)$, then a two-step process with quadratic formula (10.25) takes place. If $t \gg 2/(Nu_2)$, then we have the same result, but obtained from formula (10.12). Of course, the latter is possible only if $2/(Nu_2) \ll t_*$, which is equivalent to the requirement

$$N \gg N_{tun}, \quad N_{tun} = \frac{1}{\sqrt{u_2}}. \tag{10.28}$$

(b2) Suppose that $t \gg t_*$. In this case, we can use the approximation $P_1(t) \approx R = \sqrt{u_2}$. Condition (10.22) is equivalent to condition (10.28). That is, if this condition holds, we have a genuine two-step process. If the opposite of condition (10.28) takes place, then we have a tunneling process with the rate $R = \sqrt{u_2}$: $P_2 = 1 - exp\left(-Nu_1\sqrt{u_2}t\right)$.

All these conditions, together with conditions for disadvantageous and advantageous mutants (see below), are summarized concisely in Section 10.6.5.

10.5.8 *Disadvantageous intermediate mutants*

Disadvantageous mutants are defined by the conditions $|1-r| \gg \sqrt{u_2}$ and $r < 1$. The arguments in this case follow the same logic as those for the neutral case.

(a) Assume that $t_* \gg t_1$, figure 10.3(a). This means that for all relevant values of time, we have $P_1(t) = u_2 rt$, and $t_1 = \sqrt{\frac{2}{ru_1u_2N}}$ from equation (10.23). Given that $t_* = 1/(1-r)$, condition $t_* \gg t_1$ is equivalent to

$$u_1 N \gg \frac{(1-r)^2}{ru_2}. \tag{10.29}$$

Using inequality (10.22), we can see that the behavior is different for different values of time. If $t \ll 2\rho/(rNu_2)$, we have a two-step process with a quadratic dependence,

$$P_2(t) \approx N^2 \rho u_1 u_2 t^2/2. \tag{10.30}$$

For time values with $t \gg 2\rho/(Nu_2)$, $P_2(t)$ is defined by formula (10.26), the same as in the case of neutral mutants. Although in the case of the disadvantageous mutants the two time-dependencies are not the same, we observe that the time-threshold, $2\rho/(rNu_2)$, is extremely small (less than one time-unit), and thus does not play a practical role. We can say that formula (10.26) describes all the relevant regimes in this case, and the two-step process is not observed.

In the neutral process we noticed that case (a) coincides with the nearly-deterministic regime. It is not the case with the disadvantageous mutants. A nearly-deterministic regime is defined by the inequality (9.14) for disadvantageous mutants, but the behavior in this regime is very different, as described at the end of Section 9.4.2. As long as $|r-1| > u_1$, the mutants of type "A" are maintained at a selection-mutation balance (at level $Nu_1/(1-r)$). It is shown in Section 10.6.3 that in the deterministic regime, disadvantageous intermediate mutants produces double-hit mutants exactly at the same rate as predicted by the tunneling formula.

(b) Next, we consider the case $t_* \ll t_1$, figure 10.3(b). To determine the value of t_1, we replace $P_1(t)$ by $R = ru_2/(1-r)$, and obtain from (10.23) that $t_1 = (1-r)/(ru_1u_2N)$. Condition $t_* \gg t_1$ is equivalent to the opposite of condition (10.29). Depending on the time-scale of interest, there could be two separate cases: $t \ll t_*$ and $t_* \ll t \ll t_1$.

(b1) Suppose that $t \ll t_*$. As in case (a), we have $P_1(t) \approx ru_2t$. If $t \ll 2\rho/(Nru_2)$, then a two-step process with quadratic formula (10.30) takes place. If $2\rho/(Nu_2) \ll t \ll t_*$, which is only possible if

$$N \gg N_{tun}, \quad N_{tun} = \frac{\ln\left(\frac{(1-r)^2}{r^2u_2}\right)}{|\ln r|}, \tag{10.31}$$

then the probability of double-mutant creation is described by formula (10.26).

(b2) Finally, we assume that $t \gg t_*$. In this case, we can use the approximation $P_1(t) \approx R = ru_2/(1-r)$. Condition (10.22) is equivalent to condition (10.31). If this condition holds, we have a genuine two-step process. If the opposite of condition (10.28) takes place, then we have a tunneling process with the rate $R = u_2r/(1-r)$: $P_2 = 1 - exp\left(-Nru_1u_2t/(r-1)\right)$.

10.5.9 *Advantageous intermediate mutants*

In this case, it is the superlinear behavior that defines the growth of $P_1(t)$ before saturation, see figure 10.2. Therefore, we will use figure 10.3 with t_c instead of t_*. We start with case (b).

(b) Assume that $t_c \ll t_1$, figure 10.3(b). To determine the value of t_1, we replace $P_1(t)$ by $R = (r-1)/r$, and obtain from (10.23) that $t_1 = r/(u_1N(r-1))$. The value of t_c is given in equation (10.20). Condition $t_c \gg t_1$ is then equivalent to

$$u_1N \ll \frac{r}{\ln\left(\frac{(r-1)^2}{r^2u_2}+1\right)}. \tag{10.32}$$

Depending on the time-scale of interest, there could be two separate cases: $t \ll t_c$ and $t_c \ll t \ll t_1$.

(b1) Suppose that $t \ll t_c$. In this case, $P_1(t)$ is given by equation (10.21), and we have $P_1(t) < R = (r-1)/r$. We further note that for advantageous mutants, $\rho \approx R = (r-1)/r$. Inequality (10.22) becomes

$$\frac{r-1}{r}t > \int_0^t \frac{u_2r\left(e^{(r-1)t'}-1\right)}{r-1}\,dt',$$

which holds for any values of t. This means that in this case we have a two-step process.

(b2) Next, we assume that $t \gg t_*$, such that $P_1(t) \approx R = (r-1)/r$. In this case inequality (10.22) becomes an equality. In this regime, we have a mixed regime where sometimes the fixation of a type-"B" mutant happens before the generation of a type "C" mutant, and sometimes a type-"C" mutant is generated first.

(a) Finally, we assume that $t_c \gg t_1$, figure 10.3(a). In this case the function $P_1(t)$ is approximated by equation (10.21). Even though the integral in equation (10.23) can be evaluated easily, it is not possible to resolve the equation for t_1. Numerical simulations show that the opposite of condition (10.32) should hold for this regime. The behavior in this regime is similar to case (b1) above, with exactly the same argument. The dynamics happen predominantly by a genuine two-step process.

10.6 Dynamics of loss-of-function mutations

The mathematical theory presented above confirms the existence of three different modes of double-hit mutant generation sketched in figure 10.1. The parameters that define the dynamics are the population size, N, the mutations rates, u_1 and u_2, and the fitness of the intermediate mutants, r. Below we describe each regime in detail and show under what circumstances we can expect which kind of behavior.

10.6.1 *The genuine two-step processes*

This regime is characterized by a consecutive fixation of mutants of type "B" and then mutants of type "C", figure 10.1(a). If the number of cells in the population, N, is sufficiently small (the precise conditions for this and other regime are provided in Section 10.6.5), then the dynamics can be represented by the diagram

$$A \xrightarrow{R_{A\to B}} B \xrightarrow{R_{B\to C}} C$$

with

$$R_{A\to B} = Nu_1\rho, \qquad R_{B\to C} = Nu_2. \tag{10.33}$$

This selection-mutation diagram corresponds to two consecutive diagrams for a one-step process described in Chapter 9, see diagram 9.19 and also figure 9.3(a). Here we assumed that $\frac{1-1/r_1}{1-1/r_1^N} \approx 1$. The corresponding differential equations are derived from (9.16-9.17):

$$\dot{A} = -R_{A\to B}A, \qquad A(0) = 1, \tag{10.34}$$

$$\dot{B} = R_{A\to B}A - R_{B\to C}B, \qquad B(0) = 0, \tag{10.35}$$

$$\dot{C} = R_{B\to C}B, \qquad C(0) = 0. \tag{10.36}$$

Solving this system, we obtain

$$P_2(t) = 1 - \frac{u_2 e^{-Nu_1\rho t} - u_1\rho e^{-Nu_2 t}}{u_2 - u_1\rho},$$

where $P_2(t)$ stands for the probability to have produced a two-hit mutant by time t, and is the same as $C(t)$. The function ρ depends on the fitness of the intermediate mutant,

$$\rho = \begin{cases} \frac{1-1/r}{1-1/r^N}, & r \neq 1, \\ \frac{1}{N}, & r = 1. \end{cases}$$

At short time scales, such that

$$R_{A\to B}t \ll 1, \quad R_{B\to C}t \ll 1, \tag{10.37}$$

the function $P_2(t)$ behaves quadratically with time,

$$P_2(t) \approx \frac{1}{2}N^2\rho u_1 u_2 t^2. \tag{10.38}$$

This is a signature of a two-hit process, which tells us that the loss-of-function mutation process takes places in a sequence of two rate-limiting steps.

10.6.2 *Tunneling*

For intermediate population sizes, the tunneling regime is entered. Mutants of type "B" come and go, staying at relatively low numbers, until a two-hit mutant is produced, see figure 10.1(b). In the case of "linear" tunneling, we have the following diagram:

$$A \xrightarrow{\quad R_{A\to C} \quad} C \tag{10.39}$$

where the transition rate is given by

$$R_{A\to C} = Nu_1 R, \tag{10.40}$$

Table 10.1 Tunneling rate approximations.

Interm. mutants	Conditions	Tunneling rate, R
Disadvantageous	$r < 1$, $\|1 - r\| \gg \sqrt{u_2}$	$\frac{u_2 r}{1-r}$
Neutral	$\|1 - r\| \ll \sqrt{u_2}$	$\sqrt{u_2}$
Advantageous	$r > 1$, $\|1 - r\| \gg \sqrt{u_2}$	$\frac{r-1}{r}$

the product of the population size, the mutation rate of single-mutant production, and the constant R, whose approximations in the three regimes are summarized in Table 10.1.

Diagram (10.39) corresponds to the differential equations,

$$\dot{A} = -R_{A\to C}A \qquad A(0) = 1, \tag{10.41}$$

$$\dot{C} = R_{A\to C}A, \qquad C(0) = 0, \tag{10.42}$$

and the probability for the double-hit mutant to arise by time t is given by

$$P_2(t) = 1 - e^{-R_{A\to C}t}. \tag{10.43}$$

System (10.41-10.42) is similar to the one-hit model, which we considered in the previous chapter, see equations (9.16-9.17). What essentially happens is that for intermediate population sizes, a two-step process of creating a double-hit mutant behaves like a one-step process with rate $R_{A\to C}$ that depends on all the parameters of the system. We can say that in this regime, there is only one and not two rate-limiting steps in this process.

Equation (10.43) is a good approximation on relatively long times scales (see Section 10.6.5 for the exact bounds). On shorter time-scales, we have

$$P_2(t) = 1 - exp\left(-\frac{Nu_1u_2rt^2}{2}\right). \tag{10.44}$$

We refer to this regime as "quadratic tunneling" below.

In general, the regime of tunneling is facilitated by (1) having a high mutation rate, u_2, and (2) having a relatively large population size, such that fixation of type "B" does not typically happen before a double-hit mutant is produced. The tunneling rate, $R_{A\to C}$, was first calculated in [Nowak *et al.* (2002); Komarova *et al.* (2003b)], by a method involving an analysis of transition matrices. A different method was used later by [Iwasa *et al.* (2004b)]. The method of this chapter is the closest to that used by [Moolgavkar *et al.* (1988)], but it is somewhat different in the way it uses the theory of doubly-stochastic processes. Yet another, slightly different method, is provided in Chapter 13 to generalize this model to spatial systems. This is done in order to expose the reader to a variety of methodologies.

10.6.3 *Nearly deterministic regime*

Finally, for large system sizes, the dynamics of intermediate mutant generation is almost deterministic. The population size is sufficiently large (and the first mutation rate is sufficiently high) such that the mutants are constantly produced and steadily increase in abundance, see figure 9.3(b). The double-hit mutant production happens accordingly.

The abundance of one-hit mutants, $x_1(t)$, is described by equation (9.23). The probability to create a two-hit mutant is then given by

$$P_2(t) = 1 - exp\left(-u_2 \int_0^t x_1(t')\,dt'\right).$$

In the case of neutral intermediate mutants, for relatively short times ($u_1 t \ll 1$), we have $x_1(t) \approx N u_2 t$, and

$$P_2(t) = 1 - exp\left(-\frac{N u_1 u_2 r t^2}{2}\right).$$

For disadvantageous mutants, we have $x_1(t) \approx N u_1/(r-1)$, and

$$P_2(t) = 1 - exp\left(-\frac{N u_1 u_2 t}{r-1}\right),$$

which exactly coincides with the formula derived for stochastic tunneling. This result is intuitively clear: for very large population sizes, disadvantageous intermediate mutants remain at a relatively low level, and therefore the conditions for tunneling are still satisfied.

10.6.4 *Disadvantageous, neutral and advantageous intermediate mutants*

The behavior of intermediate mutants falls under three different types. The value r, which is the fitness parameter of intermediate mutants, defines whether type "B" cells behave as neutral, positively selected, or negatively selected mutants. If

$$|1-r| \ll \sqrt{u_2},$$

then type "B" effectively behaves as a neutral mutant. If

$$|1-r| \gg \sqrt{u_2} \text{ and } r < 1,$$

type "B" is disadvantageous. If on the other hand

$$|1-r| \gg \sqrt{u_2} \text{ and } r > 1,$$

type "B" is advantageous. Interestingly, the neutrality of intermediate mutants is defined based on the comparison of the fitness difference with the mutation rate u_2, at which double-mutants are created. This is different from the conventional definition which instead involves the population size, see condition (9.13). The reason is related to the time-scale of the processes. A mutant is defined to be neutral if it behaves effectively as a neutral mutant on time-scales of interest, which in the present situation are defined by the time-scale of tunneling, which depends on the rate of mutation u_2. More precisely, the relevant time scale is defined by the inverse tunneling rate, $1/\sqrt{u_2}$, see Table 10.1 below for the rate of tunneling in the case of intermediate mutants.

10.6.5 *The role of the population size*

The three regimes depicted in figure 10.1 and described above, can be relevant at different population sizes, depending on whether the intermediate mutants are neutral, disadvantageous, or advantageous. As described above, typically for small populations sizes, the system tends to follow the genuine two-step process, figure 10.1(a). For intermediate population sizes, the effect of tunneling is observed, figure 10.1(b). Finally, for very large population sizes, we have a nearly deterministic system, figure 10.1(c). The system sizes characteristic for these processes differ in the three cases, and not all regimes are observed for all the ranges of fitness parameter r.

Neutral intermediate mutants. In this case, all the regimes are observed depending on the population size, and we have the simplest picture:

- For small populations:
$$N \ll \frac{1}{\sqrt{u_2}},$$
a genuine two-step process dominates.
- For intermediate populations,
$$\frac{1}{\sqrt{u_2}} \ll N \ll \frac{1}{u_1}, \tag{10.45}$$
the process can be described as stochastic tunneling. In particular, for times $t < 1/\sqrt{u_2}$, we have "quadratic tunneling", and for larger times - "linear tunneling.
- For large population sizes,
$$\frac{1}{u_1} \ll N, \tag{10.46}$$
a nearly deterministic process takes place.

Disadvantageous intermediate mutants. In this case, we have the following regimes:

- For small population sizes,

$$N \ll \frac{\ln\left(\frac{(1-r)^2}{r^2 u_2}\right)}{-\ln r}, \tag{10.47}$$

a two-step process is observed. Compared to the case of neutral mutants, for disadvantageous intermediate mutants the two-step process happens for much smaller population sizes, because of a very low probability that a disadvantageous mutant fixates, ρ.

- For intermediate and large population sizes,

$$\frac{\ln\left(\frac{(1-r)^2}{r^2 u_2}\right)}{-\ln r} \ll N,$$

the effect of stochastic tunneling takes place. In particular, if the population is relatively low,

$$\frac{\ln\left(\frac{(1-r)^2}{r^2 u_2}\right)}{-\ln r} \ll N \ll \frac{(1-r)^2}{r u_1 u_2}, \tag{10.48}$$

then "quadratic tunneling" is observed for times between $2\rho/(ru_2)$ and $1/(1-r)$, and "linear tunneling" takes place for larger values of t. For intermediate population sizes, with

$$\frac{(1-r)^2}{r u_1 u_2} \ll N \ll \frac{r^{-(N-1)}}{u_1},$$

only the quadratic tunneling formula is relevant. For the largest population sizes,

$$\frac{r^{-(N-1)}}{u_1} \ll N,$$

we have a linear tunneling formula.

Advantageous intermediate mutants. In this case, we have the following regimes:

- For small population sizes,

$$N \ll \frac{r}{\ln\left(\frac{(r-1)^2}{r^2 u_2} + 1\right)},$$

a two-step process is observed for short times, $t < \frac{\ln\left(\frac{(r-1)^2}{r^2 u_2}+1\right)}{r-1}$, and a mixed regime is entered for larger values of t, such that the creation of double-hit mutants may or may not occur before the fixation of type "B" mutants. In this case, the two-step formula provides an overestimation of the probability $P_2(t)$ of double-mutant creation, and the nearly-deterministic formula is an underestimation.

- For intermediate populations,

$$\frac{r}{\ln\left(\frac{(r-1)^2}{r^2 u_2}+1\right)} \ll N \ll \frac{r}{u_1},$$

  we have a two-step process.
- For larger population sizes,

$$\frac{r}{u_1} \ll N,$$

  the mutant production follows a nearly-deterministic law.

For advantageous mutants, the two-step process is the predominant one, because of a high probability of fixation of advantageous intermediate mutants.

10.7 Summary

In this chapter we summarized the main ideas of the stochastic formalism for two-hit models of carcinogenesis. Similar mathematical problems have been studied in different contexts and for different assumptions (Kimura, 1985; Barton and Rouhani, 1987; Carter and Wagner, 2002; Komarova *et al.*, 2003b; Iwasa *et al.*, 2004b,a; Weinreich and Chao, 2005; Durrett and Schmidt, 2008; Weissman *et al.*, 2009; Gokhale *et al.*, 2009; Durrett *et al.*, 2012; Durrett and Moseley, 2012). These calculations provide an understanding of the time constraints that determine the development of cancers that rely on the inactivation of tumor suppressor genes, and provides valuable information for possible evolutionary pathways to disease. The results presented here are useful stepping stones for studying more complex models that include more biological details. So far, the assumption of mean-field, or mass-action, was used where the spatial locations of individuals were not taken into account.

In the next chapters we expand and generalize these first models to adapt them to various systems, with more complexities added. In Chapter 12 we will talk about hierarchical cell populations, where stem cells and differentiated cells interact to maintain tissue homeostasis. In Chapters 13 and 14 we will describe a spatial generalization of these results. We will also see how some of these ideas can be applied to study some of the most intriguing questions of cancer initiation and progression.

Problems

Problem 10.1. Research project. *Learn about Knudson's two-hit hypothesis and discovery of the Rb gene and other tumor-suppressor genes.*

Problem 10.2. *Derive equation (10.11) from equation (10.9) by multiplying it by x^i and summing over all i.*

Problem 10.3. *In Section 10.5.1 we made assumption (10.3) to simplify the expressions for the probabilities, such that they are linear functions of i. If assumption (10.3) did not hold, would we be able to derive equation (10.11) for the probability generating function?*

Problem 10.4. *Fill in the steps in the solution of Riccati equation, (10.16), to obtain formula (10.18).*

Problem 10.5. *In the definition of neutrality in the context of double-hit mutant production, we compare the fitness difference, $|1-r|$, with $\sqrt{u_2}$. Consider the limiting behavior of expressions for t_* and R, equations (10.19), as $u_2 \to 0$, and show how it differs in three cases (a) $|1-r| \ll \sqrt{u_2}$, (b) $|1-r| \gg \sqrt{u_2}$ and $r<1$, and (c) $|1-r| \gg \sqrt{u_2}$ and $r>1$.*

Problem 10.6. Research project. *Equation (10.38) shows that for small populations, the probability of double-hit mutant creation depends quadratically on time. This is directly related to the fact that in this regime, there are two rate-limiting steps in the process of double-hit mutant creation. In the case when tunneling is important, there is only one rate-limiting step in this process. How is the number of limiting steps related to age-incidence curves curves obtained in epidemiological studies of cancer? See e.g. [Luebeck and Moolgavkar (2002)].*

Chapter 11

Microsatellite and chromosomal instability in sporadic and familial colorectal cancers

This chapter provides an example of how the theory developed in Chapters 9, 10 can be used to study various phenomena associated with carcinogenesis. It is also one of a number of chapters which investigates the relationship between carcinogenesis and genetic instability. Here, we will examine the most basic scenario: the generation of the first malignant cell. Does the presence of genetic instability result in a faster generation of the first malignant cell? We present a simple example of how stochastic models developed for two-hit processes can be applied to biological reality.

We will concentrate on cancers which are initiated via the inactivation of a tumor suppressor gene. That is, both the maternal and the paternal copy of the gene have to lose function. A particular example which will be discussed in this context is colorectal cancer. Colorectal cancer is a major cause of mortality in the Western world. Approximately 5% of the population develop the disease, and about 40% of those diagnosed with it die within 5 years. Considerable progress has been made in identifying genetic events leading to colorectal cancer. Somatic inactivation of the adenomatous polyposis coli (APC) gene is believed to be one of the earliest steps occurring in sporadic colorectal cancer. It has been observed that the frequency of APC mutations is as high in small lesions as it is in cancers. Evidence that the APC gene plays a crucial role in colorectal cancer also comes from the study of individuals with familial adenomatous polyposis coli (FAP). FAP patients inherit a mutation in one of the copies of the APC gene; by their teens, they harbor hundreds to thousands of adenomatous polyps.

The APC gene is a tumor suppressor gene which controls cell birth and cell death processes. Inactivation of only one copy of the APC gene does not seem to lead to any phenotypic changes. Inactivation of both copies of this gene appears to result in an increased cell birth to death ratio in the corresponding cell, and leads to clonal expansion and the formation of a *dysplastic crypt*. Here, we define a dysplastic crypt as a crypt that consists of cells with both copies of the APC gene inactivated. Dysplastic crypts are at risk of developing further somatic mutations which will eventually lead to cancer. The typical estimate is that an average 70-year-old has about 1-10 dysplastic crypts, but precise counts have never been published.

How can the tumor suppressor gene be inactivated? A point mutation can induce a loss of function in one copy of the gene. Both copies of the gene can be inactivated by two subsequent point mutations in the same cell: one in the maternal, and the other in the paternal allele. Each mutation would occur with the physiological mutation rate of 10^{-7} per gene per cell division.

Now consider genetic instability. As explained in Chapters 2 and 7, there are two major types of instabilities [Lengauer *et al.* (1997, 1998)], which are described briefly below:

- (i) Small scale subtle sequence changes, such as microsatellite instability (MSI) [Boland and Goel (2010); Yurgelun *et al.* (2012)]. The MSI phenotype is generated if specific MSI genes are inactivated. Both copies of an MSI gene need to be mutated. MSI basically results in an elevated point mutation rate in the context of repeat sequences called microsatellites.
- (ii) Gross chromosomal alterations, known as chromosomal instability (CIN) [Boland *et al.* (2009); Sen (2000)]. The genetic basis of CIN is uncertain, and specific scenarios will be discussed below. If one copy of the tumor suppressor gene has been inactivated by a point mutation, the other copy can be inactivated very quickly in CIN cells due to the loss of the healthy allele. This can occur through a variety of mechanisms. They include loss of the remaining chromosome and loss of part of the remaining chromosome. These processes are also called *loss of heterozygocity*, or *LOH*.

While genetic instability might speed up the loss of tumor suppressor function, the MSI or CIN phenotypes need to be generated first (for example by basic point mutations). This chapter discusses a mathematical analysis of how MSI and CIN influence the rate of tumor suppressor gene inactivation. We will apply this analysis to various scenarios which include the sporadic (spontaneous) development of colon cancer and familial colon cancers. We start with some more detailed biological facts about CIN and MSI in colon cancer and then present the mathematical analysis.

11.1 Some biological facts about genetic instability in colon cancer

Here we will study the role that CIN and MSI may play in the inactivation of the APC gene. About 13% of all colorectal cancers have MSI and most of the rest are characterized by CIN [Lengauer *et al.* (1998)]. MSI occurs in virtually all hereditary non-polyposis colorectal cancers (HNPCC), which account for about 3% of all colorectal cancers. The MSI phenotype results from defective mismatch repair. Several genes have been identified whose inactivation leads to an increased rate of subtle genetic alterations. The main ones are hMSH2 and hMLH1. Both copies of an MSI gene must be inactivated in order for any phenotypic changes to occur. HNPCC patients inherit a mutation in one of the copies of an MSI gene and normally develop colorectal tumors in their forties. Unlike FAP patients, they do not have a vastly increased number of polyps, but the rate of progression from polyp to cancer is faster.

Molecular mechanisms leading to CIN in human cancers remain to be understood. It has been proposed that CIN might be caused by mutations in genes involved in centrosome/microtubule dynamics, or checkpoint genes that monitor the progression of the cell cycle, e.g. the spindle checkpoint or the DNA-damage checkpoint [Negrini *et al.* (2010)]. For example, heterozygous mutations in the mitotic spindle checkpoint gene hBUB1 have been detected in a small fraction of colorectal cancers with the CIN phenotype. Also, the MAD2 gene seems to be transcriptionally repressed in various solid tumors. Some CIN genes might act in a dominant-negative fashion: an alteration in one allele leads to CIN.

11.2 A model for the initiation of sporadic colorectal cancers

11.2.1 *The first model of the APC gene inactivation: no instabilities*

The colonic epithelium is organized in crypts covered with a self-renewing layer of cells (figure 11.1). The total number of crypts is of the order of $M = 10^7$ in a human. Each crypt contains of the order of 10^3 cells. A crypt

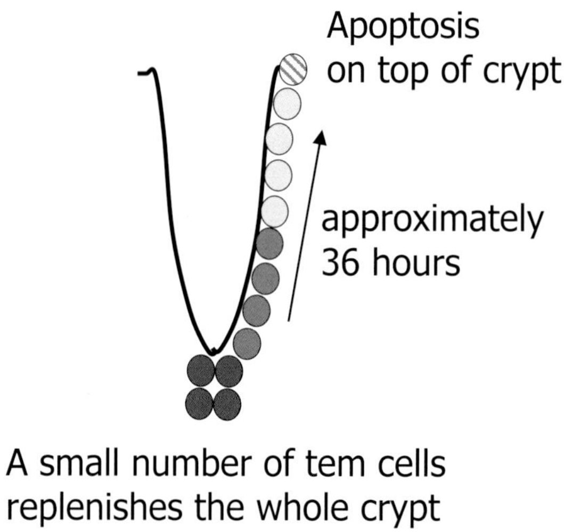

Fig. 11.1 The epithelium of the colon is organized into crypts. Each crypt contains about 10^3 cells. A small number of (stem) cells, which are thought to be located at the bottom of the crypt, divide to replenish the whole crypt. They give rise to differentiated cells which travel within 36 hours to the top of the crypt where they undergo apoptosis. Inactivation of both copies of the APC gene is believed to prevent apoptosis. The mutated cells remain on the top of the crypt, continue to divide and ultimately take over the crypt. This process gives rise to a dysplastic crypt, which represents the first step on the way to colorectal cancer.

is renewed by a small number of stem cells (perhaps $1 - 10$) [Yatabe *et al.* (2001)]. The life cycle of stem cells is of the order of $1 - 20$ days [Bach *et al.* (2000); Potten *et al.* (1992)]. Stem cells give rise to differentiated cells which divide at a faster rate, and travel to the top of the crypt where they undergo apoptosis.

We will apply the theory developed in Chapter 10 to describe the basic model of sporadic colorectal cancer initiation. Let us assume that the *effective population size* of a crypt is N; this means that N cells are at risk of developing mutations which can lead to cancer. The value of N is unknown. As explained in detail in Chapter 12, one hypothesis is that only the stem cells are at risk of developing cancer, which gives $N \sim 1 - 10$, and in this case the average turnover rate would be $\tau = 1 - 20$ days. Alternatively, we could assume that some differentiated cells are also at risk. In this case, N might be of the order of $100 - 1000$ and the average turnover rate could be less than 1 day. In this chapter, we will not consider the details of the population structure. That is, the distinction between stem cell and differentiated cell division patterns will be ignored. Chapter 12 suggests a way of incorporating this in the model.

As the first model we will concentrate only on the inactivation of the tumor-suppressor gene APC and consider the simplest network, figure 11.2. Let us denote by X_0, X_1 and X_2 the probability that the whole crypt consists of cells with 0, 1 and 2 copies of the APC gene inactivated, respectively (please note the difference in notations: in Chapter 10 the three probabilities of being in homogeneous states were denoted as A, B, and C. In the present context, as in Chapter 7, it is more convenient to adopt a different notation, as will become clear later in this chapter). The rate of change is equal to the probability that one relevant mutation occurs times the probability that the mutant cell will take over the crypt. For this model and further models, all the parameters with their respective values are summarized in Table 11.1.

Table 11.1 Parameters, notations and possible numerical values; the mutation and LOH rates are given per gene per cell division.

Quantity	*Definition*	*Range*
M	Number of crypts in a colon	10^7
N	Effective number of cells in a crypt	$1 - 100$
τ	Effective time of cell cycle, days	$1 - 20$
u	Probability of mutation in normal (non-MSI) cells	10^{-7}
$\tilde{u}$	Probability of mutation in MSI cells	10^{-4}
p_0	Rate of LOH in normal (non-CIN) cells	10^{-7}
p	Rate of LOH in CIN cells	10^{-2}
r	Relative fitness of APC$^{+/-}$ cells	≈ 1 or < 1
r_c	Relative fitness of CIN cells	?
n_m	Total number of MSI genes	$2 - 5$
n_c	Total number of CIN genes	?
u_c	Mutation rate to produce a CIN cell	$n_c u$
u_m	Mutation rate to produce a CIN cell	$n_m u$

Fig. 11.2 Mutation-selection networks of sporadic colorectal cancer initiation, in the absence of instability. (a) A two-step process. (b) A tunneling process.

In the beginning (see figure 11.2), all cells are wild type. The first copy of the APC gene can get inactivated by a mutation event. We will assume that the fitness of the resulting $APC^{+/-}$ cell, r, is the same or lower than that of the wild-type cell. Because the mutation rate per gene per cell division, $u \approx 10^{-7}$, is very small and the number of cells, N, is not large, it is safe to assume that a coarse-grained description is applicable, see Chapter 10. For realistic values of the mutation rates and populations sizes, see Table 11.1, two types of dynamics are possible: a genuine two-step process (figure 10.1(a)) and stochastic tunneling (figure 10.1(b)). Figure 11.3 shows which regime can be expected depending on the fitness of $APC^{+/-}$ cells.

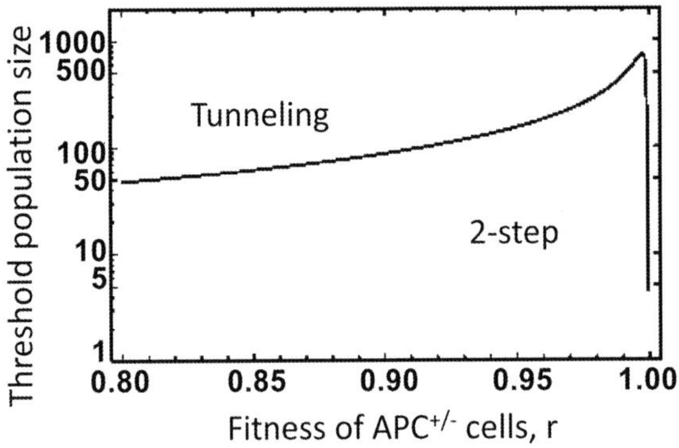

Fig. 11.3 The quantity $N_{tun} = \frac{\ln \frac{(u+p_0)r^2}{(1-r)^2}}{\ln r}$, see formula (10.47), is plotted for $u + p_0 = 10^{-6}$. For effective population sizes, N, smaller than N_{tun}, a genuine two-step process is a good approximation. For $N \gg N_{tun}$, tunneling is the dominant process.

In particular, if inactivation of one copy of the APC gene does not result in a phenotypic change, it is likely that a genuine two-step process is observed. This process is depicted in diagram 11.2(a) and described theoretically in Section 10.6.1. With the rate $N\rho 2u$, cells with one copy of the APC gene mutated will take over the crypt (state X_1). This rate of change is calculated as N times the probability (per cell division) to produce a one-hit mutant ($2u$ because either of the two alleles can be mutated) times the probability ρ of *one* mutant of type X_1 to get fixed (equal to $1/N$ if $APC^{+/-}$ cells are neutral).

Once the first allele of the APC gene has been inactivated, the second allele can be inactivated either by another point mutation (probability u per cell division) or by an LOH event (probability p_0 per cell division, see Table 11.1). We assume that mutants with both copies of the APC gene inactivated (the $APC^{-/-}$ mutants) have a large selective advantage, so that once such a mutant is produced, the probability of its fixation is close to one. Therefore, from state X_1 the system can go to state X_2 (both copies of the APC gene inactivated) with the rate $N(u+p_0)$. This rate is calculated as N times the probability per cell division to produce a two-hit mutant (u for an independent point mutation plus p_0 for an LOH event) times the probability of the *advantageous* mutant of type X_2 to take over (this is 1).

Note that the assumption that the probability of fixation of APC$^{-/-}$ mutants is one is made for simplicity. More generally, the relative fitness of type X_2 is r_1, whereas the fitness of type X_0 and X_1 is 1. Then the second rate in figure 11.2(a) should be taken to be $N\tilde{\rho}(u+p_0)$, with $\tilde{\rho} = (1-1/r_1)/(1-1/r_1^N)$. If the population size is not too large, and the relative fitness of type X_2 is much greater than 1, we have $\tilde{\rho} \to 1$, and we obtain the expression $N(u+p_0)$.

As in the examples of Section 10.6, the mutation-selection network of figure 11.2 is equivalent to a linear system of ordinary differential equations (ODE's), where the rates by the arrows refer to the coefficients and the direction of the arrows to the sign of the terms. In this chapter, we set a dimensionless time scale in terms of stem cell cycle length, τ. One (non-dimensional) time unit ($t/\tau = 1$), as before, corresponds to a generation turn-over, that is to N "elementary events" of the Moran process, where an elementary event includes one birth and one death. We have:

$$\tau \dot{X}_0 = -2uX_0,$$
$$\tau \dot{X}_1 = 2uX_0 - N(u+p_0)X_1,$$

with the constraint $X_0 + X_1 + X_2 = 1$ and the initial condition

$$X_0(0) = 1, \quad X_1(0) = 0.$$

Here, we use the fact that the intermediate mutant is neutral and that the population size is small, so that stochastic tunneling does not take place. Calculations for larger values of N can also be performed, see below.

Using $ut/\tau \ll 1$ and $N(p_0+u)t/\tau \ll 1$, we can approximate the solution for X_2 as

$$X_2(t) = Nu(u+p_0)(t/\tau)^2.$$

The quantity $X_2(t)$ stands for the probability that a crypt is dysplastic (i.e., consists of cells with both copies of the APC gene inactivated) at time t measured in days. This formula is a consequence of the fact that in the parameter regime we are considering, APC$^{-/-}$ cells are produced as a result of a genuine two-hit process. There are two steps that separate the state X_0 from the state X_2, and thus the expected number of dysplastic crypts in a person of age t is proportional to the product of the two rates and the second power of time.

The probability to have i dysplastic crypts by the age t is given by a simple binomial, $\binom{M}{i}X_2(t)^i(1 - X_2(t))^{M-i}$. The expected number of dysplastic crypts in a person of age t is then given by the following quantity,

$$\nu(t) = MNu(u + p_0)(t/\tau)^2. \tag{11.1}$$

Some estimates of the expected number of dysplastic crypts, based on equation (11.1), are given in figure 11.4, the lines marked with "Small population, neutral $APC^{+/-}$ cells".

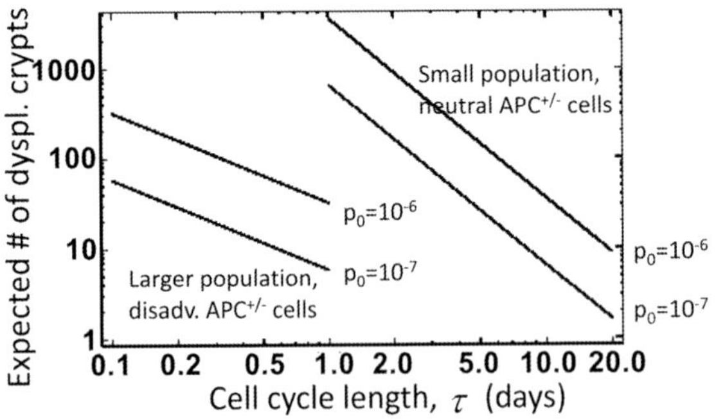

Fig. 11.4 Sporadic colorectal cancer: the expected number of dysplastic crypts, at 70 years of age, the simple model. The number of dysplastic crypts is calculated for two different values of p_0 (the rate of LOH), as a function of the effective cell cycle length, τ. The lines on the right correspond to the model where only stem cells acquire mutations ($N = 5$), and $APC^{+/-}$ cells are neutral. The lines on the left correspond to a larger effective population size ($N = 100$), and negatively-selected $APC^{+/-}$ ($r = 0.85$). The rest of the parameters are $M = 10^7$, $N = 5$, $u = 10^{-7}$, and $t = 70$ years.

For completeness, we will also consider the case where stochastic tunneling plays an important role. This can for example happen if the $APC^{+/-}$ mutants are negatively selected, and/or if the effective population size is sufficiently large, see figure 11.3. In this case the process is described by the diagram in figure 11.2(b). The type APC+/− does not have a chance to get fixated, before a double-hit mutant is created. We have

$$\tau\dot{X}_0 = -2NuRX_0, \tag{11.2}$$

$$\tau\dot{X}_2 = 2NuRX_0, \tag{11.3}$$

where R is given by $R = \sqrt{u + p_0}$ in the neutral case and $R = \frac{r(u+p_0)}{1-r}$ in the case of disadvantageous $APC^{+/-}$ mutants, Table 10.1. In this case,

$$X_2(t) = 1 - E^{-2NuRt/\tau} \approx 2NuRt/\tau,$$

a linear function of time for short time-scales. The rest of the analysis can be carried out accordingly. Calculated numbers of dysplastic crypts under the assumption that $APC^{+/-}$ cells are negatively-selected are presented in figure 11.4, the lines marked with "Large population, disadv. $APC^{+/-}$ cells". In the rest of this chapter we will only concentrate on the case where $APC^{+/-}$ mutants are neutral, and thus set $r = 1$.

One has to be careful when comparing these calculations with data. It is possible that dysplastic crypts can be lost. The model presented here gives the number of dysplastic crypts that are being *produced* over time, which could be larger than the actual number of dysplastic crypts that patients have at a particular time point. Exact measurements of the incidence of dysplastic crypts will provide important information about the crucial parameters of colorectal cancer initiation.

11.2.2 *Colorectal cancer and chromosomal instability*

Next, we include the possibility of CIN. Consider the evolutionary dynamics of a single crypt with six different types of cells. Let x_0, x_1, and x_2 denote the abundance of non-CIN cells with 0, 1, and 2 inactivated copies of APC, respectively. Let y_0, y_1, and y_2 denote the abundance of corresponding CIN cells. Figure 11.5 shows the mutation network. We assume that non-CIN cells with at least one functional APC gene have a relative reproductive rate of 1. As before, cells with two inactivated copies of APC have a reproductive rate of $r_1 > 1$, which gives them a selective advantage and enables them to take over the crypt with a high probability.

As explained above, there are two possibilities to proceed from x_1 to x_2. The second copy of APC can be inactivated by (i) another mutational event occurring with probability u or (ii) a loss of heterozygosity (LOH) event occurring with probability p_0 per cell division. Hence the resulting mutation rate is $u + p_0$. Similarly, y_1 proceeds to y_2 with a mutation rate of $u + p$, where p is the rate of LOH in CIN cells.

We assume that the crucial effect of CIN is to increase the rate of LOH, which implies $p \gg p_0$ and $p \gg u$. Intuitively, the advantage of CIN for the cancer cell is to accelerate the loss of the second copy of the APC gene. As discussed in Chapter 7, the cost of CIN is that copies of other genes are being eliminated, which might lead to nonviable cells or enhanced rates of apoptosis. Here we will assume that the reproductive rate, r_c, of CIN cells could be equal to or smaller than 1.

Whether there is any effective selection against CIN depends on the relative magnitude of the selection coefficient, $|r_c - 1|$, compared with the dominant mutational process and compared with the reciprocal of the population size. If the effective population size, N, is small, then the evolutionary dynamics are dominated by random drift, see condition of neutrality (9.13). If the rate of LOH in CIN cells is large, then the effect of selection is outpaced by LOH, Section 10.6.4. Let us define $\delta = \max(1/N, \sqrt{p})$. CIN is positively selected if $r_c - 1 > \delta$, and CIN is negatively selected if $1 - r_c > \delta$. CIN is effectively neutral if $|r_c - 1| < \delta$.

In our model, a CIN cell carries a mutation that results in an increased rate of LOH. A number of mechanisms can lead to LOH including nondisjunction of chromosomes (with or without duplication), mitotic recombination, gene conversion, interstitial deletion, or a double-strand break resulting in the loss of a chromosome arm. Here we include any such mechanism leading to an increased rate of LOH as CIN. Hence, any mutation that increases the rate of one of these processes qualifies as a CIN mutation. [Lengauer *et al.* (1997)] estimated the rate of loss of chromosomes in CIN cell lines to be $p = 0.01$ per cell division. Note, however, that even such a high rate of LOH does not imply that CIN cells must have LOH in almost all their loci, because there is conceivably selection against LOH in many loci.

We assume there are n_c genes such that a single mutation in any one of these genes can cause CIN. Hence the mutation rate from x_i to y_i is given by $u_c = 2n_c u$.

To derive equations for the probabilities in the coarse-grained approximation, as described in Chapter 10, we denote by X_i and Y_i the probability to be in homogeneous states that contain only cells of type x_i and y_i, respectively. The evolutionary dynamics are described by a linear Kolmogorov forward equation, such as (10.34-10.36) or (10.41-10.42), which can be solved analytically. It is crucial to include the phenomenon of stochastic tunneling in this description. As outlined in figure 11.5, there could be three stochastic tunnels in our system: $X_0 \rightarrow X_2$, $X_1 \rightarrow Y_2$, and $Y_0 \rightarrow Y_2$, and we can calculate the parameter conditions for individual tunnels to be "open" or "closed". It turns out there are four distinct reaction networks, figure 11.5. Each reaction network specifies a linear differential equation.

The applicability conditions for each of the networks can be derived by using methods of Chapter 10. Note that tunnel $X_0 \rightarrow X_2$ is neutral by the assumption that an inactivation of one of the two copies of the APC gene does not result in a fitness change. Tunnel $X_1 \rightarrow Y_2$ can be neutral or disadvantageous depending on the fitness of type y_1, r_c.

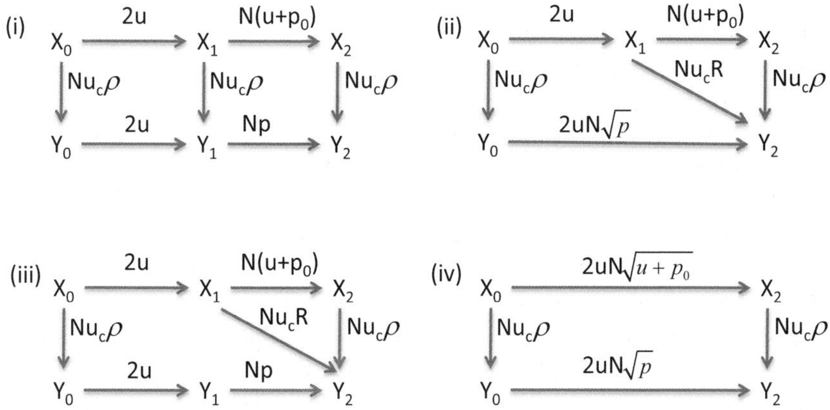

Fig. 11.5 Transition rates and stochastic tunnels of the probabilistic process describing the dynamics of early steps in colon cancer. The states X_0, X_1, and X_2 refer to the probabilities to have a homogeneous crypts of non-CIN cells with 0, 1, and 2 inactivated copies of APC, respectively. The states Y_0, Y_1, and Y_2 refer to probabilities of having homogeneous crypts of CIN cells with 0, 1, and 2 inactivated copies of APC, respectively. We assume that APC$^{+/-}$ cells are neutral. The probability of intermediate mutant fixation, $\rho = (1 - 1/r_c)/(1 - (1/r_c)^N)$ for $r_c \neq 1$ and $\rho = 1/N$ for $r_c = 1$. In the expressions for the rates, we neglected $u \ll p$ compared to p.

Let us ignore $u \ll p$ compared to p in all the expressions for simplicity. These are the generic cases, see figure 11.5:

(i) All tunnels closed. For neutral mutants y_1, such that $|1-r_c| \ll \sqrt{p}$, we have conditions $\sqrt{u+p_0} \ll 1/N$ (tunnel $X_0 \to X_2$ is closed), and $\sqrt{p} \ll 1/N$ (tunnel $Y_0 \to Y_2$ is closed). In this case tunnel $X_1 \to Y_2$ is automatically closed. For disadvantageous or advantageous mutants y_1, such that $r_c - 1 \gg \sqrt{p}$, we have an additional condition $p \ll (1-r_c)^2 r_c^{N-2}$.

(ii) Tunnels $X_1 \to Y_2$ and $Y_0 \to Y_2$ are open. For neutral or disadvantageous mutants y_1, we have conditions $\sqrt{u+p_0} \ll 1/N$ and $\sqrt{p} \gg 1$.

(iii) Only tunnel $X_1 \to Y_2$ is open. We have to require that $\sqrt{u+p_0} \ll 1/N$ and $\sqrt{p} \ll 1/N$, and also that mutants y_1 are disadvantageous, $r_c < 1$, $|1-r_c| \gg \sqrt{p}$, and $p \gg (1-r_c)^2 r_c^{N-2}$.

(iv) Channels $X_0 \to X_2$ and $Y_0 \to Y_2$ are open. In this case we have the single condition $\sqrt{u+p_0} \gg 1/N$.

In the derivation of the conditions above we used the fact that if tunnel $Y_0 \to Y_2$ is closed and y_1 is neutral, then tunnel $X_1 \to Y_2$ must also be closed, but if y_1 is disadvantageous, then tunnel $X_1 \to Y_2$ can be open. This and other similar logical arguments help reduce the number of conditions.

Instead of writing down the exact solution for the linear ODEs corresponding to the networks in figure 11.5, we can find an approximation to those solutions. The form of the approximation crucially depends on the time-scale. For example, on very long time-scales, such that $tu \gg 1$, $tp_0 \gg 1$, we have $Y_2 \to 1$ and the rest of the variables tend to zero. On very short time-scales, such that $pt \ll 1$, we can see that X_2 is quadratic in t and Y_2 is cubic in t, such that $Y_2 \ll X_2$. If however we focus on intermediate time-scales, where

$$ut/\tau \ll 1, \quad p_0 t/\tau \ll 1, \quad pt/\tau \gg 1, \tag{11.4}$$

we can see that both X_2 and Y_2 are quadratic functions of time. For example, for case (i) in figure 11.5, the system of equations reads:

$$\tau \dot{X}_0 = -(2u_N u_c \rho) X_0, \tag{11.5}$$
$$\tau \dot{X}_1 = 2uX_0 - (N(u+p_0) + Nu_c\rho)X_1, \tag{11.6}$$
$$\tau \dot{X}_2 = N(u+p_0)X_1, \tag{11.7}$$
$$\tau \dot{Y}_0 = Nu_c\rho X_0 - 2uY_0, \tag{11.8}$$
$$\tau \dot{Y}_1 = 2uY_0 + Nu_c\rho X_1 - NpY_1, \tag{11.9}$$
$$\tau \dot{Y}_2 = NpY_1 + Nu_c\rho X_2. \tag{11.10}$$

We have the following approximations for all the variables:

$$X_0 \approx 1, \quad X_1 \approx 2ut/\tau, \quad X_2 \approx 2Nu(u+p_0)(t/\tau)^2/2, \tag{11.11}$$

$$Y_0 \approx Nu_c\rho t/\tau, \; Y_1 \approx \frac{2uNu_c\rho}{Np}\left(\frac{t}{\tau}\right)^2, \; Y_2 \approx 2uNu_c\rho\left(\frac{t}{\tau}\right)^2. \tag{11.12}$$

Therefore, in this case, condition $Y_2 > X_2$ is equivalent to the condition $4uNu_c\rho > 2Nu(u+p_0)$, or equivalently,

$$u_c > \frac{u}{2\rho}\left(1 + \frac{p_0}{u}\right).$$

The same condition can be expressed in terms of the number of genes that, if mutated, can lead to CIN:

$$n_c > \frac{1}{2\rho}\left(1 + \frac{p_0}{u}\right).$$

Similar considerations apply to networks (ii-iii) of figure 11.5. In general, the condition for CIN to win can be written down by using the following combinatorial rule.

Combinatorial rule. We need to consider all paths leading from X_0 to X_2 and from X_0 to Y_2. Then we identify fast and slow leaps in the diagram. A leap is considered "slow" at a given time-frame if the corresponding rate is much smaller than τ/t. A leap is "'fast" if its rate is much larger than τ/t. In the regime identified by inequalities (11.4), leaps with rates multiplying u, u_c and p_0 are slow, and leaps corresponding to rate p are fast. For each path, we calculate the number of slow leaps. This number defines the power of t. We will only need the paths with the smallest number of slow steps. For each such path, we multiply the corresponding slow rates together; fast leaps do not count. Finally, we add all the paths together and divide by the factorial of the number of slow leaps.

In networks (i-iii) of figure 11.5, the probability $X_2(t)$ is given by

$$X_2(t) = uN(u + p_0)(t/\tau)^2.$$

The probability $Y_2(t)$ is different for the different networks, and can be calculated by similar methods.

The crucial question is whether the system reaches X_2 before Y_2 or vice versa. In the first case, APC inactivation occurs in a non-CIN cell. In the second case, APC inactivation occurs in a CIN cell, and hence CIN causes inactivation of APC (because $p \gg u$). The answer can be stated in a concise form provided there is a separation of time scales: for the relevant time scale t of human life, we have $ut/\tau \ll 1$, $p_0 t/\tau \ll 1$, and $pt/\tau \gg 1$. The condition for CIN to cause inactivation of the second APC gene can then be expressed as the number of CIN genes exceeding a certain threshold value,

$$n_c > \left(1 + \frac{p_0}{u}\right)\frac{1}{\mathcal{K}}, \qquad (11.13)$$

where $\mathcal{K} = 2\rho$ for case (i), $\mathcal{K} = \rho N\sqrt{p} + R$ in case (ii), and $\mathcal{K} = \rho + R$ in case (iii). Case (iv) can be treated similarly.

11.3 Sporadic colorectal cancers, CIN and MSI

Let us now consider the possibility of developing two types of genetic instability during cancer initiation. Starting from a population of normal cells, three different events can occur: (i) inactivation of the first copy of the APC gene, (ii) mutation of one copy of one of n_c CIN genes, and (iii) inactivation of the first copy of one of n_m MSI genes.

We use the notation X_i, Y_i and Z_i, respectively, for the probability that a crypt consists of normal cells, CIN cells or MSI cells with i copies of the APC gene inactivated, see Table 11.2. Figure 11.6 shows the mutation-selection network of colorectal cancer initiation including CIN and MSI. All the transition rates are calculated as the relevant mutation rate times the probability that the mutant will take over the crypt.

Table 11.2 The three major classes of homogeneous states.

Quantity	*Definition*	*Point mutation rate*	*Rate of LOH*
X_0, X_1, X_2	non-CIN, non-MSI	u	p_0
Y_0, Y_1, Y_2	CIN	u	$p > p_0$
Z_0, Z_1, Z_2	MSI	$\tilde{u} > u$	p_0

In this model we will assume that (a) CIN is neutral, that is, $r_c = 1$, and that (b) the population is small enough such that stochastic tunneling does not occur. This significantly simplifies the description. An interested reader can expand this framework to include stochastic tunnels, as was done in the previous section in the case of CIN only.

As before, we are interested in the probability to find the crypt in the state X_2, Y_2 and Z_2 as a function of t. In other words, we want to know the probability for the dysplastic crypt to have CIN (Y_2), MSI (Z_2) or no genetic instability (X_2). The diagram of figure 11.6 corresponds to a system of 11 linear ODE's describing the time-evolution of the probabilities to find the system in any of the 12 possible homogeneous states. An exact solution can be written down but it is a very cumbersome expression, so we will make the same approximation as for equations (11.11-11.12). Taking the Taylor expansion of the solution in terms of ut/τ and $N(u+p_0)t/\tau$, we obtain the following result:

$$X_2(t) = Nu(u+p_0)(t/\tau)^2, \qquad Y_2(t) = 4n_c u^2 (t/\tau)^2. \tag{11.14}$$

The rate $\tilde{u}$ is neither fast nor slow, so the solution for Z_2 is more compli-

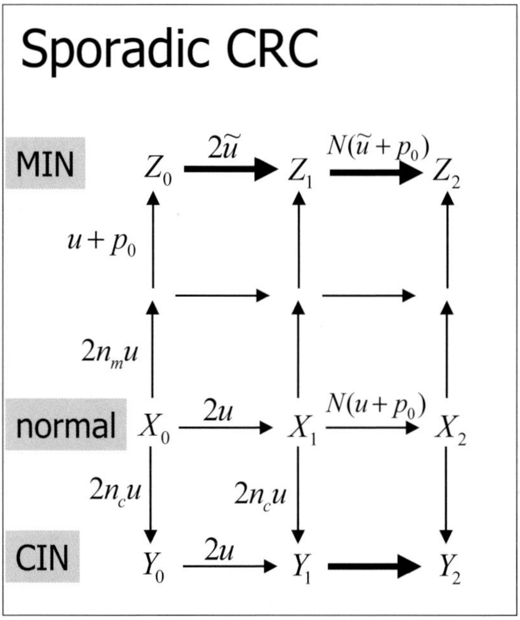

Fig. 11.6 Mutation-selection network of sporadic colorectal cancer initiation including CIN and MSI. From the initial wild-type state, X_0, the crypt can change to state X_1 as in figure 11.2(a), acquire a CIN mutation (the arrow down) or an MSI mutation (the arrow up). The line $X_0 \to X_1 \to X_2$ is identical to the process in figure 11.2(a) of developing a dysplastic crypt with no genetic instabilities. The bottom row of the diagram corresponds to CIN cells acquiring the first, and then the second, mutation (loss) of the APC gene; the second copy can be lost by a point mutation or by an LOH event whose rate is much larger for CIN cells than it is for normal or MSI cells. The state Y_2 corresponds to a CIN dysplastic crypt. The top row is the development of an MSI dysplastic crypt. The MSI phenotype is characterized by an increased point mutation rate, $\tilde{u}$. The state Z_2 is an MSI dysplastic crypt. Thick arrows denote faster steps. Note that it takes only one leap (down) to go to a CIN state from a state with no genetic instability, because CIN genes are assumed to be dominant-negative. It takes two steps to acquire MSI (up) because both copies of an MSI gene need to be inactivated before any phenotypic changes happen.

cated. We have

$$Z_2(t) = \frac{n_m u(u+p_0)}{(ab\tilde{u})^2(a-b)}\Big(2(b^3E_a - a^3E_b + a^3 - b^3 + ab(b^2-a^2)\tilde{u}t/\tau) + a^2b^2(a-b)\tilde{u}^2(t/\tau)^2\Big), \tag{11.15}$$

where

$$a = 2, \quad b = N(\tilde{u} + p_0)/\tilde{u}, \quad E_x = e^{-x\tilde{u}t/\tau}.$$

Note that if the $\tilde{u}$-steps are fast (i.e., if $\tilde{u}t/\tau \gg 1$), the limit of this expression is given by $Z(t) = n_m u(u + p_0)(t/\tau)^2$. In the opposite limit where $\tilde{u}t/\tau \ll 1$, we have $Z(t) = n_m N u(u + p_0)\tilde{u}(\tilde{u} + p_0)(t/\tau)^4/6$.

The key idea of this analysis is to identify how many slow (rate-limiting) steps separate the initial state (X_0) from the state of interest. The slow steps in our model are the ones whose rates scale with u or p_0. The step from Y_1 to Y_2 is fast, because it is proportional to the rate of LOH in CIN cells, p, which is much larger than u and p_0. The steps with the rate $\tilde{u}$ are neither fast nor slow. For all possible pathways from the initial state to the final state of interest, we have to multiply the slow rates together times the appropriate power of t/τ, and divide by the factorial of the number of slow steps. Summing over all possible paths we will obtain the probability to find the crypt in the state in question.

Applying this rule, we can see that, as before, $X_2(t)$ and $Y_2(t)$ are both quadratic in time, because it takes two rate-limiting steps to go from X_0 to X_2 and from X_0 to Y_2. The state Z_2 is separated from X_0 by two rate-limiting steps and two "intermediate" steps (whose rate is proportional to $\tilde{u}$), so the quantity $Z_2(t)$ grows as the forth power of time for $\tilde{u}t/\tau \ll 1$ and as the second power of time in the opposite limit.

The probability that a crypt is dysplastic at time t is given by $S(t) = X_2(t) + Y_2(t) + Z_2(t)$. Therefore, the expected number of dysplastic crypts in a person of age t, $\nu(t)$, is given by

$$\nu(t) = MS(t).$$

Of these dysplastic crypts, $MY_2(t)$ have CIN and $MZ_2(t)$ have MSI. This suggests that the fraction of CIN cancers is at least $Y_2(t)/S(t)$ and the fraction of MSI cancers is at least $Z_2(t)/S(t)$. The actual values may be higher because in our model, only the very first stage of cancer development is considered. At later stages of progression from a dysplastic crypt to cancer, there are more chances for cells to acquire a CIN or an MSI mutation.

Some numerical examples of such computations are presented in figure 11.7, where the logarithm of the expected number of dysplastic crypts is plotted against τ, for three different values of n_c, the number of CIN genes.

Figure 11.8(a) presents relative fractions of dysplastic crypts with CIN (gray lines) and MSI (black lines) as functions of n_c. Larger values of n_c

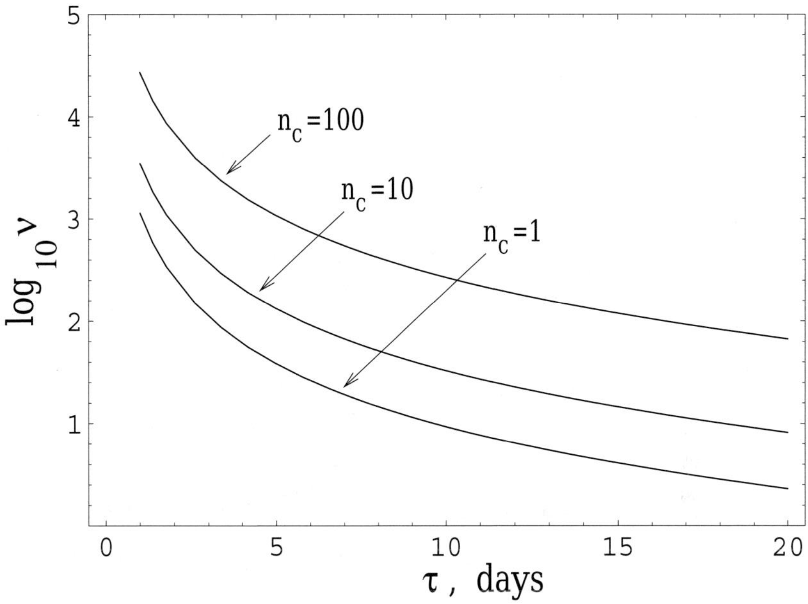

Fig. 11.7 Sporadic colorectal cancer: the logarithm of the expected number of dysplastic crypts as a function of τ at 70 years of age, in the model with CIN and MSI. The three curves correspond to three different values of n_c. $M = 10^7$, $N = 5$, $u = 10^{-7}$, $\tilde{u} = 10^{-4}$, $p_0 = 10^{-7}$, $n_m = 3$ and $t = 70$ years.

lead to an increased percentage of dysplastic crypts with CIN. The fraction of MSI crypts as predicted by this model is quite low (for $n_c = 10$ we get only 0.1% of dysplastic crypts with MSI). This could mean that MSI develops at later stages of cancer. However, there is indirect evidence that the replication error phenotype precedes, and is responsible for, APC mutations in MSI cancers (Huang *et al.*, 1996). At the same time, according to observations, 13% of all sporadic colorectal cancers have MSI (Lengauer *et al.*, 1998). Our model is consistent with this data if we assume higher rates of MSI induction in a cell. These could be a consequence of epigenetic mechanisms of MSI gene inactivation. DNA methylation of the hMLH1 gene is found at a high frequency in sporadic MSI tumors (Ahuja *et al.*, 1997; Kane *et al.*, 1997; Cunningham *et al.*, 1998).

In the diagram of figure 11.6 this means that the rates from X_0 to the MSI type (vertical arrows), $2n_m u$ and $u+p_0$, should be replaced by $2n_m u_{met}$ and $u_{met} + p_0$, respectively, where u_{met} is the rate of methylation per gene per cell division. In terms of our equations, we need to replace u by u_{met}

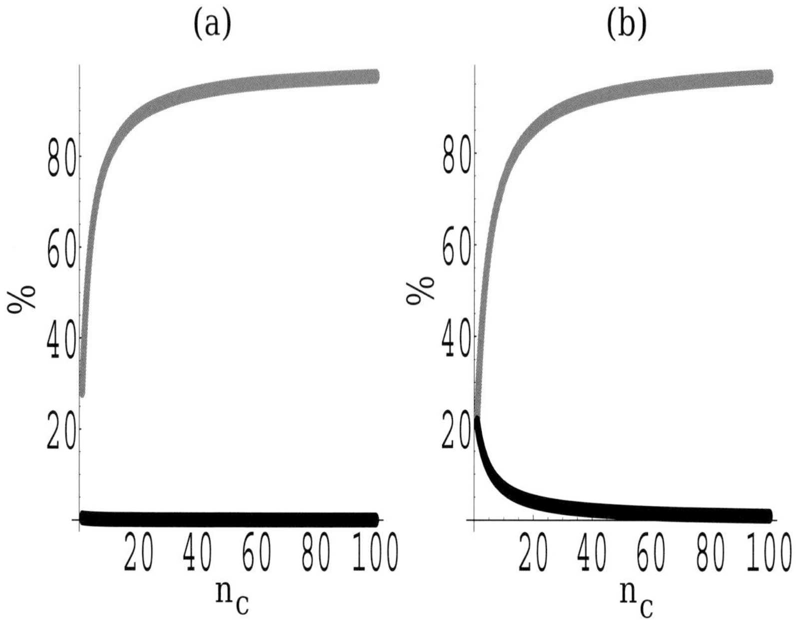

Fig. 11.8 Sporadic colorectal cancer: fractions of crypts with different instabilities, as a function of n_c; gray lines correspond to CIN and black lines to MSI. (a) The inactivation of MCI genes is by mutation, with the rate $u = 10^{-7}$. (b) The inactivation of MCI genes is by methylation, with the rate $u = 10^{-6}$. The rest of the parameters is as in figure 11.7.

in the expression for $Z_2(t)$, equation (11.15). De-novo methylation rates have not been quantified; in figure 11.8(b) we assume that u_{met} is larger than the basic mutation rate, u ($u_{met} = 10^{-6}$), and plot the proportion of MSI crypts. We can see that the expected fraction of MSI crypts predicted by our model becomes larger. Figure 11.8(b) should be compared with the observation that 13% of dysplastic crypts are MSI. We can see that this holds for values of n_c of the order of 10. As n_c becomes larger than about 20, the predicted fraction of CIN crypts becomes much higher, and the fraction of MSI crypts much lower than expected.

The calculations presented here were performed under the assumption that chromosomal instability does not have a cost. In other words, the CIN phenotype is neutral with respect to the wild type. Perhaps a more realistic model should include a possibility that CIN phenotype is disadvantageous compared to the wild-type, as presented in the previous section. In this case, the expected number of CIN dysplastic crypts will be lower. For example, if the relative disadvantage of a CIN cell is 10%, then the fraction of CIN dysplastic crypts will be reduced by 20%.

In our model, we assume that CIN is generated by means of a mutation in any of n_c dominant-negative CIN genes. In other words, a genetic hit in either of the two copies of a CIN gene will lead to the acquisition of the CIN phenotype. Alternatively, it could happen that the CIN phenotype requires the inactivation of both copies, like MSI genes or tumor suppressor genes. In terms of figure 11.6, this would mean that we have two steps separating the wild type (X_0) from the CIN phenotype (Z_0). If we assume that the CIN phenotype is neutral, the fraction of CIN dysplastic crypts in figure 11.8 would be negligible. This means that in this case, the CIN phenotype must be very advantageous in order to show up early in carcinogenesis.

11.4 FAP

The framework developed in this chapter allows us to study familial cancers in a systematic manner, by modifying the basic mutation-selection diagram of figure 11.6. In FAP patients, one allele of the APC gene is inactivated in the germ line. In terms of our model this means that all crypts start in state X_1. The corresponding mutation-selection network is found in figure 11.9(a).

Again, the mutation-selection diagram can be converted into a system of ODEs. The solutions are given by

$$X_2(t) = N(u + p_0)t/\tau, \qquad Y_2(t) = 2n_c ut/\tau,$$

and

$$Z_2(t) = \frac{n_m u(u + p_0)[2 - 2\tilde{u}bt/\tau + (\tilde{u}bt/\tau)^2 - 2E_b]}{(\tilde{u}b)^2}.$$

In the limit where $\tilde{u}t/\tau \to \infty$, we have $Z_2(t) = n_m u(u + p_0)(t/\tau)^2$. If $\tilde{u}t/\tau \ll 1$, then $Z_2(t) = n_m N u(u + p_0)(\tilde{u} + p_0)(t/\tau)^3/3$. $X_2(t)$ and $Y_2(t)$ are linear functions of time (there is one rate-limiting step), whereas $Z_2(t)$ grows slower than the second power of time (two rate-limiting steps plus one 'intermediate' step).

Some predictions of the model are shown in figure 11.10. The expected number of dysplastic crypts and the fraction of CIN crypts are calculated for $t = 16$ years. As the number of CIN genes, n_c, increases, we expect more dysplastic crypts, and a larger fraction of crypts with CIN. According to our model, the expected number of dysplastic crypts grows linearly with

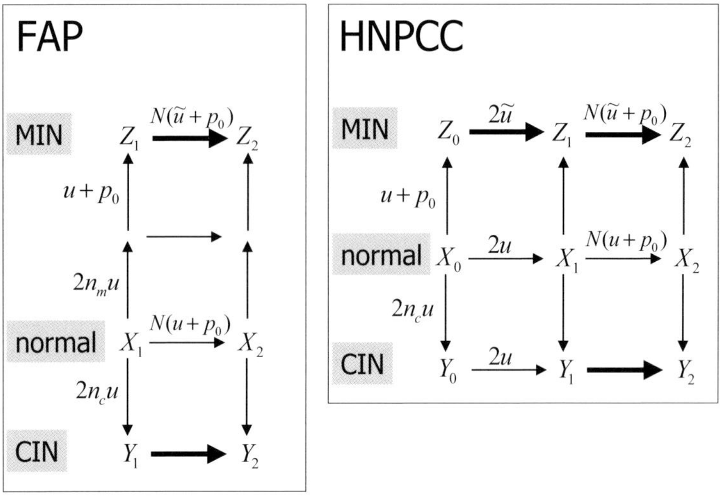

Fig. 11.9 (a) Mutation-selection network of FAP initiation. We start with the type X_1 because the first copy of the APC gene is inactivated in the germ line. (b) Mutation-selection network of HNPCC initiation. One mutation of an MSI gene is inherited, and therefore it takes only *one* step (inactivation of the second copy of the MSI gene, arrows up) to develop the MSI phenotype.

time, and by the age of 16 years is expected to be in the thousands to tens of thousands, see figure 11.10. This should be compared with the observation that patients with FAP have hundreds to thousands of polyps by age 16.

The number of polyps in FAP patients does not grow linearly with time. Instead, most polyps appear 'suddenly' in the second decade of life. These observations are consistent with the predictions of our model. It is believed that polyps result from dysplastic crypts by means of further somatic mutations and clonal expansions. Therefore, the number of polyps is expected to be a higher than linear power of time, which looks like a steep increase in the number of lesions after a relatively non-eventful period. Also, the number of dysplastic crypts ($10^3 - 10^4$ in our model) is expected to be much larger than the number of polyps ($10^2 - 10^3$) consistent with the expectation that not all dysplastic crypts progress to polyps.

Another prediction of this model is that the fraction of MSI crypts in patients with FAP is negligible. This is consistent with an experimental study where MSI was found in none of the 57 adenomas from FAP patients [Keller *et al.* (2001)].

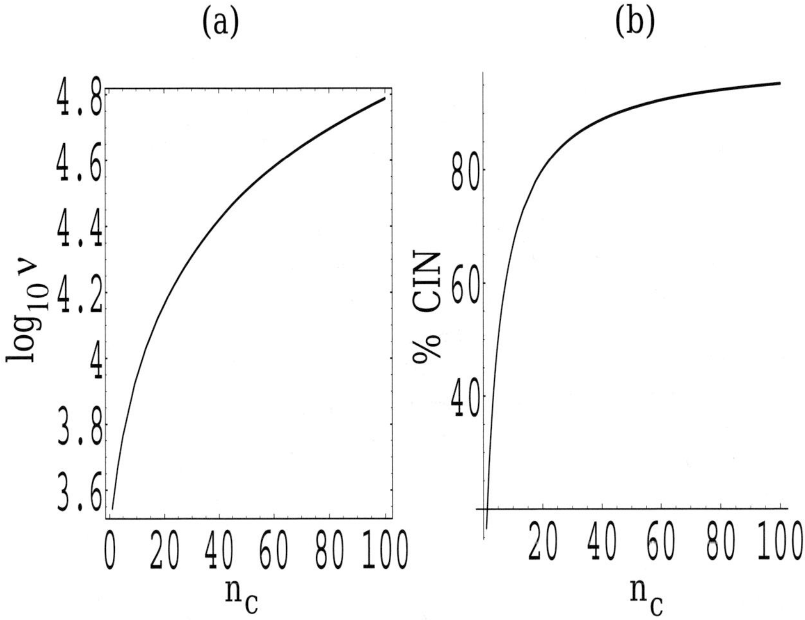

Fig. 11.10 FAP: (a) the logarithm of the expected number of dysplastic crypts and (b) the fraction of CIN crypts, at 16 years of age, as functions of n_c. Parameter values are as in figure 11.7. In (b), the fraction of MSI is nearly zero.

Finally, we note that the logical possibility exists that the second copy of the APC gene in FAP patients may be inactivated by an epigenetic event, just like the second copy of an MSI gene can be silenced by methylation. Experimental investigations [Menigatti *et al.* (2001)] however suggest that this is unlikely: out of the 84 FAP tumors, only 1 exhibited hypermethylation of the APC gene.

11.5 HNPCC

Patients with HNPCC inherit one mutation in an MSI gene. The corresponding mutation-selection network is presented in figure 11.9(b). The solutions for X_2 and Y_2 in this case are identical to those for sporadic colorectal cancers and are given by equations (11.14). The solution for Z_2 is as follows:

$$Z_2(t) = \frac{(u + p_0)[a^2 E_b - b^2 E_a + (a - b)(\tilde{u}abt/\tau - (a + b))]}{ab\tilde{u}(a - b)};$$

in the limit where $\tilde{u}$ is a fast rate we have $Z_2(t) = (u + p_0)t/\tau$. In the opposite limit, where $\tilde{u}$ is a slow rate, $Z_2(t) = N(u+p_0)\tilde{u}(\tilde{u}+p_0)(t/\tau)^3/3$. If we assume that the second copy of the MSI gene is silenced by methylation, we need to replace u by u_{met} in the expression for $Z_2(t)$.

The solutions for X_2 and Y_2 in this case are quadratic in time (two rate-limiting steps), and the quantity $Z_2(t)$ grows slower than linear but faster than quadratic, because it requires one rate-limiting and two intermediate steps (note that we are talking about linear and quadratic functions of an argument smaller than one). In figure 11.11 we present the expected number of dysplastic crypts and the fraction of MSI crypts, calculated for $t = 40$. We have varied the rate at which inactivation of the second copy of an MSI gene happens. This rate, u_{met}, ranges from 10^{-7}, which corresponds to the methylation rate equal to the mutation rate (or to inactivation happening by a point mutation), to 10^{-5}, where we assume that the methylation rate is a hundred times higher than the mutation rate.

Our model predicts that the majority of dysplastic crypts in HNPCC patients are expected to have MSI. However, we do not find that 100% of dysplastic crypts will contain MSI. On the other hand, we know that virtually all tumors in HNPCC patients have MSI. This might suggest that selection for MSI also happens at later stages of carcinogenesis: dysplastic crypts with MSI might have a faster rate of progression to cancer than dysplastic crypts containing CIN or normal cells.

Finally we note that the total number of dysplastic crypts in HNPCC patients, as predicted by our model, is of the order 10 at age 40, which is only slightly larger than the expected number of dysplastic crypts in normal individuals and is not nearly as high as in the case of FAP (of the order $10,000$, figure 11.10. This is also consistent with observations.

11.6 Summary

In this chapter we applied the tools developed in Chapter 10 to study the dynamics of colorectal cancer initiation. We calculated the rate of dysplastic crypt formation as a consequence of inactivating both alleles of the APC tumor suppressor gene. This can either happen in normal cells or in cells that have already acquired one of the two genetic instabilities, MSI or CIN. If the rate of triggering genetic instability in a cell is high and if the cost of genetic instability is not too large, then inactivation of APC will frequently occur in cells that are genetically unstable. In this case,

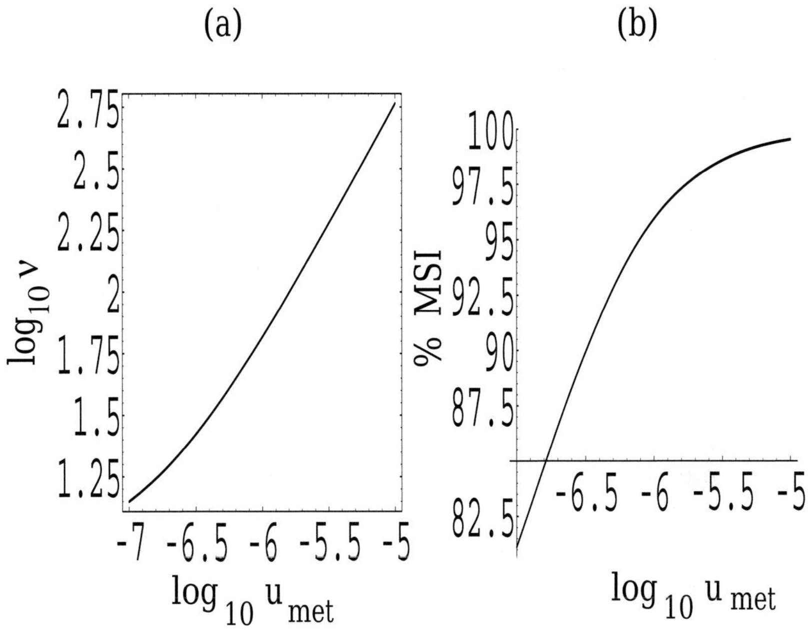

Fig. 11.11 HNPCC: (a) the logarithm of the expected number of dysplastic crypts and (b) the fraction of MSI crypts, at 40 years of age, as functions of $\log_{10} u_{met}$. Parameter values are as in figure 11.7, with $n_c = 10$.

genetic instability is the first phenotypic modification of a cell on the way to cancer.

It is interesting to compare the two types of instability, MSI and CIN. MSI, being associated with subtle changes in the genome, is probably less of a liability for the cell than CIN. In other words, CIN cells are more likely to produce non-viable offspring than MSI cells. At the same time, it may be possible that CIN is easier to trigger (for instance, if it requires a change in a single allele of many genes). Our analysis shows that if inactivation of MSI genes (either by point mutation or by methylation) occurs at a sufficiently fast rate - around 10^{-6} per cell division, then MSI can precede APC inactivation in a significant number of cases. Regarding CIN, the crucial questions are (i) how many dominant CIN genes can be found in the human genome, (ii) how fast are CIN genes inactivated, and (iii) what are the costs of CIN. A more detailed analysis of costs and benefits of CIN is given in Chapters 6 and 7.

Our calculations have shown that important insights could be derived by carefully monitoring the incidence rate of dysplastic crypts in patients as function of age. With or without early genetic instability, the abundance of dysplastic crypts should grow approximately as a second power of time. The two rate limiting steps can either refer to two mutations of APC, or one mutation of APC and one CIN mutation. In the case of CIN, LOH of the second allele of APC is not rate limiting. Hence, two rate limiting steps for the inactivation of a tumor suppressor gene can be compatible with an additional genetic instability mutation.

We have also examined the dynamics of familial colorectal cancer initiation. FAP patients inherit one inactivated copy of the APC gene, and as a result, dysplastic crypt formation involves only one rate limiting step. This could be the inactivation of the remaining copy of the APC gene by a mutation or an LOH event, or a CIN mutations (followed by a fast loss of the remaining APC copy by an LOH event). Our model predicts that the fraction of MSI dysplastic crypts in FAP patients is close to zero. A significant number of dysplastic crypts will contain CIN. This is consistent with experiments observations.

In HNPCC patients, one MSI mutation is inherited, and progression to dysplasia contains one rate-limiting step (the loss of the remaining copy of the MSI gene), followed by two "intermediate" steps (the inactivation of the APC gene as a loss-of-function mutation which happens at an elevated rate because of MSI).

An interesting possibility that we did not explore here is that dysplastic crypts can be lost and replaced by normal crypts, or that unstable crypts could have an advantage/disadvantage compared to stable crypts. In this case, many dysplastic crypts could be produced, but only a part of them is retained so that the actual number of dysplastic crypts stays low. Also, the relative fraction of CIN and MSI crypts can change. The competitive dynamics of crypts in a colon should be investigated experimentally in more detail, to inform further modeling efforts.

Problems

Problem 11.1. Research project. *Learn about details of MSI. In which cancers is it observed apart from colorectal cancer? How prevalent is it? What are the mechanisms of MSI?*

Problem 11.2. *In figure 11.5, there are three possible tunnels: $X_0 \to X_2$, $X_1 \to Y_2$, and $Y_0 \to Y_2$. Four configurations of "open" and "closed" tunnels* *are presented. Show that no other configurations are possible assuming that* $r_c \leq 1$ *and* $r = 1$.

Problem 11.3. *Write down the expressions for* $Y_2(t)$ *for networks (i)-(iv) of figure 11.5 using the combinatorial rule described in this chapter, under assumption (11.4).*

Problem 11.4. *Derive the condition for CIN to cause inactivation of the second APC gene, similar to condition (11.13), for the case of diagram (iv) in figure 11.5.*

Problem 11.5. *If we allowed "costly CIN" (that is,* $r_c < 1$*), then instead of one diagram of figure 11.6, there would be several possibilities, similar to figure 11.5. How would these different possibilities modify the analysis of Section 11.3? Without performing calculations, can you predict how the assumption* $r_c < 1$ *would change the expected percentage of CIN crypts?*

Problem 11.6. Numerical project. *The mutation spectrum of the APC gene is far from random (one reason being that the APC gene is long and multi-functional). The type of the second APC mutation may depend on where the first APC mutation took place, see e.g. [Lamlum* et al. *(1999); Segditsas and Tomlinson (2006)]. Our model is well suited to take this into account. Here is a simple way to differentiate between two kinds of point mutations. Let us assume that the total probability of a point mutation is* u *(as in the basic model), and there are two kinds of mutations. (i) With probability* u_1*, a mutation happens such that the second allele can only be inactivated by a point mutation. (ii) With probability* u_2*, a mutation happens which can be followed by another point mutation or an LOH event. We have* $u_1 + u_2 = u$*. Incorporate these two scenarios in the calculations. How does this change the results?*

Chapter 12

Evolutionary dynamics in hierarchical populations

12.1 Introduction

In Chapter 10 we explored a mathematical framework that allows modeling of cancer initiation by loss-of-function mutations. A prominent example of such a mechanism is the inactivation of tumor suppressor genes. The model considered in Chapter 10 assumed a very simple setting: a constant population of N cells was considered homogeneous, in the sense that the only phenotypic differences among cells were caused by mutations that affected their fitness. This is a crude simplification of reality. Many tissues are characterized by a hierarchical architecture, where, even in the absence of mutations, cells differ phenotypically, and different cell types play a different role. Typically, cell populations consist of stem cells and various classes of daughter cells that differ by their degrees of differentiation. Such hierarchical architecture is a feature of many tissues, especially those which are characterized by frequent self-renewal, such as blood and epithelium (MacKey, 2001; Yatabe *et al.*, 2001).

Tissue stem cells (or undifferentiated cells) are distinguished by their ability to divide or self-renew indefinitely, and generate all the cell types of the organ, potentially regenerating the entire organ from a few cells. Cellular differentiation is the process by which a less specialized cell becomes more specialized. This process takes place inside organs, for example, when stem cells divide and create differentiated daughter cells during tissue repair and during normal cell turnover. Differentiated (or transit-amplifying, TA) cells differ from stem cells by size, shape, metabolic activity, and responsiveness to signals. These changes happen due to highly controlled modifications in gene expression, and do not involve a change in the DNA.

Thus, different cells can have very different physical characteristics despite having the same genome.

In this context, the concept of cell lineages is often discussed. Cell lineages consist of classes of cells that differ by their properties such as the degree of differentiation and the capability of self-renewal. In the simplest case a lineage can be comprised of stem cells and differentiated cells. Several intermediate classes of daughter cells can also be present, giving rise to multistage cell lineages. In self-renewing tissues such as the epithelial tissue, cell lineages are considered to be basic units that ensure the correct functioning of the organ. Divisions, deaths, and differentiation events of cells in a lineage are subject to regulation. The specific mechanisms of control are complex and tissue-specific, and they are beginning to be described in the literature (Binetruy *et al.*, 2007; Gan *et al.*, 2007; Tothova and Gilliland, 2007; Discher *et al.*, 2009).

12.2 Types of stem cells divisions

Two basic models of stem cell divisions are discussed in the literature, see figure 12.1. The asymmetric model suggests that the homeostatic control of the stem cell pool is maintained at the level of single cells, whereby each stem cell produces a copy of itself plus one differentiated cell [Knoblich (2008); Fuchs *et al.* (2004); Zhong and Chia (2008); Ho (2005)]. From the engineering prospective, this model has the advantage of keeping the stem cell population level steady. An obvious disadvantage is its inability to replenish the stem cell pool in case of injury. This problem is naturally solved by the symmetric model, which maintains homeostatic control at the population level, rather than at the individual cell level. There, stem cells are capable of two types of symmetric divisions: a proliferation division resulting in the creation of two stem cells, and a differentiation division resulting in the creation of two differentiated cells [Zhang *et al.* (2009); Loeffler and Roeder (2002); Marshman *et al.* (2002); Clayton *et al.* (2007)]. Differentiation/proliferation decisions are though to be under control of numerous signals emanating from the surrounding tissue and the stem cells themselves [Liu *et al.* (2000); Simmons *et al.* (2003); Alvarez-Buylla and Lim (2004); Saha *et al.* (2006); Lien *et al.* (2006); Adams and Scadden (2007); Dehay and Kennedy (2007); Orford and Scadden (2008); Nusse (2008); Spiegel *et al.* (2008); Saha *et al.* (2008); Sen *et al.* (2008); Guilak *et al.* (2009); Lavado *et al.* (2010); De Graaf *et al.* (2010); Li and Clevers

(2010); Salomoni and Calegari (2010); Hsieh (2012); Ordóñez-Morán and Huelsken (2012)].

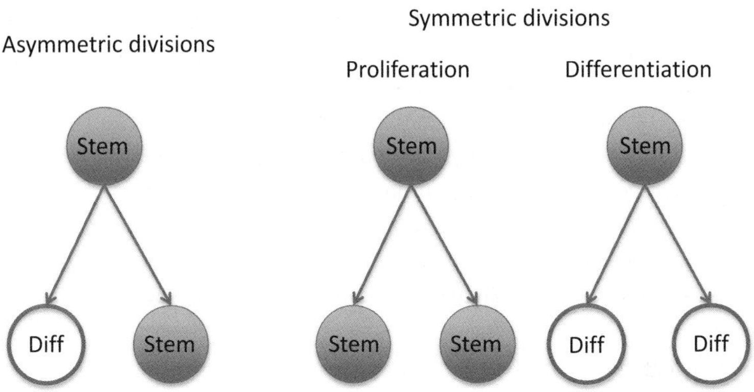

Fig. 12.1 Symmetric and asymmetric stem cell divisions. In the asymmetric division model, a stem cell produces one differentiated cell and one stem cell. In the symmetric division model, a stem cell produces two stem cells (a proliferation event) or two stem cells (a differentiation event).

The ability of stem cells to divide asymmetrically to produce one stem and one non-stem daughter cell is often considered to be one of the defining characteristics of "stemness". On the other hand, there is ample evidence suggesting that adult stem cell can and do divide symmetrically [Morrison and Kimble (2006); Shen *et al.* (2004)].

Uncovering division patterns of stem cells has been subject of intense research in the last fifteen years. Some of the first quantifications of the division strategies in vitro come from the work of [Yatabe *et al.* (2001)] who tracked methylation patterns in the dividing cells of the colon crypts. The analysis of the complex methylation patterns revealed that crypts contain multiple stem cells that go through "bottlenecks" during the life of the organism, which suggests that symmetric divisions are part of the picture. Another piece of evidence comes from experiments with chimeric mice to determine the dynamics of polyclonality of crypts. Crypts that are initially polyclonal eventually become monoclonal, which suggests that symmetric divisions must occur [Spradling *et al.* (2001); Nicolas *et al.* (2007)]. By means of radiotherapy-induced mutations, the study of [Campbell *et al.* (1996)] suggests that a significant fraction of the somatic mutations in human colon stem cells are lost within one year.

An important advance in quantification of symmetric vs antisymmetric divisions became possible with the invention of inducible genetic labeling [Klein and Simons (2011)]. This technique provides access to lineage-tracing measurements, from which the fate of labeled cells and their clones can be tracked over time. By means of the quantitative analysis of long-term lineage-tracing data [Klein *et al.* (2010); Clayton *et al.* (2007)], it has been shown that the rate of stem cell replacement is comparable to the cell division rate, implying that symmetric cell divisions contribute significantly to stem cell homeostasis [Lopez-Garcia *et al.* (2010); Snippert *et al.* (2010)]. The paper by [Simons and Clevers (2011b)] provides a review of the recent evidence of symmetric divisions in mammalian intestinal stem cells, spermatogenesis [Klein *et al.* (2010)] and epithelial tissues such as hair follicles [Doupé *et al.* (2010)].

As explained in Chapter 10, many cancerous transformations start off by an inactivation of a tumor-suppressor gene. We ask the following question: from the point of view of two-hit mutant generation, what type of stem cell divisions is advantageous for the organism? What frequency of symmetric vs asymmetric divisions can maximally delay the stochastic generation of a dangerous mutant? To this end, we consider a continuous range of strategies with mixed type divisions and explore how the frequency of symmetric vs asymmetric divisions affects the generation of mutations.

12.3 The set-up

Stem cell dynamics and evolution have been the subject of mathematical research, e.g. [Dingli *et al.* (2007); Dingli and Pacheco (2011); Traulsen *et al.* (2010); Dingli and Pacheco (2010); Michor (2008); Kern and Shibata (2007); Calabrese *et al.* (2004); Sottoriva *et al.* (2011)]. Here, we consider a non-spatial stochastic model of stem cells and differentiated cells with a constant total population. The stem cells are capable of both symmetric and asymmetric divisions (see figure 12.1). The relative proportion of symmetric divisions can vary and is denoted by the symbol σ, where $\sigma = 1$ means that all divisions are symmetrical, and $\sigma = 0$ means that stem cells only divide asymmetrically. The symmetric divisions can be of two types, proliferation and differentiation. The type of symmetric division is defined by a regulatory mechanism which assures an approximately constant level of stem cells. In this model we explicitly consider cells of two types: stem cells (S) and TA cells, D. Implicitly included are terminally differentiated cells. The total population is denoted by $N = S + D$ and remains constant. An important parameter is $\lambda = N/D$, which defines the proportion of stem cells with respect to TA cells: $S/D = \lambda - 1$.

We assume that the TA cells are removed from the system (by becoming terminally differentiated cells), and that both stem cell and TA cells have a chance to divide. We will refer to the process of terminal differentiation of TA cells as "death" because mathematically a removal from the compartment is equivalent to a death event.

Mutations are included as follows. Each time a division happens, there is a probability, u_1, that one of the daughter cells is a one-hit mutant with fitness r (while the fitness of all wild-type cells is given by 1). The fitness parameter defines the relative probability of the given cell-type to be chosen for division. As before, we consider a range of fitness values, r, such that the one-hit mutants can be disadvantageous compared to wild-type cells, neutral, or slightly advantageous. When a one-hit mutant divides, it has the probability u_2 to give rise to a two-hit mutant. Two-hit mutants are transformed cells which have a potential to give rise to a cancerous tissue transformation. We investigate how the timing of such mutant production depends on the tissue architecture, and specifically, on the symmetry of stem cell divisions.

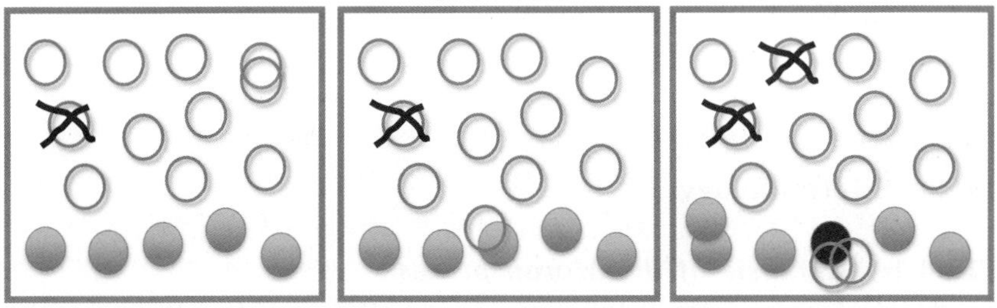

Fig. 12.2 A schematic representation of updates of the generalized Moran process. Filled circles represent stem cells and empty circles - differentiated cells (unlike in figure 9.1, mutants are not shown). (a) Following a differentiated cell death, a differentiated cell divides. (b) Following a differentiated cell death, a stem cell divides asymmetrically. (c) Two differentiated cell death events are balanced by two symmetric divisions of stem cells, one proliferation event and one differentiation event. The dividing stem cell shown in black represents the differentiation event whereby a stem cell is replaced by two differentiated cells.

In order to study this process analytically, we designed the following slightly simplified proess which is a generalization of the Moran process described in Chapters 9 and 10 that includes two types of cell populations and different types of cell divisions. At each update, a daughter cell is chosen for death at random, figure 12.2. Then a cell (a stem cell or a differentiated cell) is chosen to divide, according to its fitness, in the same way as described in Chapter 9. If a differentiated cell is chosen for division, it divides in a usual manner, and this concludes the update, figure 12.2(a). If however a stem cell is chosen for division, we proceed as follows. (1) With probability $1-\sigma$, the stem cell divides asymmetrically, which concludes this step. This is depicted in figure 12.2(b). (2) With probability σ, the stem cell divides symmetrically by differentiation, which is followed by a proliferation of another randomly chosen stem cell. Finally, another daughter cell is chosen for death, which concludes this step.

This version of the Moran process is slightly different from the process that we initially described. In the generalized Moran process, the numbers of stem cells (S) and differentiated cells (D) are kept constant at every step. This is a simplification that allows for analytical tractability (see below). In the numerical simulations the number of stem and differentiated cells fluctuates around a mean value, but despite this difference, the analytical formulas derived here are in an excellent agreement with the simulations.

Note that in order to keep S constant, the symmetric stem cell divisions have to come in pairs (one proliferation and one differentiation event), and must be combined with two cell death events, figure 12.2(c). Therefore, an update involving symmetric divisions counts as two (and not one) elementary updates.

12.4 Methodology

12.4.1 *Analysis of the Moran process*

Let us denote by j_* the number of single-mutant stem cells and by j the number of single-mutant differentiated cells. The updates can be envisaged as a Markov process in the space (j_*, j), where $j_*, j \geq 0$, with an additional state E denoting the generation of a double-mutant cell. Below we will use the condition that mutants are drifting at low numbers, $j_* \ll S$ and $j \ll D$. We have the following probabilities (cf. the probabilities listed in Section 10.5.1 for the original, homogeneous

Moran process):

- The probability that the number of mutant differentiated cells increases by one can be approximated as follows:
$$P_{j_*,j\to j_*,j+1} = \frac{r(j+(1-\sigma)j_*)}{N}(1-u_2)+\frac{D}{N}u_1+(1-\sigma)\frac{S}{N}\frac{u_1}{2}+\frac{u_1 S\sigma}{2N},$$
which is (i) a death of a wild-type differentiated cell (probability ≈ 1), followed by either a faithful division of a mutant differentiated cell, or a faithful asymmetric division of a mutant stem cell; (ii) a division of a wild-type differentiated cell with a mutation; (iii) an asymmetric division of a wild-type stem cell with a mutation happening in the differentiated daughter cell; (iv) a symmetric division of a wild-type stem cell with a mutation (times $1/2$ by association with the symmetric division process). Here and below, when calculating various transition probabilities, the terms associated with symmetric divisions require a factor $1/2$ because symmetric updates involve two death and two division events and thus count as two updates.
- The probability that the number of mutant differentiated cells decreases by one is
$$P_{j_*,j\to j_*,j-1} = \frac{j}{D},$$
which is the probability that a mutant differentiated cell dies followed by a faithful division of any w.t. cell (≈ 1).
- The probability that the number of mutant differentiated cells increases by two, and the number of mutant stem cells decreases by one:
$$P_{j_*,j\to j_*-1,j+2} = \frac{\sigma r j_*}{2N}(1-u_2),$$
which is only possible for a symmetric update, when two w.t. differentiated cells die (probability ≈ 1) followed by a mutant stem cell differentiating without a further mutation (probability $\frac{\sigma r j_*}{N}(1-u_2)$), followed by a w.t. stem cell proliferating without a mutation (probability ≈ 1); the factor $1/2$ comes from the symmetric update.
- The probability that the number of mutant stem cells increases by one is
$$P_{j_*,j\to j_*+1,j} = \frac{(1-\sigma)S}{N}\frac{u_1}{2}+\frac{\sigma}{2}\left(\frac{Su_1}{N}+\frac{rj_*}{N}(1-u_2)\right),$$

which is (i) the probability that following a death of a wild-type differentiated cell (≈ 1), a wild-type stem cell divides asymmetrically with a mutation in the stem cell offspring, (ii) a wild-type stem cell proliferates with a mutation (u_1), or (iii) a mutant stem cell proliferates without a further mutation ($\frac{rj_*}{S}(1-u_2)$).

- The probability to create a double-hit mutant is

$$P_{j_*,j\to E} = \frac{rju_2}{N} + (1-\sigma)\frac{rj_*u_2}{N} + \sigma\frac{rj_*u_2}{N},$$

which is (i) the probability that a mutant differentiated cell divides with a mutation, (ii) a mutant stem cell divides asymmetrically with a mutation, or (iii) a mutant stem cell undergoes either a differentiation or a proliferation event with a mutation.

Let us define by $\varphi_{j_*,j}(t)$ the probability to have j_* mutated stem cells and j mutated differentiated cells at time t. The Kolmogorov forward equation for this function is given by

$$\begin{aligned}\dot\varphi &= \varphi_{j_*,j-1}\left[r(j-1+(1-\sigma)j_*)(1-u_2) + Du_1 + \frac{Su_1}{2}\right]\\ &+ \varphi_{j_*,j+1}\lambda(j+1)\\ &+ \varphi_{j_*+1,j-2}\frac{\sigma r(j_*+1)}{2}(1-u_2)\\ &+ \varphi_{j_*-1,j}\left[Su_1/2 + \frac{\sigma}{2}r(j_*-1)(1-u_2)\right]\\ &- \varphi_{j_*,j}(r(j_*+j) + Nu_1 + \lambda j). \qquad (12.1)\end{aligned}$$

Let us define the probability generating function,

$$\Psi(y_*,y;t) = \sum_{j_*,j}\varphi_{j_*,j}(t)y_*^{j_*}y^j.$$

The probability to be in one of the states (j_*,j) is given by $\Psi(1,1;t)$. Therefore, the probability to transit to state E is $P_2(t) = 1-\Psi(1,1;t)$. The probability generating function satisfies the following first order PDE, whose derivation is similar to that of Section 10.5.2:

$$\begin{aligned}\frac{\partial\Psi}{\partial t} &= \frac{\partial\Psi}{\partial y_*}\left([\frac{r\sigma}{2}(y_*^2+y^2) + r(1-\sigma)y_*y](1-u_2) - ry_*\right)\\ &+ \frac{\partial\Psi}{\partial y}(r(1-u_2)y^2 + \lambda(1-y) - ry)\\ &- u_1\Psi\left(\left(D+\frac{S}{2}\right)(1-y) + \frac{S}{2}(1-y_*)\right). \qquad (12.2)\end{aligned}$$

We have

$$P_2(t) = 1 - exp\left(-u_1 \int_0^t \left\{\left(D + \frac{S}{2}\right)(1 - y(t')) + \frac{S}{2}(1 - y_*(t'))\right\} dt'\right), \tag{12.3}$$

where

$$\dot{y}_* = \left[\frac{r\sigma}{2}(y_*^2 + y^2) + r(1-\sigma)y_*y\right](1 - u_2) - ry_*, \tag{12.4}$$

$$\dot{y} = r(1 - u_2)y^2 + \lambda(1 - y) - ry, \tag{12.5}$$

$$y_*(t) = y(0) = 1. \tag{12.6}$$

Equation (12.3) states that one-hit mutants in differentiated cells are produced by divisions of differentiated cells at the rate u_1D and by divisions of stem cells at the rate $u_1S/2$. The factor $1/2$ comes from the fact that in asymmetric divisions, only a half of mutations will be in the differentiated cells, and in symmetric divisions which consist of pairs differentiation/proliferation, only half of the time a mutation will happen upon differentiation. Mutations in stem cells are produced by the divisions of stem cells at rate $u_1S/2$. The ordinary differential equations describe the dynamics of lineages that start from one differentiated mutant (equation for $\dot{y}$) or from one stem cell mutant (equation for $\dot{y}_*$). The dynamics of differentiated mutants is independent of σ.

Let us first solve equation (12.5), which informs us about the probability of creating a double-hit mutant in a differentiated cell. This Riccati equation can be solved by standard methods, see Section 10.5.3, and the growth of the quantity $1 - y$ proceeds in the following stages:

- The linear growth stage, where $1 - y \approx ru_2t$, as long as $t \ll t_*$ (to be defined).
- The saturation stage, where $1 - y \approx R$, as long as $t \gg t_*$.

The constant R obtained from the stable fixed point of equation (12.5) is given by the equation

$$y = 1 - \frac{\lambda + r - \sqrt{(\lambda + r)^2 - 4r\lambda(1 - u_2)}}{2r(1 - u_2)},$$

and can be approximated by concise expressions as shown below. Given the solution for y, equation (12.4) can also be analyzed. The function $1 - y_*$ increases monotonically and reaches saturation at $1 - y_* = 1$, after characteristic time t_{**}. To find that time-scale, we substitute the constant approximation for the function y, to obtain

$$y_*(t) \approx exp(-t/t_{**}).$$

There are several regimes where the expressions take a particularly simple form (see also summary in Table 12.1).

Regime (2A). Let us assume that $|\lambda - r| \gg \sqrt{u_2}$, $r < \lambda$, and $\sigma \ll u_2$. In this case, we have

$$R = \frac{ru_2}{\lambda - r}, \quad t_* = \frac{1}{\lambda - r}, \quad t_{**} = \frac{\lambda - r}{\lambda r u_2} \gg t_*.$$

There are therefore three distinct regimes defined by the behavior of the functions $y(t)$ and $y_*(t)$.

(1) If $t \ll t_*$, we have $1 - y = ru_2 t$ and $1 - y_* = ru_2 t + (rt)^2 u_2/2$. In this case we have

$$P_2^{lin}(t) = 1 - exp\left(-\frac{Nu_1u_2rt^2}{2} - \frac{Su_1u_2r^2t^3}{12}\right),$$

where the second term in the exponent is typically smaller than the first, and the behavior is thus indistinguishable for the usual homogeneous Moran process at early times.

(2) If $t_* \ll t \ll t_{**}$, we have $1 - y = ru_2/(\lambda - r)$ and $1 - y_* = \lambda r u_2 t/(\lambda - r)$. In this case we have

$$P_2^{inter}(t) = 1 - exp\left(-\frac{(D + S/2)u_1u_2rt}{\lambda - r} - \frac{Su_1u_2\lambda rt^2}{4(\lambda - r)}\right). \quad (12.7)$$

(3) Finally, if $t \gg t_{**}$, we have $1 - y = ru_2/(\lambda - r)$ and $1 - y_* = 1$, and

$$P_2^{sat}(t) = 1 - exp\left(-R_{A\to C}t\right), \quad R_{A\to C} = \frac{(D + S/2)u_1u_2r}{\lambda - r} + \frac{Su_1}{2}. \quad (12.8)$$

This regime becomes unimportant if for $t = t_{**}$ we can show that the quantity in the exponent is much larger than one. We have

$$P_2^{sat}(t_{**}) = P_2^{inter}(t_{**}) = 1 - exp\left(-\frac{(D + S/2)u_1}{r} + \frac{Su_1(\lambda - r)}{\lambda r u_2}\right),$$

and this quantity is very close to 1 for example if $u_1 \sim u_2$ and $(\lambda - r)S \gg 1$.

Regime (2B). Let us assume that $|\lambda - r| \ll \sqrt{u_2}$ and $\sigma \ll \sqrt{u_2}$. In this case, we have

$$R = \sqrt{u_2}, \quad t_* = \frac{1}{2\lambda\sqrt{u_2}}, \quad t_{**} = \frac{1}{\lambda\sqrt{u_2}} \sim t_*.$$

There are therefore only two regimes defined by the behavior of the functions $y(t)$ and $y_*(t)$.

(1) If $t \ll t_*$, we have as in the previous case, $1-y = ru_2t$ and $1-y_* = ru_2t + (rt)^2u_2/2$. The probability of double-hit mutant production is thus given by

$$P_2^{lin}(t) = 1 - exp\left(-\frac{Nu_1u_2\lambda t^2}{2} - \frac{Su_1u_2\lambda^2t^3}{12}\right),$$

where the second term in the exponent is typically smaller than the first, and the behavior is thus indistinguishable for the usual homogeneous Moran process at early times.

(2) If $t_* \ll t$, we have $1-y = \sqrt{u_2}$ and $1-y_* = 1$. In this case we have

$$P_2^{sat}(t) = 1 - exp\left(-R_{A\to C}t\right), \quad R_{A\to C} = (D + S/2)u_1\sqrt{u_2} + \frac{Su_1}{2}. \tag{12.9}$$

Regime (1A). Let us assume that $|\lambda - r| \gg \sqrt{u_2}$, $r < \lambda$, and $\sigma \gg u_2$. The quantity $1 - y_*(t)$ behaves as a linear function,

$$1 - y_*(t) = \frac{\lambda ru_2t}{\lambda - r},$$

for $t_* \ll t \ll t_{**}$, where

$$t_* = \frac{1}{\lambda - r}, \quad t_{**} = \frac{1}{r}\sqrt{\frac{\lambda - r}{2u_2\lambda\sigma}}. \tag{12.10}$$

For $t \gg t_{**}$, the quantity $1 - y_*(t)$ tends to a constant,

$$1 - y_*(t) = \sqrt{\frac{2u_2\lambda}{\sigma(\lambda - r)}}. \tag{12.11}$$

Note that the initial behavior of the function $1 - y_*(t)$ does not depend on σ. This means that for relatively short times ($t \ll t_{**}$), the mutant generation in stem cells proceeds in the same way for symmetric and asymmetric divisions. The length of this regime and the level of saturation however are both functions of σ. It is easy to see that both t_{**} and the saturation level increase as σ decreases. This means that the rate of mutant accumulation becomes higher for asymmetric divisions.

Regime (1B). Let us assume that $|\lambda - r| \ll \sqrt{u_2}$ and $\sigma \gg \sqrt{u_2}$. Now, the linear stage for $1 - y_*(t)$ is defined as

$$1 - y_*(t) = r\sqrt{u_2}t,$$

and it occurs for the times $t_* \ll t \ll t_{**}$, where

$$t_* = \frac{1}{2\lambda\sqrt{u_2}}, \quad t_{**} = \frac{1}{\lambda\sqrt{2\sigma}u_2^{1/4}}.$$

For $t \gg t_{**}$, the quantity $1 - y_*(t)$ tends to a constant,

$$1 - y_*(t) = \sqrt{\frac{2}{\sigma} u_2^{1/4}}. \tag{12.12}$$

Calculations for regimes (1C) and (2C) are performed in a similar manner.

The approximations obtained here and summarized in Table 12.1 are illustrated in figure 12.3. Predictions of equation (12.3), have been compared with stochastic numerical simulations, and found to be in excellent agreement with them, see below.

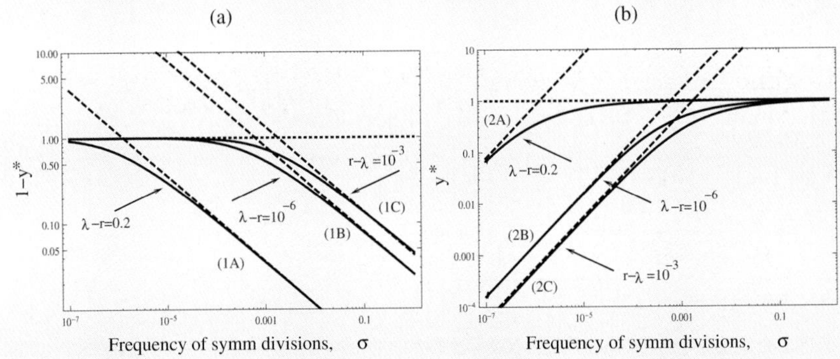

Fig. 12.3 The six different approximation regimes (Table 12.1) for the steady-state solution of system (12.4-12.5). Plotted is the quantity (a) $1 - y_*$ and (b) y_* as a function of the frequency of symmetric divisions, σ, for three different values of λ (solid lines), together with the approximations given by the formulas in Table 12.1. Approximations ($1A$), ($1B$), and ($1C$) are best demonstrated in panel (a), where the quantity $1 - y_*$ is plotted. Approximations ($2A$), ($2B$), and ($2C$) are best demonstrated in panel (b), where the quantity y_* is plotted. The other parameters are $u_1 = u_2 = 10^{-7}$, $r = 1.1$.

12.4.2 *Numerical simulations*

In the generalized Moran process described above, the number of stem and differentiated cells is kept constant at every update. To relax this rigid assumption, we can use the following stochastic process. Suppose that the population consists of four types of cells: stem cells (wild-type, i_*, and one-hit mutants, j_*), and TA cells (wild-type, i, and one-hit mutants, j). We have $i + i_* + j + j_* = N$, where N is a constant total population size. The dynamics proceed as a sequence of updates.

At each update, one TA cell is randomly removed from the population, and replaced with an offspring of another cell, thus keeping the total population size constant.

The process of division is modeled as follows. All cells (stem or TA cells) have a probability to divide. A cell is chosen for division based on its fitness. The fitness of mutated cells is given by r and the fitness of wild-type cells is 1. Let us use the notation $\mathcal{N} = i + i_* + r(j + j_*)$. Then the probability that a wild-type stem cell is chosen for division is given by $i_*/\mathcal{N}$; the probability that a mutated stem cell is chosen for division is given by $rj_*/\mathcal{N}$; the probability that a wild-type TA cell is chosen for division is given by $i/\mathcal{N}$; and the probability that a mutant TA cell is chosen for division is given by $rj/\mathcal{N}$.

If a wild-type TA cell divides, it creates another wild-type TA cell with probability $1 - u_1$, and it creates a one-hit mutant TA cell with probability u_1. If a mutant TA cell divides, it creates a one-hit mutant TA cell with probability $1 - u_2$, and it creates a two-hit mutant with probability u_2. In case of such an event, the process stops.

Divisions of stem cells can be either symmetric (with probability σ) or asymmetric (with probability $1-\sigma$). Asymmetric divisions result in a creation of a TA cell. If a wild-type stem cell is dividing asymmetrically, then with probability $1-u_1$ no mutations happen, and a one-hit mutant will be created with probability u_1. In case of such an event, with probability 1/2 the TA daughter cell will get a mutation, and with probability 1/2 it will be the stem cell that acquires a mutation. Similarly, a one-hit mutant stem cell that divides symmetrically will create a two-hit mutant with probability u_2, in which case the process stops.

Symmetric divisions can be of two types: a differentiation, which results in a replacement of the dividing stem cell with two TA cells, or a proliferation which results in a creation of a stem cell. The probability of proliferation is taken to be $p = \frac{(i_*+j_*)^{10}}{S^{10}+(i_*+j_*)^{10}}$, where S is a constant parameter which measures the expected number of stem cells in the system. The probability of proliferation is given by $1 - p$. Again, when a wild-type stem cell divides, with probability $1-u_1$ both daughter cells are wild-type, and with probability u_1 one of the daughter cells is a one-hit mutant. If a one-hit mutant stem cell divides, both daughter cells are one-hit mutants with probability $1 - u_2$, and with probability u_2 the process stops because a double-hit mutant is created.

The decision trees for stem cells are shown in figure 12.4, for wild-type stem cells (a) and for mutated stem cells (b). Stem cells are denoted by light circles with "S" and TA cells by shaded circles with "D". One-hit mutants are marked with a star.

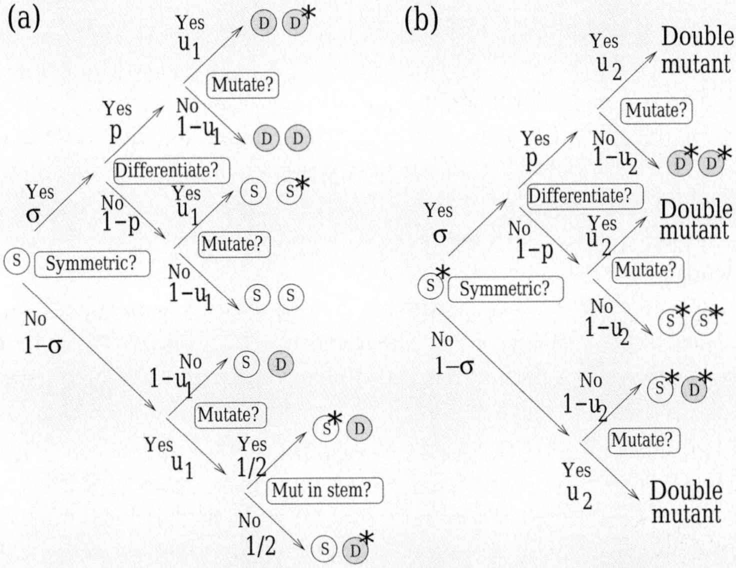

Fig. 12.4 **Stem cell division decision trees for the numerical algorithm.** (a) Divisions of wild-type stem cells. (b) Divisions of mutant stem cells. Stem cells are denoted by light circles with an "S" and TA cells by shaded circles with a "D". One-hit mutants are marked with a star.

These updates were performed repeatedly until either a double-hit mutant was created, or the maximum number of time-steps was reached. Numerical simulations of this process were compared with the analytical predictions obtained for the generalized Moran process.

12.5 Generation of mutations in a hierarchical population

12.5.1 *Tunneling rates*

Our theoretical results on the rate of double-hit mutant formation provide a generalization of the expression for the tunneling rate obtained for a simpler, homogeneous process in Chapter 10. Let us concentrate on the results for the tunneling rates - the rates at which the stem cell system of a given size produces double-hit mutants (assuming that one-hit mutants drift at relatively low levels). As before, we denote the tunneling rate as $R_{A\to C}$. We have

$$R_{A\to C} = R^{stem}_{A\to C} + R^{TA}_{A\to C} = \frac{Nu_1}{2}\left[\left(1-\frac{1}{\lambda}\right)(1-y_*) + \left(1+\frac{1}{\lambda}\right)(1-y)\right], \tag{12.13}$$

where quantities y and y_* satisfy the system

$$0 = \left[\frac{r\sigma}{2}(y_*^2+y^2) + r(1-\sigma)y_*y\right](1-u_2) - ry_*, \tag{12.14}$$

$$0 = r(1-u_2)y^2 + \lambda(1-y) - ry. \tag{12.15}$$

The time to produce double-hit mutants is distributed exponentially with the mean

$$T = \frac{1}{R_{A\to C}}.$$

Formula (12.13) describes the generation of double-hit mutants in the stem cells (the first term on the right) and in TA cells (the second term of the right). Several limiting cases are presented in Table 12.1, to be compared with Table 10.1.

Table 12.1 Important limiting cases for the tunneling rate (formula (12.13)).

Regime*	Description†	Conditions	$1-y_*$	$1-y$
$(1A)$	$r<\lambda$, s+a	$\sigma \gg u_2$, $\lvert\lambda-r\rvert \gg \sqrt{u_2}$, $r<\lambda$	$\sqrt{\frac{2u_2}{\sigma(1-r/\lambda)}}$	$\frac{ru_2}{\lambda-r}$
$(1B)$	$r\approx\lambda$, s+a	$\sigma \gg \sqrt{u_2}$, $\lvert\lambda-r\rvert \ll \sqrt{u_2}$	$\sqrt{\frac{2\sqrt{u_2}}{\sigma}}$	$\sqrt{u_2}$
$(1C)$	$r>\lambda$, s+a	$\sigma \gg \lvert\lambda-r\rvert$, $\lvert\lambda-r\rvert \gg \sqrt{u_2}$, $r>\lambda$	$\sqrt{\frac{2}{\sigma}\left(\frac{r}{\lambda}-1\right)}$	$1-\frac{\lambda}{r}$
$(2A)$	$r<\lambda$, a	$\sigma \ll u_2$, $\lvert\lambda-r\rvert \gg \sqrt{u_2}$, $r<\lambda$	$1-\frac{\sigma(1-r/\lambda)}{2u_2}$	$\frac{ru_2}{\lambda-r}$
$(2B)$	$r\approx\lambda$, a	$\sigma \ll \sqrt{u_2}$, $\lvert\lambda-r\rvert \ll \sqrt{u_2}$	$1-\frac{\sigma}{2\sqrt{u_2}}$	$\sqrt{u_2}$
$(2C)$	$r>\lambda$, a	$\sigma \ll \lvert\lambda-r\rvert$, $\lvert\lambda-r\rvert \gg \sqrt{u_2}$, $r>\lambda$	$1-\frac{\sigma}{2(r/\lambda-1)}$	$1-\frac{\lambda}{r}$

*The notations for the six different regimes refer to figure 12.3.
† "s+a" stands for symmetric and asymmetric, "a" for asymmetric.

12.5.2 *Double-hit mutants are produced slower under symmetric compared to asymmetric divisions*

An important question is how the fraction of symmetric divisions (σ) affects the rate of double-mutant production. We can see that the production of double-mutants by non-stem cells does not depend on σ, the frequency of symmetric divisions. On the other hand, the production by stem cells is crucially affected by this parameter. Our formulas show clearly that the rate of tunneling grows as σ decreases, and it is the highest when $\sigma = 0$, the case of purely asymmetric divisions. This means that in order to minimize the rate of double-hit mutant formation, one needs to maximize the share of symmetric divisions. In figure 12.5 we plot the quantity

$$\frac{R^{stem}_{A\to C}(\sigma = 1)}{R^{stem}_{A\to C}(\sigma = 0)}, \tag{12.16}$$

for different percentages of stem cells. We can see that for realistic ranges of the mutation rates, the difference is at least 10-fold, and can be as high as 10^4-fold, with the symmetrically dividing stem cells producing double-hit mutants slower than asymmetrically dividing cells.

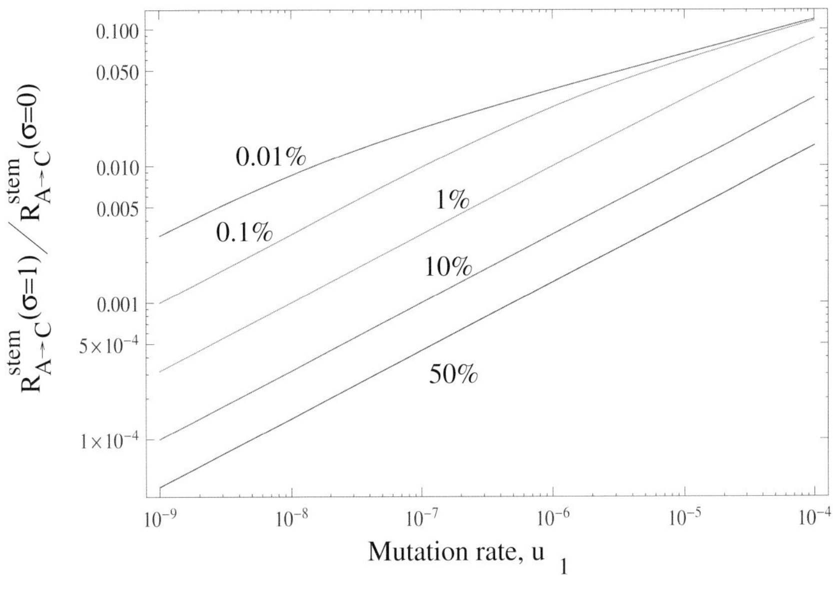

Fig. 12.5 The reduction in the rate of double mutant production in stem cells with symmetric divisions compared to stem cells with asymmetric divisions only. Plotted is the quantity in formula (12.16) as a function of the mutation rate, u_1. The percentage of the stem cells in the whole population (S/N) is marked next to the lines. The other parameters are $u_2 = u_1/2$, $r = 1$.

Figure 12.6 compares the analytical findings for the double-hit mutant production dynamics with the numerical simulations. We ran the stochastic numerical model for a fixed number of time-steps, and recorded whether or not a double-hit mutant has been generated. Repeated implementation of this procedure produced a numerical approximation of the probability of double-hit mutant generation, which is plotted (together with the standard deviations) as a function of σ, the probability of symmetric divisions, for three different values of λ, which measures the fraction of stem cells. Clearly, the probability of mutant generation in the course of a given time-interval is a decaying function of σ.

12.5.3 *Comparison with the homogeneous model*

Another result that follows from our computations is the comparison of the double-mutant production in a hierarchical (stem cells plus TA cells) model compared with the conventional, homogeneous model described in Chapter 10. It turns out the hierarchical model with purely asymmetric divisions always produces mutants faster than the homogeneous model. For the hierarchical model with purely symmetric divisions the result depends on the fitness of one-hit mutants. For disadvantageous one-hit mutants whose fitness satisfies $r < 1$, $|1 - r| \gg \sqrt{u_2}$, the hierarchical model with purely symmetric divisions produces double-mutants faster, and for neutral and advantageous mutants, it produces double-hit mutants slower than the homogeneous model. In figure 12.6 we can see that for $r = 1$ (neutral one-hit mutants), hierarchical models with a sufficiently large values of σ are characterized by slower double-hit mutant generation compared to the homogeneous model (the horizontal line).

Figure 12.7 shows additional results of simulations (together with our analytical calculations), where for three different values of r (one-hit mutant fitness) the probability of double-hit mutant generation is plotted as a function of λ. The values $\lambda \to 1$ correspond to a vanishingly low fraction of stem cells in the system, while $\lambda = 2$ corresponds to 50% of all cells being stem cells. We show purely symmetric ($\sigma = 1$) and purely asymmetric ($\sigma = 0$) cases. For fixed mutation rates and populations sizes, the homogeneous model is characterized by only one parameter, r, which is the fitness of one-hit mutants. The probability of double-hit mutant generation strongly depends on whether these intermediate mutants are disadvantageous ($r < 1$), neutral ($r \approx 1$), or advantageous ($r > 1$). In contrast to the homogeneous model, the hierarchical model contains two additional parameters, λ (the ratio of TA cells and the total population) and σ (the probability of symmetric divisions). We can see that these two parameters affect the probability of double-hit mutant generation at least as strongly as the fitness r does. The influence of σ is clear: the more the fraction of

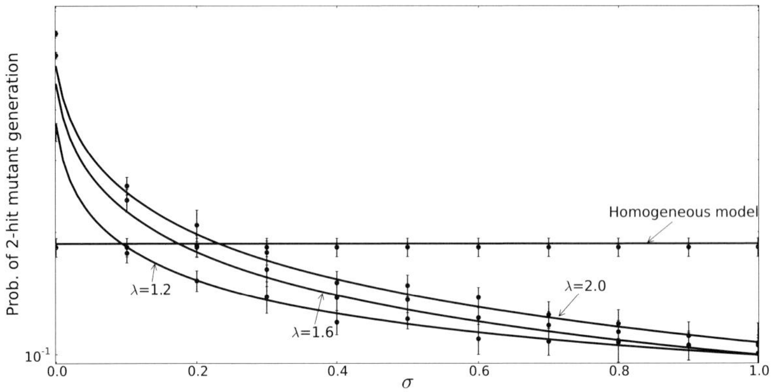

Fig. 12.6 The probability of double-hit mutant generation as a function of σ, the probability of symmetric stem cell divisions. The results of numerical simulations are presented as points (standard deviations are included). Analytical results are given by solid lines (formula (12.3)). The horizontal line represents the calculations for the homogeneous model. We ran 10 batches of 1000 runs. The parameters are $r = 1.0$, $u_1 = 0.00001$, $u_2 = 0.002$, $N = 500$.

symmetric divisions, the slower double-hit mutants are produced. Next, we examine the role of the stem cell to TA cell ratio.

12.5.4 *The optimal fraction of stem cells*

Let us consider an optimization problem for the tissue design, with the goal to delay the production of double-hit mutants. What is the optimal fraction of stem cells that the population should maintain? Analysis of the tunneling rates for a hierarchical model with purely symmetric divisions suggests that the optimal fraction of stem cells depends on the fitness of the one-hit mutants. If the one-hit mutants are disadvantageous ($r < 1$, $|1 - r| \ll \sqrt{u_2}$), then the tunneling rate grows with the parameter λ. In other words, in order to minimize the rate of double-mutant production, one would need to keep the stem cell pool as small as possible.

For neutral and advantageous intermediate mutants, where the symmetric division model gives rise to the lowest double-mutant production rate compared to the homogeneous model and the hierarchical model with asymmetric divisions, this rate is minimized for a particular fraction of stem cells. This fraction is defined by the mutation rate u_2 in the neutral case, and by the fitness of the intermediate mutants in the case of weakly advantageous mutants. For neutral one-hit mutants ($|1 - r| \ll \sqrt{u_2}$), the optimal value of λ is given by

$$\lambda_{opt} = 1 + 2u_2^{1/3}, \tag{12.17}$$

and for weakly advantageous mutants with $1 < r < \lambda$, $|r - 1| \gg \sqrt{u_2}$, we have

$$\lambda_{opt} = \frac{r}{2 - r}. \tag{12.18}$$

For example, for the biologically most relevant case of neutral one-hit mutants, the optimal fraction of stem cells is approximately 1% of the total population, assuming $u_2 = 10^{-7}$.

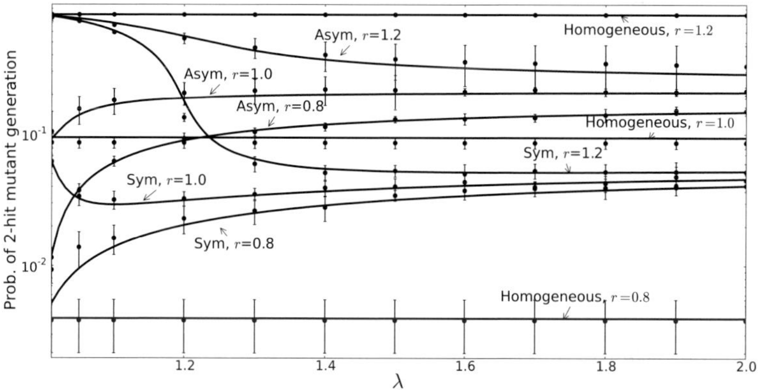

Fig. 12.7 The probability of double-mutant generation as a function of λ, the ratio of TA cells to the total number of cells. As in figure 12.6, the results of numerical simulations are presented as points (standard deviations are included), and the analytical results are given by solid lines (formula (12.3)). The horizontal lines represent the calculations for the homogeneous model. We ran 10 batches of 1000 runs. Plotted is the probability of double-mutant generation as a function of λ, for purely symmetric ($\sigma = 1$) and purely asymmetric ($\sigma = 0$) models, for three different values of r. The parameters are $u_1 = 10^{-5}, u_2 = 10^{-4}, N = 1000$.

These results are illustrated in figure 12.8. In this plot, we can see that for $r = 0.8$ the probability of having a doubly mutated cell (after a given time-span) is an increasing function of λ, as predicted. For the case of $r = 1$, the numerical simulation in figure 12.8 shows that $\lambda_{opt} \approx 1.1$ (compared with $\lambda_{opt} = 1.093$ predicted by formula (12.17)). For the case $r = 1.2$, formula (12.18) gives $\lambda_{opt} \approx 1.5$, which approximately coincides with the numerical optimum. In the case of advantageous mutants, however, the minima of λ are very shallow.

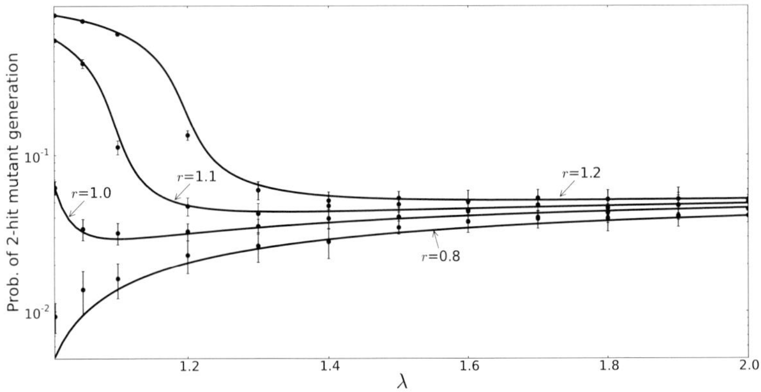

Fig. 12.8 The probability of double-hit mutant generation in the symmetric division model. The case of symmetrically dividing stem cells, same as in figure 12.7.

12.5.5 *Do mutations in TA cells produce double-mutants?*

Let us compare the relative contributions to the double-mutant production rate coming from stem cells and TA cells, equation (12.13):

$$R^{stem}_{A\to C} = \frac{Nu_1}{2}\left(1-\frac{1}{\lambda}\right)(1-y_*), \quad R^{TA}_{A\to C} = \frac{Nu_1}{2}\left(1+\frac{1}{\lambda}\right)(1-y). \tag{12.19}$$

The contribution from the TA cells grows as the fraction of TA cells increases. In figure 12.9 we plot the fraction of stem cells (given by $1-1/\lambda$) that corresponds to $R^{stem}_{A\to C} = R^{TA}_{A\to C}$. We can see that for the mutation rates around 10^{-7}, this fraction is about 0.1% for disadvantageous intermediate mutants, about 0.5% for neutral mutants, and about 15% for advantageous mutants. This means that as long as the fraction of stem cells in the population is lower than these threshold values, TA cells contribute *more* to the production of double-hit mutants than stem cells. This threshold fraction grows for larger mutation rates, making it easier for TA cells to contribute significantly to the double-hit mutant production. An analytical approximation for the threshold value of λ can be found for small values of mutation rates, such as

$$\lambda_c = \begin{cases} 1 + r\sqrt{\frac{2\sigma u_2}{1-r}}, & r < 1, \text{ Regime (1A)}, \\ r - \frac{(r-1)^2}{2\sigma}, & r > 1, \text{ Regime (1C)}. \end{cases} \quad (12.20)$$

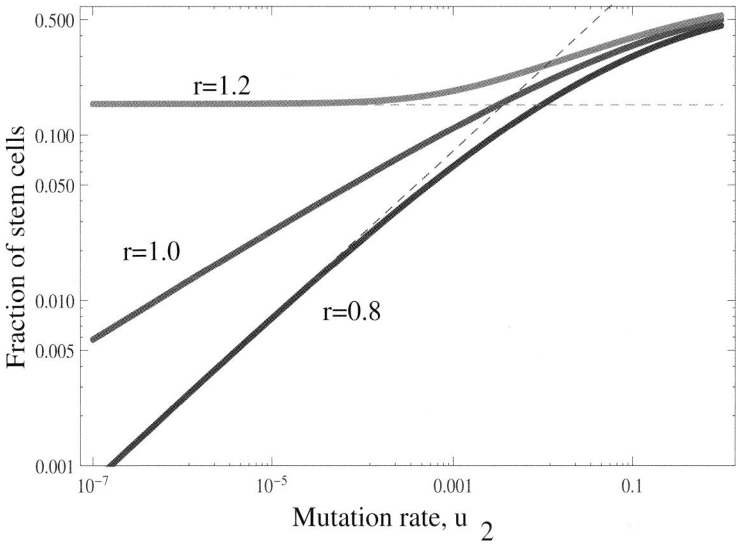

Fig. 12.9 The threshold fraction of stem cells corresponding to stem and TA cells contributing equally to double-hit mutant production. The quantity $1 - 1/\lambda_c$, which corresponds to $R^{stem}_{A\to C} = R^{TA}_{A\to C}$, is plotted as a function of the mutation rate, u_2, for three different values of r, and $\sigma = 1$. For the fraction of stem cells above these values, stem cells have a higher contribution to the rate of double-mutant production compared to the non-stem cells. Thin dashed lines show the approximations of equation (12.20).

Next we address the question of optimization assuming that only mutations acquired by stem cells are dangerous and can lead to further malignant transformations. In this case, the rate of mutant production is given by $R^{stem}_{A\to C}$, equation (12.19). It is easy to show that this quantity is maximized by asymmetric divisions only ($\sigma = 0$), and it is minimized by symmetric divisions of stem cells ($\sigma = 1$). Thus the message does not change if only stem cell mutations are assumed to contribute to carcinogenesis.

12.6 Biological discussion

In this chapter we found that symmetrically dividing stem cells are characterized by a significantly lower rate of two-hit mutant generation, compared to asymmetrically-dividing cells. This is especially important in the context of tumor-suppressor gene inactivation, which is one of the more common patterns of carcinogenesis. This provides an evolutionary framework for reasoning about stem cell division patterns.

In the literature, both types of stem cell divisions have been reported in various tissues. It has also been reported that the same stem cells are capable of both symmetric and asymmetric divisions. Whether a cell divides symmetrically or asymmetrically depends on factors such as the polarized organization of the dividing cell as well as the cell cycle length [Huttner and Kosodo (2005)]. In Drosophila germ stem cells, cell division is asymmetric or symmetric depending on whether the orientation of the mitotic spindle is perpendicular or parallel to the interface between the stem cell and its niche [Yin *et al.* (2006)]. Similarly, mammalian stem cells have been reported to employ both symmetric and asymmetric divisions to regulate their numbers and tissue homeostasis [Noctor *et al.* (2004); Morrison and Spradling (2008)]. A switch from a symmetric mode of divisions to the asymmetric model has also been reported to take place in development (see [Egger *et al.* (2011, 2010)] in the context of Drosophila).

The fact that the rate of double-hit mutant production is the lowest for symmetrically dividing cells does not in itself explain or predict any aspects of the tissue architecture. It however provides an alternative hypothesis for the observation that in mammalian tissues, symmetric patterns of stem cell division seem to be very common. The force of selection that comes from the cancer-delaying effect of such an architecture can be thought to have helped shape the observed division patterns. On the other hand, in more primitive organisms such as Drosophila, asymmetric stem cell divisions seem to dominate adult homeostasis (following the predominantly symmetric division patterns of development). Since cancer delay does not provide an important selection mechanism in the context of Drosophila, we can argue that this could help explain the observed differences.

12.6.1 *Symmetric divisions can have a cancer-delaying effect*

The mathematical result obtained here is that symmetrically dividing cells appear to delay double-hit mutant production compared to an equivalent system with asymmetrically dividing stem cells. What is the intuition behind this finding? Double-mutants are generated by means of mutations that happen in singly-mutated cells. To understand this process, let us focus on the dynamics of single mutants. In particular, we concentrate on singly-mutated stem cells, because the fates of single mutations in TA cells are identical in the two models. What happens to a singly-mutated stem cell under the different division patterns?

If stem cells divide asymmetrically, then a mutation acquired in a stem cell will remain in the system indefinitely, because at every cell division, a new copy of the mutant stem cell will be generated. On the other hand, a mutant stem cell generated under the symmetric division model has a very different and much less certain fate. Each division of a mutant stem cell can result either in (1) elimination of the mutation from the stem cell compartment as a result of a differentiation, or (2) creation of an additional mutant stem cell as a result of a proliferation event. Superficially, it might look like the two processes might balance each other out. This intuition is however misleading. A lineage of mutant stem cells starting from a single mutant stem cell is much more likely to die out than to persist and expand. In fact, only $1/K$ of all such lineages will expand to size K. Half of the lineages will differentiate out after the very first division. Statistically there will be occasional, rare long-lived lineages, but the vast majority will leave the stem cell compartment after a small number of divisions. The production of those "lucky" long-lived mutants is not enough to counter-balance the great majority of the dead-end lineages that quickly exit the stem cell compartment. This is illustrated in figure 12.10, which plots the "weight" (the net size of a lineage over time, T) of a typical symmetrically dividing mutant stem cell, X^{sym}, divided by the weight of a typical asymmetrically dividing mutant stem cell, X^{asym}. The latter quantity is simply given by T, and the former quantity is a stochastic variable. We can see that the weight of symmetrically dividing mutant lineages is always lower than that of asymmetrically dividing lineages, which means that the former will have a lower probability to produce double-mutant offspring. We conclude that the uncertainty of fate of single mutant stem cells is the reason for the statistically longer time it takes for the symmetrically dividing stem cell model to produce a double-hit mutant.

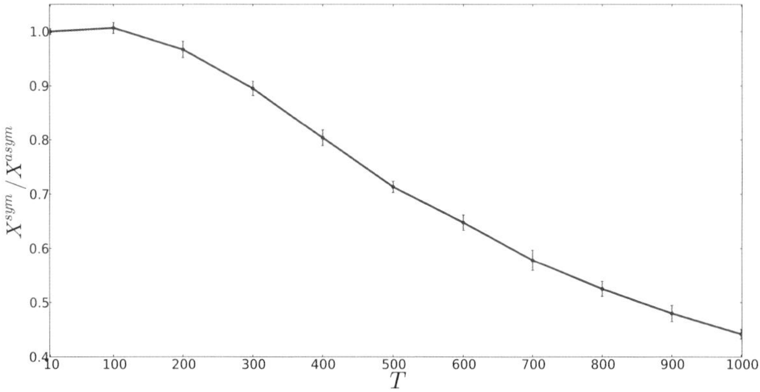

Fig. 12.10 Why are symmetrically dividing stem cells produce mutants slower? The weight of a typical symmetrically dividing mutant stem cell lineage, X^{sym}, relative to the weight of an asymmetrically dividing mutant stem cell lineage, $X^{asym} = T$, is plotted as a function of the number of stem cell divisions, T. Here, $S = 20$, $N = 1000$, and 20 batches of $10,000$ simulations were performed to calculate the mean and the standard deviation.

Interestingly, the above argument can be made in a similar manner for disadvantageous, neutral, or advantageous mutants. In any of those cases, an asymmetrically dividing mutant stem cell remains in the population indefinitely. In the model with symmetric divisions, whenever a mutant stem cell is chosen for division, its probability to proliferate is similar to its probability to differentiate (in order to keep the homeostasis), and thus the dynamics of each lineage are independent of its fitness (except that the frequency of updates is determined by the fitness of mutants; this is why the fitness parameter r factors out of equations (12.14) and (12.4)).

We note that the effect of double-hit mutant production delay caused by symmetric divisions compared to asymmetric divisions is very significant. The difference in the tunneling rate which characterizes the time-scale of the process can be as high as $1,000$-fold for tissues with 10% of stem cells and the mutation rate of 10^{-7} per gene per cell division.

12.6.2 *Can TA cells create double-hit mutants?*

The model studied in this chapter tracks single- and double-hit mutant production in both stem and TA cells. It is interesting to compare which mechanism (through stem cell single mutants or TA cell single mutants) contributes more to the double-mutant production. It turns out that as long as the fraction of stem cells is smaller than a threshold (or equivalently, if the fraction of the TA cells is larger than a threshold), non-stem cells contribute equally or more to the production of double-mutants. This threshold fraction depends on (1) the mutation rate and (2) the fitness of intermediate, one-hit mutants. For example, if the intermediate mutants are neutral and the mutation rate is 10^{-7} per gene per cell-division, then the threshold fraction of stem cells is about 0.5% of the total population. In other words, mutations originating in non-stem cells are significant if stem cells comprise less than 0.5% of the total population. This number is much higher if the intermediate mutants are advantageous, or if the mutation rate responsible for the second hit is higher. For $u_2 = 10^{-3}$, non-stem cells are the driving force behind double-mutant production as long as stem cells comprise less than about 10% of the total population. This scenario is realistic in the presence of genetic instability, where inactivation of a tumor suppressor gene is likely to occur through a small-scale mutation of the first copy of the gene followed by a loss of heterozygocity event inactivating the second copy. The latter can happen a rate as high as 10^{-2} per cell division per chromosome [Lengauer *et al.* (1997)].

The arguments presented above clarify some aspects of the long-standing debate about the origins of cancer, see also [Komarova and Wang (2004)]. It is sometimes argued that TA cells are unimportant for cancer initiation, for the following (quantitative) reason, unrelated to biological evidence. Intuitively, it seems that double-hit mutants cannot be created among TA cells, because all one-hit mutants in the TA compartment will be washed away before they have a chance to acquire the second hit. As John Cairns writes, "...there are 256 exponentially multiplying cells that divide twice a day and are being replenished continually by the divisions of a single stem cell, none of these 256 cells will ever be separated from the stem cell by more than eight divisions, and the replication errors made in those eight divisions are destined, of course, to be discarded", [Cairns (2002)]. The computations in this chapter demonstrate that under some realistic parameter regimes, double-hit mutants can be created in the TA compartment, and TA cells statistically can contribute equally or more to double-hit mutant production compared to stem cells. The simple reason for this is as follows. Even though TA cells are short-lived, and getting a second mutation in a singly-mutated TA cell is unlikely, there are many more TA cells than stem cells. The low chance of double-mutant generation in a single TA cells can be outweighed by the fact that TA cells are a large majority, and single probabilities add up to create a significant effect.

12.6.3 *Cancer stem cell hypothesis*

The question discussed above is purely mathematical, and deals with the simple possibility to acquire two hits in the TA compartment. A related biological question is whether mutations occurring in the TA compartment can lead to further carcinogenic transformations, which brings us to the cancer stem cell hypothesis [Jordan *et al.* (2006); Nguyen *et al.* (2012)]. While the concept of the cancer stem cell remains controversial [Vermeulen *et al.* (2008); Gupta *et al.* (2009)], here we do not intend to argue for or against this theory. Moreover, we refrain from making specific interpretations of this theory with regards to the exact origins of cancer. It has been argued that there is a distinction between the broader concept of the cancer stem cell on the one hand, and the narrower concept of normal stem cell becoming cancerous [Nguyen *et al.* (2012)]. While the cancer stem cell hypothesis states that cancer is maintained by a small fraction of cells with stem-like properties, without making a specific assumption of how those cells are generated, the more narrow theory argues that mutations generated among non-stem cells cannot be cancer-initiating, because (at least, some) cancers originate via the creation of a cancer stem cell, which is a modified stem cell that retains some characteristics of “stemness”.

In the light of this latter hypothesis, let us analyze the process of double-hit mutant production that occurs via mutations in stem cells only. Will our results change if only stem cell mutations can lead to carcinogenic transformation? To accommodate this assumption in our model, we must only use the first term in equation (12.13). It turns out that in this case, the message remains exactly the same: symmetrically dividing stem cell systems are characterized by a slower production of double-hit mutants compared to asymmetrically dividing stem cells. The universality of this result is explained above: the fate of mutations originating in the differentiated compartment is identical under the two models, and the only difference comes from the fates of mutant stem cells.

12.7 Summary

This chapter examined how different tissue stem cell division patterns influence the rate at which mutations are accumulated in the stem cells, and thus how they determine the risk of cancer development. Tissue stem cells have been observed to divide asymmetrically where division gives rise to one stem cell and one differentiating cell. On the other hand, evidence suggests that symmetric stem cell division can also occur, where the division fate can be determined stochastically, generating either two stem or two differentiating daughter cells. The models suggest that the stem cell division pattern can have a significant influence on the rate at which mutants are generated. Importantly, symmetric division patterns lead to a slower accumulation of mutations and thus to a lower risk of carcinogenesis. Hence, it can be hypothesized that the observed occurrence of symmetric stem cell divisions in humans have evolved as a strategy to delay the onset of cancer.

Problems

Problem 12.1. Research project. *Find out more about strategies of stem cells in the context of symmetric and asymmetric stem cell divisions, see e.g. [Simons and Clevers (2011b)].*

Problem 12.2. *In the analysis of solutions of system (12.4-12.5), different limiting cases correspond to different magnitude of the expression* $\lambda - r$. *Show that the steady-state level,* R, *takes on different values in the following cases: (a)* $|\lambda - r| \ll \sqrt{u_2}$, *(b)* $|\lambda - r| \gg \sqrt{u_2}$ *and* $r < \lambda$.

Problem 12.3. *Find the minimum of the tunneling rate* $R_{A\to C}$, *equation (12.13), as a function of* λ, *in different regimes, and derive the results of Section 12.5.4.*

Problem 12.4. Research project. *Find out more about the stem cell hypothesis and the distinction between the broader concept of the cancer stem cell on the one hand, and the narrower concept of normal stem cell becoming cancerous [Nguyen* et al. *(2012)].*

Problem 12.5. *Assume that only mutations acquired by stem cells are dangerous and can lead to further malignant transformations. In this case, the rate of mutant production is given by* $R^{stem}_{A\to C}$, *equation (12.19). Show* *that this quantity is maximized by asymmetric divisions only ($\sigma = 0$), and it is minimized by symmetric divisions of stem cells ($\sigma = 1$).*

Chapter 13

Spatial evolutionary dynamics of tumor initiation

13.1 Introduction

We have developed a stochastic methodology that describes the accumulation of tumor-initiating mutations, Chapters 9, 10. These models assumed perfect mixing in the population of cells. There was no information about spatial locations, and no spatial dynamics. This may not be a shortcoming if we talk about cancers of the blood, such as leukemia. In a discussion of solid tumors, however, spatial considerations must play a role. Many spatial mechanistic models of cancer spread have been proposed, see e.g. [Bellomo and Preziosi (2000); Chaplain (1996); Araujo and McElwain (2004); Bellomo *et al.* (2004); Chaplain and Lolas (2005); Anderson and Chaplain (1998a); Byrne and Chaplain (1996); Shochat *et al.* (1999); Bartoszyński *et al.* (2001); Ledzewicz and Schättler (2002); Hanin (2002); Gatenby and Maini (2003); Swierniak *et al.* (2003); Gaffney (2004)], and more recently, [De Pillis *et al.* (2006); Chaplain *et al.* (2006); Hinow *et al.* (2006); Byrne *et al.* (2006); Gerlee and Anderson (2007); Enderling *et al.* (2007, 2006); Deisboeck *et al.* (2008b); Anderson and Quaranta (2008); Deisboeck *et al.* (2008a); Marciniak-Czochra and Kimmel (2007); Rejniak (2007); Bellomo *et al.* (2008); Macklin *et al.* (2009); Lowengrub *et al.* (2010); Stolarska *et al.* (2009); Hinow *et al.* (2009); Swierniak *et al.* (2009); Cristini and Lowengrub (2010); De Matteis *et al.* (2013)]. These are examples of a large body of literature on spatial models of cancer, such as partial differential equation models, hybrid models, and multi-scale models. These models are briefly reviewed in Section 3.6.

In this chapter we consider much simpler, one-dimensional spatial models, and concentrate on the evolutionary aspects of carcinogenesis. The relative simplicity of the formulation allows an analytical understanding.

This chapter is rather technical and can be skipped without interrupting the logic of the book. The following chapter (Chapter 14) discusses two-dimensional spatial models and provides more biological applications.

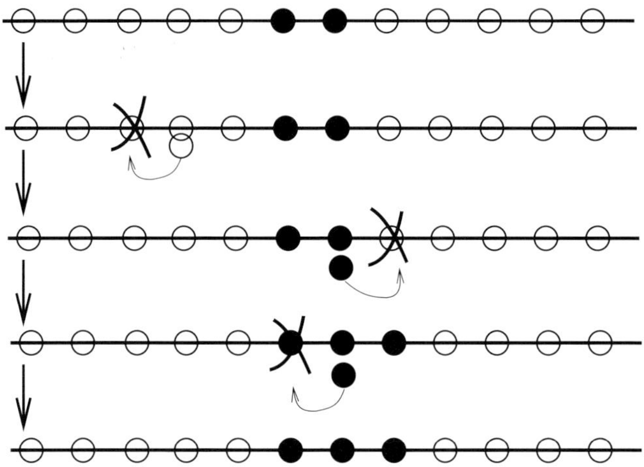

Fig. 13.1 The Moran process generalized to the one-dimensional space: a cell is chosen for death at random, and is immediately replaced by a division of one of the two neighboring cells (chosen proportional to their fitness).

13.2 1D spatial Moran process

As a first attempt to include spatial considerations in the stochastic evolutionary dynamics of malignancy, we have designed a one-dimensional spatial generalization of the mass-action Moran birth-death process (Komarova, 2006a). The cells are aligned along a regular grid, at locations $1, 2, \ldots, N$, see figure 13.1. As before, we assume that the total number of cells does not change. Cells are randomly chosen for death. Each cell death is followed by a cell division of one of its two neighboring cells, which places its daughter cell at the empty slot. Cell death occurs randomly and the division probability is proportional to the relative fitness of the cells.

In this chapter we study the dynamics of mutations that occur in such spatial models, both in the context of oncogenes and tumor suppressor genes. Let us first consider the two-species process depicted in diagram (9.1), which includes only one type of mutations. The wild-type cells (type "A" cells) mutate with probability u_1 per cell division, to create type-"B" cells. Type "B" cells are assumed to reproduce faithfully.

In order to appreciate the difference in the mutant dynamics between the space-free and the spatial models, let us for the time being ignore de-novo mutations and concentrate on the process of the spread of existing mutants.

Let us assume that there is a mutant cell with the relative fitness r at position j. If any cell at position $1, \ldots, j-2$ or $j+2, \ldots, N$ dies, then there can be no change in configuration. A change can occur only in two cases:

- Death occurs at position j, in which case the mutant disappears.
- Death occurs at position $j+1$ or $j-1$. Then the number of mutants can increase by one if the mutant cell is chosen for division.

Similarly, if we have several mutant cells at sequential positions from i through j, then a change of the number of mutants can only happen if death occurs at positions $i-1$, i, j or $j+1$ (see figure 13.1). In this model, a mutant colony which originated as one cell can only occupy a set of adjacent slots. A change in the position of this set can only be caused by cell death at its boundary.

We can see that the dynamics of mutant spread is very much affected by the spatial constraints imposed by the 1D geometry of this model. Only a small subset of elementary updates (namely, only the updates that happen at the interface between the mutants and wild-type cells) can lead to a change in cell configuration. The difference in the mutant growth between the non-spatial and the spatial models can be illustrated by looking at the spread of advantageous mutants. In this case, the situation is similar to the difference between the exponential growth and the surface growth described in Chapter 4. In the former case, the mutants can grow significantly faster, because a death anywhere in the population can lead to a division of a mutant. In the latter case, growth can only occur by a change near a boundary (or a surface).

This feature of the spatial process leads to the differences in the dynamics of mutants in the spatial model compared to the space-free model in the context of both gain-of-function and loss-of-function mutations. Similar tendencies are also observed in more realistic 2D models studied in Chapter 14.

In the following sections we will first concentrate on a two-species process without mutations, and calculate the probability of invasion of a type-"B" mutant. Then we proceed to study a three-species system, similar to the loss-of-function process described in Chapter 10.

13.3 Two-species dynamics

13.3.1 *Preliminaries*

We can characterize the states of the system by positions of the leftmost and the rightmost mutants, i and j, such that

$$1 \leq i \leq j \leq N. \tag{13.1}$$

The transition matrix is given by the following. If the left boundary of the mutant domain is at 1 (or the right boundary is at N), then these boundaries cannot move anymore. Otherwise, if the number of mutants is larger than one, then the probabilities to expand the mutant domain to the left and right are given by $P_{i,j\to i-1,j} = P_{i,j\to i,j+1} = \frac{1}{N}\frac{r}{1+r}$, and the probabilities to reduce the domain on left and right are given by $P_{i,j\to i+1,j} = P_{i,j\to i,j-1} = \frac{1}{N}\frac{1}{1+r}$. Finally, if there is only one mutant (that is, $i = j$), then we have $P_{i,i\to i-1,i} = P_{i,i\to i,i+1} = \frac{1}{N}\frac{r}{r+1}$, and the probability to lose the mutant is $\frac{1}{N}$. All the rest of the elements of the transition matrix are equal to zero.

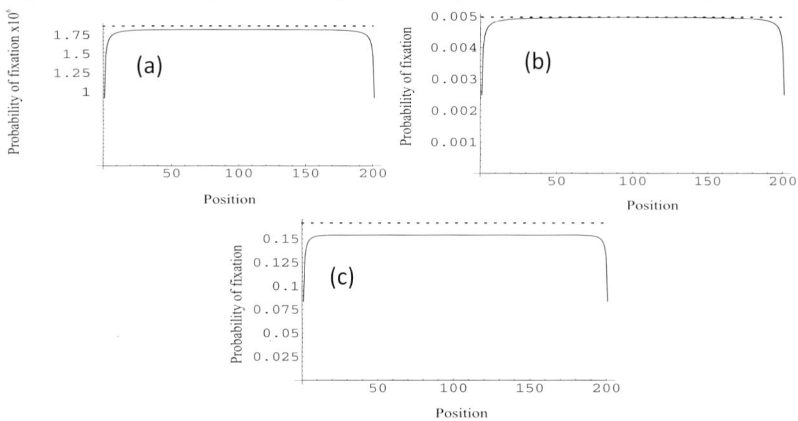

Fig. 13.2 Numerical solutions of system (13.2-13.5), where the probability of absorption u_{ii} is plotted as a function of the position of the mutant, i, in a system of $N = 200$ cells. (a) Disadvantageous mutants, $r = 0.95$; (b) neutral mutants, $r = 1$; (c) advantageous mutants, $r = 1.2$. The dotted horizontal line corresponds to the value $\rho(r)$, the probability of fixation for the space-free system.

We can envisage the dynamics as a 2-dimensional Markov random walk inside domain (13.1), with an additional absorbing state which can be reached from the diagonal $i = j$; this state corresponds to the extinction of the mutant and is denoted by $\tilde{E}$. The other absorbing state is the fixation of the mutant, $(0, N)$. The random walk is governed by the matrix above. We can set i to be the horizontal and j the vertical coordinate of the position of the walker, and then the above probabilities can be referred as $P_{ij}^{\rightarrow}$, $P_{ij}^{\leftarrow}$, $P_{ij}^{\uparrow}$ and $P_{ij}^{\downarrow}$.

13.3.2 *Probability of mutant fixation*

We can calculate the probability of absorption in $(0, N)$ starting from a state (i, j), which we call u_{ij}. We have the following system of equations,

$$\begin{aligned} u_{ij} &= u_{i-1,j}P^{\leftarrow} + u_{i+1,j}P^{\rightarrow} + u_{i,j-1}P^{\downarrow} + u_{i,j+1}P^{\uparrow} \qquad (13.2)\\ &\quad + u_{ij}[1 - (P^{\leftarrow} + P^{\rightarrow} + P^{\downarrow} + P^{\uparrow})], \quad 1 < i < j < N, \\ u_{1j} &= u_{1,j+1}P^{\uparrow} + u_{1,j-1}P^{\downarrow} + u_{1j}[1 - (P^{\uparrow} + P^{\downarrow})], \quad 1 < j < N \\ u_{iN} &= u_{i-1,N}P^{\leftarrow} + u_{i+1,N}P^{\rightarrow} + u_{iN}[1 - (P^{\leftarrow} + P^{\rightarrow})], \quad 1 < i < N, \\ u_{jj} &= u_{j-1,j}P^{\leftarrow} + u_{j,j+1}P^{\uparrow} + u_{jj}[1 - (P^{\leftarrow} + P^{\uparrow} + P^{\tilde{E}})], \quad 1 < j < N, \\ u_{11} &= u_{12}P^{\uparrow} + u_{11}[1 - (P^{\uparrow} + P^{\tilde{E}})], \qquad (13.3)\\ u_{NN} &= u_{N-1,N}P^{\leftarrow} + u_{NN}[1 - (P^{\leftarrow} + P^{\tilde{E}})], \qquad (13.4)\\ u_{1N} &= 1. \qquad (13.5) \end{aligned}$$

The quantities u_{ii} are probabilities of fixation starting from one mutant at position i.

The results for this model must be compared with the probabilities of fixation in the mass-action model, equations (9.8) and (9.9). Numerical solutions for the probabilities of absorption show that quantities u_{ii} are symmetric one-hump functions which are flat except for narrow boundary regions near $i = 1$ and $i = N$, see figure 13.2.

In order to find expressions for the "inner" values of fixation probabilities, we note the following. If point (i, j) is sufficiently far away from the boundary, then the boundary effects are not felt and u_{ij} only depends on $|j - i|$ rather than on the initial position of the mutant interval. In order to solve the problem away from the boundary, we can use the periodic boundary conditions, which is equivalent to replacing equations (13.3) and (13.4) with the following:

$$u_{11} = u_{12}P^{\uparrow} + u_{11}[1 - (P^{\uparrow} + P^{\tilde{E}}/2)], \tag{13.6}$$

$$u_{NN} = u_{N-1,N}P^{\leftarrow} + u_{NN}[1 - (P^{\leftarrow} + P^{\tilde{E}}/2)]. \tag{13.7}$$

Now, quantities u_{ij} do not depend on the position of the mutant interval, but only on its length. Let us denote by π_j the probability that the mutant will reach fixation starting from a mutant interval of length $j+1$. We have a self-consistent system of equations for the probabilities π_i,

$$\pi_i(P^{\uparrow} + P^{\downarrow}) = P^{\uparrow}\pi_{i+1} + P^{\downarrow}\pi_{i-1}, \quad 0 < i < N-1, \tag{13.8}$$

$$\pi_0(P^{\uparrow} + P^{\tilde{E}}/2) = P^{\uparrow}\pi_1, \tag{13.9}$$

$$\pi_{N-1} = 1. \tag{13.10}$$

This system can be solved by setting $\pi_i \propto \alpha^i$, and finding $\alpha = 1$ and $\alpha = P^{\downarrow}/P^{\uparrow} = 1/r$. Therefore, we have

$$\pi_i = A + B/r^i,$$

and the constants A and B are found from the boundary conditions,

$$A = \left(1 - \frac{r+1}{r^{N-1}(3r-1)}\right)^{-1}, \quad B = \frac{r^{N-1}(r+1)}{1 + r + r^{N-1} - 3r^N}.$$

The probability to reach fixation starting from only one mutant cell is given by $\rho_{space} \equiv \pi_0 = A + B$. We have,

$$\rho_{space} = \begin{cases} \frac{2r^{N-1}(1-r)}{1+r+r^{N-1}-3r^N}, & r \neq 1, \\ \frac{1}{N}, & r = 1. \end{cases} \tag{13.11}$$

For large values of N, we can approximate the fixation probabilities as follows:

$$r < 1: \quad \rho_{space} = \frac{2(1-r)}{r+1}r^N, \tag{13.12}$$

$$r > 1: \quad \rho_{space} = \frac{2(r-1)}{3r-1}. \tag{13.13}$$

These expressions are further discussed in Section 13.5.1, where we compare them with the probability of mutant fixation in the space-free model.

13.4 Three-species dynamics

Next, let us formulate the dynamics for a three-species model, diagram (10.1), in a one-dimensional space. Again, we will use a homogeneous state approximation and describe the behavior of the system by means of equations (10.34-10.36) or (10.41-10.42). The applicability conditions for this approximation are now somewhat more restrictive and they are derived in section 13.4.2. The rate constants $R_{A\to B}$ and $R_{B\to C}$ can be calculated in the same way as for the non-spatial model. We have (cf. formula (10.33)),

$$R_{A\to B} = Nu\rho_{space} = Nu\frac{2r^N(1-r)}{1+r+r^N-3r^{N+1}}, \tag{13.14}$$

where ρ_{space} is the probability of successful fixation of a mutant starting from one cell of type "B". Approximations for neutral and disadvantageous mutants are given by formulas (13.45) and (13.44). Similarly, we calculate the second rate in the two-step process, which is the same as in the mass-action model, equation (10.33),

$$R_{B\to C} = Nu_2. \tag{13.15}$$

Finally, we need to find the tunneling rate, $R_{A\to C}$.

13.4.1 *Calculating the tunneling rate by the doubly-stochastic approximation*

In order to find the rate of tunneling, we again use the doubly-stochastic approximation, see (10.12, 10.15):

$$P_2(t)=1-exp\left(-\int_0^t Nu_1(1-x(t'))\,dt'\right)=1-exp\left(-Nu_1\int_0^t P_1(t')\,dt'\right). \tag{13.16}$$

In this approach, the key is to calculate the probability $P_1(t) = 1 - x(t)$ of creating a double mutant starting with one cell of type "B" (and no further mutations). Each (independent) lineage of type "B" spreads as a one-dimensional connected spot. The size of the spot is given by the random variable i. The state $i = 0$ is the extinction of the mutant. The state $i = E$ corresponds to the creation of a mutant of type "C". The probability $P_1(t)$ is the probability to acquire a second mutation among the lineage of a single cell of type "B".

Let us introduce a short-hand notation,

$$\tilde{r} = r/(r+1).$$

Then for the dynamics within a lineage, the transition probabilities are given by: $P_{i\to i+1} = \tilde{r}(1-u_2)$ for $0 \le i \le N-1$, $P_{i\to i-1} = 1/(r+1)$ for $1 \le i \le N$, $P_{1\to 0} = 1/2$, $P_{N\to N} = N(1-u_2)$, $P_{i\to E} = 4\tilde{r}u_2 + (i-2)u_2$ for $3 \le i \le N$, $P_{2\to E} = 4\tilde{r}u_2$, and $P_{1\to E} = 2\tilde{r}u_2$, where time is measured in terms of generations.

In what follows we will simplify the problem so that the transition probabilities are:

$$P_{i\to i+1} = \lambda, \quad P_{i\to i-1} = \mu, \quad P_{i\to E} = \beta i. \tag{13.17}$$

These probabilities represent a "model" of the real situation rather than an approximation. Indeed, in equations (13.17), with

$$\lambda = \tilde{r}(1-u_2), \quad \mu = 1/(r+1), \quad \beta = 3\tilde{r}u_2,$$

we neglect several subtleties that we discovered for spatial propagation of mutants. For instance, we ignore the fact that $P_{1\to 0} \ne P_{i\to i-1}$ for $i > 1$. We also ignore the fact that the probability for exiting into state E from state i is not exactly proportional to i: for $i < 3$ it does not depend on i, and for larger i it has a constant (in i) term.

Next we find the quantity $P_1(t)$, equation (10.15). Let us denote by $\xi_i(t)\Delta t$ the probability to be absorbed in E during the interval $(t, t+\Delta t)$ starting from state i at $t = 0$. We have

$$\xi_1 = \frac{dP_1(t)}{dt}. \tag{13.18}$$

On the other hand, we have the following equations for ξ_i:

$$\dot{\xi}_i = \lambda\xi_{i+1} + \mu\xi_{i-1} + \beta i\xi_E - (\lambda+\mu+\beta i)\xi_i, \quad 2 \le i,$$

with the boundary condition

$$\dot{\xi}_1 = \lambda\xi_2 + \mu\xi_0 + \beta\xi_E - (\lambda+\mu+\beta)\xi_1,$$

with $\xi_0 = \xi_E = 0$, and the initial condition

$$\xi_i(0) = i\beta. \tag{13.19}$$

Let us take the Laplace transform of the ODE, where

$$\mathcal{L}\xi_i(t) = f_i(s).$$

Using $\mathcal{L}\dot{\xi}_i = sf_i - \xi_i(0)$, we obtain the following system of equations,

$$[\lambda f_{i+1} + \mu f_{i-1} - (\lambda + \mu + s)f_i] - f_i\beta i = -i\beta, \quad i > 1, \quad (13.20)$$

$$[\lambda f_2 - (\lambda + \mu + s)f_1] - f_1\beta = -\beta. \quad (13.21)$$

While a direct solution of this system is difficult, we note that equation (13.20) is similar to an inhomogeneous discrete Airy equation. Let us denote

$$h_i = f_i - 1,$$

and set

$$\epsilon = \lambda - \mu.$$

The quantity ϵ measures the difference between the "birth rate" and the "death rate" of the one-hit mutants. We have for the function h_i,

$$\lambda(h_{i+1} - 2h_i + h_{i-1}) + \epsilon(h_i - h_{i-1}) - sh_i = \beta i h_i, \quad (13.22)$$

$$\lambda(h_2 - h_1) - h_1(\lambda - \epsilon + \beta + s) = \lambda - \epsilon. \quad (13.23)$$

Using the continuous limit, we obtain the system

$$\lambda h'' + \epsilon h' - sh = \beta x h, \quad (13.24)$$

$$\lambda h'(0) - h(0)(\lambda - \epsilon + \beta + s) = \lambda - \epsilon; \quad (13.25)$$

for the second boundary condition we use the boundedness of the solution for large x. This system can be solved exactly in terms of the Airy function Ai and its derivative. We have,

$$h(x) = \frac{2e^{-\frac{\epsilon x}{2\lambda}}(\epsilon - \lambda)Ai[K_s(x)]}{(2\beta - \epsilon + 2(\lambda + s))Ai[K_s(0)] - 2\beta(\lambda/\beta)^{2/3}Ai'[K_s(0)]}, \quad (13.26)$$

where

$$K_s(x) = \frac{\epsilon^2 + 4\lambda(s + \beta x)}{2(\beta/\lambda)^{2/3}\lambda^2}.$$

In order to obtain $P_1(t)$, the mutation probability in each lineage, we would need to find $f_1(s)$ (we start from one cell of type "B"), and evaluate $\mathcal{L}^{-1}[f_1(s)/s]$. Indeed, the cumulative probability $P_1(t)$ is related to the function $\xi_1(t)$ by equation (13.18). Therefore their Laplace transforms differ by a factor s:

$$\mathcal{L}(\xi_1) = \mathcal{L}\left[\frac{dP_1}{dt}\right] = s\mathcal{L}[P_1] - P_1(0), \quad \text{where } P_1(0) = 0.$$

Therefore, we have

$$\mathcal{L}[P_1] = \mathcal{L}(\xi_1)/s = f_1/s, \tag{13.27}$$

and

$$P_1(t) = \mathcal{L}^{-1}\left[\frac{f_1}{s}\right].$$

Next, we note that much like in the calculations of Section 10.5.3, the probability $P_1(t)$ grows from zero and as $t \to \infty$ saturates at a nonzero level, see equation (10.19). It is shown below that it is the limiting value $R = \lim_{t\to\infty} P_1(t)$ that is of interest, because it gives the main contribution to probability (13.16) in some important cases. This value can be found without having to perform the inverse Laplace transform. By the final value theorem of Laplace transform, we have

$$\lim_{t\to\infty} P_1(t) = \lim_{s\to 0} s\mathcal{L}[P_1(s)] = \lim_{s\to 0} f_1(s) = f_1(0),$$

because of equation (13.27). Finally, we note that

$$f_1(0) = 1 + h_1(0) = 1 + h(0)|_{s=0} + O(1/N),$$

where $h(x)$ is given by equation (13.26). Therefore, we have

$$\lim_{t\to\infty} P_1(t) \equiv R \approx 1 + h(0)|_{s=0}. \tag{13.28}$$

In the next section we investigate when this limiting value gives the main contribution to the probability $P_2(t)$ of double-mutant generation, and derive three important limiting cases.

13.4.2 *Limiting cases and the tunneling rate approximations*

Let us find the solution $h(0)$, equation (13.26), in the limit of small values of s. We have

$$h(0) = \frac{2(\epsilon-\lambda)Ai[K_s(0)]}{(2\beta-\epsilon+2(\lambda+s))Ai[K_s(0)] - 2\beta(\lambda/\beta)^{2/3}Ai'[K_s(0)]}, \tag{13.29}$$

$$K_s(0) = \frac{\epsilon^2+4\lambda s}{2(\beta/\lambda)^{2/3}\lambda^2}. \tag{13.30}$$

The contribution corresponding to $s = 0$ will yield the saturation constant, R (equation (13.28)), and the correction corresponding to the first order in s will inform us on the time-scale on which the function $P_1(t)$ reaches saturation. We will expand expression (13.29) in the Taylor series in terms of small s, and consider three limiting cases, compare with results in Section 10.5.4:

(i) Disadvantageous intermediate mutants. Let us suppose that $\epsilon \gg (\beta\lambda^2)^{1/3}$, and $\epsilon < 0$, which is equivalent to conditions

$$r < 1, \quad |1-r| \gg 2(3u_2)^{1/3}.$$

Then as $\beta \to 0$, the argument of the Airy function, $K_s(0)$ (equation (13.30)), tends to infinity, and we can use the large-argument expansion of the Airy function:

$$Ai(z) \sim \frac{e^{-\frac{2}{3}z^{-3/2}}}{2\sqrt{\pi}z^{1/4}}.$$

We obtain in the limit of small β,

$$1 + h(0) = \frac{(\lambda-\mu)^2+\lambda^2}{(\lambda-\mu)^2\mu}\beta - \frac{s}{\lambda-\mu} + O(s^2).$$

To obtain the time-scale of saturation we ask: how small should s be in the above expression in order for the second term to be negligible compared to the first term? If $s \ll \beta(\lambda^2+(\lambda-\mu)^2)/(\mu(\lambda-\mu))$, then the value of $h(0)$ can be approximated by its $s=0$ value, and therefore for $t \gg t_*$ with

$$t_* \sim \frac{(\lambda-\mu)\mu}{(\lambda^2+(\lambda-\mu)^2)\beta} = \frac{1-r^2}{3ru_2(r^2+(r-1)^2)},$$

we have

$$\begin{aligned} P_1(t) \approx R &= \frac{(\lambda-\mu)^2+\lambda^2}{(\lambda-\mu)^2\mu}\beta \\ &= 3ru_2\frac{(1-r)^2+r^2}{(1-r)^2}. \end{aligned} \qquad (13.31)$$

(ii) Neutral intermediate mutants. If $\epsilon \ll (\beta\lambda^2)^{1/3}$, which is equivalent to

$$|1-r| \ll 2(3u_2)^{1/3},$$

then we have

$$1 + h(0) = \left(\frac{3\beta}{\lambda}\right)^{1/3}\frac{\Gamma(2/3)}{\Gamma(1/3)} + \left(\frac{9\beta}{\lambda^2}\right)^{1/3}\left(\frac{\Gamma(2/3)}{\Gamma(1/3)}\right)^2 s + O(s^2).$$

For times $t \gg t_*$ with

$$t_* \sim \left(\frac{3}{\beta^2\lambda}\right)^{1/3}\frac{\Gamma(2/3)}{\Gamma(1/3)} = \frac{\Gamma(2/3)}{\Gamma(1/3)}\frac{2}{(3u_2^2)^{1/3}},$$

we have

$$\begin{aligned} P_1(t) \approx R &= \left(\frac{3\beta}{\lambda}\right)^{1/3}\frac{\Gamma(2/3)}{\Gamma(1/3)} \\ &= (9u_2)^{1/3}\frac{\Gamma(2/3)}{\Gamma(1/3)}. \end{aligned} \qquad (13.32)$$

(iii) Advantageous intermediate mutants. If $\epsilon \gg (\beta\lambda^2)^{1/3}$ and $\epsilon > 0$, which is equivalent to conditions

$$r > 1, \quad |r-1|gg2(3u_2)^{1/3},$$

then we can use the same type of a large-argument expansion as in (i), and obtain

$$t_* \sim \frac{(2\lambda-\mu)\mu}{\lambda(\lambda-\mu)^2} = \frac{(2r-1)(r+1)}{(r-1)^2 r},$$

such that with $t \gg t_*$,

$$P_1(t) \approx R = \frac{\lambda-\mu}{\lambda} = \frac{r-1}{r}. \tag{13.33}$$

Tunneling rate approximations derived here are summarized in Table 13.1.

13.4.3 *When is tunneling important?*

For the tunneling rate approximation to be relevant, we need to show that (a) the system is not too large such that individual mutant lineages do not accumulate too fast, and (b) the system is not too small, such that tunneling takes place faster than fixation of the intermediate mutant.

In order to satisfy the first of these conditions (case (a) above), let us estimate expression (13.16). We will use roughly the same approach as in Section 10.5.4, but omit some details to get the main picture. The function $P_1(t)$ approaches its saturation level around the time t_* (the expressions for these values in different limits are given in Section 13.4.2). Again, we denote by t_1 the time-scale when the expression in the exponent of (13.16) becomes of the order one,

$$u_1 N \int_0^{t_1} P_1(t')\,dt' \sim 1.$$

If condition $t_* \ll t_1$ holds, then for times $t > t_*$, the creation of double-hit mutants is well approximated by the constant tunneling rate,

$$R_{A\to C} = N u_1 R, \tag{13.34}$$

$$R = \left[1 - \frac{1}{2\lambda}\left(\lambda+\mu+\beta-\sqrt{(\lambda+\mu+\beta)^2-4\lambda\mu}\right)\right]. \tag{13.35}$$

If the opposite is true, that is, if $t_* \gg t_1$, then although for $t \gg t_*$ the tunneling formula above works, the probability $P_2(t)$ is exponentially close to 1, and therefore the tunneling approximation does not provide a useful tool.

Condition $t_* \ll t_1$ is equivalent to

$$Nu_1 \ll \frac{1}{Rt_*}, \tag{13.36}$$

which can be rewritten as follows in the three limiting cases:

$$\text{Disadvantageous mutants}: \quad Nu_1 \ll \frac{1-r}{r+1}, \tag{13.37}$$

$$\text{Neutral mutants}: \quad Nu_1 \ll \frac{1}{2}\left(\frac{u_2}{3}\right)^{1/3}\left(\frac{\Gamma(1/3)}{\Gamma(2/3)}\right)^2, \tag{13.38}$$

$$\text{Advantageous mutants}: \quad Nu_1 \ll \frac{(r-1)r^2}{(r+1)(2r-1)}. \tag{13.39}$$

The second condition (case (b) above) requires that typically, mutants of type "C" are generated before fixation of type "B" occurs. This is equivalent to condition

$$R_{A\to B} \ll R_{A\to C},$$

or

$$\rho_{space} \ll R, \tag{13.40}$$

where the fixation rates are calculated in (13.11, 13.12, 13.13), and the tunneling rates are given by (13.31, 13.32, 13.33). Condition (13.40) leads to condition $N \gg N_{tun}$. For disadvantageous mutants of type "B", N_{tun} is given by

$$N_{tun} = \log\left[\frac{3}{2}\frac{u_2 r(r+1)[(r-1)^2+r^2]}{(r-1)^3}\right] / \log r. \tag{13.41}$$

In figure 13.3 we present the comparison of N_{tun} for the space-free (see formula (10.31)) and spatial models, for a fixed value of u_2, and different values of r. In the case of neutral intermediate mutants, we have

$$N_{tun} = \frac{\Gamma(1/3)}{\Gamma(2/3)}\frac{1}{(9u_2)^{1/3}}, \tag{13.42}$$

to be compared with formula (10.28) for the mass-action model. We can see that N_{tun} is smaller for the spatial model compared to that for the mass-action model.

Finally, in the case of advantageous mutants, the rate of intermediate mutant fixation does not become smaller than the tunneling rate. This situation is similar to the regime of advantageous intermediate mutants described in Chapter 10.

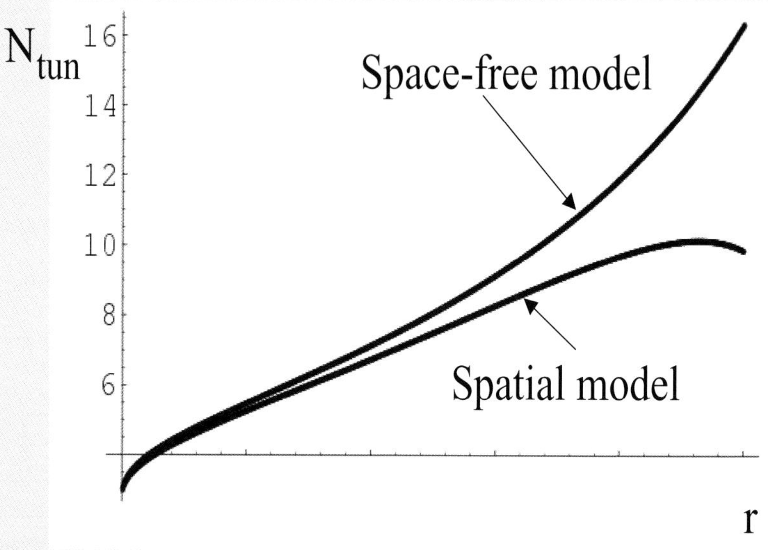

Fig. 13.3 The quantity N_{tun} for disadvantageous mutants, for the spatial and space-free models, calculated from formulas (13.41) and (10.31) respectively. For this graph, $u_1 = 10^{-3}$; the range of r is from zero to 0.75, so that the mutant is disadvantageous in the spatial model.

13.5 Dynamics of mutant generation

13.5.1 *Gain-of-function mutations: a two-species problem*

Gain-of-functions mutations were described in Chapter 9, and a mathematical formalism describing their dynamics was developed for mass-action (space-free) models. Here we generalize these findings to the 1D spatial spatial process. We consider the dynamics of fixation of mutants of type "B", diagram (9.1). Gain-of-function mutations are essential when talking about oncogene activation, in which case we assume that the fitness of type-"B" mutants is larger than that of the wild type cells, $r > 1$. It is however important to also study the cases with $r = 1$ and $r < 1$ (neutral and negatively selected mutants). Such mutations can correspond to an inactivation of one copy of a tumor suppressor gene, or to the activation of CIN genes which can result in a fitness reduction, see Chapters 7 and 11.

In the spatial problem with natural boundary conditions, the probability of eventual fixation of a given mutant clone depends on the position of the originating cell. If the original mutant is produced near the domain's boundary, its offspring has a relatively smaller chance to expand and invade the population compared with clones originating away from the boundary. For mutants originating in the bulk of the domain, the probability of fixation approaches the value given by equation (13.11).

We can compare this quantity with the fixation probability, ρ, in the mass-action model (equation (9.11)),

$$\rho_{space} = \rho \frac{2r(1 - r^{N-1})}{1 + r + r^{N-1} - 3r^N}. \qquad (13.43)$$

In particular, for neutral mutants such that $|r - 1| \ll 1/N$, we have

$$\rho_{space} = \rho = 1/N.$$

For large values of N, we obtain, in the case of disadvantageous mutants ($r < 1$, $|r - 1| \gg 1/N$),

$$\rho_{space} = \frac{2r}{1 + r}\rho < \rho, \qquad (13.44)$$

and in the case of advantageous mutants ($r > 1$, $|1 - r| \gg 1/N$) we have

$$\rho_{space} = \frac{2r}{3r - 1}\rho < \rho. \qquad (13.45)$$

To summarize, the probability that the offspring of one cell of type "B" will reach fixation is given by $1/N$, as long as the fitness of type "B" is equal to that of type "A". If the fitness of type "B" is smaller or larger than that of type "A", then the probability of fixation is always bigger in the mass-action model compared to the spatial model. This is illustrated in figure 13.4.

13.5.2 *Loss-of-function mutations: a three-species problem*

Loss-of-function mutations are characteristic of tumor suppressor gene inactivation, see Chapter 10, and are described by diagram (10.1). The general pattern of double-hit mutant generation in space is similar to what we obtained in the space-free model, Section 10.6.5. Depending on the population size, the dynamics will follow different scenarios. For small populations, we

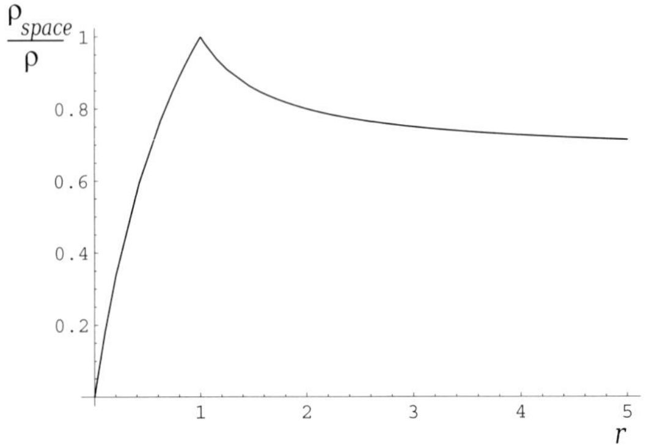

Fig. 13.4 The quantity ρ_{space}/ρ, formula (13.43), as a function of r for $N = 1000$. This quantity tells us how much less likely a mutant fixation is in the spatial model compared to the mass-action model.

have a genuine two-step process (figure 10.1(a)), where the acquisition of a double-hit mutant requires two distinct rate-limiting steps. The two rates are given by

$$R_{A\to B} = Nu\rho_{space}, \quad R_{B\to C} = Nu_2.$$

For intermediate populations, we enter the regime of stochastic tunneling (figure 10.1(b)), where intermediate mutants do not get a chance to become fixated, because a double-hit mutant is produced. In the tunneling process, the single rate is given by

$$R_{A\to C} = Nu_1R,$$

where the approximations for the value R are given in Table 13.1.

Table 13.1 Tunneling rate approximations in the spatial model.

Interm. mutants	Conditions	Tunneling rate, R
Disadvantageous	$r < 1$, $\|1-r\| \gg 2(3u_2)^{1/3}$	$3ru_2\frac{(1-r)^2+r^2}{(1-r)^2}$
Neutral	$\|1-r\| \ll 2(3u_2)^{1/3}$	$(9u_2)^{1/3}\frac{\Gamma(2/3)}{\Gamma(1/3)}$
Advantageous	$r > 1$, $\|1-r\| \gg 2(3u_2)^{1/3}$	$\frac{r-1}{r}$

Finally, for the largest populations we have a nearly-deterministic regime, figure 10.1(c).

13.5.3 *Definition of neutrality*

An interesting issue is the definition on neutrality in different models. It is the same for spatial and mass-action descriptions in the regime where a two-step process dominates (or if we have only two types, "A" and "B" in the system). In this case mutants with fitness satisfying $|1-r| \ll 1/N$ can be considered neutral. Indeed, in the expansion of the fixation probability of a mutant in terms of r around the value $r = 1$, the highest order term is given by $1/N$, and the next term is $(N-1)(r-1)/2N$ without spatial effects, and $(N-1)^2(r-1)/2N^2$ in the spatial model. The smallness of the second term compared to the first term is a criterion of neutrality.

The meaningful definition of neutrality changes in the system where double-mutants are produced. There, it is not the time-scale of fixation, but rather the rate of second mutation acquisition which is the dominant factor. Now, the definition is different in the spatial model compared to that in the mass-action model. In the latter case, neutral mutants were defined by the condition

$$|1-r| \ll \sqrt{u_2}. \tag{13.46}$$

In the spatial case, we have

$$|1-r| \ll 2(3u_2)^{1/3}. \tag{13.47}$$

That is, a larger region of fitnesses around $r = 1$ qualifies as neutral.

13.5.4 *Three-species dynamic: a comparison with the space-free model*

Let us compare the behavior of the mass-action and the spatial models.

Consider the tunneling rates, Tables 13.1 and 10.1. In the case of disadvantageous mutants, the tunneling rate for the spatial model is always larger than that for the mass-action model. It has the same order of magnitude in terms of small u_2. Regarding the case of neutral mutants, it is interesting that the rate of tunneling in the spatial model has a *larger order of magnitude* than that in the mass-action model. In both cases, tunneling happens faster in the spatial model compared to the mass-action model. In the case of advantageous intermediate mutants, the tunneling rate is the same in the spatial and the mass action model.

Finally, we turn to the two-step process. It is characterized by two coefficients, $R_{A\to B}$ and $R_{B\to C}$. While the value of $R_{B\to C}$ is the same for both spatial and non-spatial models, the rate $R_{A\to B}$ is larger in the mass-action model (unless $r = 1$).

As we can see, the two-step dynamics tend to happen faster in the mass-action model, and tunneling proceeds faster in the spatial model. Depending on the parameters, the two-step or the tunneling regime may be more important. If the dynamics occur predominantly by a genuine two-step process (which tends to be the case for smaller population sizes), then the mass-action model produces two-hit mutants faster. If on the other hand the tunneling process contributes the most, as is the case for larger populations, then the spatial model will produce two-hit mutants faster.

In figure 13.5 we present a numerical simulation where some of these effects are demonstrated. We ran stochastic spatial and non-spatial models in the regime where tunneling played an important role (compared to a two-step process, see the conditions below). The mutant "B" was taken to be neutral ($r = 1$). The initial condition was the *all "A"* state, and the simulations were stopped as soon as the first mutant of type "C" was created. For each model, we performed 10,000 runs and found out numerically the distribution of generation times of double-hit mutants. One can see that

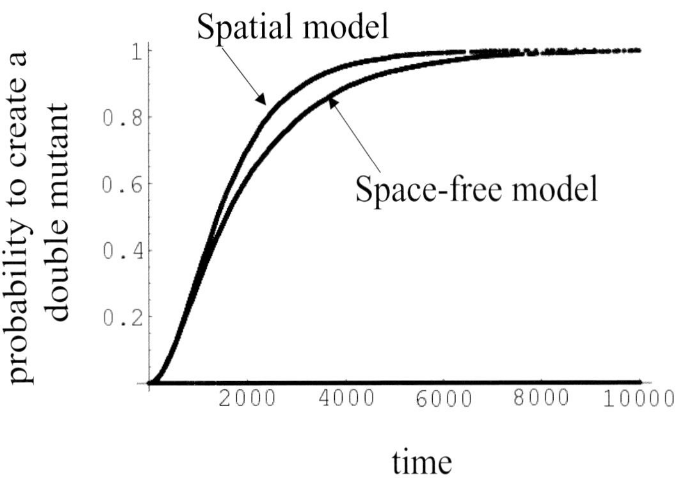

Fig. 13.5 Cumulative probability distribution function for the generation of a mutant of type "C", in spatial and non-spatial models (numerical results). Here, $r = 1$, $N = 100$, $u = 0.005$ and $u_1 = 0.02$. The simulation was performed for a discrete-time Markov process.

for the spatial model, the generation of a double-hit mutant occurs earlier, that is, at a higher rate, than that for the space-free model.

The difference in the predictions between the two models can be quite substantial. For instance, if we take $u_1 = u_2 = 10^{-5}$ and $N = 10^4$, after 100 generations the probability to have started a colony of type "C" is given by $p_{m-a} = 3 \cdot 10^{-2}$ in the mass-action model and $p_{sp} = 2 \cdot 10^{-1}$ in the spatial model. For parameter values $u_1 = u_2 = 10^{-6}$, and the same colony size, we have after 1000 generations: $p_{m-a} = 10^{-2}$ and $p_{sp} = 10^{-1}$. We can see that the two models result in an order-of-magnitude difference in their predictions. One could envisage the following experiment where these calculations can be verified. Suppose we have a cell culture where space is a limiting factor such that the cells are in competition with each other. Various degrees of cellular mixing can be attained by putting cells in different environments. Suppose all cells are of type "A" at the beginning, and type "C" cells are marked for an easy detection (e.g. with GFP, the green fluorescent protein). After a fixed number of generations, we can determine the number of cultures which developed clones of type "C" cells. This should be compared with the quantities p_{m-a} and p_{sp} to see which model gives a better fit.

We have shown that the tunneling rate is higher for the spatial model than it is for the mass-action model. This finding is of general interest and will be discussed further in Chapter 14 in the context of more advanced spatial models. The reason for this accelerated rate of double-mutant production compared to the mass-action model is the very slow dynamics of single mutant cells. Let us look at the rate at which the number of mutants of type "B" can decrease/increase in the two models. We need to compare the spatial rates, equations (13.17), with the mass-action rates given by equations (10.4, 10.5). We notice that for the mass-action model, these rates increase with the number of mutants. In the spatial model, they are constant. Indeed, in the spatial model, cells of type "B", once produced, form mutant "islands", i.e., joint sets of points of type "B". The number of mutants can only change if a death occurs at one of the two boundaries of such islands. In the mass-action model, a change in the mutant number can occur upon death of any cell, followed by a reproduction of any cell of the other type. Therefore, in the spatial model, once an island of type "B" mutants is created, it tends to linger for a long time, serving as a platform for creating a double mutant.

13.6 Outlook

A lesson that we learn from the simple model analyzed in this chapter is that spatial considerations are important. Therefore, one must be cautious and instead of using the simplest, mass-action model, examine carefully whether or not the spatial structure should be included. For instance, when modeling the initiation of cancers of the blood, spatial considerations do not enter in the same way as in solid tissues. In solid tumors, spatial consideration must be taken into account, thus making the spatial model the tool of choice.

An important area of applications of multistage models of the kind presented here, is the comparison with epidemiological data. Starting with early studies of Knudson [Knudson (1971)] and Moolgavkar and colleagues [Moolgavkar (1978); Moolgavkar *et al.* (1988)], until more recent works [Luebeck and Moolgavkar (2002); Little and Wright (2003); Little and Li (2006); Meza *et al.* (2008)], two-hit and multi-hit models have been compared with age-incidence data on various cancers. Usually, the goal of such research is to derive unknown system parameters from fitting the model results with the age-incidence curves. There, the difference between the spatial and non-spatial models can become very important.

In Chapter 14 we will extend the discussion of this chapter to two-dimensional settings, and to a larger number of intermediate steps. The complex role of space in the mutant dynamics will be explained in more detail. We will also depart from the rigidity of the Moran process and consider an alternative process (contact process) to explore the robustness of the results.

Problems

Problem 13.1. *Derive inequalities (13.44) and (13.45) from expression (13.43).*

Problem 13.2. *In this chapter, cells can reproduce by placing their offspring in the adjacent location. More generally, one could assume that cells can reproduce by placing their offspring within a certain radius. The approach described here cannot be easily generalized to such models. What essential assumption breaks down if we allow reproduction into locations not immediately adjacent to the dividing cell?*

Problem 13.3. Research project. *Elegant "Isothermal theorem" for graphs proven in [Lieberman* et al. *(2005)] states that, if each node of a graph has the same number of neighboring nodes, i.e. a symmetric bidirectional graph, then the invasion probability is equivalent to that of the fixation probability of the traditional Moran process in the absence of migration. Formula (13.43) is an apparent contradiction with this statement. Can you resolve this paradox? Hint: pay attention to the order of the events in an elementary update.*

Chapter 14

Complex tumor dynamics in space

14.1 Introduction

In Chapters 9, 10, 12, and 13 we concerned ourselves with the most common elementary patterns of cancer initiation and progression: gain-of-function mutations and loss-of-function mutations. We developed a very detailed theory for probabilities and timing of invasion of advantageous mutants, in one- and two-step processes. We examined the associated dynamics and showed that depending on the system size and mutation rates, different patterns of mutant emergence can be observed. A two-step process is characterized by a consecutive fixation of the intermediate, one-hit mutant followed by the rise of the two-hit, advantageous mutant. Stochastic tunneling is the effect whereby the two-hit mutant is created before fixation of the intermediate mutant occurs. We looked at the homogeneous populations (Chapters 9 and 10), hierarchical populations (Chapter 12), and spatial populations (Chapter 13).

Chapter 13 provides the simplest, 1D generalization of the Moran process, see also [Komarova (2006c)]. Similar results for 2D and 3D models were obtained in (Durrett *et al.*, 2012; Durrett and Moseley, 2012), and the effect of migration on these evolutionary processes was studied in [Thalhauser *et al.* (2010)]. All of this work considered Moran processes, which are birth-death processes in a constant population. In particular, these models assume that a death of an individual is immediately balanced by the birth of another, thus not allowing for variation in the population size or for empty space. In addition, these models were studied in the context of two sites that need to be mutated in order to attain the advantageous trait, and only the situation of the nearest-neighbor interactions was considered.

The present chapter explores the role of spatial interactions, comparing different population growth processes in 2D. In addition to the Moran process, we also consider a different type of model, called the contact process, where the population size is determined by the birth and death rates and where it is possible to have empty, unoccupied space. This relaxes the very rigid requirement of a microscopic balance between births and deaths in the Moran model. The Moran process can be considered as a limiting case of the contact process where the division rates are extremely high compared to death rates. We examine the dynamics for different numbers of sites that need to be mutated to attain the advantageous phenotype, and explore the whole range of spatial restriction, from the nearest-neighbor model to the mass-action scenario. We examine the cases of disadvantageous, neutral, and advantageous intermediate mutations.

In this chapter we look at cancer as an evolutionary process that happens in a spatial setting, and relate it to the well-known concept of *complex traits*. Some broad notions from evolutionary biology that can be useful are reviewed next.

14.2 Complex traits and fitness valleys

Complex traits depend on the interactions between several different genetic loci (Whitlock *et al.*, 1995). The simplest case of a complex trait that we have considered so far was the inactivation of tumor suppressor genes, where two separate mutations were required to achieve the advantageous phenotype. In general, cancerous phenotypes require the presence of many genetic (or epigenetic) alterations, each of which might in itself be disadvantageous. One of the central questions is to understand how the complex (advantageous) phenotype evolves by gradually acquiring the intermediate, possibly disadvantageous alterations.

In ecology, this process is termed "crossing the fitness valley". Graphically this is represented in figure 14.1(a). Starting from the wild-type cells, several intermediate mutations may induce changes that lead to fitness reduction. The corresponding cells will be selected against, making it difficult for the cell lineages to survive long enough to acquire further mutations that would eventually confer a fitness advantage. When the intermediate mutants are disadvantageous, the term "fitness valley" is used (Whitlock *et al.*, 1995; Weinreich and Chao, 2005; Weissman *et al.*, 2009). For neutral mutants, we have a fitness "plateau", figure 14.1(b), and for slightly advantageous intermediate phenotypes we can talk about the fitness "foothill", figure 14.1(c). Even though it is easier to envisage that evolution can proceed toward the foothill, if the advantage of intermediate phenotypes is small, the process may take a very long time.

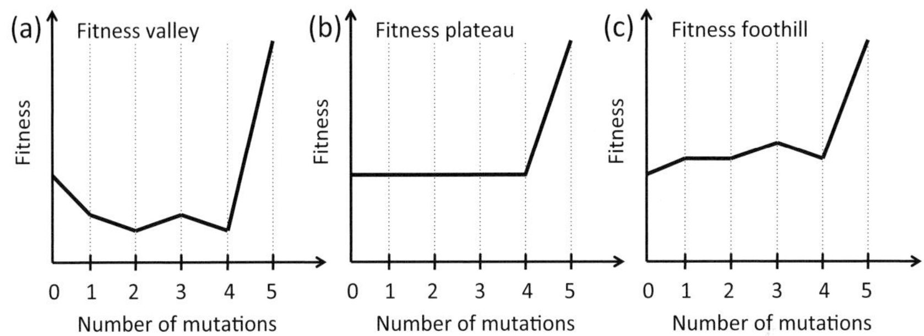

Fig. 14.1 A schematic representing the concepts of a (a) fitness valley, (b) fitness plateau, and (c) fitness foothill.

Scenarios where intermediate mutants are disadvantageous or neutral are abundant and have been discussed in the previous chapters in the context of inactivation of tumor-suppressor genes. Scenarios where intermediate mutants are slightly advantageous compared to the wild-type, and the complex trait is even more advantageous (evolutionary foothills) are also common in nature. An interesting example is the evolution of p53 loss, leading to accelerated and uncontrolled cell growth and the formation of tumors [de Vries *et al.* (2002)]. While complete loss of p53 function has been mainly attributed to an inactivation of both copies of the gene, a dominant negative effect has also been documented, where a single mutated copy of the gene can reduce overall p53 function to a certain degree through disruption of p53 tetramer formation, which is required for function [de Vries *et al.* (2002)].

14.3 The Moran process

We first consider the Moran process, as this allows a greater degree of mathematical insight. We will restrict our attention to a process in a 2D square grid of size N. In this process, which is a generalization of the model described in Chapter 13, individuals die at random, independently of their phenotype, and are immediately replaced by the progeny of one of the nearby individuals, selected randomly from the square neighborhood of a fixed size, M. The probability to be selected for reproduction is proportional to the fitness of each phenotype.

In order to model the process of mutation accumulation, we assume that the genome of each cell contains m sites, such that if all m sites are mutated, the cell acquires the advantageous (complex) phenotype, while any smaller subset of mutations confers fitness r to the cell. Mutations can happen in any order, and mutations happen with a given (and equal) probability for each out of m sites upon reproduction. As before, back-mutations are not taken into account. Figure 14.2(a) shows an example of a mutation diagram for $m = 3$ sites. The population size stays constant (equal to N) after each update, and there are no empty spots on the grid.

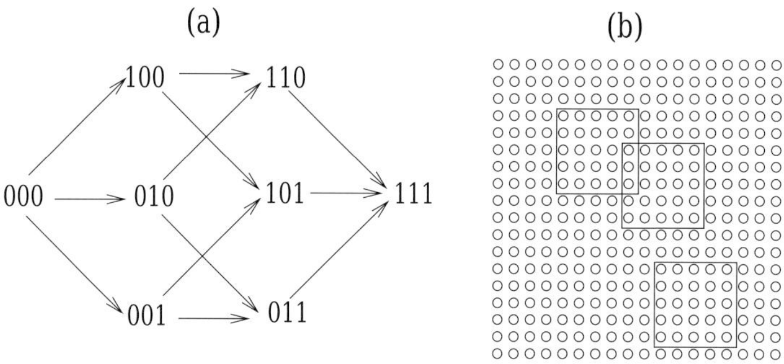

Fig. 14.2 The simulation setup. (a) The combinatorial mutation diagram for $m = 3$ sites. (b) The concept of neighborhood, illustrated with the neighborhoods of radius 2.

We will investigate the role of spatial interactions by varying the neighborhood size where the individuals can place their offspring. If for a given individual, the neighborhood only includes the 8 surrounding spots on the square grid, this corresponds to the nearest neighbor situation, or a "radius 1 neighborhood". If the neighborhood is as big as the whole population, then this is the mass-action situation. An example of a neighborhood of radius 2 is shown in figure 14.2(b). The question we ask is how the rate of m-hit mutant production depends on the neighborhood size.

14.3.1 *Spatial restriction accelerates evolution*

Numerical simulations of this process are presented in figure 14.3(a), assuming that two loci need to be mutated to acquire the advantageous phe-

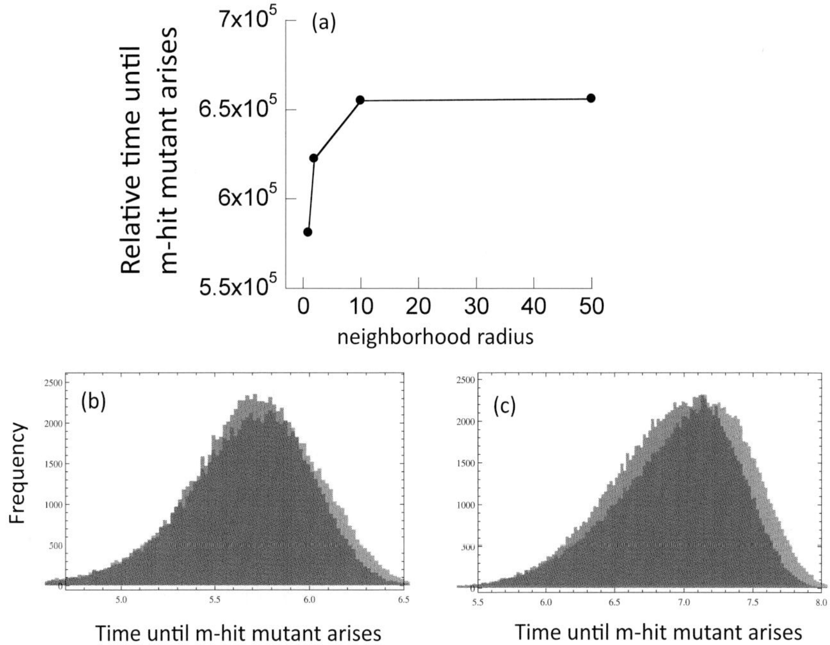

Fig. 14.3 Effect of the neighborhood radius on the time at which the m-hit mutant arises in the Moran process. (a) The average time of emergence for 4 neighborhood radii: 1, 2, 10, 50. Averages are based on at least 10^5 iterations of the simulation. The chosen parameters were: grid size $N = 50 \times 50$, $m = 2$, intermediate mutants were neutral, $u = 10^{-4}$. (b) Histograms showing the distribution of outcomes, corresponding to the data presented in part (a). For simplicity only two radii are compared: 1 and 50. (c) Same type of histogram, but with $u = 10^{-5}$, showing a greater difference.

notypes ($m = 2$). There, for each neighborhood size, we ran the process at least 10^5 times, stopping when the m-hit mutant was generated, and recording the waiting times. An example of mean waiting times as a function of the neighborhood radius is shown in figure 14.3(a). The exact distribution of waiting times for different neighborhood sizes are shown in the form of histograms in figures 14.3(b,c). We can see that even if the differences in the mean waiting times between the mass-action and the nearest neighbor model are smaller than the widths of the distributions, these distributions are clearly distinct, and the means are significantly different (with the p-value in both the T-test and Mann-Witney U-test less than 10^{-10}). In fact, for the examples presented in figures 14.3(b,c), the sample size of only 2500 and 1000 points respectively, consistently yields p-values smaller than 0.05 for the Mann-Witney U-test.

We observe (figure 14.3(a)) that the waiting times increase monotonically with the neighborhood size. In other words, tight spatial interactions lead to the fastest m-hit mutant production. The results hold for both neutral and disadvantageous intermediate mutants. An intuitive explanation for this phenomenon that was proposed in Chapter 13 evokes the concept of mutant islands. Under mass-action rules, all types are mixed randomly (figure 14.4(a)). On the other hand, if reproduction is only allowed within a small neighborhood, spatial structures tend to form, where the nearby individuals are likely to have identical genotypes (figure 14.4(b)). For example, in the case of $m = 2$, where one-hit mutants are neutral or disadvantageous, once a one-hit mutant is generated, its clone (if it has a chance to form) will be located in the vicinity of the original de-novo mutation. It can be argued that such localized clones are on average longer-lived than spatially dispersed clones. This is because a dead mutant is more likely to be replaced by the progeny of another mutant than a wild-type individual, if most of its neighbors are mutants. In turn, longer-lived clones of intermediate mutants are more likely to produce m-hit mutants, which speeds up the process of m-hit mutant generation.

How tight the mutant islands are depends on the neighborhood size. For very small neighborhoods, the dynamics are the most localized, and the islands are more pronounced than in systems with larger neighborhood sizes.

14.3.2 *Dependence on parameters*

The extent to which spatial restriction accelerates the emergence of the m-hit mutant is intimately related to the exact scenario of m-hit mutant generation. In Chapters 10 and 13 we described three regimes, see figure 10.1 in the case when $m = 2$. A genuinely two-step process, stochastic tunneling, or a nearly deterministic process are observed depending on the population size, N, relative to other parameters of the system.

In the present chapter, instead of varying the system size, N, we varied the mutation rate, u. Figure 14.5 shows the relative mean waiting time for the mass-action system compared to that for the nearest-neighbor system, for different mutation rates. For each mutation rate, we calculated the mean waiting time for the m-hit mutants in the mass-action model, and

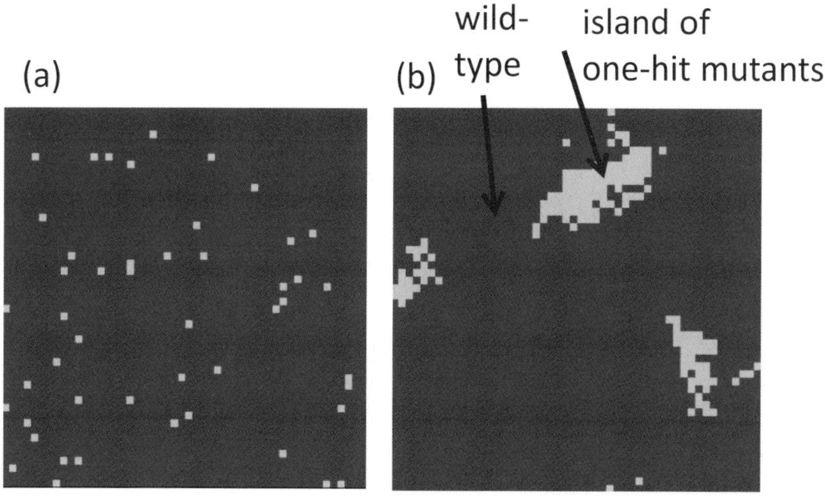

Fig. 14.4 Spatial configuration of the Moran process for the case of (a) mass-action and (b) a neighborhood radius of one. With spatial restriction, islands of intermediate mutants are observed. The chosen parameters were: grid size=50, $m = 2$, intermediate mutants were neutral, $u = 10^{-4}$. Dark color indicates wild-type, light color mutants.

divided it by the mean waiting time in the nearest-neighbor setting. This ratio was plotted for several different values of the mutation rate, u. Panels (a,c,e) correspond to neutral intermediate mutants, and panels (b,d) to disadvantageous mutants. We can see a common pattern where for very high mutation rates, the spatial interactions do not have much influence on the rate of m-hit mutant generation (the ratio of waiting times is close to one). For intermediate mutation rates, the spatially-restricted systems produce mutants significantly faster (the ratio > 1). Then for low mutation rates, the difference between spatial and non-spatial systems either stops changing with u (for disadvantageous intermediate mutants), or it decreases (for neutral intermediate mutants). The notions of "high", "intermediate", and "low" mutation rates are defined relative to other parameters, in particular, the population size N, the intermediate mutant fitness, r, and the number of sites, m. Here we describe the observed trends:

High mutation rates. For relatively large mutation rates, the difference between the spatial and mass-action systems is very small. This is because for very high mutation rates, new mutants are produced very frequently,

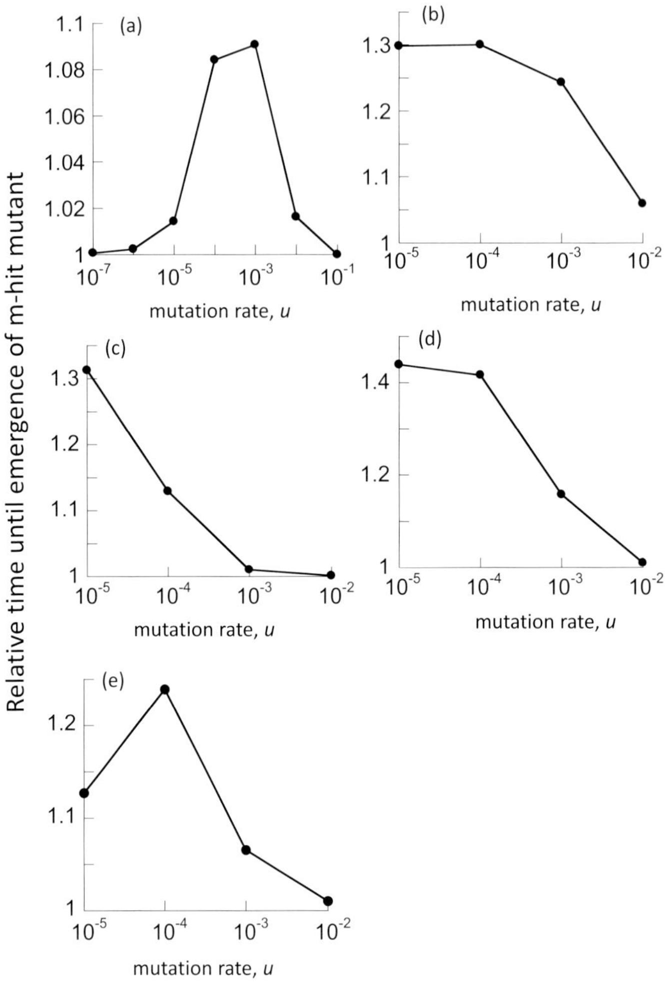

Fig. 14.5 Time to emergence of m-hit mutant in the Moran process. Compared are mass-action relative to extreme spatial restriction, where the neighborhood radius is one. The time to emergence for radius=1 was set to unity, and the time observed for mass-action was scaled accordingly. This number is plotted against the mutation rate, u. (a) $m = 2$, grid size $N = 10 \times 10$, neutral intermediate mutants. (b) $m = 2$, $N = 10 \times 10$, intermediate mutants have 10% fitness cost. (c) $m = 2$, $N = 50 \times 50$, neutral intermediate mutants. (d) $m = 2$, $N = 50 \times 50$, intermediate mutants have 10% fitness cost. (e) $m = 4$, $N = 50 \times 50$, neutral intermediate mutants.

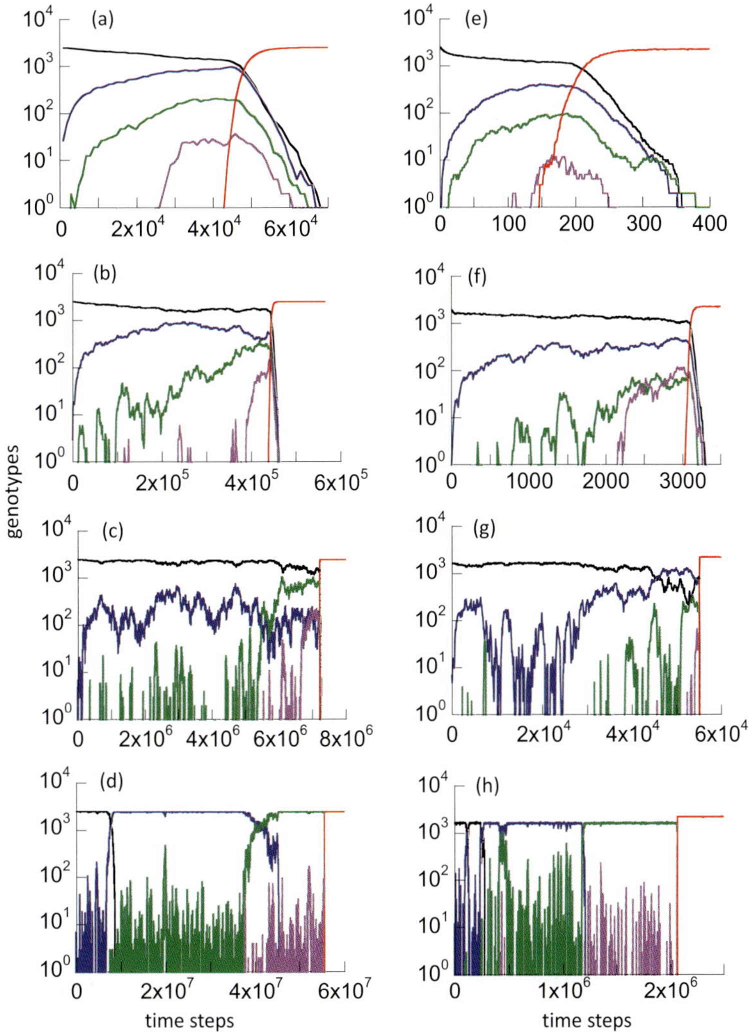

Fig. 14.6 Time series simulation showing the evolutionary dynamics for (a-d) the Moran process and (e-h) the contact process, assuming $m = 4$, a 50×50 grid and neutral intermediate mutants. Wild-types are shown in black, intermediate 1-hit, 2-hit, and 3-hit mutants are shown in blue, green and pink, respectively. The intermediate mutants have fitness $r = 1$. The advantageous 4-hit mutant is shown in red. For the Moran process, the mutation rates were (a) $u = 10^{-2}$, (b) $u = 10^{-3}$, (c) $u = 10^{-4}$, (d) $u = 10^{-5}$. For the contact process, $L/D = 3$, and (e) $u = 10^{-2}$, (f) $u = 10^{-3}$, (g) $u = 10^{-4}$, (h) $u = 10^{-5}$.

and are accumulated mostly by de-novo production rather than by mutant reproduction (the nearly-deterministic regime of Chapters 9 and 10). In this case, the role of mutant islands is negligible, and the waiting time is not changed by the neighborhood size. Figure 14.6(a) shows a typical time-series of mutant generation for high mutation rates, in the case of neutral intermediate mutants. We can see that intermediate mutants are generated constantly and experience a steady climb. This behavior is similar to the behavior of advantageous mutants. Even though the mutations are neutral, the presence of a one-way mutation process at a high rate effectively makes those mutants behave like advantageous mutants.

This behavior is observed if the mutation rate is significantly larger than a threshold, $u_c^{(1)}$. This threshold for the case of $m = 2$ can be determined from the theory in Chapter 10: $u_c^{(1)} \sim \frac{1}{N}$ for neutral intermediate mutants when $|1 - r| \ll \sqrt{u}$, see inequality (10.46) with $u_1 = u_2$. For disadvantageous intermediate mutants, we have a selection-mutation balance defined by production and selection, and the threshold value $u_c^{(1)}$ is given by the upper bound in the inequality (10.48): $u_c^{(1)} \sim \sqrt{\frac{1}{N}(1 - r)/r}$, where we set $u_1 = u_2$.

Intermediate mutation rates. As the mutation rates become smaller than $u_c^{(1)}$, the accelerating role of space in m-hit mutant generation becomes more pronounced, because de-novo mutant generation is less frequent now, and the mutant island effect becomes more important. Figures 14.6(b) and (c) show typical time-series in this regime, where the m-hit mutant is formed by stochastic tunneling, with no intermediate mutants reaching fixation. A similar picture is observed for disadvantageous intermediate mutants.

For even smaller values of u, there is another change in behavior. The system behaves differently for neutral and disadvantageous intermediate mutants.

Low mutation rates, neutral intermediate mutants. In the case of neutral intermediate mutants, when $u \ll u_c^{(2,neut)}$ (which is defined later), the role of space is again diminished, figure 14.5(a,e); figure 14.5(c) does not show this regime because simulations for lower mutation rates for the given parameter values are exceedingly long. We can see that the difference between the spatial and non-spatial model for very low mutation rates is smaller than that for intermediate mutation rates. The explanation again is given by the time-series, figure 14.6(d). For very small mutation rates, sequential fixation takes place for one or more consecutive mutants. The reason is that it is unlikely to mutate the next locus before the preceding mutant has reached fixation. The total waiting time now consists of two parts: the time waiting for fixation (one or more) and the time waiting for tunneling through the rest of intermediate mutations. While the tunneling rate is accelerated by spatial constraints, fixation events take somewhat longer in the nearest-neighbor model compared to mass action, simply because the spread of mutants happens via "surface growth" where the expansion can only take place at the outside rim of the mutant island, compared to the bulk growth of the mass-action model. The two effects act in the opposite ways, and we can see that the mass action waiting time for such small mutation rates is not too different from that of the nearest-neighbor model.

Let us denote by R_i the tunneling rate starting from a homogeneous population of type i, and by $\rho^{i\to j}$ the probability for a mutant of type j to reach fixation in a homogeneous population of mutants of type i, starting from one cell. The characteristic time of mutant fixation was found in formula (9.12), and the time of tunneling is given by the inverse R_i. If tunneling is the dominant process, which happens when

$$Nu_0\rho^{0\to 1} \gg R_0, \tag{14.1}$$

we can estimate the time until an m-hit mutant is produced as

$$T \approx \frac{1}{R_0}.$$

If condition (14.1) is reversed, the fixation of the first mutant is typically followed by a tunneling through the rest of the mutations, and we have

$$T \approx \frac{1}{Nu_0\rho^{0\to 1}} + \frac{1}{R_1}.$$

Since tunneling rates are accelerated by spatial interactions, and fixation rates are slowed down, we can see that lowering mutation rates will decrease the accelerating effect of space. In the limit of very small mutation rates, the dynamics proceed as a sequence of consecutive fixations of intermediate mutants, and the total mean time to m-hit mutant generation can be approximated as

$$T \approx \sum_{i=0}^{m-1} \frac{1}{Nu_i\rho^{i\to i+1}},$$

the sum of inverse consecutive fixation rates. Condition (14.1) allows us to calculate the threshold value, $u_c^{(2,neut)}$, when the role of space becomes negligible. We have in the neutral case, $\rho^{0\to1} = 1/N$, and the tunneling rate is given by $\prod_{i=0}^{m-1} u_i^{1/2^i}$ (Iwasa *et al.*, 2004a). In the symmetric case where all the mutation rates are the same and given by u, the expression for $u_c^{(2,neut)}$ is given by $u_c^{(2,neut)} = N^{-\frac{2^{m-1}}{2^{m-1}-1}}$. In particular, for $m = 2$, we simply have $u_c^{(2,neut)} = 1/N^2$.

Low mutation rates, disadvantageous intermediate mutants. In the case where $r < 1$ and $|1-r| \gg \sqrt{u}$, fixation of intermediate mutants is very unlikely. There, the difference between spatial and non-spatial rates of m-hit mutant production does not decrease for small values of u, but it merely stops increasing. In other words, starting from some threshold value, $u_c^{(2,dis)}$, decreasing u will not lead to further growth in the ratio of waiting times. This threshold value can be estimated if we impose the condition that the time-scale of new mutant production (given by $1/(Nu)$) is significantly larger than a typical life-span of a disadvantageous mutant (given approximately by $1/(1-r)$, see [Weissman *et al.* (2010)]). We then have the value $u_c^{(2,dis)} = \frac{1-r}{N}$.

14.4 The contact process

Next, we consider a model that is arguably more realistic than the Moran process described above. In the Moran process, each death is immediately followed by a reproduction event, which makes the model analytically tractable, but imposes a very rigid constraint on the timing of events. In what follows we describe a model that is more widely used in ecological and evolutionary simulations, and which is a type of a contact process studied in different contexts (Liggett, 1999).

In a square 2D grid of size N, nodes can be unoccupied or occupied with cells of different types. Each time-step consists of $\mathcal{N}$ elementary updates, where $\mathcal{N}$ is the total number of occupied sites. At each elementary update, we pick an individual at random. With probability $0 < D < 1$ this individual is removed, and with probability $L = 1 - D$ it attempts reproduction. Reproduction proceeds as follows. A random site in the neighborhood of size M of this individual is picked, and if it is occupied, the reproduction is aborted, and the update is complete. If the site is empty, the offspring (possibly with a mutation) of the individual is placed at the site, which completes the update. The Moran process described earlier can be considered the limit of the contact process where $L \gg D$ (except the time-steps proceed uniformly in our version of the Moran process, and they happen with an exponentially-distributed time-steps in this limit of the contact process). In the case where $L \gg D$, the grid is completely filled with live cells at all times, and the dynamics are driven by the death events, each of which is immediately followed by a division event.

14.4.1 *The steady-state density of cells*

Before we explore the question of the rate of fitness valley crossing in the contact process, we need to gain an understanding of the model's basic properties. Let us set the mutation rates to zero and simply observe the spatio-temporal behavior of individuals on the grid. Let us suppose that $L > D$. In the mass-action model, the process reaches a quasi-stationary state where the individuals are distributed around the grid with the equilibrium density of

$$\nu = 1 - \frac{D}{L}.$$

That is, there are typically $N(1 - D/L)$ individuals, and the probability of each site to be occupied is given by ν (in other words, they are distributed uniformly throughout the grid). This result can be derived very easily by assuming a uniform distribution of individuals and equating the probability of death with the probability of successful reproduction.

Things are significantly more complicated in a spatially-restricted model ($M < N$). In this case, it is known that the probability of reproduction L has to be greater than a threshold value, $L_c > D$, for the system to reach the quasi-steady state, but the exact value of L_c is not theoretically known. Also, the equilibrium density has not been calculated (apart from the limiting values as $M \to N$ (Bramson *et al.*, 1989)). In the regime of interest where $M \ll N$, no estimates of the equilibrium density are available.

The complication comes from the macroscopic structures forming in the system in the case of spatially-restricted interactions (figure 14.7). In figure 14.8 we show numerically obtained steady-state numbers of individuals as a function of the neighborhood size, M. We can see that the resulting density is lower than the mass-action density $\nu_{m.a.}$. An attempt to describe this behavior was made by the pair approximation method, which takes local interactions into account by considering pair correlations of neighboring sites [Boots and Sasaki (2002); Tomé and de Carvalho (2007)]. Unfortu-

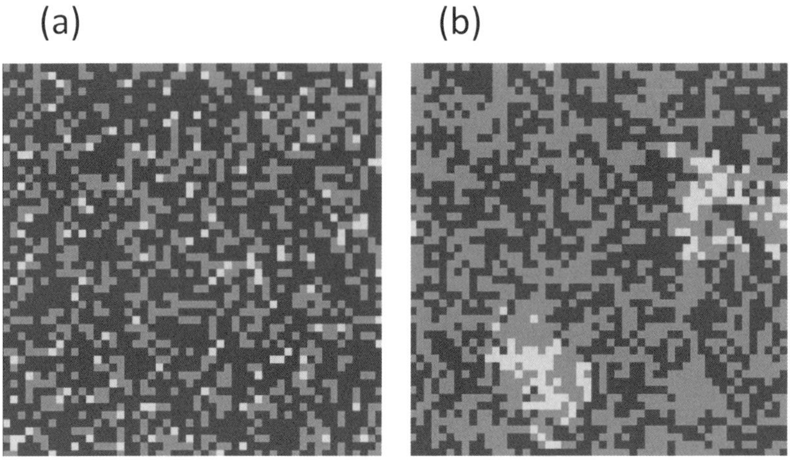

Fig. 14.7 Spatial configuration of the contact process model for the case of (a) mass-action and (b) a neighborhood radius of one. With spatial restriction, islands of intermediate mutants are observed. Individuals are not evenly distributed, but form macroscopic structures. The chosen parameters were: grid size 50×50, $m = 2$, intermediate mutants were neutral, $u = 10^{-4}$, $L/D = 3$. Darkest color indicates wild-type, lightest color mutants, and intermediate gray color shows empty space.

nately this approximation crudely underestimates the difference between the spatial behavior and the mass-action behavior.

To capture the spatial effects more effectively, we studied numerically generated quantity $(1-\nu)L/D$ as a function of the neighborhood size, M. For the mass-action system, this quantity is simply 1. For spatially-structured system, this quantity grows as M decreases. We noticed that this quantity is very well described by a simple inverse function of M, which yielded the following empirical formula:

$$\mathcal{N} = N\left(1 - D/L\left(1 + \frac{c}{M}\right)\right), \tag{14.2}$$

where c does not depend on M (that appears to depend on the parameter D/L, for example, $c \approx 4$ for $D/L = 1/3$ and $c \approx 4.5$ for $D/L = 1/2$). It was further empirically found that the density of individuals in the neighborhood of a given individual is on average given by

$$\nu_{loc} = 1 - D/L\left(1 + \left(\frac{c_1}{M}\right)^{\alpha}\right), \tag{14.3}$$

where c_1 and α are both M-independent (with $c_1 \approx 1.6$ and $\alpha \approx 0.9$ both for $D/L = 1/3$ and $D/L = 1/2$).

We observe the following two important trends:

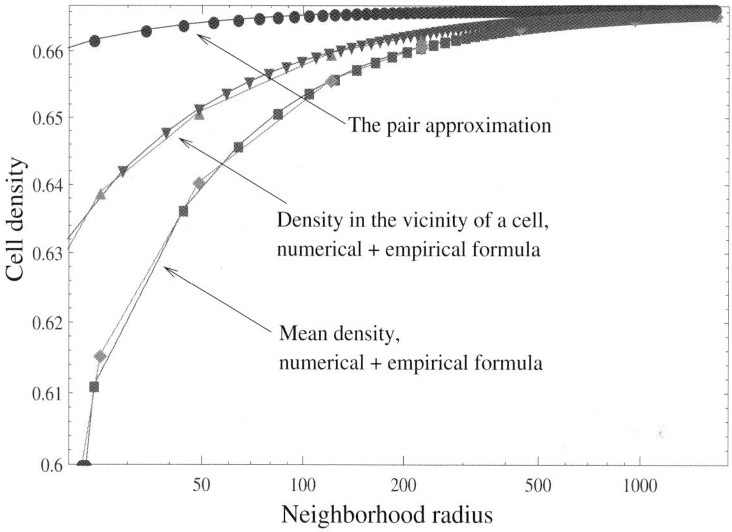

Fig. 14.8 Numerically calculated densities of cells in the contact process, plotted as functions of the neighborhood radius. We present the mean density (together with the empirical formula 14.2), and the mean density of cells in a neighborhood of nonempty spots (together with empirical formula (14.3). Also, plotted is the pair-approximation of the mean density. The parameters are $N = 41 \times 41$, and $L/D = 3$.

(i) The density in the vicinity of an occupied spot is greater than the mean density. The distribution of cells throughout the grid is no longer uniform, and the probability of finding an occupied site in the vicinity of a given individual is higher than the mean density. This is a quantification of the clustering effect, which can also be observed by simply examining a typical spatial distribution of individuals in a spatially-structured system, figure 14.7(b).

(ii) The mean number of individuals and the mean number of neighbors of an individual both increase with the neighborhood size, M, and approach $N(1 - D/L)$ as $M \to N$ (the mass-action limit). This can be viewed as a direct consequence of macroscopic structures. For smaller values of M, the density becomes non-uniform, with the local density higher than the average density. This makes individuals land on other occupied spots and thus reduces the number of successful events per time-step. The number of successful divisions at equilibrium must be matched by the deaths, which are proportional to the total population.

Therefore, the total population in spatial systems must be lowered compared to the mass-action system.

The importance of these trends for our evolutionary question becomes clear once we realize that the equilibrium number of individuals as well as the number of neighbors of a given individual define the time-scale of the contact process. This is explained below, after we talk about the empirical observations for this model.

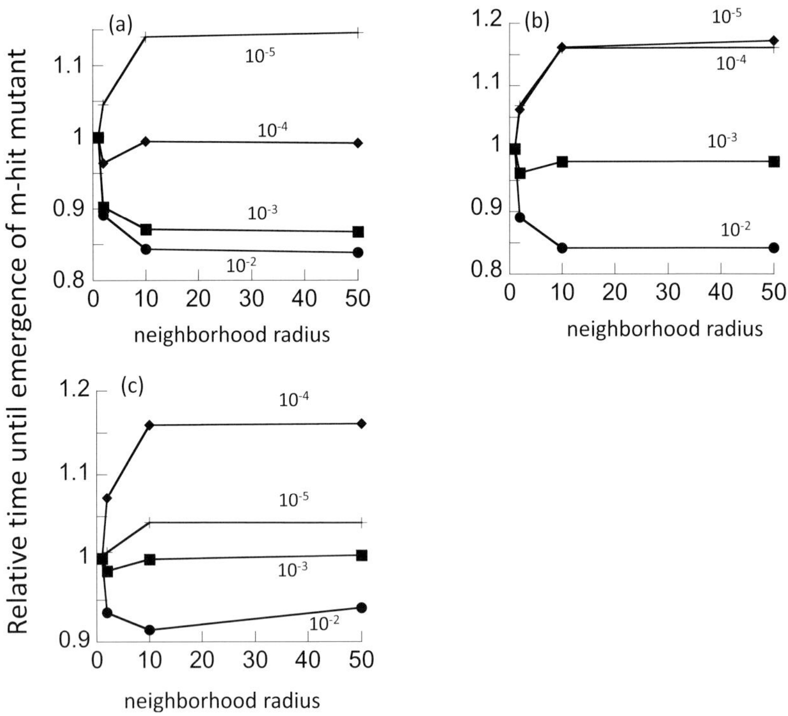

Fig. 14.9 Time to emergence of m-hit mutant in the contact process. Compared are neighborhood radii of 1, 2, 10, and 50. The time to emergence for neighborhood radius 1 was set to unity, and the waiting time observed for the other radii was scaled accordingly. Different curves are shown for different mutations rates u, as indicated in the plots. For all plots, $L/D = 3$. (a) $m = 2$, grid size is 50×50, neutral intermediate mutants. (b) $m = 2$, grid size is 50×50, intermediate mutants have 10% fitness cost. (c) $m = 4$, grid size is 50×50, neutral intermediate mutants.

14.4.2 *Complex effects of spatial restriction*

The effect of spatial restriction on the waiting time for the emergence of the m-hit mutant is more complicated than that in the Moran process model. In the current setting, spatial restriction can either accelerate or delay the generation of the m-hit mutant, depending on parameters. In some cases, there is an optimal neighborhood size that maximizes the rate of emergence of the m-hit mutant. That is, evolution works fastest for intermediate-range interactions. Figure 14.9 presents some typical results. In this figure, for a given mutation rate u, we calculated the mean waiting time for the m-hit mutant emergence for several neighborhood radii, and divided it by the mean waiting time for the nearest neighbor scenario (the radius 1 neighborhood model). These relative mean waiting times were plotted as functions of the neighborhood radius, for different mutation rates. In panels (a) and (b), we have $m = 2$ and the intermediate mutants are neutral and disadvantageous respectively. In panel (c), we have $m = 4$ with neutral intermediate mutations. We can see that the waiting time decays monotonically with the neighborhood radius e.g. for $u = 10^{-2}$ in figures 14.9(a,b) and $u = 10^{-3}$ in figure 14.9(a). It increases monotonically for $u = 10^{-5}$ in figures 14.9(a,b) and for $u = 10^{-4}$ in figures 14.9(b,c). The waiting time as a function of the neighborhood radius experiences an intermediate minimum for $u = 10^{-4}$ in figure 14.9(a), for $u = 10^{-3}$ in figure 14.9(b,c) and for $u = 10^{-2}$ in figure 14.9(c).

To explain this complicated behavior, we note that there are two different mechanisms that govern the spatio-temporal dynamics of m-hit mutant generation.

(1) The formation of mutant islands is facilitated by tight spatial interactions (figure 14.7). This is exactly the same trend as observed and explained in the context of the Moran model, and it results in the increase of the tunneling rate for the nearest-neighbor model.
(2) In contrast to the Moran model, the population development over time in the contact process is nonuniform and is defined by the equilibrium number of individuals. For smaller values of M, the total number of individuals is smaller, and thus the total number of events is also lower. Therefore, the rate of evolution (measured e.g. by the rate of tunneling) for the mass-action model is faster compared to the nearest-neighbor model.

Combining the two opposite effects, we can describe the mean time of fitness valley crossing as

$$\frac{T_{Moran}}{\tau}, \tag{14.4}$$

where T_{Moran} grows with M and $1/\tau$ decays with M. Under some parameter regimes, the resulting function can be shown to possess an intermediate minimum, which corresponds to the evolutionary optimum for the fitness valley crossing. It can also be a monotonically increasing or decreasing function of M, depending on the parameters.

14.4.3 *Parameter dependencies*

The mutation rate is an important parameter in determining the exact effect of spatial restriction on the rate at which the m-hit mutant is generated. Figure 14.9 shows the effect of spatial restriction on the time it takes to generate the m-hit mutant, for different mutation rates. We observe the following patterns:

- For large mutation rates, the crossing of the fitness valley happens the fastest in the mass-action model, because the tunneling rate is largely independent of the neighborhood size (as we learned in the Moran process), and the time-scale of the events, τ, is faster for mass action, as explained in the previous section.
- For intermediate mutation rates, the two trends trade-off and the mean time for fitness valley crossing experiences a minimum for intermediate values of M.
- For small mutation rates, spatial restriction decreases the time until the m-hit mutant emerges. Evolution occurs slowest for the mass-action scenario and fastest for the nearest neighbor scenario. As the mutation rate decreases, the magnitude of this effect rises.
- However, if the mutation rate is decreased below a threshold, the m-hit mutants are not generated by tunneling anymore. Instead, the sequential fixation of intermediate mutants occurs. While the nearest-neighbor model still allows for fastest evolution, the effect is less pronounced in this parameter region.

The evolutionary pathways are shown as time series in figure 14.6(e-h), demonstrating evolution through tunneling for lower mutation rates, and through sequential fixation for higher mutation rates.

The difference in the rate of evolution in the spatially restricted and the mass-action scenarios is generally smaller than that for the Moran process (figure 14.9 compared to figure 14.5). The reason is that in the contact process the forces that accelerate evolution in a spatial setting are offset by the slower dynamics inherent in the spatial situation, where growth involves the formation of macroscopic structures. In general, the maximal difference between the spatial and mass-action settings did not exceed 15-20% in our simulations. Hence, the difference is less pronounced than in the Moran process model, where a difference of up to 40-50% was observed between nearest neighbor and mass-action settings for disadvantageous intermediate mutants.

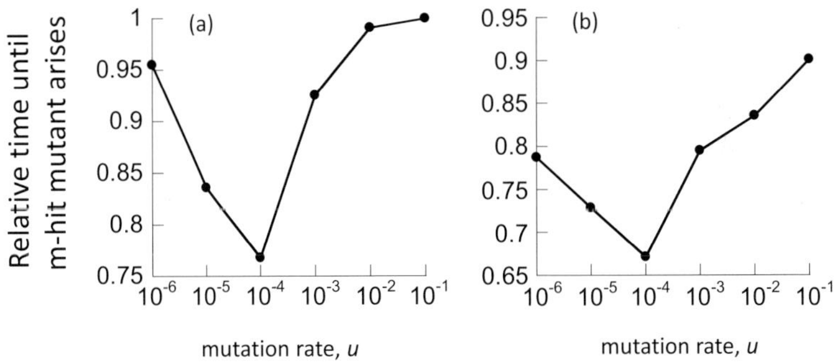

Fig. 14.10 Time to emergence of m-hit mutant in (a) the Moran process and (b) the contact process, assuming that intermediate mutants are advantageous compared to the wild-type, and that the m-hit mutant is even more advantageous. Compared are mass-action relative to extreme spatial restriction, where the neighborhood radius is one. The time to emergence for radius=1 was set to unity, and the time observed for mass-action was scaled accordingly. This number is plotted against the mutation rate, u. The other parameters were assigned the following values: $m = 2$, grid size $N = 50 \times 50$, intermediate mutants have 10% fitness advantage. For (b) $L/D = 3$.

14.5 Advantageous intermediate mutants

The analysis so far concentrated on situations where intermediate mutants are either neutral or disadvantageous, and similar patterns were observed for the two cases. Here, we investigate the evolutionary dynamics assuming that intermediate mutants are advantageous compared to the wild-type, and that the m-hit mutant is even more advantageous. Results are qualitatively similar for the Moran process (figure 14.10(a)) and the contact process (figure 14.10(b)). In these plots, for each mutation rate we calculated the mean waiting time for the m-hit mutants in the mass-action model, and divided it by the mean waiting time in the nearest-neighbor setting. We can see that in sharp contrast to the patterns seen for neutral and disadvantageous mutants, for advantageous intermediate mutants nearest-neighbor interactions always slow down the emergence of the m-hit mutant (figure 14.10). The degree to which this happens depends on the mutation rate, u (in relation to other system parameters). The effect is highest for intermediate mutation rates, and the difference is small for low and high mutation rates. Below we describe the observed patterns for the Moran process, and give explanations.

High mutation rates. Advantageous intermediate mutants are selected for and grow (almost) deterministically rather than drifting. For very high mutation rates, the production of the intermediate mutants by de-novo mutations from wild-type also contributes to this growth, and in fact is the dominant force as long as $uN \gg r$, see figure 14.11(a). As explored in previous sections, mutant growth by production from wild-type is not influenced by the spatial configuration, explaining the small difference between the spatial and mass action simulations for high mutation rates. For the parameters in figure 14.10(a), we can see that the difference between the mass-action and the nearest-neighbor model is small for $u \gg 4 \cdot 10^{-4}$.

Intermediate mutation rates. In this regime, replication contributes significantly to the growth of m-hit mutants, figure 14.11(b). As seen in figure 14.10, for intermediate mutation rates, systems with spatial restrictions produce m-hit mutants slower than the mass-action model. The reason for that, interestingly, is the existence of mutant islands. Advantageous intermediate mutants in the spatial systems tend to expand, but they can do so only by growing on the outer rims of the mutant islands (the so-called surface growth, see e.g. [Wodarz *et al.* (2012)]). On the contrary, in the mass-action systems, the mutants experience a bulk growth, leading to a faster (exponential) expansion. The existence of mutant islands serve to accelerate the production of m-hit mutants for the neutral and disadvantageous case, because of its protective role. For advantageous intermediate mutants, these islands constrain the growth, slowing down the evolution compared to the mass-action system.

Low mutation rates. For very low mutation rates, fixation of the intermediate mutant often occurs before the generation of the m-hit mutant, see figure 14.11(c). As explored above, this reduces the influence of the neighborhood radius on the rate of evolution, explaining the reduced difference at low mutation rates.

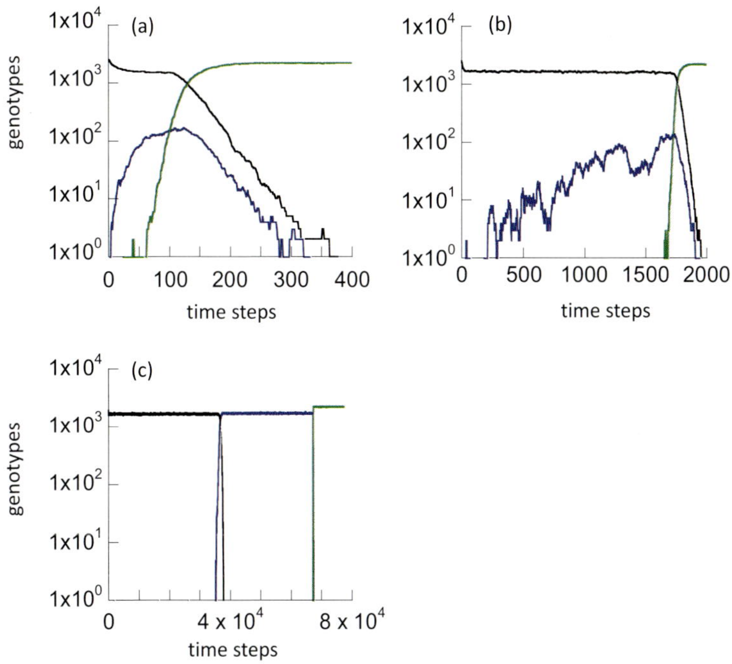

Fig. 14.11 Time series simulations showing the evolutionary dynamics for the Moran process, assuming that intermediate mutants are advantageous and the m-hit mutant is even more advantageous. Plots are shown for three mutation rates: (a) 10^{-2}, (b) 10^{-4}, and (c) 10^{-6}. Wild-types are shown in black, intermediate mutants (i.e. 1-hit mutants) in blue, and m-hit mutants in green. The simulations were done with $m = 2$, a 50×50 grid, and a 10% fitness advantage of intermediate mutants compared to wild-types.

In the contact process (figure 14.10(b)), the same effects take places as described above for the Moran process. The difference is that in the contact process, there is an additional effect whereby the dynamics in the mass-action model always happens faster than in the spatially-restricted systems. As explained before, this happens because of the uniform distribution of individuals across the grid in the mass-action model, which increases the probability of successful reproduction events. Since this effect combines in a positive way with the effects described for the Moran process, the overall effect in the contact process is more significant compared to the Moran process. In the contact process, we observe that the evolution happens faster in the mass-action system across all mutation rates, with the effect being the strongest for the intermediate values of u.

14.6 Summary and discussion

In this chapter we saw that the effects of spatial structure on the rate of cancer evolution are far from straightforward.

- In the Moran process, spatial interactions speed up the evolution of the complex trait for neutral and disadvantageous intermediate mutants, consistent with previous work (Chapter 13, (Komarova, 2006c; Durrett *et al.*, 2012; Durrett and Moseley, 2012)).
- For advantageous intermediate mutants in the Moran process, spatial interactions slow down the evolution of the complex trait. This is the opposite effect of space compared to the case of neutral and disadvantageous intermediate mutants.
- In the contact process for neutral and disadvantageous intermediate mutants, spatial interactions can either accelerate or slow down the rate of evolution, depending on the mutation rate. It is also possible to have an intermediate optimal interaction radius (between mass-action and nearest neighbor interactions), which minimizes the time until the advantageous trait emerges.
- For advantageous intermediate mutants in the contact process, spatial interactions speed up the evolution of the complex trait, even more so than in the Moran process.

The two types of models (the Moran and the contact processes) correspond to different biological scenarios, which we discuss next.

Moran and contact processes: comparisons and applications. The Moran process practically assumes a high degree of regulation that keeps the population at a constant level and avoids the generation of empty space that is not filled with live individuals. This situation can apply to populations of tissue cells, or perhaps even tissue stem cells, whose homeostasis is tightly controlled by various feedback-mechanisms (Lander *et al.*, 2009). In this situation, spatial restriction always minimizes the time until the m-hit mutant is generated. This confirms the results obtained in the previous studies and is especially pronounced if the intermediate mutants are disadvantageous. It is actually a biologically surprising result. Tissue cells are for the most part characterized by a high degree of spatial restriction, and this can facilitate the emergence of aberrant cells that could potentially lead to the development of cancer. It is likely that spatial restrictions of cellular interactions are vital for other processes that determine the functioning of the tissue and the organism, and that the accelerated evolution of aberrant cells is a side-effect of this architecture. As shown in [Thalhauser *et al.* (2010)], migration and motility result in dynamics that converge towards mass-action scenarios, and could thus slow down the rate of evolution in tissues. One can hypothesize that migration of cells in tissues could hence be an adaptive response to slow down the accumulation of aberrant mutations.

On the other hand we can envisage situations where populations are regulated less tightly, and where a death of a cell is not immediately followed by a reproduction event. In this case the Moran process is not a realistic description anymore. For this reason we also studied the contact process, which is a stochastic birth-death process that is characterized by a stochastic (long-lived) quasi-steady state with a nearly constant density of cells, resulting from the balance of death and division events. Hence, at equilibrium, the entire space is not filled with individuals, but contains unoccupied space. Interestingly, this makes a lot of difference because macroscopic structures can form, and successful reproduction events tend to occur only at the surface of those structures. This counters the effect that spatial restriction accelerates evolution through the formation of mutant islands that are protected from being replaced by wild-types. This has several consequences, in the context of neutral or disadvantageous intermediate mutations.

(1) Increasing the interaction radius can monotonically increase the rate of evolution, it can monotonically decrease it, or there can be an optimum interaction radius that minimizes the time until the m-hit mutant is generated. Which pattern is observed depends largely on the mutation rate (compared to some characteristic constant that depends on the population size, the number of sites and the fitness of the intermediate mutants). Spatial restriction tends to slow down evolution for higher mutation rates, and accelerate it for lower mutation rates. Which pattern is in fact observed depends not only on the mutation rate, but also on other factors such as the number of loci involved.

(2) Another consequence is that because of trading-off forces in the dynamics, the extent of the difference observed between spatial and nonspatial settings is relatively small. No more than a 15-20% difference was observed in the simulations. Nevertheless, this can still be enough to provide significant selection pressure that can shape the life-style of organisms, i.e., whether they adopt a sedentary life-style characterized by significant

spatial restrictions and limited movement, or whether they adopt a life style characterized by extensive movement and migration. An interesting example are bacteria, which can either show planktonic growth or sessile growth characterized by the formation of biofilms (Jefferson, 2004). According to the model explored here, sessile growth could be favored if mutation rates are relatively low because in this case, populations can adapt faster through accelerated crossing of fitness valleys. If, however, mutation rates are higher, faster adaptation could be promoted by a planktonic life-style. Note, however, that many other factors are likely to influence the life-style of bacteria and organisms in general, and that the rate at which organisms are able to cross fitness valleys and adapt is only one of many dimensions that play a role.

When we turn to slightly advantageous intermediate mutations, we discover that in some sense, spatial interactions play the opposite role compared to the situation with neutral or disadvantageous mutants. In both Moran and contact processes, the speed of evolution grows monotonically with the neighborhood size, and the nearest-neighbor model corresponds to the lowest rate of m-hit mutant generation. Interestingly, this effect increases in the contact process compared to the Moran process. For all intermediate mutant fitnesses, the effect of space is the most pronounced when stochastic tunneling is the dominant mode of m-hit mutant generation.

The dynamics of mutant islands. To explain our observations, we studied the formation of "mutant islands". These are spatially-correlated colonies of mutants, which arise in models with spatial restrictions, and are the tightest for the nearest-neighbor model. Clearly, in these models, neighboring cells tend to have similar genotypes, which gives rise to these islands of intermediate mutants. The abundance of the intermediate mutant is the main correlate of the rate of m-hit mutant generation in the tunneling regime. We discovered that mutant islands promote the persistence of mutant colonies when they are disadvantageous or neutral. In this case, the generation of m-hit mutants relies on the survival of intermediate mutant colonies, which are typically transient and short-lived, and which in the spatial setting are facilitated by the protective islands. In contrast to this, mutant islands effectively slow down the growth of mutant colonies when they are advantageous. In the mass-action scenario, advantageous intermediate mutants experience a bulk (exponential) growth, while they can only expand along the island surface in spatial settings.

These arguments explain the opposite influence of space on the m-hit mutant generation in the case of neutral/disadvantageous intermediate mutants on the one hand, and advantageous intermediate mutants on the other.

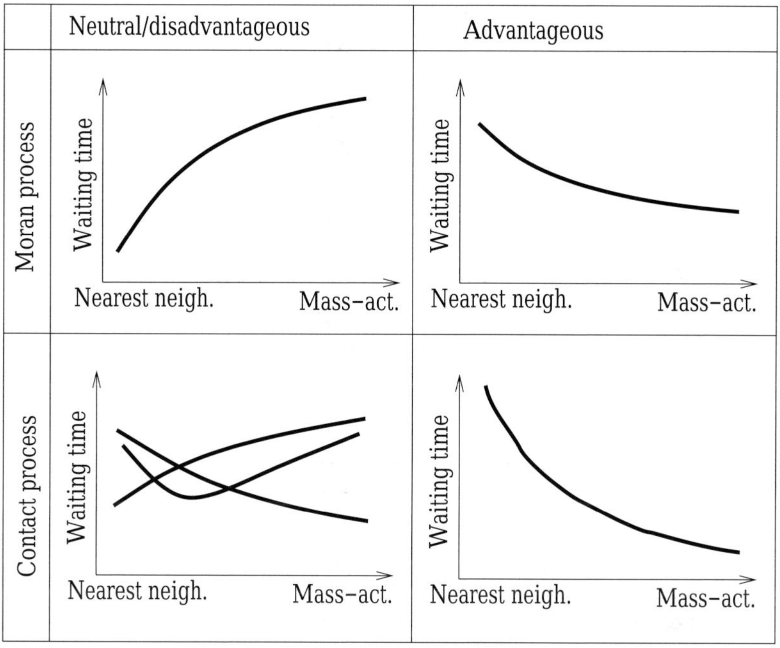

Fig. 14.12 A schematic summarizing the complex role of space in the generation time of m-hit mutants. The top row represents the Moran process, and the bottom row the contact process. The left column corresponds to neutral/disadvantageous intermediate mutants, and the right column to the advantageous intermediate mutants. In each graph, the horizontal axis represents the degree of mixing, from the nearest neighbor model to the mass-action model. The curves show whether the waiting time increases or decreases (or is non-monotonic) as a function of the neighborhood size. It also shows the relative size of the effect, when comparing the Moran process with the contact process.

In the case of the Moran process, the influence of space on the rate of m-hit mutant generation can be explained completely by the mutant island arguments. The contact process involves an additional effect, which is related to the existence of empty space in that model, and the dependence of the individuals' spatial distribution on the neighborhood size. It turns out that in the spatially restricted situation, the distribution is very different from that in the mass-action situation, resulting in a smaller number of successful divisions per time unit. We will refer to this effect as the "time-scale effect".

How do the two effects (the mutant island effect and the time-scale effect) trade-off to influence the rate of m-hit mutant generation in the contact process? Do they add up or do they work in the opposite directions? In the case of neutral/disadvantageous mutants, the time-scale effect slows down the evolution while the mutant island effect tends to speed it up. Therefore, (1) the effect of space is weaker in the contact process compared to the Moran process, and (2) space could speed up evolution, slow it down, or have a nonmonotonic effect depending on the neighborhood size.

For advantageous intermediate mutants, the two effects work in the same direction: the slowing down of the time-scale of evolution by spatial constraints adds to the effect of mutant islands. As a result, (1) the effect of space is stronger in the contact process compared to the Moran process, and (2) space always slows down evolution in this setting. All these effects are summarized schematically in figure 14.12.

An important message that follows from our analysis is that the effect of spatial restriction on the rate at which fitness valleys are crossed is not at all straightforward but depends on a multitude of factors. This is in contrast to previous notions that space accelerates the evolution of complex phenotypes. Of particular importance are the laws according to which populations grow in spatially restricted settings. It is currently not possible to say whether the contact process presented here is a sufficiently realistic description of spatial growth in experimental or natural populations. This will need to be examined by experiments, tracking both the number of individuals, as well as the spatial patterns the develop over time. Cell cultures or experiments with microorganisms, which can be fluorescently labeled (Wodarz *et al.*, 2012; Hofacre *et al.*, 2012), would be a suitable system to examine this. Indeed, it is likely that spatial population growth is characterized by different laws in different settings, and that the effect on the rate of fitness valley crossing varies for different spatial population growth laws. This can be investigated by further computational models, once more experimental information is available about spatial population growth laws in different organisms and different settings.

Problems

Problem 14.1. Research project. *Learn about the general concept of complex phenotypes and fitness value crossing in evolutionary biology.*

Problem 14.2. *In Chapter 10 we characterized three different scenarios of two-hit mutant generation, figure 10.1: a two-step process (or more generally, sequential fixation), stochastic tunneling, and a nearly-deterministic process. Identify which processes are observed in figures 14.6(a-d) for neutral intermediate mutants and* $m = 4$.

Problem 14.3. *Similar to the previous questions, but for figure 14.11. In this case, intermediate mutants are advantageous, and* $m = 2$. *Use theory of Chapter 10 to predict which regime is expected for the three different values of* u.

Problem 14.4. *In figures 14.5(a) and (e) it is shown that for very small mutation rates the difference betweem spatial and non-spatial dynamics decreases. This is related to the general trend that as* u *decreases, we enter the regime of sequential fixation of mutants. Explain why we do not see this trend in figures 14.5(b), (c) and (d). Hint: by using theory developed in Chapter 10, find the value of* u *small enough such that sequential fixation is observed for parameter values of figures 14.5(b), (c) and (d); take* $u_1 = u_2$.

Chapter 15

Stochastic modeling of cancer growth, treatment, and resistance generation

15.1 Introduction

In this chapter we develop some useful mathematical tools that allow us to model the growth of cancer, the generation of resistance, and the effect of multi-drug treatment. The methodology is stochastic, based on the description of a cellular colony as a birth-death process with mutations.

We start by describing the growth of a cellular colony as a birth-death process with mutations. Several different mutations can be included, and the resulting cell types are characterized by the mutations that they have accumulated. The general problem of mutation accumulation by a growing cell colony goes back to Luria and Delbrück, who developed the so-called "fluctuation analysis" of mutations in bacterial cultures in 1943 [Luria and Delbrück (1943)]. Since then the distribution of the number of mutants in growing populations of cells has been studied by many authors (see the review by [Zheng (1999)] and references therein). Even though the fluctuation analysis of Luria and Delbrück was originally designed for bacterial populations, it has since been widely applied to cancer genetics [Kendal and Frost (1988); Jaffrézou *et al.* (1994); Dewanji *et al.* (2005)].

This chapter is not a comprehensive review of the Luria-Delbrück analysis, but instead it provides a toolbox sufficient to study certain aspects of cancer. Specifically, we are interested in generation of drug resistance in growing tumors, under different treatment regimes. Several applications of the theory developed here will be considered in Chapter 16.

15.2 The basic model of cancer growth and generation of mutations

15.2.1 *The concept: a birth-death process with mutations*

The aim is to describe the dynamics of birth, death, and mutations in a colony of cells [Komarova and Wodarz (2005); Komarova (2006b)]. We assume that there are four different processes that can take place in the colony, see figure 15.1. A death will lead to the number of cells of the given type decreasing by one. Faithful reproduction will increase the number of cells of the given type. Reproduction with a mutation will result in an increment in the number of cells of the mutant type. Transformation will decrease the number of cells of a given type and simultaneously increase the number of cells of the mutant type. Such transformations can be caused by environmental factors or can be a direct consequence of treatment by mutagenic drugs.

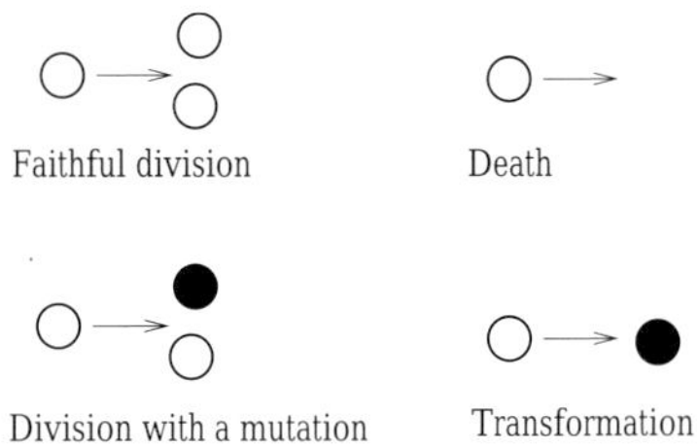

Fig. 15.1 The four basic processes: death, faithful and unfaithful division, and transformation. For simplicity, only two cell types are included: wild type and transformed cells. Wild type cells are depicted as empty circles, and transformed cells as filled circles.

Because of mutations, there may be cells of different types in a colony. Each cellular type (phenotype) in this model is characterized by the types of mutations that it has acquired. Figure 15.2 illustrates the mutation network in the case where $m = 3$ different mutations are tracked. It follows that with m different mutations, the total of

$$n = 2^m - 1$$

genotypes can exist in the system. Each phenotype can be expressed as a binary number of length m, where 1 stands for mutation and 0 for the absence of mutation in a given site.

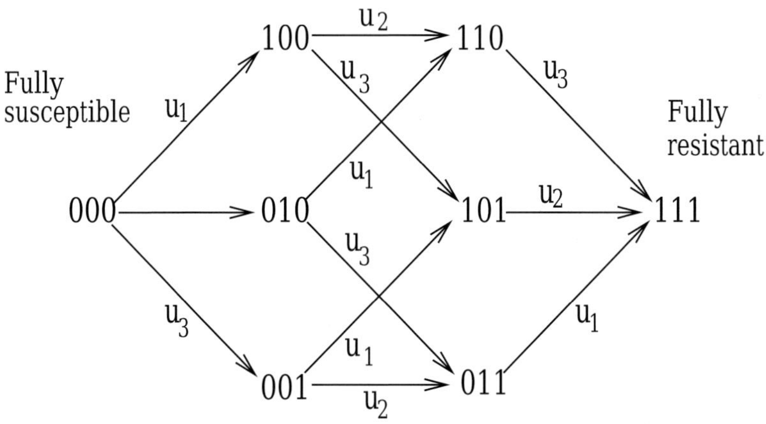

Fig. 15.2 Mutation diagram corresponding to three drugs. The binary number, s, for each node describes the genotype with respect to three possible mutations, where 0 is "not mutated" and 1 is "mutated". In the context of drug resistance to $m = 3$ drugs, 0 stands for "susceptible" and 1 for "resistant".

15.2.2 *Summary of all the probabilities*

We will model the dynamics by using a continuous time, discrete state space Markov birth-death process on a combinatorial mutation network. For an infinitesimal time increment, Δt, the probability for a cell to undergo each of the four basic processes (death, faithful division, mutation and transformation) is proportional to Δt; the probability of two events occurring simultaneously (within Δt) is neglected. Different events will occur with different probabilities (or rates), depending on whether treatment is applied. We denote by L_s the birth rate of type s, and by d_s its death rate.

Here is a formal description of all the processes [Komarova and Wodarz (2005); Komarova (2006b)]. Each phenotype "A^s" with some $0 \leq s \leq 2^m$, is represented as a node of a mutation network with arrows coming in and out, see figure 15.3. We assume that in time interval Δt, the following events can occur with each phenotype "A^s":

- With probability $L_s(1-\sum_j u_j^{s,out})\Delta t$ a cell of type "A^s" reproduces, creating an identical copy of itself.
- For each outgoing arrow, with probability $L_s u_j^{s,out}$ a cell of type "A^s" reproduces with a mutation, creating a cell of type "$A_j^{s\,next}$", for all j.

- With probability $d_s \Delta t$ a cell of type "A^s" dies.
- For each outgoing arrow, with probability $K_s \alpha_j^{s,out}$ a cell of type "A^s" is transformed into a cell of type "A$_j^{s\,next}$", for all j.

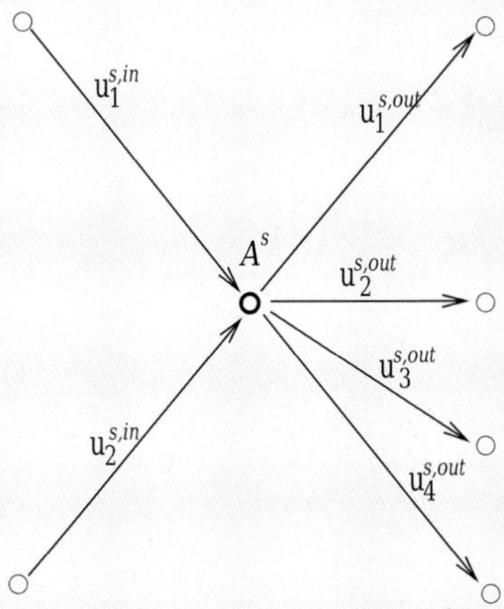

Fig. 15.3 An example of a vortex in a mutation diagram.

We start with a given number of cells of type "A^0", and follow the process until the first cell of type "A^n" has been created. We would like to calculate the probability, $P_n(t)$, that a cell of type "A^n" has been created as a function of time.

15.2.3 *Stochastic description: the example of one mutation*

Before we introduce the general methodology, let us consider a specific example of a one-drug treatment. The states of the system are characterized by integer-valued vectors, (i_0, i_1), where i_0 is the number of cells susceptible to the drug and i_1 is the number of cells resistant to the drug. Let us denote by $\varphi_{i_0,i_1}(t)$ the probability to be in state (i_0, i_1) at time t. Let us consider the processes of cell divisions, death, and mutations from type 0 to type 1 with probability u per cell division. This function

φ_{i_0,i_1} satisfies the following Kolmogorov forward equation:

$$\begin{aligned}\varphi_{i_0,i_1} &= L_0(1-u)\varphi_{i_0-1,i_1}(i_0-1) + L_1\varphi_{i_0,i_1-1}(i_1-1)\\ &\quad + L_0u\varphi_{i_0,i_1-1}i_0\\ &\quad + d_0\varphi_{i_0+1,i_1}(i_0+1) + d_1\varphi_{i_0,i_1+1}(i_1+1)\\ &\quad + -\varphi_{i_0,j_0}(L_0+L_1+D_0+D_1), \end{aligned} \tag{15.1}$$

where the terms on the first line represent divisions of both types of cells, the second line corresponds to mutations creating resistant cells, the third line is death of both types of cells, and the last line corresponds to the possibility of no change (the negative term). To proceed, it is convenient to define the probability generating function,

$$\Psi(\xi_0,\xi_1;t) = \sum_{i_0,i_1} \varphi_{i_0,i_1}(t)\xi_0^{i_0}\xi_1^{i_1}.$$

In order to derive an equation satisfied by this function, we multiply equation (15.1) by $\xi_0^{i_0}\xi_1^{i_1}$ and sum over i_0 and i_1. Then the left hand side becomes $\partial\Psi/\partial t$. On the right hand side we have several types of terms. For example, the first term can be written as

$$\sum_{i_0,i_1}\varphi_{i_0-1,i_1}(i_0-1)\xi_0^{i_0}\xi_1^{i_1} = \sum_{k,i_1}\varphi_{k,i_1}k\xi_0^{k+1}\xi_1^{i_1},$$

where $k = i_0 - 1$. Next we notice that this can be rewritten as

$$\xi_0^2\sum_{k,i_1}\varphi_{k,i_1}k\xi_0^{k-1}\xi_1^{i_1} = \xi_0^2\frac{\partial}{\partial\xi_0}\sum_{k,i_1}\varphi_{k,i_1}\xi_0^{k}\xi_1^{i_1} = \xi_0^2\frac{\partial}{\partial\xi_0}\Psi.$$

Similarly, the other terms can be rewritten in terms of partial derivatives of the function Ψ with respect to ξ_0 or ξ_1. We obtain the following equation:

$$\frac{\partial\Psi}{\partial t} = \frac{\partial\Psi}{\partial\xi_0}(L_0(1-u)\xi_0^2+d_0-\xi_0(L_0+d_0-L_0u\xi_1)) + \frac{\partial\Psi}{\partial\xi_1}(L_1\xi_1^2+d_1-\xi_1(L_1+d_1)). \tag{15.2}$$

In the next section we will generalize this methodology to multiple mutations, and then show how it can be used to answer important questions related to drug treatments of cancer. Before we go on, we will explain the general usefulness of the approach outlined here.

The probability function, $\varphi_{i_0,i_1}(t)$ contains all the "microscopic" information about the stochastic process of interest. In fact, it contains "too much" information, and often one could make shortcuts to extract only the "useful" information needed to answer a specific question. One such shortcut is demonstrated by the method outlined above and generalized below. The calculations leading to equation (15.2) demonstrate the convenience of the probability generating function description: instead of an infinite system of coupled equations (15.1) for the original probability function, $\varphi_{i_0,i_1}(t)$, we obtained a single first order PDE for the probability generating function. This PDE can be solved by the method of characteristics (as reviewed below). The solutions can be related to important characteristics of the process, such as the probability of cancer treatment success/failure.

15.2.4 *The probability generating function description*

In the general case of m drugs, let us introduce the function

$$\varphi_{i_0,\ldots,i_n}(t),$$

the probability to have i_s cells of type A^s at time t, where $0 \le s \le n = 2^m$ are binary numbers. We can write down the Kolmogorov forward equation,

$$\dot{\varphi}_{i_0,\ldots,i_n} = \sum_{s=0}^{n} Q\{A^s\}, \tag{15.3}$$

where $Q\{A^s\}$ is the contribution obtained from considering probabilities of reproduction and death of cell-type "A^s",

$$\begin{aligned} Q\{A^s\} = {} & \varphi_{\ldots,i_s-1,\ldots}(i_s-1)L_s\left(1-\sum_j u_j^{s,out}\right) + i_s L_s \sum_j \varphi_{\ldots,i_s,\ldots,i_j-1,\ldots} u_j^{s,out} \\ & + \varphi_{\ldots,i_s+1,\ldots}(i_s+1)d_s + K_s \sum_j \alpha_j^{s,out} \varphi_{\ldots,i_s+1,\ldots,i_j-1,\ldots}(i_s+1) \\ & - \varphi_{\ldots} i_s(L_s + d_s + K_s \sum_j \alpha_j^{s,out}). \end{aligned} \tag{15.4}$$

We used the following short-hand notations: $\varphi_{\ldots}$ stands for $\varphi_{i_0,\ldots,i_n}$, and the only explicit subscripts indicate the indices which are different from $(i_0,\ldots,i_n)$. In equation (15.4), the first term is faithful reproduction, the second term represents all possible mutations, the third term is death, the fourth term is transformation of cells, and the last term comes from the probability of no change.

It is convenient to define the probability generating function, $\Psi(\xi_0, \ldots, \xi_n; t)$,

$$\Psi(\xi_0, \ldots, \xi_n; t) = \sum_{i_0, \ldots, i_n} \varphi_{i_0, \ldots, i_n} \prod_{s=0}^{n} \xi_s^{i_s}. \tag{15.5}$$

This expression can be viewed as a transformation from discrete variables $i_0, \ldots i_n$ to continuous variables $\xi_0, \ldots, \xi_n$.

Let us multiply equation (15.3) by $\prod_{s=0}^{n} \xi_s^{i_s}$ and sum over all indices to obtain the equation for the generating function. The rule for rewriting various types of terms in terms of the generating function can be summarized as follows:

- Terms with $\varphi_{\ldots} i_s$ give $\xi_s \frac{\partial}{\partial \xi_s}$;
- Terms that multiply $\varphi_{\ldots i_s - 1 \ldots}(i_s - 1)$ give $\xi_s^2 \frac{\partial}{\partial \xi_s}$;
- Terms that multiply $\varphi_{\ldots i_s + 1 \ldots}(i_s + 1)$ give $\frac{\partial}{\partial \xi_s}$;
- Terms like $\varphi_{\ldots i_s \ldots i_j - 1 \ldots} i_s$ give $\xi_s \xi_j \frac{\partial}{\partial \xi_s}$;
- Terms with $\varphi_{\ldots i_s + 1 \ldots i_j - 1}(i_s + 1)$ give $\xi_j \frac{\partial}{\partial \xi_s}$.

It is convenient to introduce the following shorthand notations,

$$u^{s,out} = \sum_j u_j^{s,out}, \quad \alpha^{s,out} = \sum_j \alpha_j^{s,out}.$$

Then the function $\Psi(\xi_0, \ldots, \xi_n; t)$ satisfies the following hyperbolic partial differential equation:

$$\frac{\partial \Psi}{\partial t} = \sum_s \frac{\partial \Psi}{\partial \xi_s} \Bigg[\xi_s^2 L_s \left(1 - u^{s,out}\right) + d_s + \xi_s L_s \sum_j \xi_j u_j^{s,out}$$
$$+ K_s \sum_j \xi_j \alpha_j^{s,out} - \left(L_s + d_s + K_s \alpha^{s,out}\right) \xi_s \Bigg]. \tag{15.6}$$

The fact that this is a first order equation follows from the assumption of linearity of the underlying birth-death process. Nonlinear processes lead to higher order equations.

15.2.5 *The method of characteristics*

The first-order PDE (15.6) can be solved by the standard method of characteristics, see e.g. [Gockenbach (2010)], Chapter 8. The equations for characteristics for equation (15.6) are given by:

$$\dot{\xi_s} = L_s\left(1 - u^{s,out}\right)\xi_s^2 + \left[L_s \sum_j u_j^{s,out}\xi_j - \left(L_s + d_s + K_s\alpha^{s,out}\right)\right]\xi_s$$

$$+ K_s \sum_j \xi_j \alpha_j^{s,out} + d_s), \quad 0 \le s \le n. \tag{15.7}$$

Let us assume that at $t = 0$, we have M_0 wild-type (non-mutated) cells, so that $\varphi_{M_0,0,\ldots,0}(0) = 1$. From definition (15.5) we have

$$\Psi(\xi_0, \ldots, \xi_n; 0) = \xi_n^{M_0}.$$

Suppose that we want to obtain expression for $\Psi(\bar{\xi}_0, \ldots, \bar{\xi}_n; \bar{t})$ by the method of characteristics, where $\bar{\xi}_0, \ldots, \bar{\xi}_n$ and $\bar{t}$ are some fixed values. For time $\bar{t}$, we have

$$\Psi(\bar{\xi}_0, \ldots, \bar{\xi}_n; \bar{t}) = \xi_n(\bar{t})^{M_0},$$

where the function $\xi_n(t)$ is a solution of system (15.7) which satisfies the initial conditions,

$$\xi_j(0) = \bar{\xi}_j, \quad 0 \le j \le n. \tag{15.8}$$

This method can for instance be used to find the probability of non-production of m-hit mutants, see also Chapter 10. The quantity

$$\Psi(0, 1, \ldots, 1; t) = \xi_0(t)^{M_0}$$

has the meaning that by time t, no cell which contains all m mutations has been produced. This is an important quantity when we talk about the process of carcinogenesis, as well as resistant mutant generation, as described in the next section.

15.3 Application to cancer treatment and generation of resistance

Here we apply the stochastic models developed above for multi-drug resistance and investigate the dependence of treatment outcomes on the initial tumor load, mutation rates, the turnover rate of cancerous cells, and the treatment strategy. The main goal is to elucidate the general principles of the emergence and evolution of resistant cells inside the tumor, before and after the start of treatment.

The first stochastic model of drug resistance was created by Goldie and Coldman [Goldie and Coldman (1979)], who developed a whole new approach to mathematical treatment of resistance in their subsequent work, see e.g. [Goldie and Coldman (1983a,b); Coldman and Goldie (1985); Goldie and Coldman (1985, 1998)]. A number of important theoretical and numerical results have been obtained by the authors for the case of one and more drugs. Since this ground-breaking work, a lot of mathematical models of drug resistance in cancer have been proposed. Several models, including stochastic branching models for stable and unstable gene amplification and its relevance to drug resistance, were explored by [Kimmel and Axelrod (1990); Harnevo and Agur (1991, 1993); Axelrod *et al.* (1994); Kimmel and Stivers (1994)]. Methods of optimal control theory were used to analyze drug dosing and treatment strategies [Cojocaru and Agur (1992); Kimmel *et al.* (1998); Coldman and Murray (2000); Swierniak and Smieja (2001); Murray and Coldman (2003); Smieja and Swierniak (2003)] (for a review of the optimal control theory in chemotherapy see [Swan (1990)]). Models for tumor growth incorporating age-structured cell cycle dynamics, in application to chemotherapy scheduling, have been developed by [Gaffney (2004, 2005)]. Mechanistic mathematical models developed to improve the design of chemotherapy regimes are reviewed in [Gardner and Fernandes (2003)]. [Jackson and Byrne (2000)] extended an earlier PDE model of [Byrne and Chaplain (1995)] to study the role of drug resistance and vasculature in tumors' response to chemotherapy; in this class of spatial models, the tumor is treated as a continuum of different types of cells, which include susceptible and resistant cells. Another class of models is based on the Luria-Delbrück mutation analysis [Kendal and Frost (1988); Jaffrézou *et al.* (1994); Chen *et al.* (2000)].

The stochastic model described here follows the tradition of [Goldie and Coldman (1983a, 1998)], and takes this classical work a step further. In particular, it has been possible to include a nonzero death rate for cancer cells and still obtain analytical results. Other extensions include the studies of cross-resistance, quiescence/cycling transitions, and a variety of different treatment strategies, which are considered in Chapter 16.

15.3.1 *The framework*

The methodology developed in the previous section is a very useful modeling framework for studying treatment of tumors with single and multiple drugs. In this case, each cellular type (phenotype) is characterized by its resistance properties. For instance, if we consider one drug, then a cell can be either resistant or susceptible to this drug. The system has only two types, and the state space consists of vectors (i, j), where $i, j \geq 0$ are integer numbers denoting the number of susceptible and resistant cells respectively. We assume that resistance to the drug is acquired by means of a mutation. This process is reflected by the following (very simple) network, $x_0 \rightarrow^u x_1$. Here u is the mutation rate, that is, the probability to create a resistant mutant upon division, and the index k in notation x_k characterizes susceptibility, with $k = 0$ meaning susceptible, and $k = 1$ not susceptible (resistant) to the drug.

We can generalize this process of resistance generation to the case of m different drugs [Komarova and Wodarz (2005); Komarova (2006b)]. Each cell can acquire resistance to each of the drugs, by means of a certain mutation. There are m types of mutations, with rates $u_1, \ldots, u_m$. Each mutation event corresponding to rate u_i leads to a phenotype resistant to drug i. We assume that resistance to one drug does not imply resistance to another drug (in general, this is not always true and cross-resistance is often observed, which is addressed in Chapter 16). In order to develop resistance to all m drugs, a cell must accumulate m mutations. The total number of types is $2^m - 1$, and the system's state is characterized by an integer-valued vector of length 2^m. In particular, there are $\binom{m}{k}$ phenotypes resistant to $k \leq m$ drugs. We label each phenotype by a binary number of length m, where "1" indicates resistance to the drug corresponding to its position and "0" indicates susceptibility. For example, if $m = 3$ (see figure 15.2), then type 000 is fully susceptible, phenotype 111 is fully resistant, and type 101 is resistant to drugs number 1 and 3 and it is susceptible to drug number 2.

In the context of drug treatment, the death rate of cells, d_s, is comprised of their *natural death date*, D_s, which is equal to the death rate in the absence of treatment, and possibly a *drug-induced death rate*, H_s. We have

$$d_s = D_s + H_s.$$

For the wild-type cancer cells we assume that the corresponding rates satisfy $L_0 > D_0$, so that the colony grows in the absence of treatment. The ratio $0 \leq D_0/L_0 < 1$ defines the *turnover* of cancer cells. The values $D_0/L_0 \ll 1$ correspond to low turnover, low death cancers, whereas values $D_0/L_0 \approx 1$ describe extremely high-turnover, slow-growth cancers.

In addition to the natural death rate, cells can be subject to an additional, drug-induced death rate, H_s. There are many ways to model drug-induced death rates. For instance, we could assume that each drug a cell is susceptible to increases its death rate by a certain amount, so the more drugs a cell is resistant to, the higher its death rate. To model partial resistance, we could assume that even "maximally resistant" cells have a residual drug-induced death rate. However, here we will adopt a simpler framework. We assume that if a cell is resistant to *all* the drugs applied, then its drug-induced death rate is simply zero. On the other hand, if a cell is susceptible to at least one of the drugs, then its drug-induced death rate is equal to a constant, H, which has the meaning of the intensity of therapy. The more complicated scenarios described above can also be incorporated in the model, see e.g. [Katouli and Komarova (2010)].

15.3.2 *Treatment regimes*

Let us denote by t_* the time when treatment begins. The timing of treatment, that is, the value t_*, is related to the tumor size at the start of treatment, N. In the simplest case we can assume a deterministic relationship,

$$N = M_0 e^{(L_0 - D_0)t_*}, \tag{15.9}$$

where all the types in the absence of treatment are assumed to divide and die with the same rates given by L_0 and D_0. This is an approximation, which is made throughout this chapter and also Chapter 16. Its consequences are explored in detail in [Komarova *et al.* (2007)].

In the simplest case, we envisage a one-stage treatment, which begins at a given time-point, t_*. More complex processes consisting of several stages are considered in Chapter 16.

When applying the method of characteristics in the context of cancer treatment (Section 15.2.5), one needs to keep in mind that the coefficients in system (15.7) are time-dependent. The reason for this is that the death rate of cells, d_s, depends on whether treatment is applied or not. In the simplest case, we have the following 2-stage process: for $0 < t < t_*$ no treatment is applied, such that $d_s = D_s$. We call this regime the pre-treatment stage. At time $t = t_*$, therapy starts and continues for all values of $t \geq t_*$, such that $d_s = D_s + H_s$. This is the treatment stage. We would like to calculate the function $\Psi(\bar{\xi}_0, \ldots, \bar{\xi}_n; t_* + \bar{t})$, that is, the characteristic function at time $\bar{t}$ after the start of therapy. In this case we have two sets of equations for characteristics of type (15.7), one with "pre-treatment coefficients" and the other with "treatment coefficients".

In order to solve this problem, we can employ the following simple but superficially counterintuitive algorithm. First, we solve the system of ODEs for characteristics with *treatment* coefficients, with the initial conditions given by $\xi_0(0) = \bar{\xi}_0, \ldots, \xi_n(0) = \bar{\xi}_n$, and obtain the solution $\xi_0(\bar{t}), \ldots, \xi_n(\bar{t})$. Next, we use these values as the initial condition for the system with *pre-treatment* coefficients, and get the solution of that at time t_*, $\xi_0(t_*), \ldots, \xi_n(t_*)$. The result is then given by

$$\Psi(\bar{\xi}_0, \ldots, \bar{\xi}_n; \bar{t}) = \xi_n(t_*)^{M_0}.$$

The reason for this time-reversal is the fact that the equations for characteristics are used to trace the trajectories back to the initial conditions. This is consistent with the standard 1st order PDE techniques.

15.3.3 *Probability of extinction and treatment success*

As explained above, the function $\Psi(0, 1, \ldots, 1; t)$ has the meaning of the probability that at time t, no cells of type "A^n" exist. We will call this quantity "the probability of non-production" (of resistance); it is given by

$$P_{non-prod}(t) = \Psi(0, 1, \ldots, 1; t) = \xi_0(t)^{M_0},$$

where M_0 is the type cells, and functions $\xi_i(t)$ satisfy system (15.7) with initial conditions

$$\xi_0 = 0, \quad \xi_i = 1, \quad 1 \leq i \leq n.$$

The probability of extinction, that is, the probability that there are zero cells of each type at time t, can be defined as

$$P_{ext} = \Psi(0, 0, \ldots, 0; t) = (\xi_0(t))^{M_0},$$

where $\xi_n(t)$ is the solution of the system (15.7) with different initial conditions,

$$\xi_0(0) = \xi_1(0) = \ldots = \xi_n(0) = 0.$$

Note that in the regime of treatment, the probability of extinction, as $t \to \infty$, coincides with the long-time limit of the probability of non-production of a resistant mutant,

$$\lim_{t\to\infty} P_{ext} = \lim_{t\to\infty} P_{non-prod}.$$

The function $\Psi(0, \ldots, 0; t)$ is a monotonically increasing function, which equals zero for $t = 0$, and steadily approaches a horizontal asymptote. The function $\Psi(0, 1, \ldots, 1; t)$ is not necessarily monotonic. It starts at one for $t = 0$, can drop to a minimum and then climb up and approach the same asymptote from below. In other scenarios, it can be a monotonically decreasing function of time (e.g. for $D = 0$). The reason for a non-monotonic behavior is this. As treatment starts at $t = 0$, there are no mutants, and then some resistant mutants may be produced quickly, while the population of cells is still large. Later on, the susceptible population decreases dramatically, so no new resistant mutants are produced, but there is a natural death rate for the mutants which may lead to accidental extinction of the mutant colony before it reaches a significant size. Once the resistant colony grows, the chance of its spontaneous extinction approaches zero, and thus the probability to have a resistant mutant stabilizes at a constant level.

The quantities P_{ext} and $P_{non-prod}$ are related to the probabilities of treatment success and failure. The probability of treatment failure can be defined as

$$P_{failure}(t) = 1 - \xi_0^{M_0}.$$

To define the probability of treatmnet success, we note that the quantity $\varphi_{0,\ldots,0}(t) = \Psi(0, \ldots, 0; t)$ is the probability of having zero cells of all types at time t. This probability includes (i) the scenario where the colony goes extinct spontaneously, and (ii) the scenario where the tumor grows and is subsequently treated successfully. The latter process has the meaning of the probability of treatment success. For small mutation rates, the probability that scenario (i) occurs can be approximated as $(D_0/L_0)^{M_0}$. We have,

$$\varphi_{0,\ldots,0}(t) = (D_0/L_0)^{M_0} + (1 - (D_0/L_0)^{M_0})P_{success}(t),$$

where M_0 denotes the initial number of wild-type cells, and L_0 and D_0 are division and death rates of susceptible (wild-type) cells before treatment starts. Thus we have

$$P_{success}(t) = \frac{\xi_0^{M_0}(t) - (D_0/L_0)^{M_0}}{1 - (D_0/L_0)^{M_0}}. \qquad (15.10)$$

15.3.4 *Symmetric coefficients*

All the resistant types can be separated into classes such that in each class k, all the types are resistant to exactly k drugs (and susceptible to $m - k$ drugs). For each k, the class consists of all variables ξ_s such that the binary numbers s contain exactly k nonzero entries. Therefore, we can denote by ξ_k, with $0 \le k \le m$, the class of variables describing resistance to k drugs. Let us suppose that within each class, the birth and death rates are equal, and also that all mutation rates are equal to each other. Because of this symmetry assumption, it does not matter which k drugs we pick. The total number of distinct equations in this case is not $n = 2^m + 1$, but $m + 1$:

$$\dot{\xi}_k = L_k(1-(m-k)u)\xi_k^2 + [(m-k)L_k u\xi_{k+1} - (L_s + d_k)]\xi_k + d_k, \quad 0 \le k \le m. \tag{15.11}$$

We can further simplify the description by assuming a fully symmetrical case, where all the rate coefficients are the same for all types and all drugs.

15.4 Example: the case of two drugs

Let us suppose that the treatment is a combination of $m = 2$ drugs. Therefore we can distinguish four phenotypes, "A^{00}", "A^{10}", "A^{01}" and "A^{11}". The Kolmogorov forward equation is given by:

$$\begin{aligned}
\dot{\varphi}_{i_{00},i_{10},i_{01},i_{11}} = & \Big[\varphi_{i_{00}-1,i_{10},i_{01},i_{11}} L_{00}(i_{00}-1)(1-u_1-u_2-u_{12}) + \\
& \varphi_{i_{00},i_{10}-1,i_{01},i_{11}} L_{10}(i_{10}-1)(1-u_2-u_{12}) + \\
& \varphi_{i_{00},i_{10},i_{01}-1,i_{11}} L_{01}(i_{01}-1)(1-u_1-u_{12}) + \varphi_{i_{00},i_{10},i_{01},i_{11}-1} L_{10}(i_{11}-1)\Big] \\
& + \Big[i_{00}(u_1\varphi_{i_{00},i_{10}-1,i_{01},i_{11}} + u_2\varphi_{i_{00},i_{10},i_{01}-1,i_{11}} + u_{12}\varphi_{i_{00},i_{10},i_{01},i_{11}-1}) \\
& + i_{10}(u_2\varphi_{i_{00},i_{10},i_{01},i_{11}-1} + u_{12}\varphi_{i_{00},i_{10},i_{01},i_{11}-1}) \\
& + i_{01}(u_1\varphi_{i_{00},i_{10},i_{01},i_{11}-1} + u_{12}\varphi_{i_{00},i_{10},i_{01},i_{11}-1})\Big] + \\
& \Big[d_{00}\varphi_{i_{00}+1,i_{10},i_{01},i_{11}} + d_{10}\varphi_{i_{00},i_{10}+1,i_{01},i_{11}} + \\
& d_{01}\varphi_{i_{00},i_{10},i_{01}+1,i_{11}} + d_{11}\varphi_{i_{00},i_{10},i_{01},i_{11}+1}\Big] -
\end{aligned}$$

$$\varphi_{i_{00},i_{10},i_{01},i_{11}}\left[i_{00}(L_{00}+d_{00})+i_{10}(L_{10}+d_{10})\right.$$

$$\left.+i_{01}(L_{01}+d_{01})+i_{11}(L_{11}+d_{11})\right]. \quad (15.12)$$

In this master equation, the first term in square brackets on the right hand side comprises all the processes of faithful cell division, the second term in square brackets includes all the mutation events, the third one represents all the cell death events, and the fourth term corresponds to no change in the system's state.

15.4.1 *Equations for the moments*

From the master equation, information about all the moments can be extracted. In particular, the equations for the mean number of cells in each class can be written. Let us denote

$$x_{00}(t)=\sum \varphi_{i_{00},i_{10},i_{01},i_{11}}(t)i_{00}, \quad x_{10}(t)=\sum \varphi_{i_{00},i_{10},i_{01},i_{11}}(t)i_{10},$$

$$x_{01}(t)=\sum \varphi_{i_{00},i_{10},i_{01},i_{11}}(t)i_{01}, \quad x_{11}(t)=\sum \varphi_{i_{00},i_{10},i_{01},i_{11}}(t)i_{11},$$

where the summation is performed over all the four indices. Then we have:

$$\dot{x}_{00}=[L_{00}(1-u_1-u_2)-d_{00}]x_{00}, \quad (15.13)$$

$$\dot{x}_{10}=[L_{10}(1-u_2)-d_{10}]x_{10}+L_{00}u_1x_{00}, \quad (15.14)$$

$$\dot{x}_{01}=[L_{01}(1-u_1)-d_{01}]x_{01}+L_{00}u_2x_{00}, \quad (15.15)$$

$$\dot{x}_{11}=[L_{11}-d_{11}]x_{11}+L_{10}u_2x_{10}+L_{01}u_1x_{01}. \quad (15.16)$$

These equations can be obtained directly from the master equation; for example, the first equation is nothing but equation (15.12) multiplied by i_{00} and summed over all the indices. The initial conditions can be written as

$$x_{00}(0)=M_0, \quad x_{10}(0)=x_{01}(0)=x_{11}(0)=0. \quad (15.17)$$

In other words, we assume that at time zero, there are M_0 fully-susceptible cells, and no mutants are initially present. The deterministic equations obtained in this way can help one reason about the expected dynamic of the colony growth and resistance generation. Equations of this type are used, for example, in [Komarova and Wodarz (2007); Katouli and Komarova (2011)]. However, they cannot address questions of the probability of treatment success. To quantify the likelihood of a successful treatment outcome, we need to use the stochastic approach, which is described next.

15.4.2 Equations for the characteristics

The probability generating function is defined in accordance with equation (15.5),

$$\Psi(\xi_{00},\xi_{10},\xi_{01},\xi_{11};t)=\sum_{s=0}^{n}\varphi_{i_{00},i_{10},i_{10},i_{11}}(t)\xi_{00}^{i_{00}}\xi_{10}^{i_{10}}\xi_{01}^{i_{01}}\xi_{11}^{i_{11}}.$$

The equations for the characteristics are as follows:

$$\dot{\xi}_{11}=L_{11}\xi_{11}^2-(L_{11}+d_{11})\xi_{11}+d_{11}, \tag{15.18}$$
$$\dot{\xi}_{10}=L_{10}(1-u_2)\xi_{10}^2+[L_{10}u_2\xi_{11}-(L_{10}+d_{10})]\xi_{10}+d_{10}, \tag{15.19}$$
$$\dot{\xi}_{01}=L_{01}(1-u_1)\xi_{01}^2+[L_{01}u_1\xi_{11}-(L_{01}+d_{01})]\xi_{01}+d_{01}, \tag{15.20}$$
$$\dot{\xi}_{00}=L_{00}(1-u_1-u_2)\xi_{00}^2 + [L_{00}(u_1\xi_{10}+u_2\xi_{01})-(L_{00}+d_{00})]\xi_{00}+d_{00}, \tag{15.21}$$

with general initial conditions (15.8). Solutions of this system with the initial conditions

$$\xi_{00}=\xi_{10}=\xi_{01}=\xi_{11}=0 \tag{15.22}$$

are used to calculate the probability of treatment success. Equations (15.18-15.21) can be solved recursively. Let us assume the absence of treatment ($d_s=D_s$). We make the change of variables,

$$\xi_s=-\frac{\dot{X}_s}{L_s\left(1-u^{s,out}\right)X_s},\quad 0\le s\le n \tag{15.23}$$

and obtain a Riccati-type equation for X_s:

$$\ddot{X}_s+\left[L_s\left(1-\sum_j u_j^{s,out}\xi_j\right)+D_s\right]\dot{X}_s+L_s\left(1-u^{s,out}\right)D_sX_s=0. \tag{15.24}$$

Note that the solution, X_s, depends of the functions ξ_j, the variables downstream from the node s. For $D_s\neq 0$, only the first of the equations, the one for X_{11}, can be solved analytically because there are no variables downstream from node (11), and we have a Riccati equation with constant coefficients. All the rest of the equations in general have to be solved numerically.

One way to get some analytical insights is to suppress the dynamics of type "A^{11}", such that the equation for ξ_{11} becomes

$$\dot{\xi}_{11} = 0. \tag{15.25}$$

In this case the quantity

$$1 - \xi_s(t)$$

stands for the probability to create at least one mutant of type "A^s" by time t; note that this mutant may have died away by time t, or it may have created offspring: because of equation (15.25) we do not distinguish between these scenarios. From initial conditions (15.22) we have $\xi_{11}(t) = 0$, and the two equations for the variables corresponding to resistance to one drug can be solved analytically. These variables correspond to the indices s that contain only one nonzero entry. We have,

$$\xi_{01} = -\frac{b_1 + Ab_2 e^{(b_2-b_1)t}}{L_{01}(1-u_1)(1+Ae^{(b_2-b_1)t})}, \tag{15.26}$$

where

$$A = -\frac{b_1 + L_{01}(1-u_1)}{b_2 + L_{01}(1-u_1)},$$

and $b_1 > b_2$ are roots of the quadratic equation,

$$b^2 + (L_{01} + D_{01})b + L_{01}(1-u_1)D_{01} = 0. \tag{15.27}$$

Similarly, ξ_{10} is obtained by changing $u_1 \to u_2$, $L_{01} \to L_{10}$, and $D_{01} \to D_{10}$.

To get the answer for ξ_{00}, these expressions should be substituted in equation (15.21) to continue the recursion. Only a solution by numerical integration is possible, unless further simplifying assumptions are made.

15.5 Mutant production before and during treatment

15.5.1 *General theory*

At what stage are mutants predominantly generated: before the start of treatment, or after therapy is applied? Let us introduce two quantities,

$P^{\uparrow}(N)$ and $P^{\downarrow}(N)$, in order to be able to characterize and compare the production of mutants before therapy and during therapy.

To simplify the description, we will assume the symmetry in coefficients described in Section 15.3.4, so that the number of equations for m drugs is $m+1$.

Pre-existence of resistant mutants. Let us suppose that we can switch off the mutation rate during therapy, and calculate the probability of non-extinction (which is the same as the probability to have resistant mutants as $t \to \infty$). Let us find the probability to have mutants in the limit when $t \to \infty$. We have the following systems of equations. Before start of treatment,

$$\dot{\xi}_m = L\xi_m^2 - (L+D)\xi_m + D, \tag{15.28}$$

$$\dot{\xi}_i = L(1-iu)\xi_i^2 + (iLu\xi_{i+1} - (L+D))\xi_i + D, \quad 0 \le i < m, \tag{15.29}$$

and upon the start of therapy,

$$\dot{\xi}_m = L\xi_m^2 - (L+D)\xi_m + D, \tag{15.30}$$

$$\dot{\xi}_i = L(1-iu)\xi_i^2 - (L+D+H)\xi_i + D + H. \quad 0 \le i < m. \tag{15.31}$$

In order to solve this general problem, we need to implement the method described in Section 15.2.5. We are interested in the quantity

$$P^{\uparrow}(N) = \lim_{\bar{t}\to\infty} \Psi(0, 1, \ldots, 1, t_* + \bar{t}).$$

The meaning of $P^{\uparrow}$ is the probability to develop resistance, as $t \to \infty$, if mutations only happen before the start of treatment.

First we solve system (15.30-15.31) and find its limiting behavior. Then we use the steady-state values for all the variables as the initial condition for system (15.28-15.29). The quantity describing pre-existence of resistant mutants can be calculated as

$$P^{\uparrow}(N) = 1 - \xi_0(t_*)^{M_0}, \quad t_* = \frac{1}{L-D}\ln(N/M_0).$$

This corresponds to the probability to have at least one fully resistant mutant as $t \to \infty$, given that the size of the colony at the start of treatment is N, and no further mutants are produced during treatment.

Generation of mutants during treatment. In order to characterize the role of the treatment phase in the generation of resistance, we will switch off the mutation rates during the growth (pre-treatment) phase, and turn them back on once treatment starts. The simplest way in which the problem can be formulated is as follows. We consider only the treatment phase, and start from N susceptible cells (no pre-existence). We have

$$\dot{\xi}_m = L\xi_m^2 - (L+D)\xi_m + D, \tag{15.32}$$

$$\dot{\xi}_i = L(1-iu)\xi_i^2 + [iLu\xi_{i+1} - (L+D+H)]\xi_i + (D+H), \quad 0 \le i < m, \tag{15.33}$$

$$\xi_i(0) = 0, \quad 0 \le i \le m. \tag{15.34}$$

The quantity characterizing the generation of mutants during therapy is given by

$$P^{\downarrow}(N) = 1 - \lim_{t\to\infty} \xi_0(t)^N.$$

This function has the meaning of the probability to have created viable mutants in the course of therapy, starting from N fully-susceptible cells.

We can calculate the limiting behavior of solutions of system (15.32-15.34), as time goes to infinity, by using the method described in Section 15.3.3. We have, under the assumption that $u \ll (L-D)$, that

$$\lim_{t\to\infty} \xi_m = \frac{D}{L}, \tag{15.35}$$

$$\lim_{t\to\infty} \xi_i = 1 - \frac{i!(L-D)L^{i-1}u^i}{(D+H-L)^i}, \quad 0 \le i < m. \tag{15.36}$$

If the total number of drugs used is m, then the probability of mutant generation with no pre-existence is given by

$$P^{\downarrow}(N) = 1 - \left(1 - \frac{m!(L-D)L^{m-1}u^m}{(D+H-L)^m}\right)^N. \tag{15.37}$$

We observe that the larger the initial size, the lower is the probability of treatment success.

In the rest of this section we will calculate and compare the quantities $P^{\uparrow}$ and $P^{\downarrow}$. If it turns out that $P^{\uparrow}(N) > P^{\downarrow}(N)$, then we can conclude that resistance is generated at a higher intensity before therapy, and the contribution of the therapy phase is less important. The opposite result, $P^{\uparrow}(N) < P^{\downarrow}(N)$, would tell us that most mutants are generated after therapy begins. The biological consequences of the results are discussed in more detail in the next chapter.

15.5.2 *The case of one drug*

Mutant production during treatment. The expression for the probability to create resistance during therapy is obtained from equation (15.37):

$$P^{\downarrow}(N) = 1 - \left(1 - \frac{(L-D)u}{H-(L-D)}\right)^N. \tag{15.38}$$

In the analysis below it is convenient to use the following threshold value of treatment intensity, H:

$$H_c = 2(L-D). \tag{15.39}$$

For $H = H_c$, we have

$$P^{\downarrow}(N) = 1 - (1-u)^N \approx 1 - e^{-Nu}. \tag{15.40}$$

For values of H larger than the threshold, $H > H_c$, the corresponding probability to create resistance is lower.

Mutant production before treatment. In the case of one drug, we have only two equations in system (15.28-15.29). The equation for $\xi_m = \xi_1$ can be solved exactly. However, the equations for ξ_0 will contain non-constant coefficients and cannot be solved analytically (unless $D = 0$). Therefore, instead of solving systems (15.28-15.29, 15.30-15.33) directly, we will use a different method.

Let us first suppress the dynamics of the resistant mutant ($\dot{\xi}_1 = 0$) and calculate the probability of mutant generation during the pre-treatment phase. We have $\xi_1 = 0$, and thus the equation for ξ_0 reads

$$\dot{\xi}_0 = L(1-u)\xi_0^2 - (L+D)\xi_0 + D.$$

This Riccati equation can be solved to yield

$$\xi_0 = -\frac{\beta_1 + A\beta_2 e^{(\beta_2-\beta_1)t}}{a(1+Ae^{(\beta_2-\beta_1)t})},$$

with

$$\beta_{1,2} = \frac{1}{2}\left(-(L+D) \pm \sqrt{(L+D)^2 - 4LD(1-u)}\right),$$

and

$$A = -\frac{\beta_1 + L(1-u)}{\beta_2 + L(1-u)}.$$

Taking the highest order terms in u, we obtain,

$$\xi_0(t) = 1 - \frac{(e^{(L-D)t} - 1)Lu}{L - D} \approx 1 - \frac{NLu}{M_0(L - D)}, \tag{15.41}$$

and the probability to have created a mutant is simply

$$P_{create} = 1 - \xi_0^{M_0} \approx \frac{LuN}{L - D}.$$

This quantity has the meaning of the probability to have produced at least one resistant mutant by the time the colony has reached size N (this mutant and all its progeny may or may not be present at this point).

This is a very intuitive result, as the probability to have created a mutant is given by the total number of cell divisions, $\mathcal{N}$, times the mutation rate, u. The total number of cell divisions from one cell to N cells is roughly given by $\mathcal{N} = NL/(L - D)$, which results in the formula for P_{create} above.

Now, each resistant mutant created during the pre-treatment phase, will give rise to a lineage of progeny, with the growth rate L and the death rate D, which is not affected by the presence of the drug. The probability for a lineage starting from one cell to survive (not to go extinct as $t \to \infty$) is given by $P_{survive} = 1 - D/L$. Therefore, the probability to generate resistance during the pre-treatment phase is given by

$$P^{\uparrow}(N) = P_{create}P_{survive} = Nu.$$

This result was obtained by intuitive reasoning, and it obviously holds only when $Nu < 1$. A more rigorous derivation for the quantity $P^{\uparrow}$ is possible, and is presented next.

Let us consider one equation,

$$\dot{x}_1 = Lx_1^2 - (L + D)x_1 + D, \quad x_1(0) = 1,$$

where the quantity $1 - x_1(t)$ has the meaning of the probability that a one-hit colony survives until time t. We have,

$$1 - x_1(t) = \frac{L - D}{L - De^{-(L-D)t}}. \tag{15.42}$$

Using doubly-stochastic processes, we can calculate the probability that the colony has at least one mutant by time t,

$$1 - exp\left\{-uL \int_0^t N(t')[1 - x_1(t - t')]\, dt'\right\},$$

where the multiplier $uN(t')$ reflects the production of mutants and the quantity $(1-x_1)$ corresponds to the survival probability of each mutant lineage. We set $N(t) = M_0 e^{(L-D)t}$. Integration gives

$$1 - exp\left\{-Nu\frac{L}{D}\left(Dt + \ln\frac{Le^{-Dt} - De^{-Lt}}{L-D}\right)\right\}. \qquad (15.43)$$

This quantity corresponds to the probability to have at least one resistant mutant by the time when therapy starts. Note that in the limit $D \to 0$ we obtain

$$P^{\uparrow} = 1 - e^{-u(N(t)-M_0)},$$

which coincides exactly with the result by [Goldie and Coldman (1983a)]. Equation (15.43) is a generalization of the formula by [Goldie and Coldman (1983a)] for nonzero death-rates.

In order to calculate the quantity $P^{\uparrow}(N)$, we need to find the limiting behavior of the probability to have a mutant as $t \to \infty$, and also to ignore the production of new mutants during the treatment stage. Instead of the time-dependent formula (15.42) we can use its limiting value at $t \to \infty$, $1 - D/L$, so we have

$$P^{\uparrow}(N) = 1 - e^{\left\{-M_0 uLe^{(L-D)t}\int_0^t e^{-(L-D)t'}\left(1-\frac{D}{L}\right)dt'\right\}} = 1 - e^{-Nu}. \quad (15.44)$$

We can see that this is exactly equal to expression (15.40).

Probability of treatment success. Let us solve the two-stage problem for the case of one drug, $m = 1$. Therapy starts at the time, t_* (and the colony size reaches the value N). Before start of treatment, we have system (15.28-15.29), and during treatment, we have equations (15.32-15.33).

In order to solve this problem, let us use the method of Section 15.2.5. We first need to find the limiting values of ξ_i under the treatment conditions (system (15.32-15.33)), as given by formulas (15.35-15.36), and use these as the initial conditions for equations (15.28-15.29), in the interval $0 \le t \le t_*$, where t_* is the time when treatment starts. The quantities $[\xi_i(t_*)]^{M_0}$ are the probabilities of treatment success with therapy starting at time t_*.

Equation (15.28) for $\xi_1(t)$ with initial condition (15.35) can be solved exactly to give

$$\xi_1(t) = \frac{D}{L}.$$

The equation for ξ_0 is a constant coefficient Riccatti equation,

$$\dot{\xi}_0 = L(1-u)\xi_0^2 + (Du - (L+D))\xi_0 + D, \quad \xi_0(0) = 1 - \frac{(L-D)u}{D+H+L}.$$

We can write down the exact solution:

$$\xi_0 = \frac{Ae^{(D[1-u]-L)t}L + D(1-u)}{L(1-u)[1+Ae^{(D[1-u]-L)t}]}, \quad A = \frac{(1+u)(L-D)(D+H-L[1+u])}{Lu[H+(D-L)u]}.$$

The limiting behavior is given by

$$\lim_{t_* \to \infty} \xi_0 = \frac{D}{L}.$$

In the case where $Nu < 1$, we can take the limit $u \to 0$ and find an approximate formula for $\xi_0(t)$,

$$\xi_0 \approx 1 - \left(\frac{He^{(L-D)t}}{H+D-L} - 1 \right) u.$$

Setting

$$e^{(L-D)t} = \frac{N}{M_0},$$

and neglecting 1 compared to this quantity, we can write down the probability of treatment success,

$$P_{success}(N) \approx \left(1 - \frac{HNu}{(H+D-L)M_0} \right)^{M_0}. \qquad (15.45)$$

We can see that larger values of D correspond to a higher probability of treatment success. If therapy is very strong such that $H \gg L - D$, we have a very simple formula,

$$P_{success}(N) \approx \left(1 - \frac{N}{M_0} u \right)^{M_0},$$

i.e., treatment success does not depend on D for strong therapies.

15.5.3 *The case of two drugs*

Mutant production during treatment. From expression (15.37) we obtain in the case $m = 2$,

$$P^{\downarrow}(N) = 1 - \left(1 - \frac{2(L-D)Lu^2}{(H-(L-D))^2} \right)^N.$$

Although Nu may or may not be a small quantity, it is safe to assume that $Nu^2 \ll 1$, which allows the following approximation:

$$P^{\downarrow}(N) = \frac{2(L-D)Lu^2N}{(H-(L-D))^2}. \tag{15.46}$$

Mutant production before treatment. In the case of two drugs, let us calculate the probability of double-mutant creation as a function of the tumor size. Again, we will suppress the dynamics of the double mutants, such that $\xi_0 \equiv 0$, and only keep track of the creation process. We will make the approximation of a doubly-stochastic process, whereby generation of each one-hit mutant leads to a birth-death process, all of which are independent and identically distributed (see also [Moolgavkar *et al.* (1988); Iwasa *et al.* (2004b)]). We will assume that the total population size changes according to the deterministic exponential law, $N(t) = M_0 e^{(L-D)t}$ (see [Komarova *et al.* (2007)] for extended discussion). Generalizing the notion of a filtered Poisson processes described e.g. in [Parzen (1962)], we have,

$$\Psi(0,1;t) = \exp\left[-2Lu\int_0^t N(t')\,(1-\Phi(0,1;t-t'))\,dt'\right], \tag{15.47}$$

where $LuN(t')dt'$ is the probability to create a one-hit mutant in the time-interval $(t, t+dt)$, and $1-\Phi(0,1;t-t')$ is the probability that the lineage resulting from that mutant will give rise to the production of a double-mutant in the time from the creation of the lineage, t', to the current time, t. The latter probability is given by $1-\xi_1(t-t')$, see formula (15.41). The factor 2 in the exponent comes from the two possibilities of acquiring two hits. We have,

$$P_{create} = 2\left(\frac{Lu}{L-D}\right)^2 N\left(\ln\frac{N}{M_0}-1\right). \tag{15.48}$$

The method of a doubly-stochastic process is a good approximation as long as the one-hit mutants are "rare". Multiplying P_{create} by the probability of each double-mutant to survive, $P_{survive} = 1-D/L$, we obtain

$$P^{\uparrow}(N) = \frac{2Lu^2N}{L-D}\left(\ln\frac{N}{M_0}-1\right), \tag{15.49}$$

that is, this quantity now depends on the turnover rate, D/L. Note that formula (15.49) breaks down as $D \to L$ as the doubly-stochastic approximation is not applicable in this regime anymore; at this moment we do not have a method to handle the regime $D \approx L$.

15.6 Outlook

In this chapter we developed a mathematical formalism which allows us to calculate the probability of resistance generation, in the setting of single- or multi-drug treatment. This framework only includes the basics. For example, we only considered a one-phase treatment with a combination therapy in the absence of cross-resistance. The main point is to show the reader how to approach the basic problem. More complicated scenarios can be studied by using similar methods.

In Chapter 16 we will demonstrate how this framework can be applied to approach several important questions in drug treatments of cancer. We will consider multi-stage therapies, include the existence of quiescent cells, and examine the problem of cross-resistance.

Problems

Problem 15.1. *Derive the partial differential equation (15.2) from the Kolmogorov forward equation (15.1), in the case of a one-drug treatment.*

Problem 15.2. *The same in the case of 2-drug treatments: Derive the partial differential equation and the equations for characteristics, (15.13-15.16), from Kolmogorov forward equation (15.12). How does the description simplify in the case of symmetric coefficients, $L_s = L$, $D_s = D$, etc.?*

Problem 15.3. *In the general case of m-drug treatments: Derive the partial differential equation (15.6) from the Kolmogorov forward equation (15.4). Show that the rewrite rules listed in Section 15.2.4 hold.*

Problem 15.4. *Apply the method of characteristics outlined in Section 15.2.5 to the PDE in the case of one-drug treatment, (15.2). Find the probability of treatment success assuming that at $t = 0$, there are M_0 susceptible mutants, and treatment starts at time t_*.*

Problem 15.5. *Show that the probability of spontaneous colony extinction starting from M_0 cells in the absence of treatment is given by $(D_0/L_0)^{M_0}$, where L_0 and D_0 are division and death rates of the cells. This result was used in the derivation of formula (15.10). (Hint: this is similar to the gambler's ruin problem.)*

Problem 15.6. ***Research project.*** *Determining the number of mutations* *in a growing colony of cells is related to the famous Luria-Delbrk̈ distribution. Find out about the history and major benchmarks in the efforts to solve this problem.*

Problem 15.7. ***Research project.*** *In this book, approximation (15.9) was used, which postulates a deterministic connection between the total cell number and the time elapsed. What are the consequences of this assumption (see [Komarova* et al. *(2007)])?*

Problem 15.8. ***Numerical project.*** *Use any programming language to set up a simulation that calculates the probability of treatment success, given a treatment with* m *drugs that begins at time* t_* *(with approximation (15.9)). Assume that at* $t = 0$*, the colony consists of* M_0 *susceptible mutants. First solve the system for the charactetristic equations under the treatment conditions (for infinitely long treatment, the result can be found from the stable equilibrium of the system for characteristics). Then use this result as the initial condition for the system with pre-treatment coefficients.*

positive cells of about 0-35%. The molecular response quantifies the number of BCR-ABL mRNAs and is the most sensitive method to monitor the treatment response. A complete molecular response is defined by the absence of detectable BCR-ABL mRNAs, measured by reverse transcriptase polymerase chain reaction (RT-PCR). A major molecular response occurs if a 3-log reduction of the BCR-ABL/BCR level occurs compared to the median pre-treatment levels.

The initial treatment approach involved the use of chemotherapeutic agents [Bolin *et al.* (1982)] which normalized the blood cell count but failed to alter disease progression to blast crisis. Treatment outcomes

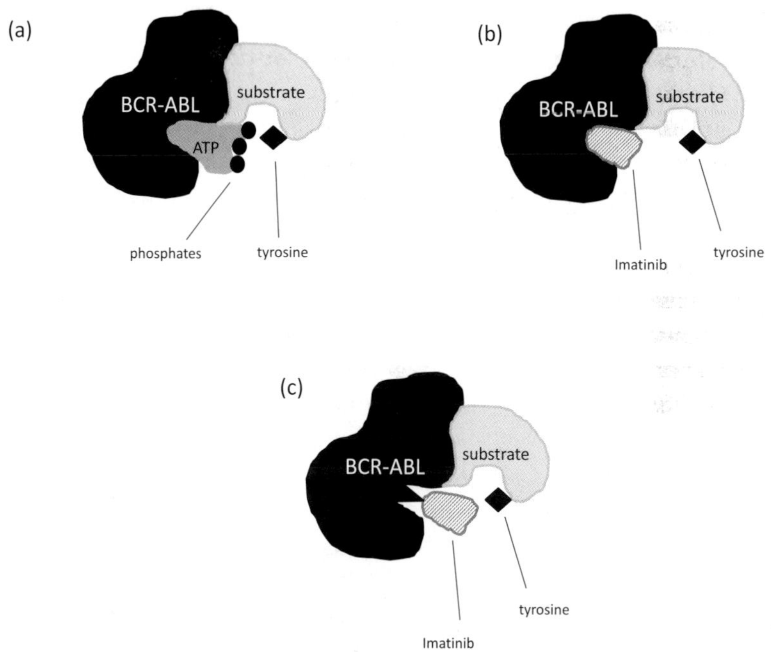

Fig. 16.1 Schematic explaining the concept behind the tyrosine kinase inhibitor imatinib in the treatment of CML. (a) BCR-ABL tyrosine kinase binds ATP and transfers phosphates from ATP to tyrosine residues on substrates. This leads to downstream effects that cause unregulated cellular behavior. (b) Tyrosine kinase inhibitors, such as imatinib, block ATP from binding to BCR-ABL tyrosine kinase, thus preventing unregulated behavior. (c) A point mutation in the ATP binding site of the BCR-ABL tyrosine kinase changes the shape of the binding site (indicated by a triangle), thus blocking the drug from binding.

were significantly improved by the introduction of recombinant interferon alpha (rIFN-α) therapy [Group (1997)]. Hematopoietic stem cell transplants also offered benefits and could even lead to cures, although they are associated with a high risk of morbidity and mortality [An *et al.* (2010)]. The breakthrough in CML treatment came with the development of small molecule inhibitors of the tumorigenic BCR-ABL tyrosine kinase. The first such small molecule inhibitor was imatinib mesylate, also referred to as imatinib, STI-571, or Gleevec [Deininger *et al.* (2005); Deininger and Druker (2003)]. The principle according to which this drug works is as follows. The BCR-ABL gene product is a constitutively active kinase that binds ATP and transfers phosphate from ATP to tyrosine residues, thereby leading to aberrant cellular behavior. Imatinib is designed to specifically block the binding of ATP to the BCR-ABL tyrosine kinase, thereby preventing continued functioning of the tumor cells (figure 16.1).

The use of imatinib in the chronic phase of the disease leads to very good treatment results, demonstrating impressive hematological, cytogenetic, and molecular responses. Drug treatment typically leads to a biphasic decline in the BCR-ABL transcript number [Michor *et al.* (2005b); Roeder *et al.* (2006)]. Initially a fast phase of decline is observed, followed by a slower phase. These dynamics can be explained with a number of different hypotheses [Michor *et al.* (2005b); Roeder *et al.* (2006); Komarova and Wodarz (2007); Wodarz (2010)]. Patients with complete cytogenetic and molecular responses also showed prolonged progression-free survival, indicating that the course of the disease had been altered [An *et al.* (2010)]. However, in patients that were treated during blast crisis, the outcome was less impressive. A major obstacle for treatment success was resistance to imatinib [Shannon (2002); Shah *et al.* (2004); Tauchi and Ohyashiki (2004); Kantarjian *et al.* (2006); Branford *et al.* (2003); Volpe *et al.* (2009); Bozic *et al.* (2012)]. Most commonly, resistance is conferred by point mutations that prevent the binding of the drug to its target (figure 16.1). In a minority of cases, gene amplification events can also lead to drug resistance. The consequent increase in the level of BCR-ABL protein levels allows ongoing oncogenic activity in the presence of the drug. Subsequent, second generation inhibitors have been developed to broaden the therapeutic approach and to address problems arising from drug resistance. These include the drugs dasatinib and nilotinib [Leitner *et al.* (2011); Kantarjian *et al.* (2011); Weisberg *et al.* (2007)].

There are many point mutations that confer resistance against a particular inhibitor. Many of them specifically confer resistance to one (e.g.

imatinib) but not the other inhibitors (e.g. dasatinib and nilotinib). However, one of the most common mutations found in blast crisis patients, the T315I mutation [O'Hare *et al.* (2006)], confers resistance to imatinib, dasatinib, and nilotinib. This phenomenon is referred to as "cross-resistance". It can considerably complicate treatment options. A number of drugs are under development that are aimed at overcoming the T315I mutation [An *et al.* (2010); Giles *et al.* (2007)]. Examples of such drugs include ponatinib, a pan-BCR-ABL inhibitor, and the aurora kinase inhibitor MK-0457, which have documented activity against cells containing the T315I mutation.

16.3 The computational framework

In order to understand how resistant mutants are generated during cancer progression and treatment, we use the mathematical framework that was introduced in the previous chapter. The following provides a basic summary, and the basic concepts are also explained schematically in figure 16.2. Cancerous cells are described by a stochastic birth-death process with a positive net proliferation rate. If we denote the growth rate of cells as L and the death rate as D, the condition $L > D$ corresponds to clonal expansion. We further assume that cancer is detected when the colony reaches a certain size, N, at which moment therapy starts (treatment size). The effect of therapy is modeled by the drug-induced death rate, H, which shifts the balance of birth and death such that the colony shrinks. That is, the net cell death rate is now larger than the birth rate, $D + H > L$. If all cancerous cells were susceptible to the drug, then therapy would inevitably lead to eradication of cancer. However, in the course of cancer progression, mutations can lead to the generation of cell types which are resistant to the drug. This is assumed to occur with a probability u upon cell division. Before the tumor is treated, the mutant will behave identically compared to the wild-type. During therapy, however, the resistant phenotype will proliferate while the wild-type will be killed with a rate H. For simplicity, we assume that mutant cells which are not resistant to all drugs in use are killed with the same rate as wild-type cells. Alternatively, it can be assumed that such mutants are partially resistant (i.e., are affected less than the wild-type but more than the fully resistant phenotype [Schabel Jr *et al.* (1983, 1979)]. However, analysis indicates that this does not alter our results significantly.

More than one drugs can be used simultaneously or in sequence in order to reduce the probability of treatment failure due to drug resistance. The rational behind using multiple drugs is quite simple. In the absence of cross-resistance, in order to become resistant to m drugs, the cell has to accumulate m mutations. Much of the discussion below is devoted to the analysis of multiple drug treatments, and what protocols can optimize the likelihood of treatment success.

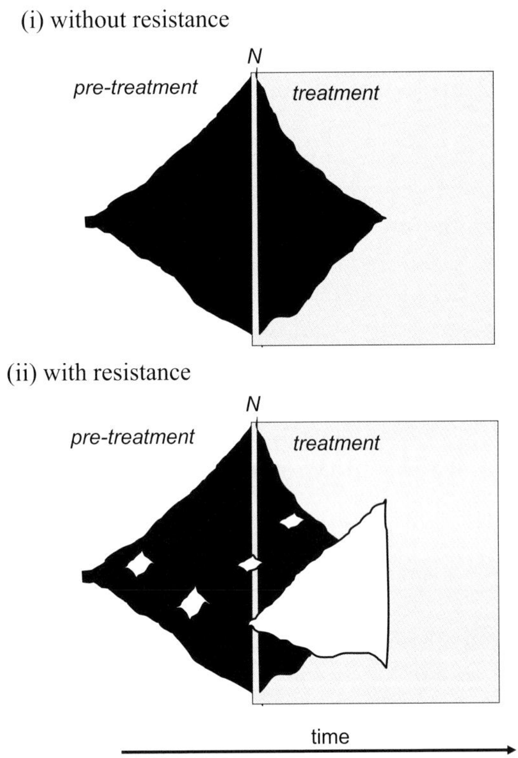

Fig. 16.2 Schematic representation of the assumptions which underlie the modeling framework. (i) Without resistance, the cancer grows exponentially, and treatment starts at tumor size N. Upon treatment, the tumor size shrinks until it is driven extinct. (ii) If mutations can occur, resistant cell clones are generated. This can occur both before and during treatment. As therapy is applied, a resistant cell clone can expand while the wild-type declines. Therefore, treatment fails to drive the cancer extinct. Treatment phases are indicated by shading.

16.4 When do resistant cells emerge?

This is a fundamental question which has important implications for the development of treatment strategies with the aim to prevent failure due to resistance. On the one hand, the growth of a tumor involves many cell divisions during which mutations can occur. In addition, many cancers are characterized by some form of genetic instability which accelerates the rate at which cells acquire mutations during this growth phase. Hence, it is possible that by the time the tumor reaches a certain size, there will be one or more cells resistant to therapy. Upon start of treatment, these cells may proliferate and grow to large numbers, preventing the eradication of the tumor. The computational framework of Section 15.5 has allowed us to quantify the probability that at least one cell is resistant before the start of treatment. On the other hand, once therapy starts, cells can also acquire mutations. While the death rate of the tumor cells is now larger than their rate of division, cell divisions, and thus mutations, can still occur. The probability that at least one resistant mutant is generated during the phase of treatment is also calculated in Section 15.5.

The upshot of these calculations is that the treatment phase does not contribute significantly to the generation of drug resistance. If resistance poses a problem during treatment, the model suggests that resistant cells must have pre-existed before the start of therapy. The situation is slightly different depending on whether treatment occurs with a single drug, or whether several drugs are used in combination:

(1) If the cancer is treated with a single drug, then there is a parameter region in which the generation of resistant cells is more likely to occur during the treatment phase than during the growth phase before therapy. This occurs if the efficacy of treatment is weak relative to the growth rate of the cancer. In our notation explained above, it occurs if $H < 2(L - D)$. However, we argue that this is not relevant for practical purposes. This condition means that the number of cell divisions during treatment is higher than the number of cell divisions during the growth phase before treatment. In other words, the time it would take to eradicate the tumor by drugs in the absence of resistance is larger than the age of the tumor upon start of therapy. This seems like an unrealistic parameter regime.

(2) If two or more drugs are used, the treatment phase becomes even less significant in the context of the generation of resistant cells. That is,

in all parameter regions, resistance is more likely to pre-exist than to be generated during therapy. The reason lies in the dynamics of the intermediate mutants. During the growth phase, a cell with a single resistant mutation will undergo clonal expansion, which facilitates the generation of further mutations. During the treatment phase, a cell with a single resistant mutation has a negative growth rate (as it is susceptible to one or more drugs). This makes it unlikely that additional mutations can be attained before the clone is extinct.

Because the treatment phase can be ignored, we note that the chances of treatment failure due to resistance are not influenced significantly by the efficacy of treatment (assuming that treatment is strong enough to remove the cancer in the absence of resistance).

16.5 Cancer turnover and the evolution of resistance

Let us consider the number of cell divisions which occur during the growth phase until the tumor has reached size N. This is roughly given by $\nu = NL/(L-D)$. We can see that if $D = 0$ or $D \ll L$, the number of cell divisions is approximately given by $\nu \approx N$. On the other hand, if D is close to L ($D \approx L$), many more cell divisions are required to reach size N, since a high death rate cancels the effect of cell divisions. For convenience, we will call the scenario where $D \approx L$ a high-turnover cancer. In contrast, we will call the scenario where $D = 0$ or $D \ll L$ a low-turnover cancer.

How does the turnover rate of the cancer influence the emergence of resistant cells during the growth phase and thus the pre-existence of resistance? The answer depends on how many drugs are used to treat the cancer.

(1) If the cancer is treated with a single drug, then the probability that a resistant mutant exists before therapy is not dependent on the turnover rate of the cancer. That is, high-turnover and low-turnover cancers behave in exactly the same way as far as the pre-existence of mutants is concerned. An intuitive explanation is as follows. A higher turnover cancer requires more cell divisions to reach size N, and thus more mutants are created. At the same time, however, the death rate of the mutants is also increased. The two effects cancel each other out. Similar results were also observed in related and earlier mathematical models by Goldie and Coldman who did pioneering work in this field of research.

(2) If two or more drugs are used, the probability that resistant mutants pre-exist does depend on the natural death rate of the tumor cells, D. In other words, the dynamics are different for high-turnover and low-turnover cancers. The higher the turnover rate of the cancer cells, the higher the probability that a resistant mutant exists when the cancer has reached size N. The larger the number of drugs used, the stronger this dependency. To explain this, consider the process of mutant generation. In the case of one drug, the increase in mutant production is canceled out exactly by the increase in mutant death as the turnover rate of the tumor cells is increased. This does not hold for two or more drugs. Now, an increase in the turnover rate of the tumor cells increases the production rate of resistant mutants more than it increases the death rate of the mutant cells. The net effect is that a resistant mutant is more likely to be present at the time of treatment if the turnover rate of the tumor cells is higher. In general, if the number of drugs is increased, a higher natural death rate of tumor cells, D, contributes increasingly to the production of resistant mutants and thus to treatment failure.

This gives rise to the important insight that cancers which are characterized by a high turnover rate (i.e., the death rate of cells is close to their division rate) might be difficult to control with combination therapy. This is discussed further in the next section.

16.6 Combination therapy and the prevention of resistance

The combination of several drugs together seems like an obvious strategy to prevent treatment failure as a result of resistance. If a one cell has to accumulate a sufficient number of mutations in order to become fully resistant, it is less likely that a resistant cell will exist upon start of treatment. If there is no cross-resistance between different drugs, then a cell has to acquire m mutations in order to become resistant to m drugs (see figure 15.2). Combination therapy has shown great success in the context of HIV infection [Maldarelli *et al.* (2007)]. In a typical HIV infected patient, we can expect that viruses are present which are resistant to one or two drugs. However, it is extremely unlikely that a virus exists which is resistant to three drugs [Ribeiro *et al.* (1998)]. This provides the rationale for using a combination of three drugs to achieve long-term suppression of the virus by drug therapy. In the following, we discuss how combination therapy affects the chance of treatment failure in the context of cancers treated with targeted small molecule drugs.

We ask at which tumor size N the probability of treatment failure reaches a threshold value, which we denote by δ. This means that if we

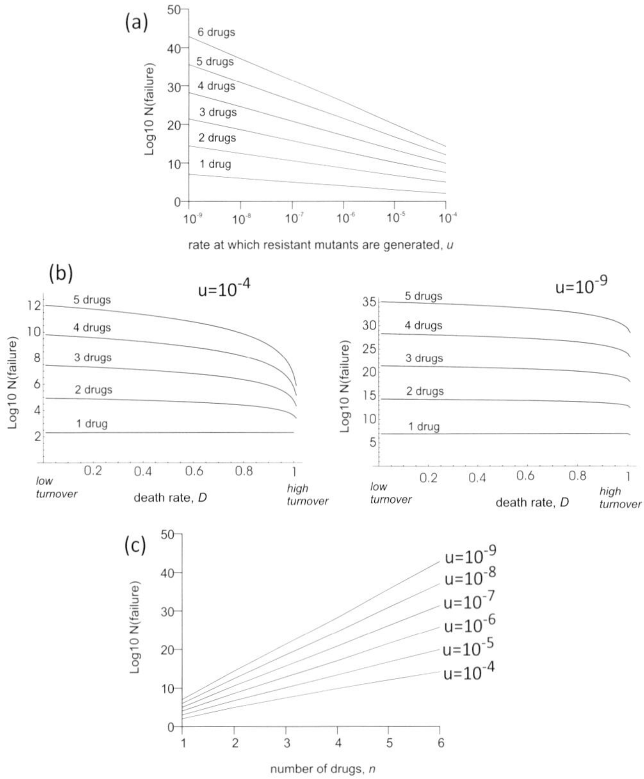

Fig. 16.3 Log tumor size, N, at which treatment failure is observed, depending on the parameters of the model. (a) Dependence on the rate at which resistant mutants are generated, u. The higher the value of u, the lower the tumor size at which treatment fails. The larger the number of drugs, the stronger this dependency. (b) Dependence on the natural death rate of tumor cells, D. The higher the value of D (i.e., the higher the turnover of the cancer), the lower the tumor size at which treatment fails. The higher the number of drugs, and the higher the rate at which resistant mutants are generated, u, the more pronounced this trend. (c) Dependence on the number of drugs, n. Increasing the number of drugs increases the tumor size at which treatment fails. The higher the mutation rate, however, the lower the advantage gained from adding further drugs. Baseline parameter values were chosen as follows: $L = 1, \delta = 0.01$.

start treatment at tumor size N, failure will be observed in a fraction δ of the patients, while treatment will be successful in a fraction $1 - \delta$ of patients. For now we assume that an acceptable goal is to treat 99% of patients successfully, that is $\delta = 0.01$. In other words, if more than 1% of patients shows resistance, we consider the treatment strategy a failure. So we ask at which tumor size treatment failure is expected to occur. In particular, we ask how the number of drugs used in combination influences the tumor size when resistance is observed. According to the model, this depends on the mutation rate, u, and the turnover rate of the tumor cells (value of D relative to L) (see figure 16.3).

(1) The higher the rate at which resistance mutations are acquired, u, the less the effect of adding another drug, and the more difficult it becomes to treat (figure 16.3). Consider the most optimistic scenario when $D = 0$ (table 16.1). Assume that cancers can reach up to sizes of 10^{13} cells [McKinnell (1998)]. Then, the physiological point mutation rate, $u = 10^{-9}$, requires two drugs, values of u between $u = 10^{-8}$ and 10^{-7} require three drugs, u between 10^{-6} and 10^{-5} require four drugs, and $u = 10^{-4}$ requires six drugs (table 16.1). By extrapolation, 10 drugs are needed if $u = 10^{-3}$, and about 30 drugs are needed if $u = 10^{-2}$. Therefore, if resistance mutations can occur at levels which are significantly higher than the physiological mutation rate (e.g. because genetic instability promotes the generation of resistance mutations), combination therapy is unlikely to be advantageous.

(2) A high turnover rate of cancer cells also abolishes benefits which can be obtained from combination therapy (figure 16.3(b)). In the context of combination therapy, resistance arises at lower tumor sizes as the death rate of tumor cells, D, is increased. In fact, if the death rate of tumor cells, D, comes close to their division rate, L (high turnover cancer), then the effect of combining multiple drugs disappears (figure 16.3(b)). The size at which resistance arises converges to the same value, no matter how many drugs are used. In this case, the frequency with which cancers arise is low because they have a high chance to go extinct spontaneously, but when they do arise, the chances of complete tumor eradication are very slim. Because high turnover cancers are likely to grow very slowly, however, drug therapy could still increase the life-span of the patient by reducing the number of tumor cells for a prolonged period of time. Re-growth of resistant cells to large sizes would take a long time.

This leaves us with the following message: combination therapy will provide benefits for preventing treatment failure due to resistance, except in the following cases: (i) The cancer has a high turnover rate. Such cancers are likely to grow slowly. (ii) Resistance mutations against the drugs in use can be generated at rates which are several orders of magnitude higher than the physiological mutation rate.

Table 16.1 The log10 size at which resistance becomes a problem (i.e., treatment fails in more than 1% of patients), depending on the number of drugs and the rate at which resistant mutants are generated, u. If we assume that the cancers cannot grow beyond 10^{13} cells without causing death, a treatment regime can be considered acceptable if resistance only becomes a problem at sizes which are greater than 10^{13} cells (i.e., $\log_{10}$ of the size > 13). The parameter regimes where this occurs and treatment is expected to be successful are indicated by bold font in the table. The calculations assume $L = 1, D = 0$.

	1 drug	2 drugs	3 drugs	4 drugs	5 drugs	6 drugs
$u = 10^{-4}$	2.01	4.95	7.46	9.81	12.06	**14.23**
$u = 10^{-5}$	3.01	6.73	10.13	**13.36**	**16.70**	**20.02**
$u = 10^{-6}$	4.01	8.61	12.91	**17.04**	**21.49**	**25.83**
$u = 10^{-7}$	5.01	10.53	**15.75**	**20.8**	**26.17**	**31.43**
$u = 10^{-8}$	6.01	12.47	**18.62**	**24.6**	**30.90**	**37.10**
$u = 10^{-9}$	7.01	**14.42**	**21.36**	**28.23**	**35.61**	**42.86**

16.7 Parameters and CML

While the treatment of CML with the targeted drug imatinib (Gleevec) has shown significant success especially during the chromic phase of the disease, treatment of blast crisis often fails because of drug resistance [McCormick (2001)]. In accordance with our framework it has been reported that mutants might pre-exist the initiation of treatment rather than being generated during the treatment phase [Gambacorti-Passerini *et al.* (2003); Nardi *et al.* (2004)]. Data suggest that two main types of mutations confer resistance to the cells [McCormick (2001); Gambacorti-Passerini *et al.* (2003); Gorre *et al.* (2001)]: the amplification of *BCR-ABL*, or a point mutation in the target protein. Genetic instability [Loeb (2011)] is likely to promote the occurrence of gene amplifications which have been measured to occur in cancer cells at a rate of 10^{-4} per cell division [Tlsty *et al.* (1989)]. On the other hand, the point mutation rate is about 10^{-9} per base per cell division

[Loeb *et al.* (1974)]. However, the frequency of gene amplifications is much less than that of point mutations among patients [Gambacorti-Passerini *et al.* (2003)]. Part of the reason might be that *BCR-ABL* amplifications are costly to the cells in the absence of treatment [Tipping *et al.* (2001)]. Including this assumption into the modeling framework, however, shows that even if this fitness cost is very significant, amplifications should still be observed more often than point mutations. However it is thought that the level of resistance is a function of the number of extra copies of the *BCR-ABL* gene. Therefore, if a significant degree of resistance requires 2 or more amplification events (but only one point mutation event), we expect that a resistant mutant is generated faster by point mutation than by gene amplification, explaining the observed frequencies.

Thus, for prevention of drug resistance we assume that resistant mutants are generated maximally with a point mutation rate between 10^{-9} and 10^{-8} per base pair per generation. Experiments with susceptible CML cell lines have shown viability measurements (in the absence of treatment) of about 90% [Tipping *et al.* (2001)]. From this we can roughly calculate that the relative death rate of cancer cells is in the range of $D/L = 0.1-0.5$. In this parameter region, we find that a combination of three drugs should prevent resistance and ensure successful therapy even for advanced cancers (table 16.2). This assumes that the size of advanced cancers is less than 10^{13} cells, which derives from white blood cell count measurements which range from 10^5 to 10^6 per microliter of blood in blast crisis. The findings of [Nowicki *et al.* (2004)] indicate that *BCR-ABL* might increase the amount of reactive oxygen species and thus the rate of point mutations. As long as the elevation of the mutation rate is less than a hundred fold, our results remain robust (table 16.3).

In summary these calculations provide optimistic results for the treatment of CML with a combination of different targeted small molecule inhibitors.

Table 16.2 $\log_{10}$ size at which resistance becomes a problem, depending on the number of drugs and the turnover rate of the cancer cells (value of D/L).

	1 drug	2 drugs	3 drugs	4 drugs	5 drugs
$D/L = 0.1$	5.95	12.34	18.45	24.38	30.19
$D/L = 0.5$	5.95	12.13	17.99	23.69	29.26
$D/L = 0.9$	5.95	11.48	16.70	21.74	26.66

Table 16.3 Same as in table 16.2, except that we assume that resistant mutants are generated with an elevated rate of $u = 10^{-6}$.

	1 drug	2 drugs	3 drugs	4 drugs	5 drugs
$D/L = 0.1$	4.00	8.55	12.80	16.89	20.86
$D/L = 0.5$	4.00	8.31	12.37	16.20	19.93
$D/L = 0.9$	4.00	7.68	11.07	14.40	17.40

16.8 Tumor architecture and tumor stem cells

So far, we have treated the growing cancer cells as a homogeneous population without taking into account tissue architecture. It is, however, widely believed that tumors are characterized by an architecture that is similar to that found in healthy tissue. Thus, the tumor is thought to be maintained by a small number of tumor stem cells, while the bulk of the tumor is made up of more differentiated cells that cannot maintain a tumor or initiate new tumor growth [Visvader and Lindeman (2008); Reya *et al.* (2001); Dalerba *et al.* (2007)]. Further, tumor stem cells seem to share certain characteristics with their healthy counterparts. With tissue stem cells, cellular quiescence is a central process that regulates the kinetics of cellular proliferation and tissue homeostasis, especially in stem cells [Pelayo *et al.* (2006); Arai and Suda (2007); Moore and Lemischka (2006); Fuchs *et al.* (2004); Weissman (2000); Cheshier *et al.* (1999)]. If stem cells are not needed to divide and to replenish tissue cells, they temporarily stop to progress through the cell cycle until further divisions are required. While primitive cancer cells proliferate with a higher rate than healthy cells, data indicate that they can still undergo quiescence, both during tumor growth and during treatment. This has been demonstrated in CML [Holyoake *et al.* (1999, 2001)]. It has even been suggested that in CML, therapy induces quiescence in primitive cancer cells [Graham *et al.* (2002)]. Quiescent cells in turn are not affected by the drug and are therefore shielded from therapy-induced elimination [Graham *et al.* (2002)]. Stem cell dynamics and the distinction between active and quiescent tumor stem cells can be important key factors influencing the dynamics of tumor growth [Ducrot *et al.* (2011); Gyllenberg and Webb (1991)] and the evolutionary dynamics of drug-resistant cell variants. This is explored in the current section.

We use an extension of the stochastic model of Chapter 15 that includes a population of primitive, proliferating CML cells, and a population of quiescent CML cells [Komarova and Wodarz (2007); Komarova (2007)]. The cells enter a quiescent state with a rate α, and quiescent cells re-enter the cell cycle with a rate β. Quiescent cells do not divide or die and are not susceptible to any drug activity. In addition to these processes, CML cells can mutate to give rise to acquired drug resistance with a probability u, as before.

The conclusion that resistance is most likely due to the pre-existence of resistant mutants before therapy still holds in the context of quiescent tumor stem cells. Interestingly, while quiescence prolongs the time until therapy reduces the number of cells to low levels or extinction, the therapy phase is irrelevant for the evolution of drug resistant mutants. If treatment fails as a result of resistance, the mutants will have evolved during the tumor growth phase, before the start of therapy. Thus, prevention of resistance is not promoted by reducing the quiescent cell population during therapy (e.g., by a combination of cell activation and drug-mediated killing).

The probability of treatment failure as a result of resistance in the context of a single drug is not affected by quiescence parameters (figure 16.4(a)). To put this in quantitative terms, the probability to have at least one resistant mutant at size N is independent of α and β.

This is demonstrated by the following argument. Let us assume for simplicity that there is no cell death in the colony (all the arguments can be extended to nonzero death rates). In the model, mutants are generated during cell division. The probability of resistance is the same as the probability to generate mutants, which is defined by the number of cell divisions (and the constant mutation rate). It is easy to see that the total number of cell divisions until the tumor reaches size N does not depend on the quiescence parameters α and β. For instance, if there is no cell death, then the number of cell divisions to expand from one cell to N cells is exactly $N-1$, no matter what the quiescence rates are, see figure 16.5. It is of course the case that the higher the rate at which cells enter quiescence, and the lower the rate at which cells exit quiescence, the longer it takes the tumor to grow to size N. However, the actual number of cell divisions to reach size N is unchanged by quiescence. Therefore, the probability to produce resistant mutants is independent of quiescence rates.

The situation is different when considering resistance against two or more drugs. For treatment with multiple drugs, the probability of treatment failure as a result of resistance depends on the quiescence parameters (figure 16.4(b-d)). The higher the rate of entry into the quiescent state (larger α) and the lower the rate of exit from the quiescent state (lower β), the higher the probability of treatment failure. This has important implications for understanding the principles underlying potential combination treatments and for developing computational methods to predict the number of drugs needed to overcome drug resistance.

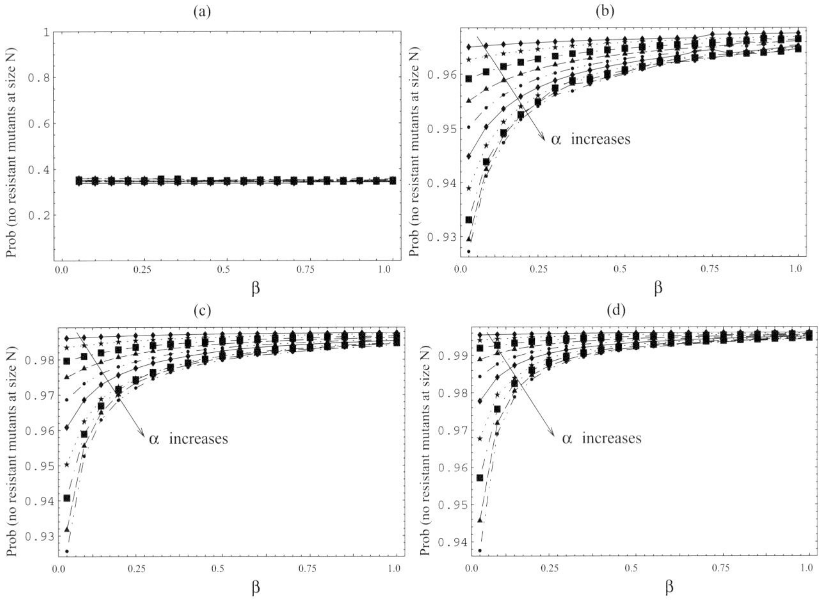

Fig. 16.4 The probability of having no fully-resistant mutants at size N for different quiescence parameters. Each figure (a)-(d) shows the probability of no resistant mutants as a function of β (the rate of cell awakening), for 10 different values of α (the rate at which cells become quiescent), $\alpha = 0.1, 0.2, \ldots,$ and 1.0. (a) Treatment with $m = 1$ drugs; all the curves corresponding to different values of α are the same. The parameters are $N = 10^7$ and $u = 10^{-7}$. (b) Treatment with $m = 2$ drugs, $N = 10^{11}$, $u = 10^{-7}$. (c) $m = 3$ drugs, $N = 10^{13}$, $u = 10^{-6}$. (d) $m = 4$ drugs, $N = 10^{13}$, $u = 10^{-5}$. In all plots, we took $M_0 = 10^3$, $L = 1$, $D = 0$. The reason we used different values of N and u for different values of m is because we chose the parameter regime corresponding to intermediate values of the probability of treatment success. When this probability is nearly 100% or nearly 0, the dependence on α and β is less apparent and less meaningful.

16.9 Short-term versus long-term treatment strategies

The previous sections examined the number of targeted drugs that need to be used in combination in order to overcome treatment failure as a result of drug resistance. If a good response to treatment is achieved, it is likely that the drug will have to be taken for substantial amounts of time in order to sustain the response. The combination of multiple drugs, however, can increase side-effects, which might not be possible to endure in the long term. The question arises whether the number of drugs can be reduced over time once the tumor burden has declined to a certain degree. This was investigated mathematically [Komarova and Wodarz (2009)] by using the mathematical framework of Chapter 15, taking into account the tumor cell tissue architecture and the occurrence of tumor stem cell quiescence. This analysis showed that one or more drugs can be removed once the number of tumor cells is reduced significantly, without compromising the chances of sustained tumor suppression. Which drug to remove first depends on the

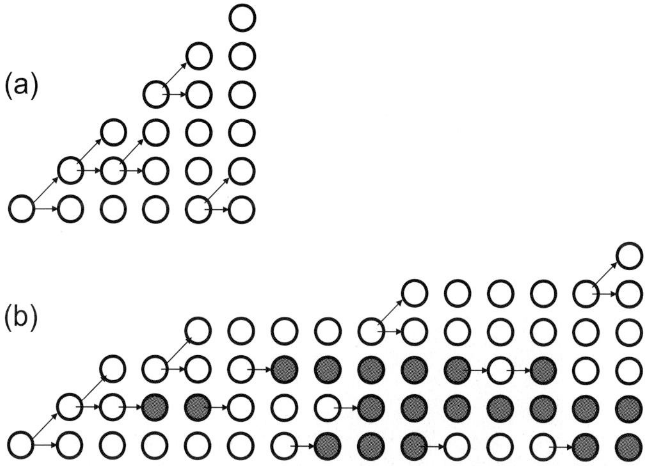

Fig. 16.5 A schematic demonstrating the number of cell divisions that is needed for a colony of cells to expand from 1 cell to N cells (in the figure, $N = 6$). Empty circles represent cycling cells, and gray circles represent quiescent cells. Columns depict states of the colony in consecutive moments of time. The changes are marked by arrows. Two arrows stemming from one cell represent a cell division. A single arrow represents either a cell becoming quiescent or a quiescent cell waking up. (a) A colony without quiescence. (b) A colony with quiescence. In both cases we can see that it takes exactly $N - 1 = 5$ cell divisions to expand to size N; however the process in (b) contains more "events" and it takes a longer time.

number of mutations in the BCR-ABL gene that confer resistance to the drugs, as well as on how effectively the drugs inhibit Bcr-Abl protein tyrosine kinase activity and inhibit tumor growth. The model further showed that the number of CML cells at which the number of drugs can be reduced does not correlate with the two phases of decline of the BCR-ABL transcript numbers. Neither does it depend much on kinetic parameters of CML growth, except for the mutation rates at which resistance is generated (figure 16.6). This is a significant finding because even without any information on most parameters, and using only the data on the number of cancer cells and the rate at which resistant mutants are generated, it is possible to predict at which stage of treatment the number of drugs can be reduced. These calculations provide a clinically useful guide for the development of long term treatment strategies that initiate with a combination of different drugs.

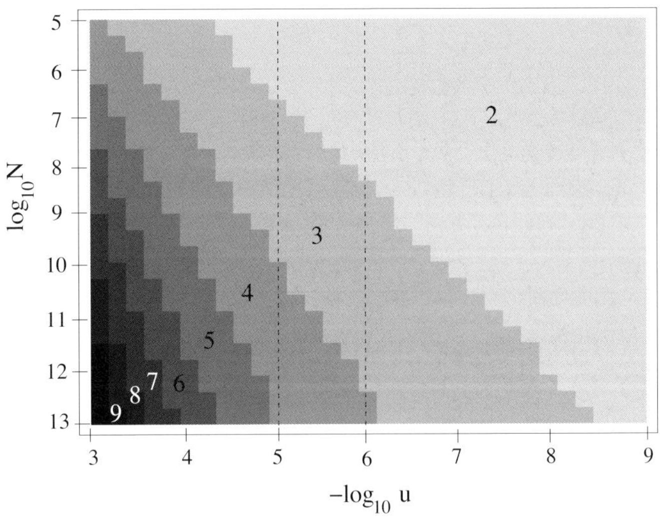

Fig. 16.6 The number of drugs needed for the probability of treatment success $1 - \delta$ with $\delta = 0.01$, as a function of u and N. Other parameters are: $L = 1$, $D = 0$, $H = 2$, $\alpha = 0.01$, $\beta = 0.05$, $M_0 = 100$. The jagged appearance of the boundaries between the domains with different values of m is due to a discrete method used to calculate it, and can be smoothed by using a simulation with a higher resolution.

16.10 Cross-resistance and combination therapy

The initial model described here has shown how the combination of three or four tyrosine kinase inhibitors can potentially overcome drug resistance even during the advanced blast crisis stage of CML. It was assumed that mutations that confer resistance against one drug do not confer resistance against any of the other drugs in use. In addition to imatinib, the second generation inhibitors dasatinib and nilotinib are alternative inhibitors of the BCR-ABL gene product. More than 50 mutations have been identified that confer CML resistance against one of the drugs (in particular imatinib), but not against the others [Rix *et al.* (2007)]. However, one particular mutation has been identified, called T315I [O'Hare *et al.* (2007)], which confers resistance to all three drugs: imatinib, dasatinib, and nilotinib. This obviously complicates the use of combination therapy. In the following, we examine the effectiveness of combining the currently used inhibitors under the assumption that the T315I mutation confers cross-resistance to all drugs, while on the order of 10-100 mutations confer resistance to only one drug in the combination. We also explore the consequences of combining existing drugs with new drugs that can potentially inhibit T315I mutant cells [O'Hare *et al.* (2007)].

Compared to the simple framework of Chapter 15, cross-resistance is characterized by the existence of mutants simultaneously resistant to more than one drug. In terms of the combinatorial mutation diagrams (see figure 15.2), this manifests itself in "short-cuts" - mutational steps that create resistance to several drugs at once, see figure 16.7. The existing medical knowledge on the types of mutations for CML drugs can be incorporated in the rates of mutations u. We assume that k_i mutations give rise to resistance to drug i without affecting resistance properties with respect to the other drugs (the index i takes values $1, 2, 3$ for the three drugs). This translates into $u_i = k_i u$, where u is the basic point mutation rate. Similarly, we denote by u_{ij} with $i \neq j$ the rate of mutations leading to cells simultaneously resistant to drugs i and j (for three drugs, there are three such rates, u_{12}, u_{13}, and u_{23}). Finally, u_{123} is the rate of generation of mutants conferring resistance to all the three drugs.

As an example relevant to CLM treatments, we assume that there are of the order $k_i = k \sim 10 - 100$ different point mutations that can confer resistance against only one of the drugs. There is also one particular mutation which can confer resistance against all the drugs in use ($k_{123} = 1$).

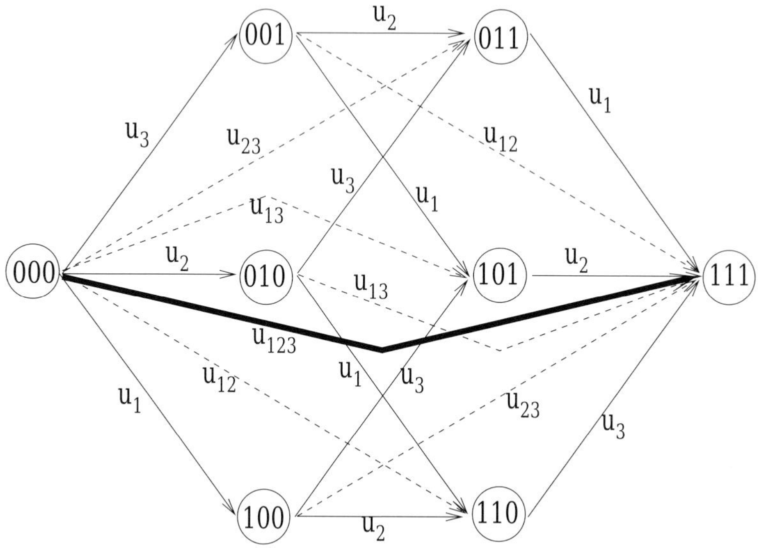

Fig. 16.7 Combinatorial mutation diagram for different resistance classes. Treatment with $m = 3$ drugs. One-drug mutations are denoted by thin solid arrows, two-drug mutations by dashed arrows, and three-drug mutations by a thick solid arrow. The mutation rates are indicated next to each arrow.

Different drugs can possess different degrees of cross-resistance. For combinations of three drugs, there are five possibilities, depending on pair-wise cross-resistance properties of the drugs and the triple-resistance properties. These five cases can be captured by means of five possible three-drug cross-resistance networks, see figure 16.8. In the figure, each drug is represented as a node of the network. Each pair of circles is connected if there exists a mutation event that confers resistance against the two drugs. Different lines (single, double, dashed) correspond to different mutations. In figure 16.8, the drugs are denoted by the numerals "1", "2", and "3". If a cross-resistance network has no connecting lines (the leftmost diagram in figure 16.8) this means that no cross-resistance takes place. If the lines in a cross-resistance network are different, this means that different mutants are resistant against each pair of drugs. If the lines are the same, then triple cross-resistance takes place (the rightmost cross-resistance network in figure 16.8). This case is exemplified by the drugs imatinib, dasatinib, and nilotinib. By setting some of the double-resistance and triple-resistance rates to zero, we can model any of the five possible cross-resistance networks by using the diagram of figure 16.7.

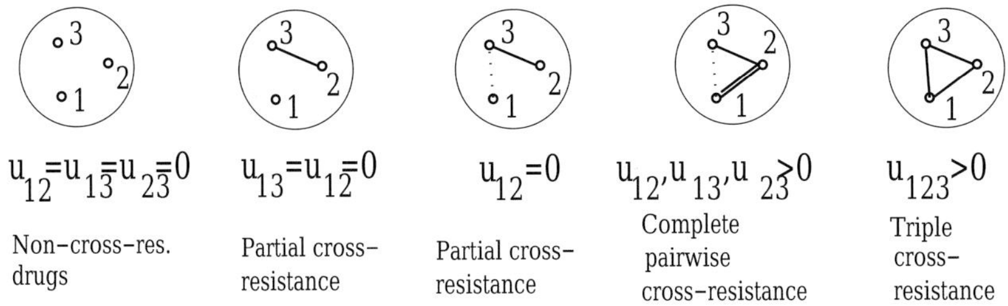

Fig. 16.8 All possible three-drug resistance networks. The number of nodes corresponds to the number of drugs used. Connected nodes correspond to the existence of a cross-resistant mutation. Identical connecting lines indicate that the same mutation confers resistance to all connected drugs. Different (single, double, dashed) lines correspond to different mutations.

First, consider the combination of two drugs. For reference, figure 16.9 shows the probability of treatment success for one and two drugs assuming the absence of cross resistance. This is compared to various cross-resistance scenarios. While the probability of treatment success is lower in the presence than in the absence of cross resistance, combining two drugs with cross-resistance clearly improves the probability of treatment success relative to the use of only one drug. The reason is that it is much more likely to acquire a mutation that confers resistance against only one drug than to acquire the doubly-resistant mutation. This is because only one specific mutation can lead to cross-resistance, while many mutations can confer resistance against a single drug. Hence, for most mutations, combination therapy will not be challenged by cross-resistance. On a qualitative level, this result does not depend on the kinetic parameters of the model, such as the division and death rates. The advantage of combining two drugs becomes insignificant if the number of mutations that confer resistance to only one drug is very low (figure 16.10(a)), or if the rate at which the doubly-resistant mutants are acquired is relatively high. If the number of mutations that confer resistance against only one drug is on the order of $50 - 100$, and if the rate at which resistant mutations are generated is on the order of $10^{-6} - 10^{-9}$, however, the model suggests that combining two drugs is advantageous to the patient, even if cross-resistance is possible.

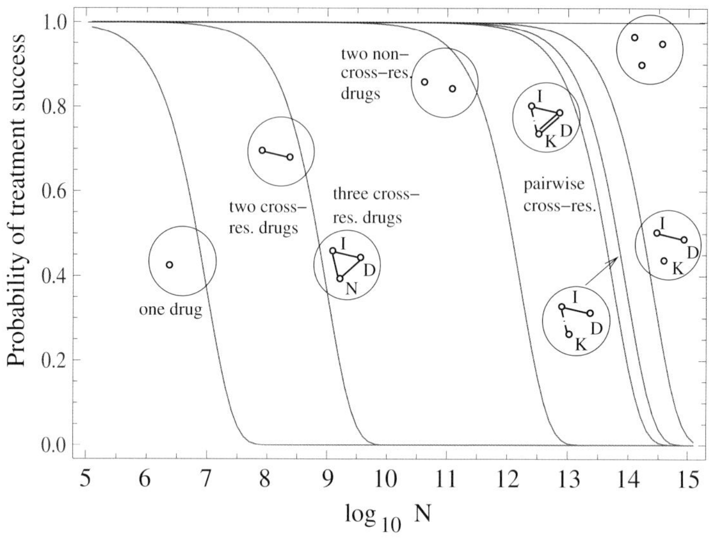

Fig. 16.9 The probability of treatment success is plotted as a function of the colony size, N. Different curves correspond to different combination treatments, with one, two and three drugs. The cross-resistance networks are presented by using connected and disconnected nodes, see figure 16.8. Simulation parameters are as follows: $u = 10^{-9}$, $k = 100$, $M_0 = 100$, $D/L = 0.5$, $H/L = 3$. The symbols "I", "D", "N", and "K" stand for "imatinib", "dasatinib", "nilotinib" and a future drug which can bind to T315 mutants.

Next, consider a combination of three drugs, i.e., imatinib, dasatinib, and nilotinib. The model shows that combining three drugs will not lead to any further advantage compared to the combination of two drugs (figure 16.9). For triple combination therapy to be advantageous, most resistant cells must harbor mutations that render them resistant against two of the drugs (but not the third one). Accumulating two separate resistance mutations, however, is a relatively rare event. It is much more likely that a cell acquires the single cross-resistance mutation. Hence, triple combination therapy does not improve the probability of treatment success compared to double combination therapy. Triple combination therapy can only provide an additional advantage if the number of mutations that confer resistance against only one drug is unrealistically high (figure 16.10(b)). Assuming reasonable values for the cellular division, death, and mutation rates, there must be at least $k = 1000$ mutations that confer resistance against only one drug for triple combination therapy to somewhat improve the chances of

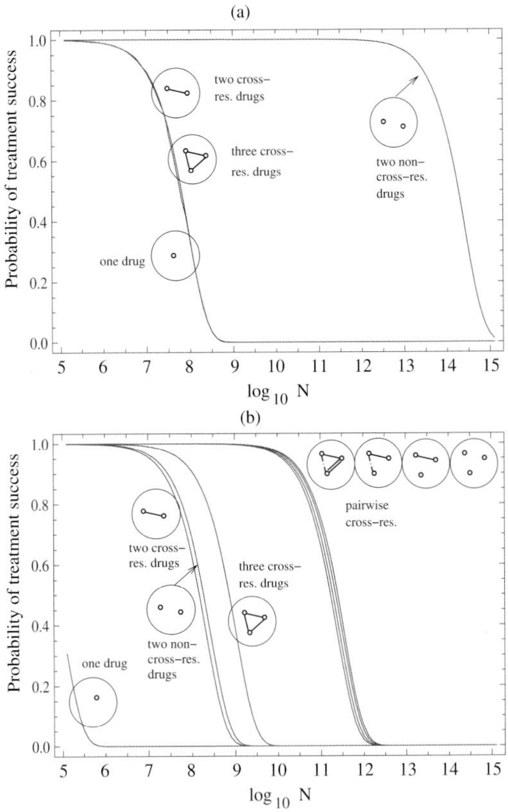

Fig. 16.10 The probability of treatment success is plotted as a function of the colony size, N. (a) The number of non-cross-resistant mutations is low ($k = 10$) and the mutation rate for cross-resistance is 10 times higher than in figure 16.9. Conclusion: combining more than one cross-resistant drugs does not improve the chances of treatment success. (b) The number of non-cross-resistant mutations is high, $k = 10^4$. Conclusion: combining three cross-resistant drugs improves the chances of treatment success compared with two cross-resistant or non-cross-resistant drugs (which in turn is better than using only one drug). The rest of parameters are as in figure 16.9.

treatment success. The improvement becomes significant for $k = 10,000$ or more mutations (figure 16.10(b)). For such high values of k we observe that the effect of cross-resistance is insignificant and treatment failure occurs as a result of mutations that confer resistance to one drug at a time. In this case combining three drugs with triple cross-resistance gives a better outcome than two drugs in the absence of cross-resistance (see figure 16.10(b)).

Again, these results do not qualitatively depend on the kinetic parameters of the model.

General cross-resistance networks have relevance for future generation treatment options. If drugs are used that show no cross resistance with imatinib, dasatinib, and nilotinib, such as VX-680, danusertib, and ponatinib [Winter *et al.* (2012); Pinilla-Ibarz and Flinn (2012); Burley (2006); Carter *et al.* (2005); Melo and Chuah (2007); Harrington *et al.* (2004); Young *et al.* (2006); Duncan *et al.* (2008)], would it be advantageous to combine such a drug with two or more drugs that do not inhibit the T315I mutation? Our calculations show that it is advantageous to combine such a future generation drug (call it drug "K") with two drugs that cannot act on the T315I mutation, say, imatinib (I) and dasatinib (D). This is seen for example in figure 16.9 where we compare the probability of treatment success for two cross-resistant drugs with that for "pairwise cross-resistant" drugs. We observe that any of the pairwise cross-resistant combinations of three drugs gives a significant improvement. Note that adding a third cross-resistant drug does not improve the chances of successful therapy.

We can examine all possible resistance networks for three-drug therapies, see figure 16.9. Let us suppose that there is some cross-resistance between drug K and, say, I. In other words, even though the T315I mutation does not confer resistance to K, there may be a different mutation which makes the cell resistant to both K and I, but not to D. In this situation, treating with I and K is obviously better than just treating with I (this follows from our previous results), but what is interesting, adding D does give an advantage in this case. Adding D will even give an advantage if there is a third mutation which confers resistance to both K and D (but not I). In other words, even though treating with three drugs characterized by triple cross-resistance does not improve the chances of treatment success compared to two such drugs, treating with three pairwise-cross-resistant drugs is advantageous compared to treating with two such drugs.

16.11 Combination versus cyclic sequential treatment

In order to overcome resistance-induced treatment failure, we have so far considered various combination therapy strategies. While combination therapies with targeted drugs are gaining some attention [Bozic *et al.* (2013); Komarova and Boland (2013)], clinicians often consider the use of cyclic drug applications, where treatment is initiated with the first drug, then the patient is switched to a second drug, and so on. Hence, it is important to determine whether such cyclic treatment regimes lead to improved treatment outcomes compared to single drug mono-therapy. This was investigated with mathematical models [Katouli and Komarova (2011)]. The stochastic framework that was developed in the previous chapter and that was used to study the effects of combination treatment can also be applied to study cyclic treatment regimes. Cyclic drug treatments are assumed to proceed as follows, see figure 16.11(a). Treatment starts at time t_*. Drug 1 is applied for a time-duration of Δt_1. Then the drug is discontinued and replaced by drug 2. After time-duration Δt_2, drug 2 is in turn replaced by drug 1. The total treatment duration is denoted by T_{treat} and consists of $2\mathcal{N}$ cycles of treatment (here the word "cycle" refers to a one-drug treatment with drug 1 or 2).

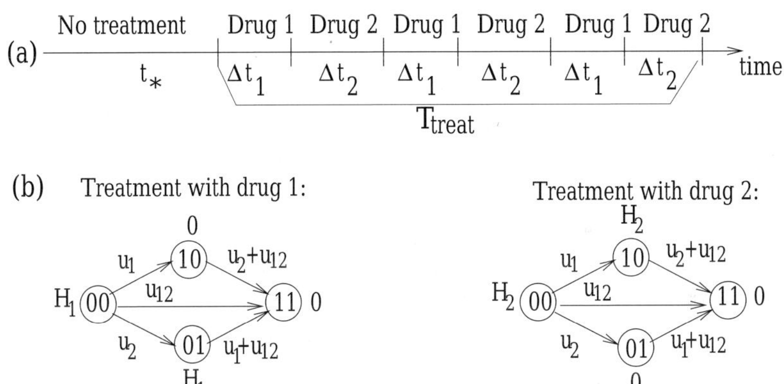

Fig. 16.11 Cyclic two-drug treatments. (a) A cyclic treatment protocol. (b) The mutation diagrams and drug-induced death rates for treatments with drug 1 and drug 2. The resistance types are represented by circles with binary indices; the drug-induced death rates are marked next to the circles; the mutation rates are indicated next to the arrows.

Mathematically, each treatment protocol corresponds to specific values of the death rates, D_s, at different moments of time:

$$d_s = D_s + H_s(t),$$

where as before, the coefficients D_s are natural death rates of the cancer cells, and $H_s(t)$ are the drug-induced cell death rates. The functions

$H_s(t)$ depend on the particular treatment strategy used. As different drugs are applied, the "strength" of each drug, which depends on the concentration of the drug in the patient's blood, changes as some smooth function of time. The exact shape of these functions, and therefore, the shape of $H_s(t)$, depends not only on the treatment strategy (that is, whether drugs are applied in combination, or cyclically), but also on the way the drugs are administered, and on how quickly they are absorbed. For example, it can be assumed that $H_s(t)$ for a susceptible class reaches a maximum sometime after the drug is taken, and decays until the next administration of the drug. However, we can simplify this picture by assuming that the functions $H_s(t)$ are piecewise constant. They are assumed to have a constant nonzero value for all the susceptible classes as long as the patient is treated with a given drug, and they become zero after the drug is discontinued. For the effects of pharmacokinetics on the dynamics of treatment see [Gaffney (2005)].

This framework, which is based on [Day (1986)], can be used to determine how different cyclic treatment strategies influence the chances of treatment success by preventing the production of fully resistant mutants. The mutation diagrams corresponding to the two different cycles (treatment with drug 1 and treatment with drug 2) are shown in figure 16.11(b). Depending on which drug is applied, different mutants are experiencing different drug-induced death rates. For example, during the cycle of the 1st drug application, mutants of type (10) (that is, mutants resistant to drug 1 and susceptible to drug 2) are growing, and mutants of type (01) are killed at rate H_2. Similarly, during the cycle of the 2nd drug application, mutants of type (10) are killed at rate H_2, and mutants of type (01) are growing.

In figure 16.12 we plot the population sizes of the two partially-resistant colonies, $x_{10}(t)$ (dashed lines) and $x_{01}(t)$ (solid lines), in the course of treatment, for some fixed values of the parameters. In figure 16.12(a), treatment starts with drug 1, and we can see that during the first cycle the colony x_{01} which is susceptible to this drug, decays exponentially, while the colony x_{10}, which is resistant to this drug, grows. In the second cycle, when drug 2 is applied, colony x_{01} grows and colony x_{10} decays. In figure 16.12(b) we present the scenario where the order of the drugs is switched (while all the parameters are kept the same). We can see immediately that the dynamics of mutant decay are sensitive to the order in which the two drugs are applied.

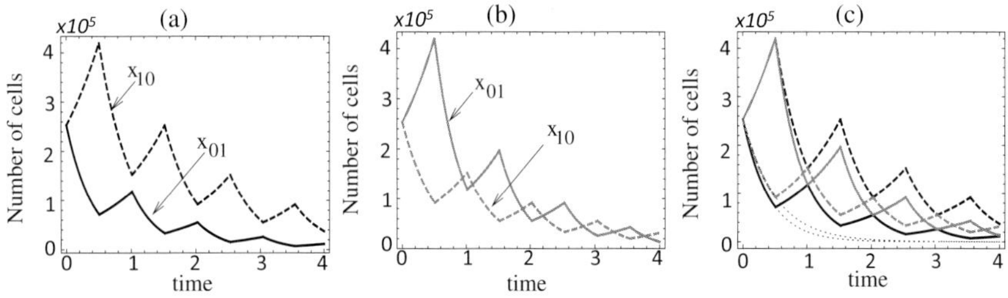

Fig. 16.12 Cyclic two-drug treatments. The deterministic dynamics of the populations of partially-resistant mutants, x_{01} (solid lines) and x_{10} (dashed lines). (a) Worst drug first (black lines), (b) best drug first (gray lines), (c) lines from panels (a) and (b) are plotted together; also, the thin dotted lines present the same populations under a combination treatment. The parameters are $H_1 = 3$, $H_2 = 3.5$, $\gamma = 1$, the total number of cells at the start of treatment is $N = 10^{11}$, $u = 10^{-7}$, the numbers of partially-resistant mutants at start of treatment are $N_{01} = N_{10} = N \log Nu$, and $\Delta t_1 = \Delta t_2 = \log N/\gamma/50 \approx 0.5$. Neither of the two treatment strategies is optimal.

Before summarizing the model results about cyclic treatment protocols, we note an important implication of our model. Combining two drugs instead of using a cyclic treatment protocol will always correspond to a larger probability of colony elimination. Figure 16.12(c) shows the dynamics of the two partially-susceptible colonies under the cyclic treatment protocols of figures 16.12(a) and (b) together with the dynamics corresponding to the combination treatment where drugs 1 and 2 are applied simultaneously and continuously (dotted lines). The drug-induced death rates of all types are always smaller (or at least not larger) under combination treatment, thus eliminating the two colonies at a faster rate.

With this in mind, the model suggests optimal cyclic treatment regimes that reduce the probability of resistance-induced treatment failure compared to singe drug mono-therapy. In this context, we need to distinguish two different characteristics of the drugs: their *potency* and their *activity spectrum*. By potency, we mean how effectively a drug kills cells that are susceptible to the drug. Further, a drug is characterized by a broad activity spectrum if it is effective against a large spectrum of mutant cells. On the other hand, a drug with a narrow activity spectrum is a more specific agent, which is active against a relatively small number of cell variants.

We find that in order for a cyclic treatment to be effective, the drugs' potencies must satisfy a certain condition, which we call the condition of "mutual strength" [Katouli and Komarova (2011)]. This condition states that the strength of the two drugs involved must lie above a certain threshold relative to the growth rate of the tumor cell population:

$$\frac{1}{H_1} + \frac{1}{H_2} < \frac{1}{L - D}. \qquad (16.1)$$

For realistic parameters, drugs which are not mutually strong will yield very poor probabilities of treatment success if applied cyclically. If the drugs under consideration are characterized by similar activity spectra, then the optimal treatment regime is to use the best drug first, but use the worst drug for longer. This "best drug rule" also applies in the presence of cross-resistance. If the drugs under consideration are characterized by similar potency but different activity spectra, then the optimal treatment regime is to use the less active drug first, and to use the more active drug for longer. In general, this framework can be used to calculate optimal cycle sequences and durations for given parameters such as drug strength, drug potencies, and mutation rates. We would, however, again, like to point out that no cyclic treatment regime can lead to better results than combination treatments, which is also supported by experimental data [Shah *et al.* (2007)].

16.12 Summary

In this chapter, we described how the mathematical methodology developed in the previous chapter can be used to gain clinically important insights into the evolution of drug resistance against small molecule inhibitors, and to design treatment regimes that overcome resistance and lead to sustained suppression of the tumor. We started from a relatively simple model that can be used to describe the blast crisis phase of CML, and found that in the context of measured parameters, a combination of 3 drugs can successfully overcome the problem of drug resistance. We subsequently explored how this framework can be extended to include a variety of other biological aspects, such as tumor tissue architecture, the dynamics of tumor stem cell quiescence and activation, and the evolution of simultaneous resistance to multiple drugs. We explored treatment strategies under various circumstances, including the long-term application of therapy. Finally, we compared combination treatments with sequential cyclic treatment approaches, and described optimal cyclic treatment approaches that can lead to better outcomes than monotherapies, although such cyclic approaches can never lead to better outcomes than combination therapies. These models have given rise to a variety of clinically important insights, and can be improved further by including more biological details.

Problems

Problem 16.1. Numerical project. *Use the code written for Problem 15.8 to investigate how tumor turnover, D/L, influences the likelihood of having resistant mutants at a given time.*

Problem 16.2. *Suppose there are i_0 cycling susceptible cells and j_0 quiescent susceptible cells in the population. Further suppose that there are i_1 and j_1 cycling and quiescent cells resistant to the drug (assuming $m = 1$). Transition rates from cycling to quiescence and back are given by α and β respectively. (a) Write down the Kolmogorov forward equation for the probability function $\varphi_{i_0,j_0,i_1,j_1}(t)$ (see Section 15.2.4). (b) Define the probability generating function to be*

$$\Psi(\xi_0, \eta_0, \xi_1, \eta_1; t) = \sum \varphi_{i_0,j_0,i_1,j_1}(t)\xi_0^{i_0}\xi_1^{i_1}\eta_0^{j_0}\eta_1^{j_1},$$

and derive the PDE it satisfies (see Section 15.2.4). (c) Write down the equations for characteristics for ξ_0, η_0, etc (see Section 15.2.5).

Problem 16.3. *Draw a mutation diagram with cross-resistance in the case of $m = 2$, similar to the one in figure 16.7. Repeat the analysis of Section 15.4 including cross-resistance.*

Problem 16.4. Research project. *Find out about the research into therapeutic agents that are active against T315 mutants.*

Problem 16.5. Research project. *Learn about the work by [Day (1986)] and the "Worst drug first" rule.*

PART 3
Advanced topics

Chapter 17

Evolutionary dynamics of stem-cell driven tumor growth

One of the first topics considered in this book were single-species growth dynamics. We described various tumor growth laws, such as exponential growth, logistic growth, Gompertzian growth, etc., and discussed various data that can be fitted by those models. The basic models that describe different tumor growth laws are often phenomenological in nature, i.e., they do not have a particular mechanistic underpinning. In this chapter, we will link different tumor growth laws to specific evolutionary processes that can occur in stem cell driven cancers. In particular, we concentrate on the evolution of escape from feedback regulatory mechanisms that normally ensure tissue homeostasis. We will show how such an evolutionary model of stem cell-driven tumors can predict the different observed patterns of tumor growth, depending on the evolutionary pathways that lead to the escape from feedback control and the consequent breakdown of tissue homeostasis.

Evidence suggests that tumors are maintained by a relatively small number of tumor stem cells or tumor initiating cells that have the potential to divide indefinitely [Clevers (2011); Petersen and Polyak (2010); Wang *et al.* (2010); Johnston *et al.* (2010)]. The rest of the tumor bulk is assumed to be made up of more differentiated cells that can only undergo a limited number of divisions [Clevers (2011); Al-Hajj *et al.* (2003)]. This notion is supported by experiments where tumor cells are transplanted into immunodeficient mice, which show that a majority of cells fail to establish new tumors and only a small, defined subset of cells is capable of sustained proliferation [Bonnet and Dick (1997); Al-Hajj *et al.* (2003); Passegué *et al.* (2003)]. In these experiments the subset of tumorigenic cells expressed markers which are also expressed in healthy tissue stem cells. Hence, these cells are thought to have certain stem characteristics that allow them to self-renew and regenerate the tumor.

Healthy human tissue is highly regulated to ensure homeostasis, with feedback loops playing a fundamental role in this regard. In particular, two types of feedback loops have been suggested to be crucial: Differentiated cells secrete factors that inhibit the division of stem cells. In addition, differentiated cells secrete factors that suppress self renewal of stem cells and instead promote cell death following terminal differentiation [Lander *et al.* (2009); Wu *et al.* (2003); McPherron *et al.* (1997)]. Strong evidence for these feedback mechanisms has been found in the olfactory epithelium [Lander *et al.* (2009); Wu *et al.* (2003)], but also in a variety of other tissues, including striated muscle, liver, bone, central nervous system, hematopoietic system, among others [McPherron *et al.* (1997); Daluiski *et al.* (2001); Yamasaki *et al.* (2003); Tzeng *et al.* (2011); Elgjo and Reichelt (2004)].

As pointed out throughout this book, tumor formation occurs through a multi-step process where cells sequentially accumulate random mutations and epigenetic changes. Different types of cancers (depending on the tissue of origin) tend to arise through different and specific mutational pathways of varying complexity [Vogelstein and Kinzler (2004); Deininger *et al.* (2000)]. Tumor initiation and progression can involve key events, such as the emergence of genetic instability that allows mutations to be accumulated faster, or the acquisition of the angiogenic phenotype that enables the formation of new blood supply, among other processes [Weinberg]. Despite this great complexity and heterogeneity in the mechanism of tumor formation, there is ample evidence that escape from feedback regulation is a key ingredient in the formation of most, if not all, stem-cell driven tumors (reviewed in [Vogelstein and Kinzler (2004)]).

17.1 The model

The dynamics of tissue regulation through feedback loops has been studied in several cell lineages including the hematopoietic system, the lymphocytic system, the olfactory epithelium and the colon crypt [Johnston *et al.* (2007); Marciniak-Czochra *et al.* (2009a); Lander *et al.* (2009); Bocharov *et al.* (2011)]. To study the escape from feedback, we introduce a basic computational model of feedback-regulated tissue homeostasis that is based on previously published and experimentally validated work [Lander *et al.* (2009)] (figure 17.1(a)). This model [Rodriguez-Brenes *et al.* (2011)], described by equation (17.1), takes into account two populations: stem cells, S, which have unlimited reproductive potential, and differentiated

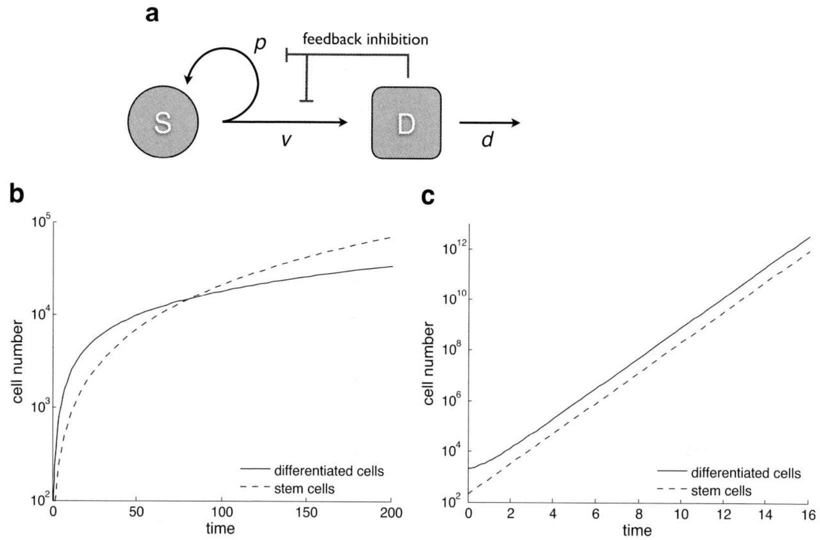

Fig. 17.1 Feedback-regulated tissue homeostasis and cell growth properties in the ordinary differential equations model. (**a**) Stem cells divide at a rate v producing either two stem cells with probability p or two differentiated cells with probability $(1-p)$. Differentiated cells die at a rate d and produce factors that inhibit self-renewal and division in stem cells. (**b**) Inhibited growth. If only differentiation feedback is lost, the population of stem cells and differentiated cells grows without bound at a slower than exponential rate. (**c**) Uninhibited growth. If both feedbacks are lost stem cells and differentiated cells grow at a rate dominated by the same exponential. Time is expressed in units of $\ln 2/v(\hat{D})$, the expect duration of one cell cycle at equilibrium.

cells, D, that eventually die (this includes all cell populations with limited reproductive potential, such as transit cells).

$$\begin{aligned} \dot{S} &= (2p(D)-1)\, v(D)\, S, \\ \dot{D} &= 2(1-p(D))\, v(D)\, S - dD. \end{aligned} \tag{17.1}$$

Stem cells divide at a rate v; this results in either two daughter stem cells with probability p; or two differentiated cells with probability $1-p$. Differentiated cells die at rate d. In accordance with data, we assume that differentiated cells secrete two types of feedback signals: one inhibits the rate of cell division, and the other reduces the probability of stem cell self-renewal, leading to cell death via terminal differentiation. The rate of cell division and the probability of self-renewal are treated as general functions of the number of differentiated cells, $v(D)$ and $p(D)$, respectively.

That is, we do not assume specific mathematical terms for these processes. This ensures that results are robust and are not dependent on particular and arbitrary mathematical expressions. We require however, that both feedbacks be decreasing functions of the number of differentiated cells, D, and go to zero if D grows without bound. Also, $0.5 < p(0) \leq 1$, otherwise the only outcome is population extinction. While asymmetric stem cell divisions (giving rise to one stem and one differentiated cell) is a possibility, its introduction does not change any of our results.

17.2 Evolutionary dynamics in ODE models

In model (17.1), only loss of the differentiation feedback can lead to uncontrolled growth, two types of which are observed: If the division feedback is still intact, we observe relatively slow, sub-exponential, growth, which we call "inhibited growth" (figure 17.1(b)), see also Chapter 4. If the division feedback is also lost, we observe faster exponential growth, which we call "uninhibited growth" (figure 17.1(c)). We further observe that with uninhibited growth, the ratio of stem cells to differentiated cells always converges to a fixed percentage. With inhibited growth, however, stem cells make up an ever increasing fraction over time. The predicted importance of losing the differentiation feedback for carcinogenesis is in line with previous modeling approaches [Johnston *et al.* (2007)].

To study the evolutionary dynamics of feedback escape, we assume that the healthy cell population is near equilibrium, and investigate the growth of mutational phenotypes from low numbers. We denote mutations that lack production of feedback signals by differentiated cells with the prefix D; and those that lack response by stem cells to these signals with the prefix S. Mutations that affect cell differentiation carry the suffix *diff-* and those affecting the division rate the suffix *div-*. The phenotypes considered are described as follows: (i) Type *Ddiff-* lacks production of differentiation-regulating signals by differentiated cells; (ii) Type *Sdiff-* lacks the stem cell response to the differentiation-regulating signals; (iii) Type *Ddiv-* lacks production of division inhibiting signals by differentiated cells; (iv) Type *Sdiv-* lacks the stem cell response to division inhibiting signals. The mutational steps, the nature and number of which are likely tissue dependent, are not modeled.

In a background of healthy tissue, types *Ddiv-*, *Ddiff-*, *Sdiv-*, are not selected for; they are selectively neutral with respect to healthy cells, and are

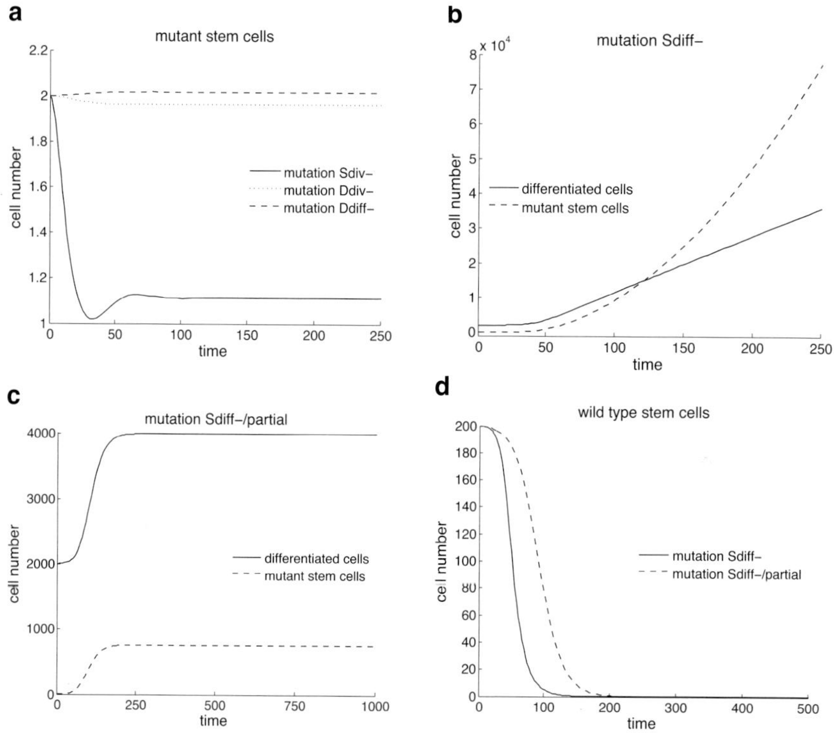

Fig. 17.2 Evolutionary dynamics of feedback loss in the ordinary differential equations model. The simulations begin at equilibrium with two stem cells carrying the specified mutation. (**a**) For populations near equilibrium mutations *Sdiv-*, *Ddiv-* and *Ddiff-* do not confer any competitive advantages over their wild type counterparts. If the mutation arises in a small number of cells the steady state number of mutant stem cells will be negligible. (**b**) Mutation *Sdiff-* results in inhibited growth in the number of mutant stem cells and differentiated cells. (**c**) Mutation *Sdiff-/partial* produces a finite increase in both the number of mutant stem cells and differentiated cells. (**d**) Mutations *Sdiff-* and *Sdiff-/partial* result in the extinction of the wild type stem cell population.

thus likely to go extinct in a stochastic setting (figure 17.2(a)). Only *Sdiff*-types, lacking a stem cell response to the differentiation-regulating signals, can have a growth advantage in a background of healthy tissue, eventually taking over the entire population and growing uncontrolled according to the "inhibited"' growth pattern (figure 17.2(b,d)). Hence, this must be the first significant step towards malignancy. Only in these *Sdiff*-type of mutants can the acquisition of the phenotype *Sdiv-* confer a selective advantage to the cells, which will grow exponentially according to the "uninhibited"

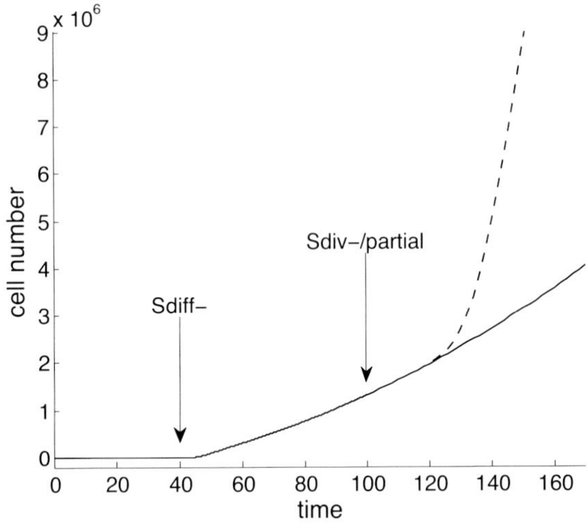

Fig. 17.3 In a healthy population at equilibrium, a stem cell acquires mutation *Sdiff-* at the time indicated by the arrow (solid line); the subsequent appearance of mutation *Sdiv-/partial* in a *Sdiff-* cell produces an acceleration in the growth of the tumor (dashed line).

growth pattern and eventually dominate the population (figure 17.3). None of the other phenotypes enjoys a selective advantage in any setting. Hence, loss of feedback control in this system requires that mutants lose the ability to respond to, rather than produce, the signals. Importantly, this can only occur via a unique sequence of events, where first the response to differentiation, then to division feedback is lost (figure 17.4). We suggest this to be a universal pathway of feedback escape among stem cell driven cancers, although the nature and number of mutation events to achieve this is certainly tissue specific.

We also consider mutations that confer only a partial loss of response to feedback signals. A mutation that only partially compromises the response to differentiation-regulating factors (denoted by *Sdiff-/partial*) will eventually take over the entire population, producing a bounded "sigmoidal" growth pattern (figure 17.2(c,d)). A mutation that produces a partial loss in the response to division rate factors (*Sdiv-/partial*) cannot invade a healthy cell population, but it will produce an acceleration in the growth rate of a tumor exhibiting inhibited growth (figure 17.3).

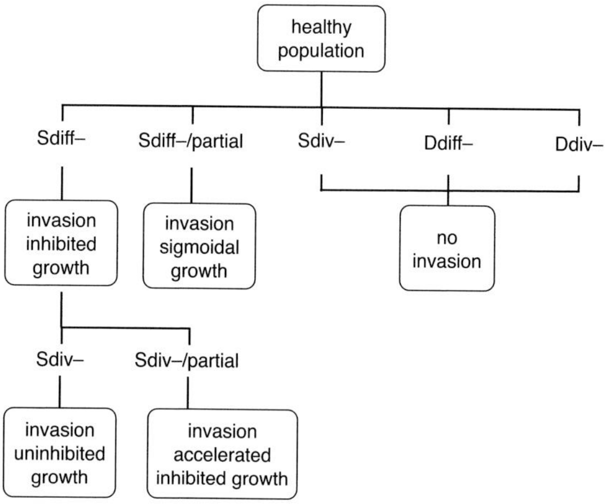

Fig. 17.4 Tumor progression towards uninhibited growth follows a unique sequence of feedback inactivations: first mutation *Sdiff-* must occur, followed by mutation *Sdiv-*.

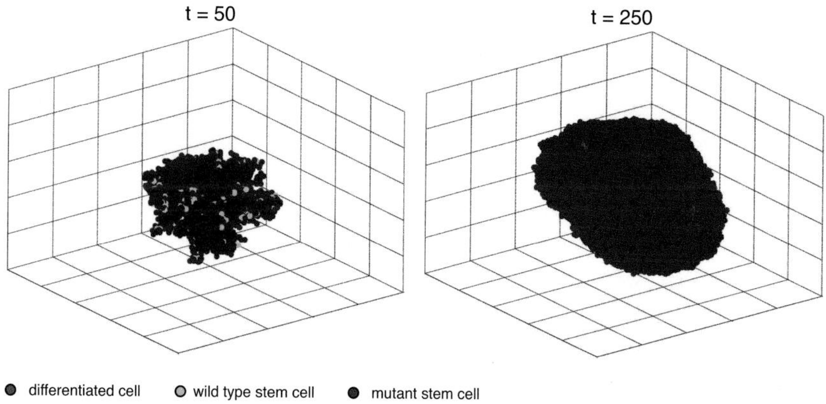

Fig. 17.5 Spatial arrangement of the cell population at two different times in the spatial model. The simulation begins with a tissue at near equilibrium with two stem cells randomly selected to carry mutation *Sdiff-* at time $t = 0$.

17.3 Evolutionary dynamics in a stochastic, spatial model

Next, we consider these dynamics in the context of a stochastic, three-dimensional rectangular lattice model corresponding to solid tumor growth with spatial structure. A lattice point can host at most one cell at any time. For a cell to divide, there must be a free lattice point adjacent to it to place the offspring. A stochastic simulation algorithm is used, where the probabilities of cell division, differentiation and death correspond to our previous non-spatial model. Our main results remain unchanged (figures 17.5 and 17.6). Again, we observe uninhibited tumor growth if both feedback loops are broken, and inhibited growth when only the differentiation feedback loop is broken. The percentage of stem cells increases progressively with inhibited growth, while it converges to a fixed percentage for uninhibited growth. However, in contrast to the non-spatial situation, the tumor growth rates are slower. Uninhibited growth is not characterized by exponential, but by cubic growth. Inhibited growth is characterized by sub-cubic growth. This difference is partly caused by surface growth dynamics in the three-dimensional model, a behavior observed in several types of solid tumors [Brú *et al.* (2003); Drasdo and Höhme (2005); Enderling *et al.* (2009a)]. Looking at the evolutionary dynamics, we find that feedback inactivation occurs via the same unique pathway as in the non-spatial model. A full analysis of this model is given in [Rodriguez-Brenes *et al.* (2011)].

17.4 Predicted versus observed tumor growth patterns

Performing extensive literature searches, we found that most growth patterns belong to one of the categories resulting from our models: The uninhibited pattern, which is exponential in the non-spatial system and cubic in the spatial system; the inhibited pattern, which is sub-exponential in the non-spatial system and sub-cubic in the spatial system; and the sigmoidal growth pattern. In the following examples, specific forms of the models were fitted to different types of published tumor growth data using least squares procedures. (i) Inhibited non-spatial growth is found in Ehrlich's ascites tumor [Laird (1964)] where cells grow sub-exponentially. (figure 17.7(a)). (ii) Inhibited spatial growth is found in A2780 human ovarian carcinoma growth in mice [Simeoni *et al.* (2004)] (figure 17.7(b)). The data show sub-cubic kinetics with a power law of 2.17 and no saturation. (iii) Uninhibited non-spatial growth is found in data from L1210 cells [Shackney (1970)], a mouse lymphocytic leukemia, where cells grow exponentially (figure 17.7(c)). (iv) Uninhibited spatial growth is found in data from spatial multicellular tumor spheroids of EMT6/Ro cells [Freyer and Sutherland

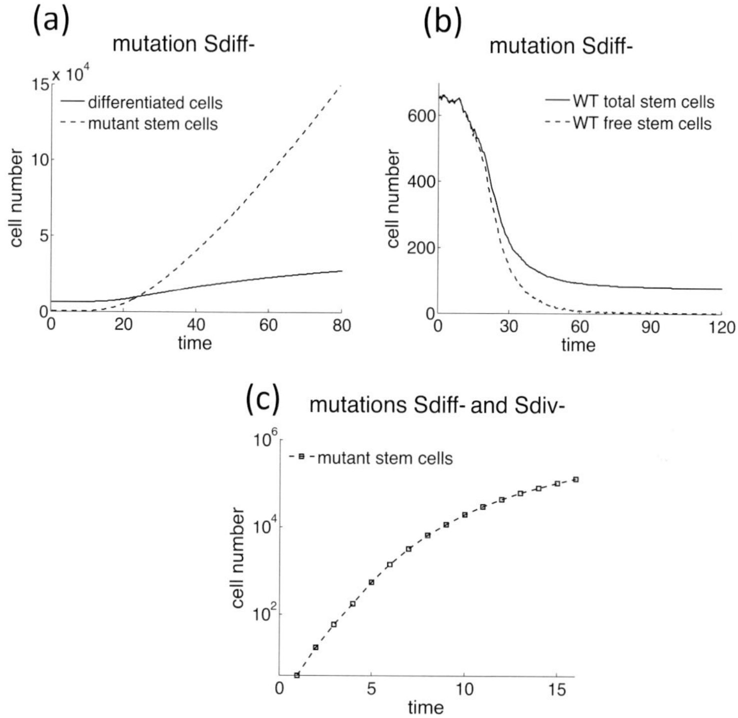

Fig. 17.6 Evolutionary dynamics in the spatial model. (**a**) The appearance of mutation *Sdiff-* results in the unlimited growth of the mutant stem cell and differentiated cell populations. (**b**) The number of wild type stem cells decreases. Note that a small number of stem cells that are trapped -and thus unable to divide-lingers in the population for a long time. The number of wild type stem cells however, becomes a negligible percentage of the entire cell population. (**c**) Cell population with stem cells carrying mutations *Sdiff-* and *Sdiv-* (simulations start with a small number of stem cells carrying both mutations). Cell growth is much faster than if only mutation *Sdiff-* is present (note different scales in **a** and **c**); but, unlike the non-spatial model, the growth is not exponential. Results represent the average of 24 runs.

(1986)], derived from a mouse mammary tumor, which show cubic growth (figure 17.7(d)). (v) Finally, a sigmoidal growth pattern is found in Jurkat cells [Reuss *et al.* (2004)] originating from a T cell human leukemia (figure 17.7(e)).

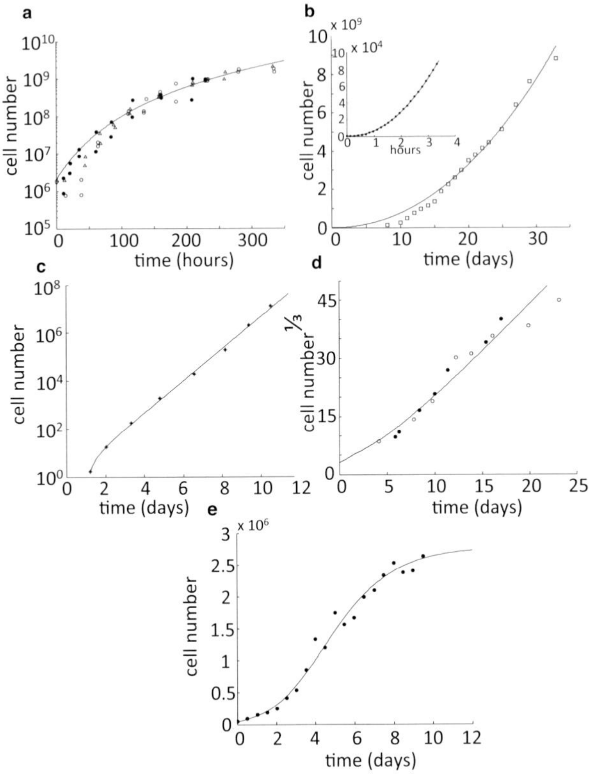

Fig. 17.7 Experimentally observed growth patterns and model fits. (**a**) Inhibited growth in the non-spatial model. Ehrlich's ascites tumor [Laird (1964)] (three experiments shown: •, △, ○). (**b**) Inhibited growth in the spatial model. Main frame: (□) A2780 human ovarian carcinoma [Simeoni *et al.* (2004)] and projection of the model using the function $y = ax^b$ (solid line). Inset: Simulation results (•) and projection of the model (simulations were not carried further in time due to computational constraints). (**c**) Uninhibited growth in the non-spatial model. (∗) L1210 a mouse lymphocytic leukemia [Shackney (1970)]. (**d**) Uninhibited growth in the spatial model. Multicellular tumor spheroids of EMT6/Ro cells [Freyer and Sutherland (1986)], a mouse mammary tumor (two experiments shown: ○, •). (**e**) Sigmoidal growth in the non-spatial model. (•) Jurkat T cell human leukemia [Reuss *et al.* (2004)]. Simulations are shown in solid lines; those corresponding to the stochastic spatial model represent the average of 24 runs.

17.5 The order of phenotypic transitions

Here, we address model predictions about the order of phenotypic transitions and the central importance of the *Sdiff*-type with previously published data. A transgenic mouse model of hepatocellular carcinoma was developed, in which it is possible to regulate the expression of the human MYC oncogene in murine liver cells, suppressing it through doxycycline treatment [Shachaf *et al.* (2004); Shachaf and Felsher (2005); Felsher (2003)]. While transgenic mice treated with doxycycline remained disease free, those with active MYC from the discontinuation of treatment developed malignant tumors that were locally invasive and able to metastasize. When MYC was subsequently inactivated, rapid tumor regression was observed that was associated with terminal differentiation into normal liver cells and apoptosis. Moreover, re-activation of MYC resulted in significant tumor re-growth and de-differentiation. These observations validate key model predictions. MYC expression influences self-renewal and differentiation of cells, and thus influences the function $p(D)$ in our model. Activation of MYC corresponds to corrupted differentiation feedback, i.e., to the *Sdiff-* phenotype in the model, while inhibition of MYC reverses this phenotype. The model predicts *Sdiff-* to be the initial and most crucial event in the evolution of feedback loss and uncontrolled growth. Even if cells have acquired other mutations that can also contribute to tumor progression, these mutations are predicted to only contribute to growth in cells that already have corrupted differentiation feedback. Hence, restoration of the differentiation feedback loop even in cells with further complex genetic alterations is predicted by the model to result in tumor regression and tissue dynamics that are characteristic of a healthy state. This same behavior is observed in the experiments where the macroscopic and malignant nature of the tumors indicate the presence of additional mutations, which are incapable of promoting growth in the absence of MYC [Shachaf *et al.* (2004)]. Figure 17.8 shows a computer simulation that successfully recapitulates the dynamics observed in the MYC regulation experiments. Similar dynamics have been observed in the context of other tumors and/or oncogenes (e.g. Myc-induced hematopoietic tumors, breast cancers and osteogenic sarcoma, or Ras-induced melanomas), although details of the results can differ in various ways (e.g. treatment leading to complete extinction of the tumor) [Shachaf *et al.* (2004); Shachaf and Felsher (2005); Felsher (2003)]. Understanding the mechanisms that lead to differences in outcome will be important future work that can be aided by computational models.

In general, many key mutations in carcinogenesis disrupt negative feedback regulation of cell division patterns. Consider the protein transforming growth factor beta (TGF-β). This protein plays a key role in tissue

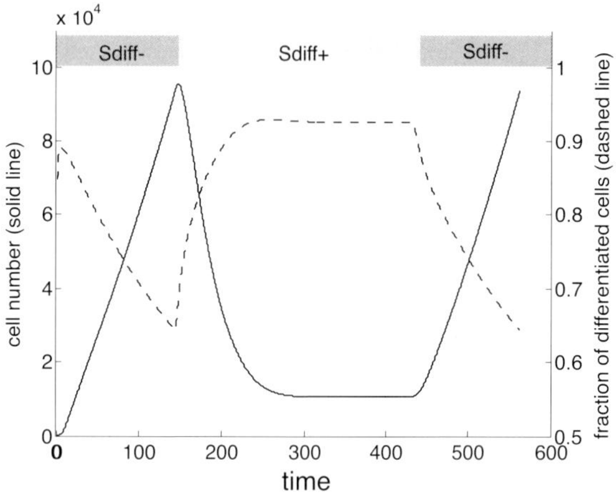

Fig. 17.8 Simulation of MYC regulation experiments [Shachaf *et al.* (2004); Shachaf and Felsher (2005); Felsher (2003)]. Upon activation of MYC (represented in the model by the *Sdiff-* phenotype) the cell population (solid line) exhibits sustained growth characterized by a reduction in the fraction of differentiated cells (dashed line). When MYC is subsequently inactivated (*Sdiff+* phenotype) the fraction of differentiated cells increases while the cell population decreases to a lower valued steady state; when MYC is re-activated the cell population rises again. See text for discussion.

homeostasis by inhibiting mitosis and promoting cell differentiation [Weinberg]. Many types of cancers must circumvent TGF-β growth inhibition to be able to thrive. In these cancers feedback escape is accomplished by inactivating the genes for the TGF-β receptors or through downstream alterations that disable the tumor-suppressive arm of the pathway [Derynck *et al.* (2001); Massagué (2000, 2001)]. Half of all pancreatic carcinomas and more than a quarter of colon cancers carry mutations that make cells irresponsive to TGF-β signals that inhibit cell division and promote differentiation [Rozenblum *et al.* (1997); Woodford-Richens *et al.* (2001)]. Mutations that affect TGF-β receptors also occur in gastric, biliary, pulmonary, ovarian, esophageal, and head and neck carcinomas [Massagué (2000)]. Another specific example of feedback escape in cancers is observed in glioblastomas that inactivate the bone morphogenetic protein 4 pathway (BMP4). There is strong evidence that glioblastomas are maintained by a small population of tumor initiating cells that have stem cell characteristics [Wu *et al.* (2008)]. In humans, naturally occurring BMP4 induces glia stem cells to differentiate, inhibiting cell proliferation [Lim *et al.* (2000)]. The relationship between BMP4 and cancer is supported by evidence of epigenetic silencing of BMP4 receptors in glioblastomas [Lee *et al.* (2008)], and by in vitro experiments that show that the addition of BMP4 causes a colony of glioblastoma multiforme cells to increase the fraction of differentiated cells and lose their tumorigenic capabilities [Piccirillo *et al.* (2006)]. Finally, in colorecal cancer the initial mutational events are defined relatively well. They are the loss of the APC tumor suppressor gene and the concomitant activation of the Wnt cascade, followed by the activation of the K-Ras oncogene. While both alterations lead to complex phenotypic changes, a common effect of both is that the cell division pattern is shifted away from differentiation and towards self-renewal, consistent with our model.

17.6 Summary

Using evolutionary computational models, we found that escape from feedback-regulated tissue homeostasis can only occur via a unique sequence of phenotypic transitions that we propose to be common among stem cell driven tumors, even if the nature and number of mutational events required to achieve this are certainly tissue specific. The resulting growth dynamics predicted by the model fall into three categories: uninhibited, sigmoidal, and inhibited. These can describe many experimental growth patterns found in the literature, which was demonstrated by fitting the model to five sets of published data. The finding of inhibited tumor growth patterns in the literature is of particular interest. Such a growth pattern, especially in the form of sub-cubic growth, could only be explained if the growing tumor is still partially subject to feedback regulation that has remained from the underlying tissue. This gives support to the notion that not only the tissue architecture, but also the regulatory mechanisms of the corresponding healthy tissue continue to operate to a certain degree in tumors, especially at early stages.

Problems

Problem 17.1. Research project. *Find out more about feedback loops in stem cell dynamics.*

Problem 17.2. Numerical project. *Modify the basic stem cell model (17.1) to include an intermediate transit amplifying population. This population, call it A, is assumed to be generated by differentiation from the stem* *cells population, S. The A population is assumed to divide, although with a slower rate than stem cells, and to differentiate into the terminally differentiated cell population D. The processes of A division and differentiation are subject to the same types of feedback inhibition as the stem cells. Thus, instead of a 2-compartment model, you will now have a 3-compartment model. Explore the dynamics of uninhibited and inhibited growth in this new model. How do they compare to the dynamics observed for model (17.1)?*

Chapter 18

Tumor growth kinetics and disease progression

In Part 2 of the book, a mathematical framework was developed to study the evolutionary dynamics of mutant generation during tumor growth. An obvious application of this framework is the generation of mutants that are resistant to drug therapy, and the theory was developed in this context. The current chapter explores implications of this mathematical framework for the evolution of tumor cells, namely, the progression of cancers towards increased malignancy.

As explored in the introductory chapters, the process of carcinogenesis involves the accumulation of multiple genetic mutations enabling cells to defy homeostatic regulation and grow in an unregulated manner [Vogelstein and Kinzler (2002); Boland and Goel (2005)]. This process continues throughout the progression of the disease. Initially, the tumor may be able to grow to a certain degree until it hits a selective barrier and growth plateaus. Further growth then requires the presence of mutants that can overcome the relevant selective barrier. This process can repeat as the cancer progresses further towards malignancy. We explore how the potential of the tumor to accumulate mutations and to progress depends on parameters describing the tumor growth kinetics.

In particular, we are interested in the effect of cell death on cancer progression. The rate of cell death is related to the turnover of the tumor cell population. The closer the death rate of cells is to their division rate, the higher the turnover. Tumor cell populations that mostly grow and do not exhibit significant amounts of cell death are considered low turnover tumors. In Chapter 16, we have shown that the turnover of the tumor cell population can have a substantial effect on the rate at which mutants are

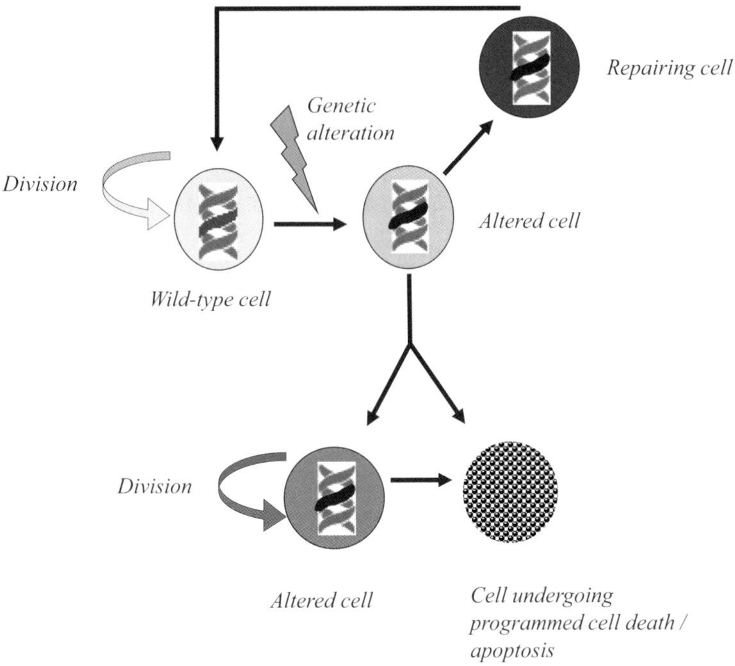

Fig. 18.1 Schematic diagram that puts the concept of programmed cell death or apoptosis into bigger context of cellular checkpoints. When the genome of cells becomes altered (which includes oncogenic mutations), two things can happen. First, the cell can try to repair the alteration, in which case it returns to its previous (wild-type) state. Repair, however, is not always possible. In this case, the cell can either undergo programmed cell death (most often apoptosis), or the altered cell can continue to divide. If the altered cell continues to divide, then apoptosis can occur during cell divisions because checkpoints are invoked that detect the altered state. In healthy cells, repair and apoptosis checkpoints are intact and function well. In tumor cells these pathways are often compromised. The checkpoints can still function relatively well in early cancer cells, and tend to be more compromised during later stages and in malignant cancer cells.

generated, with high turnover cancers exhibiting significantly faster rates of evolution.

Similar to healthy cells, cancer cells, especially at earlier stages, can respond to their altered state by undergoing programmed cell death (PCD) [Jäättelä (2004); Kroemer and Martin (2005)] (figure 18.1). At later stages of the disease, cancer cells tend to have inactivated some PCD pathways.

The most commonly discussed form of PCD is apoptosis, although several non-apoptotic modes of PCD can also be induced and are an important mechanism to remove cancerous cells [Jäättelä (2004); Kroemer and Martin (2005)]. PCD is induced by a variety of triggers including damaged DNA, and altered levels of regulatory proteins that exist in the cell as a result of oncogenic mutations [Jäättelä (2004)] (e.g. the *c-myc* or *Ras* oncogenes in a variety of cancers). Cancer cells might also die by necrosis in certain circumstances, but this is a very different mode of death and does not represent a response to stress or genetic alterations. When talking about cell death in this chapter, we refer to programmed cell death.

It is generally accepted that PCD protects against the development of tumors. Here we argue that the situation is more complex than this. It is certainly true that PCD can prevent the establishment of a tumor mass, and that a certain loss of PCD can be required for clonal expansion of transformed cells. We argue, however, that cancer progression to malignancy can be prevented by reduced levels of PCD, while it can be promoted by higher levels of programmed cell death. A reduced rate of programmed cell death correlates with fewer cell divisions during clonal expansion and thus fewer mutants present in the tumor cell population at a given size. More cell death correlates with a greater number of cell divisions during clonal expansion and thus with more mutants present in the tumor cell population when a certain size is reached. Since a pool of mutant cells is required for the tumor to overcome selective barriers (e.g. the requirement for blood supply or for overcoming growth inhibition) and to expand further, a higher level of cell death in the expanding tumor cell population can increase the chances that the tumor progresses towards malignancy instead of becoming a dead end.

18.1 Cell death and mutant generation

Consider a population of first stage tumor cells that clonally expands (i.e., cells that have just been transformed and that can divide beyond homeostatic control). Growth stops once the tumor has reached size N. Further progression towards malignancy requires the presence of mutants such that selective barriers can be overcome (figure 18.2(a)). The more mutants in the tumor cell population, the greater the likelihood that the appropriate mutants are present to overcome selective barriers. This increases the chances that the tumor progresses to malignancy and that growth does not result in a pathogenic dead end.

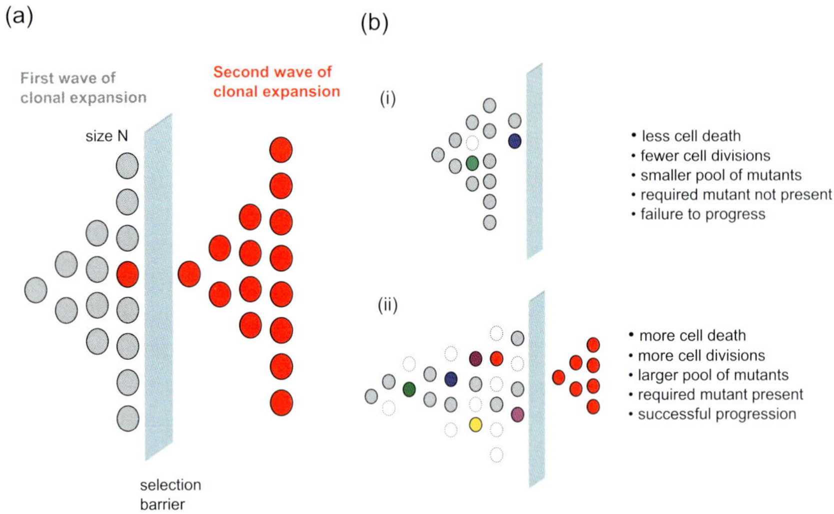

Fig. 18.2 Schematic diagram of the basic concept that underlies our arguments. (a) We consider a growing colony of cells and the generation of mutants. Upon cancer initiation, a first wave of clonal expansion occurs. This stops if a selective barrier is encountered, such as the requirement for new blood supply. The tumor size at which the selective barrier is encountered is denoted by N. In order for the tumor to progress and grow further, an appropriate mutant (indicated in red) has to be generated during this first wave of clonal expansion. If the appropriate mutant is not present, then the tumor fails to progress and becomes a dead end. The color grey depicts wild-type tumor cells, and different colors indicate different mutants. (b) Schematic representation of our central message. (i) If the death rate of tumor cells is relatively low, fewer divisions are required to reach tumor size N. Consequently, fewer mutants are present in the tumor cell population when the selective barrier is hit. This decreases the chance that the appropriate mutant required for progression (indicated in red) exists. If the required mutant does not exist, the tumor does not progress and is a dead end. (ii) If the death rate of tumor cells is higher, then more cell divisions are required to reach tumor size N. Consequently, more mutants are produced. This increases the chance that the appropriate mutant required for progression (marked in red) exists by the time the selective barrier is hit.

The mathematical framework that was used to study these processes was reviewed in detail in Chapter 15. Here, we briefly summarize the main points that are relevant for the current arguments. Suppose we have a colony of tumor cells which grows stochastically from size M_0, until it reaches size N. This clonal expansion is a birth-death process with rates of cell division and cell death defined by constants L and D, respectively. The ability of a cell to undergo PCD correlates with the death rate of the cell, D. In addition, cells can mutate with a probability u. Mutant cells are also characterized by the appropriate division and death rates. It can be assumed that mutant cells have a death rate D_1 which is higher than D. The reason is that these can be mutations which contribute to cancer progression, and therefore drive up the level of PCD. The expected numbers of wild-type and mutant cells in a colony starting from M_0 cells are given by equations

$$\begin{aligned}\dot{x}_0 &= L(1-u)x_0 - Dx_0,\\ \dot{x}_1 &= Lux_0 + (L - D_1)x_1,\\ x_0(0) &= M_0, \quad x_1(0) = 0,\end{aligned} \tag{18.1}$$

see Section 15.4.1. Let us assume for simplicity that $D_1 = D$. We denote the expected number of mutants after time t by $G_1(t)$. By solving the above system we can show that it is given by

$$G_1(t) = M_0 e^{(L-D)t}(1 - e^{-Lut}).$$

On the other hand, we can relate the time-variable t to the colony size, N:

$$N = M_0 e^{(L-D)t}.$$

Using this in the solution for x_1, we can obtain the expression for the expected number of mutants by the time the colony reaches size N. Calling this quantity $G_2(N)$, we obtain approximately

$$G_2(N) = N\left(1 - \left(\frac{N}{M_0}\right)^{-\frac{Lu}{L-D}}\right).$$

The higher the death rate, the longer it takes for the colony to reach size N. Let T be the threshold time, that is, the time remaining from the beginning of the clonal expansion to a certain upper age limit where the patient death is likely to be caused by a multitude of reasons other than cancer. Let us denote $D_c = L - \frac{1}{T}\ln\frac{N}{M_0}$. Then for $D < D_c$, the size N is reached before time T, and for $D > D_c$ it takes longer than T to reach size N. The expected number of mutants produced by the colony once it reaches size N or time T (whichever happens first) is given by

$$E_{mut} = \theta(D_c - D)G_2(N) + \theta(D - D_c)G_1(T),$$

which is a one-humped function of the death rate D (here $\theta(.)$ denotes the Heaviside function). If we assume a more general case. I.e., that the death rate of mutants $D_1 > D$ (because oncogenic mutations can induce cell death), and that it grows with D, the property of this relationship remains the same. Note that size N and time T can vary from tumor to tumor. Here we consider the ensemble average, but this does not affect the notions discussed.

How does the average number of mutant cells and thus the potential of the tumor to progress after the first wave of clonal expansion depend on the death rate of tumor cells, i.e., the rate of PCD, D (assuming that other cellular parameters, such as the division rate, remain constant)? As the death rate D increases from low to high, the average number of mutants present in the colony at size N rises. This is because more cell divisions are required for the tumor to reach size N. In other words, higher levels of cell death (PCD) increase the mutant pool and promote tumor progression. Because more cell divisions are required to reach size N, the rate of tumor growth is slowed down as the death rate D is increased relative to the division rate L. We can see that from the point of view of the tumor there is a clear trade-off: Very low death rates lead to fast growing but benign tumors (as not enough mutants are generated before the clone reaches size N). Very high death rates lead to a larger mutant pool but make the growth too inefficient. One way to capture these trends is as follows. Assume that T denotes the time or the age up to which progression of a first stage tumor can significantly contribute to mortality. If it takes less time than T to reach size N, we can count the number of mutants (at size N) as a function of the death rate. Beyond time T, the organism is too old and is more likely to die from other causes before the cancer has invaded sufficiently. In this parameter region, we have to consider the number of mutants found in the tumor at time T. This is a decreasing function of the death rate D.

Overall, there is a one-humped relationship between the number of mutant cells at the end of clonal expansion, and the death rate of tumor cells, D (i.e., we see an increase followed by a peak and a decline, see figure 18.3). As seen in figure 18.3, the number of mutants at size N always increases with higher death rates D, but when the value of D crosses a threshold, the tumor does not reach size N within the life-span of the organism. There, an increase in the death rate of tumor cells leads to a reduction in the number of mutants at time T.

Note that these arguments assume that during the growth plateau N, cell divisions stop. This has been observed to occur, for example when initial early cells undergo replicative senescence after a certain number of divisions. In this case, for further growth to occur, clonal expansion needs to give rise to a mutant cell which escapes replicative senescence and continues to divide. However, not all growth plateaus are characterized by a cessation of cell division. In some cases, such as in telomeric crisis, cell growth can stop because the death rate of the cells is increased and roughly matches the division rate of cells. Our arguments do not apply to such scenarios.

18.2 Does PCD protect against cancer?

The predominant view is that PCD protects against cancer because cell death counters clonal expansion [Jäättelä (2004)]. Loss of PCD certainly is a defining characteristic of tumor cells, and is thought to allow cancer cells to survive better and to progress towards malignancy. It is true that strong PCD (i.e., a high death rate of tumor cells) can prevent the progression of cancer because clonal expansion is so slow that the cancer will not grow to a sufficient size or accumulate a sufficient number of mutations during the life-span of the organism. Even stronger, if the death rate of tumor cells is greater than their division rate, clonal expansion can never be successful. However, loss of PCD can reduce the chances that the tumor progresses towards malignancy if the first wave of clonal expansion occurs sufficiently fast relative to the life-span of the organism. In this case, loss of PCD is actually protective for the organism. Assume that cells have inactivated some PCD pathways, such that clonal expansion can occur at a significant rate and that the tumor cell population can reach a defined size within the life-span of the organism. Now, if the cells reduced the ability to undergo PCD by too much, not enough mutants will be produced during this wave of clonal expansion, consequently the tumor has a high probability of failing to progress from a benign to a malignant form. On the other hand, if the ability of cells to undergo PCD is lost to a lesser extent (i.e., the cells can still undergo PCD to a certain degree), then the number of mutants produced during this wave of clonal expansion is significantly higher, increasing the chances that the tumor can overcome selective barriers and progress to a malignant phenotype. This conclusion is summarized schematically in figure 18.2(b). This argument can also be thought of as an optimization problem. For the organism, the optimal strategy is either to have the

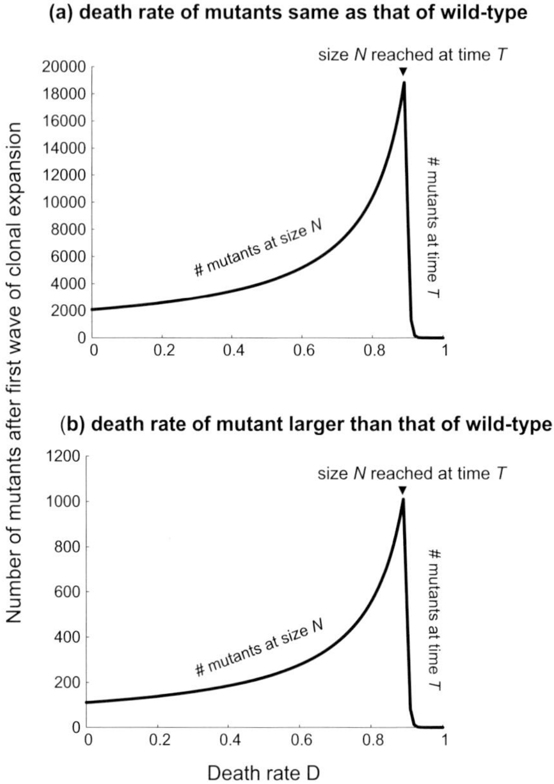

Fig. 18.3 Relationship between the number of mutants found after the first wave of clonal expansion and the death rate of tumor cells, D. (a) Mutant tumor cells are assumed to have the same death rate as wild-type cells. (b) Mutant tumor cells are assumed to have a death rate that is higher than that of wild type tumor and satisfies $D_1 = D + 0.9(L - D)$. Both scenarios show the same pattern. That is, the number of mutants after the first wave of clonal expansion is a one humped function of the death rate D. If tumor size N is reached before time T, we consider the number of mutants at size N, which is an increasing function of the death rate D. If tumor size N is reached after time T (when the organism dies of cancer-unrelated causes), then we consider the number of mutants at time T, which is a decreasing function of the death rate D. Parameters are given by: $N = 10^9$, $u = 10^{-7}$, $M_0 = 1$, $T = 20$ years, $L = L_1 = 1$, assuming on average 10 cell divisions per year.

highest rate of PCD such that a tumor never undergoes significant clonal expansion, or a relatively weak rate of PCD such that potentially harmful mutants are not present. The optimal strategy for the cancer is to have an intermediate rate of PCD. The rate of PCD has to be reduced such that initial clonal expansion occurs, but it should not be reduced too much such that the mutants necessary for progression are present.

It is a well-known fact that especially at older ages, people tend to harbor a number of harmless tumors that fail to progress [Folkman and Kalluri (2004); Volpe *et al.* (2004)]. Mutational processes are random, and hence, the exact extent to which PCD is lost can vary along a spectrum. The presence of these non-progressing tumors might be explained by the fact that the loss of PCD was too great, resulting in the failure to generate a sufficient number of mutants during clonal expansion. A well-known selective barrier is the requirement of tumors for the formation of new blood supply (angiogenesis) [Folkman *et al.* (1992)]. Tumor cells can divide until the population reaches a certain small size at which it stops growing unless angiogenic cells induce the formation of new blood supply. If the population of tumor cells has not generated many mutants during this expansion, chances are low that angiogenic cells will be found in this tumor cell population (because angiogenic cells can be created by defined mutations [Folkman *et al.* (1992)]. However, it is important to point out that this is a hypothesis at this stage, and that other mechanisms can also explain an abundance of covert tumors in aging individuals (such as a balance between proliferation and death).

In the light of this discussion, we note that some important tumor suppressor genes never tend to be lost in the initial stages of carcinogenesis, but only during later stages, just before the transition to full malignancy. The most famous example is the tumor suppressor p53 which is responsible for apoptosis and senescence, among other things [Vogelstein and Kinzler (2002); Kemp *et al.* (1993)]. According to our theory, an early inactivation of p53 would result in a low rate of cell death early on. Consequently, tumor cells would divide less and thus not create a sufficient number of mutants during clonal expansion. Consequently, such a tumor would reach a dead end. On the other hand, at later stages of carcinogenesis, p53 can be lost because lots of mutants have already been generated, and this pool is enough to allow progression towards malignancy. Inactivation of p53 would then be a very common event, as observed, because it allows the population of cancer cells to grow faster than in the presence of p53 [Blagosklonny (2002); Oren (2003)].

It is interesting to consider our arguments in the context of genetic instability (elevation of the rate of genetic alterations observed in many tumor cells). In some cancers, such as colon cancer, genetic instability is thought to arise relatively early in the course of the disease [Shih *et al.*

(2001)]. While genetic instability can promote the generation of mutant cells, it also leads to the destruction of the tumor cells' genome [Komarova and Wodarz (2004)] and this in turn can lead to apoptosis. Therefore, in this case the rate of mutation and the rate of apoptosis can be linked in cancer cells. Cells with a high degree of genetic instability might also undergo apoptosis more readily because of the deranged genomes, and thus undergo more cell divisions and create more opportunities to mutate until a certain tumor size is reached.

18.3 Cell turnover and pathology

The arguments presented here can also be applied to interpret differences in the natural history and pathology of various cancers. Here we compare two specific examples: renal cell carcinoma and pancreatic cancer.

The investigation of the natural history of renal cell carcinoma (RCC) has been difficult because most patients are treated surgically with partial or radical nephrectomy shortly after the diagnosis [Volpe *et al.* (2004); Motzer *et al.* (1996); Rathmell *et al.* (2007); Ozono *et al.* (2004); Bosniak (1995); Oda *et al.* (2001)]. As a result of improved diagnostic tools, the reported incidence of RCC has increased, and many RCCs are now detected incidentally as small tumors in asymptomatic patients. This has allowed a more detailed examination of the natural history of this cancer. Patients with such small tumors usually do not have any metastases. The potential of the tumor to grow and progress varies among patients [Volpe *et al.* (2004)]. Some tumors progress relatively fast, while others grow very slowly or not at all. Some small tumor masses were even observed to shrink. About one third of small renal masses that are presumed to be RCC grow and progress significantly [Volpe *et al.* (2004)]. Most small tumors, however, do not grow or progress at a significant rate [Volpe *et al.* (2004)]. These could be tumors that have not generated a sufficient number of mutants in the initial wave of clonal expansion, and that are consequently unable to progress further. According to our model, this is because the degree of PCD that these initial cancer cells undergo is too low. Thus, relatively few cell divisions are required for the initial growth of the cancers to the small size at which they were found. This notion is supported by autopsy reports, showing that 67-74% of RCCs are undetected until death, and that only 9-20% of undiagnosed RCCs result in malignant progression and death of the patient [Volpe *et al.* (2004)].

The picture is quite different with pancreatic cancer [Bardeesy and De-Pinho (2002); Leach (2004); Rosenberg (2000); Hingorani *et al.* (2003)]. Infiltrating ductal adenocarcinoma of the pancreas is responsible for over 95% of all exocrine pancreatic malignancies and is the fifth leading cause of cancer-related death in the USA. Moreover, diagnosis with pancreatic cancer usually has a poor prognosis, with five year survival between 1-4%. Surgery of small lesions is the only possible form of treatment because the cancer is resistant against all possible forms of available chemotherapy and radiotherapy. However, even surgical removal of lesions does not provide a strong benefit for the patient because metastases have been established even if the primary tumor is very small (<2cm) [Hingorani *et al.* (2003)]. The cancer spreads to the regional lymph nodes at an early stage of the disease, and subclinical liver metastases are present in most patients at the time of diagnosis [Rosenberg (2000)]. In the light of our framework, we hypothesize that even growth to relatively small sizes is sufficient to generate a large pool of mutants that allows the cancer to progress towards malignancy with a fast rate, including the potential of the cancer cells to metastasize at this early stage. The reason would be as follows. Although the tumor cells have lost some ability to undergo PCD such that the cancer is initiated and the cells can undergo clonal expansion, the cells retain a significant ability to undergo cell death. Consequently, a significant amount of cell death occurs during clonal expansion. Therefore, a relatively large number of cell divisions is required for the cancer to grow to a defined size, and this enables the cancer to generate a large pool of mutants that allows the cancer to progress further and to be resistant against chemo- and radiotherapies.

18.4 Conclusions

The relationship between cell death and the process of carcinogenesis and cancer progression is complex. According to the traditional view, higher levels of cell death, in particular PCD, prevent the emergence of cancer, while lower levels of cell death promote the emergence of cancer. According to the mathematical framework discussed here, some loss of cell death is required for the tumor cells to clonally expand. However, if the extent to which cells have lost the ability to undergo death is too large, then not enough mutants might be accumulated during clonal expansion, and the tumor might fail to progress, essentially becoming a dead end and remaining benign. On the other hand, if the ability to undergo cell death has been lost to a lesser extent, then clonal expansion is more likely to produce a sufficient number of mutants such that the tumor can overcome selective barriers and progress towards malignancy. These insights might help us understand genetic pathways which lead towards malignancy, and differences in the patterns of pathology that are observed among different cancers. Importantly, to go beyond theoretical arguments, this hypothesis should be tested by experiment. Two cell lines could be compared, a control population and a population of cells that have a defect in a defined apoptotic pathway. These cells should divide until a defined population size has been reached, and the distribution of mutations of certain genes should be measured. This should directly address the arguments presented here.

Problems

Problem 18.1. Research project. *This chapter only addresses growth plateaus where cells essentially stop dividing. There are other situations where the net growth stalls because the death rate matches the division rate of cells. The two types of growth plateaus are termed "division-driven growth" and "death-driven growth" in [Sorace and Komarova (2012)]. The process of mutant accumulation is very different in the two cases. Learn about these differences by reading [Sorace and Komarova (2012)].*

Chapter 19

Epigenetic changes and the rate of DNA methylation

A large portion of the book dealt with somatic evolutionary processes that contribute to carcinogenesis and cancer progression. Somatic evolution is driven by the accumulation of phenotypic changes, such as the activation of oncogenes or the loss of tumor suppressor genes. These phenotypic changes were linked to specific genetic events, such as point mutations and larger chromosomal alterations that can lead to loss of heterozygosity. Based on the details of these genetic events and their effect on the fitness of cells, models were explored that calculated the rate at which cancerous mutants are generated, which has important implications for determining the most likely evolutionary pathways towards cancer. While such genetic events are certainly fundamental to the development of cancer, epigenetic events are likely to be equally important, and probably interact with genetic changes to drive the disease [Jones and Baylin (2002); Iacobuzio-Donahue (2009); Esteller and Herman (2002); Komarova *et al.* (2008b); Goel and Boland (2012)]. As mentioned in the introductory Chapter 2, DNA methylation patterns are thought to be particularly significant [Laird and Jaenisch (1996); Esteller and Herman (2002); Ehrlich *et al.* (2002); Baylin (2005); Robertson *et al.* (2001); Baylin and Herman (2000); Wajed *et al.* (2001); Clark *et al.* (2002)]. Methylation is a covalent chemical modification that leads to the addition of a methyl group (CH3) at the carbon 5 position of the cytosine ring, which occurs mostly at CpG dinucleotide sequences. DNA methylation is mediated by a group of enzymes called DNA methyltransferases (DNMTs), some of which are responsible for de novo methylation, while others are responsible for maintenance methylation. For a detailed review of DNA methylaiton processes and their biochemical underpinnings, the reader is referred to the references cited above. Aberrant methylation patterns, i.e., deviation from "normal" methylation levels, have been associated with carcinogenesis. Hypomethylaiton, an overall decreased level of methylation throughout the genome, can contribute to the formation of tumors in a variety of ways. It can affect the CpGs in the promotes of proto-oncogenes genes, which can lead to their over expression and consequently to aberrant growth. Decreased methylation in repetitive DNA sequences can also promote the occurrence of chromosomal rearrangements, which are tightly linked with the development of cancer. On the other hand, increased levels of methylation is referred to hypermethylation, and this can also contribute to the development of cancer. In particular, it can lead to the silencing of tumor suppressor genes through transcription repression of their promoter regions, which has obvious tumorigenic properties.

With both genetic and epigenetic processes being able to contribute to major phenotypic transitions in tumorigenesis, an important question concerns the rate with which epigenetic events induce such changes, compared to the rate at which genetic events induce them. For example, is it more likely that a tumor suppressor gene becomes inactivated by genetic or epigenetic processes? This information would shed light onto the relative importance of these two mechanisms for cancer initiation and progression. Mutation rates have been quantified in a variety of settings, e.g. reviewed in [Wodarz and Komarova (2005)]. The activation of oncogenes requires specific nucleotide changes that turns on expression. Such mutations occur at the order of 10^{-9} per base pair per generation. The inactivation of one tumor suppressor gene copy occurs at a faster rate because many different mutations can potentially lead to the inactivation, hence an estimated rate of about 10^{-7} per gene per generation. In genetically unstable cells, the mutation rates are significantly elevated, e.g. it has been shown that chromosome loss in chromosomally unstable cells occurs at a rate of 10^{-2} to 10^{-3} per chromosome per generation. On the other hand, the rate at which epigenetic processes lead to the equivalent phenotypic changes is not well understood. An interesting aspect is hypermethylation. Similar to the mutator phenotypes in the context of genetic events, CpG island methylator phenotypes, or CIMP cells have been identified. These are cells that are generally characterized by relatively high levels of CpG island methylation [Issa (2004); Nosho *et al.* (2008); Toyota *et al.* (1999); Toyota and Issa (1999)]. However, the amount of CpG island methylation measured at one point in time does not tell us anything about the rate or speed at which methyl groups are added to the DNA. Do methylator phenotypes add methyl groups faster?

This turns out to be a difficult question to answer. The rate of de novo methylation has bee recently measured in select cancer cell lines in the context of different promoters and genes [Wodarz *et al.* (2013)]. In order to obtain a more detailed understanding of the differences between CIMP and non-CIMP cells, this study aimed to investigate the methylation kinetics in cells that were partially de-methylated. This was achieved by measuring the de novo methylation rate following $5-aza-2'deoxycytidine$ (5-AZA) treatment in CIMP and non-CIMP cell lines, using a combination of experimental and mathematical approaches. Another study quantified the rate of de novo methylation in CIMP cells without de-methylation, i.e., in cells characterized by relatively high degrees of methylation [Ushijima *et al.* (2005); Watanabe *et al.* (2006)]. This chapter will review these data, and discuss a mathematical framework to help interpret the results.

19.1 *De novo* methylation kinetics in CIMP and non-CIMP cells following demethylation

Based on observed patterns of hypermethylation, cell lines SW48 and RKO have been designated as CIMP cell lines in the literature [Boland and Goel (2005)]. The cell lines HT29 and HCT116 are thought to be non-CIMP cell lines [Boland and Goel (2005)]. In these cells, methylation levels were measured with respect to the 7 different sites [Wodarz *et al.* (2013)]: ALU, Line-1, APC1, RASSF2-1, HPP1, SFRP2, and MGMT. This was done using quantitative pyrosequencing assays after de-methylating the cells using 5-AZA. After treatment, the temporal process of re-methylation was examined with respect to the 7 sites. The results of these measurements are presented in figure 19.1 [Wodarz *et al.* (2013)]. There, each of the 7 plots (a-g) corresponds to a different site, which is marked on the corresponding graph. The four lines on each graph correspond to the four cell lines, which are color-coded. The measured methylation level (as percentage) is plotted against time post-treatment. The baseline methylation levels for the cell lines are presented by horizontal lines of the corresponding color.

We can see that qualitatively, the majority of remethylation curves start at a level lower than their base methylation level, and then climb up with visibly different slopes. Some of the curves reach saturation (which interestingly may or may not coincide with the measured baseline methylation level), while others continue to climb during the whole duration of the experiment. Another noticeable feature of some of the curves is the presence

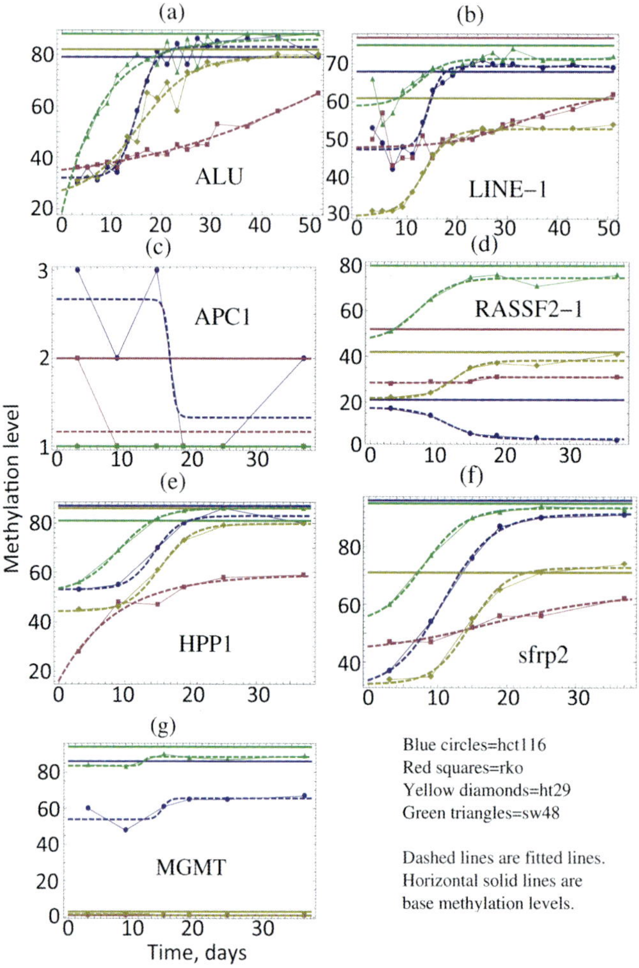

Fig. 19.1 Methylation time-series for four cell lines, for 7 sites. Experimental measurements are shown by connected points, and the fitted functions by dashed lines. The cell lines are color-coded, and the base methylation level for each cell line is presented by a horizontal line of the appropriate color. The panels (a-g) correspond to the 7 different sites.

of a time-delay. In fact, the majority of the experimental runs can be assigned to one of two different groups, as shown in figure 19.1. In one group, the methylation level starts climbing up immediately upon the cessation of 5-AZA treatment. An example of such behavior is exhibited by line SW48 (the green line in figure 19.1(a)). In the other group, there is a certain time-delay between the cessation of 5-AZA treatment and the point where the methylation process picks up. For example, the blue line (HCT116) in figure 19.1(a) climbs up between time-points approximately 10 and 25. For the first 10 days post 5-AZA treatment, the process of remethylation is slow to gain momentum.

19.2 Quantifying the *de novo* methylation kinetics

In order to quantify these data and to extract information about methylation rates, the data were fitted with the function

$$f(t) = y_0 + y_1 \tanh(mt - b). \tag{19.1}$$

This function is characterized by 4 parameters. Parameter m (days^{-1}) measures the methylation rate; the dimensionless parameter b characterizes the time-delay until the *de novo* methylation process starts to gain momentum, as described above; parameters y_0 and y_1 are related to the initial and target methylatin level, with $y_0 - y_1 \tanh(b)$ being the initial methylation level (after the 5-AZA treatment), and $y_0 + y_1$, the target methylation level. The choice of this functional form was dictated by the patterns seen in the methylation time-series and described above. Function $f(t)$ is one of many possible functions capable of accounting for the two types of methylation patterns, with and without delay. The exact mathematical form of this function is unimportant, as long as the following requirements are met: (i) the function has the ability to capture a climb (gradual or step-like) from one level to another level and (ii) has parameters which can be extracted to characterize the slope of the climb (the *de novo* methylation rate) and the amount of time delay. In the function $f(t)$ chosen here, parameter m measures the *de novo* methylation rate, and parameter b characterizes the time-delay.

Figure 19.1 presents the results of fitting the function $f(t)$ to the data, which for each case is given by a dashed line of the appropriate color. Out of the total of $4 \times 7 = 28$ experimental runs, it was possible to successfully fit 18 [Wodarz *et al.* (2013)]: these include all the cell-lines in the context of ALU, LINE-1, HPP1, and SFRP2, as well as lines HT29 and SW48 in the context of RASSF2-1. The discussion will consequently focus on these cases.

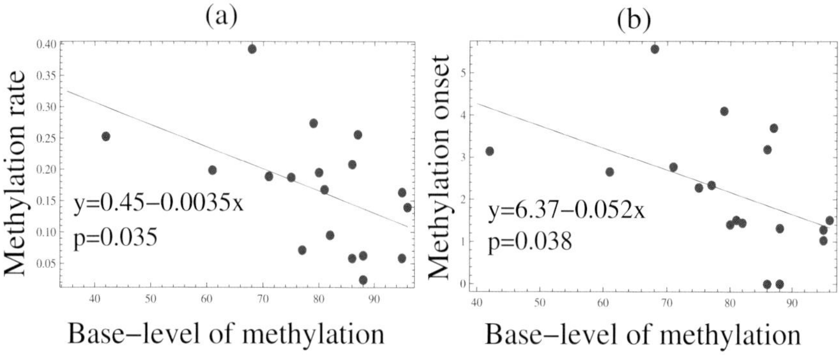

Fig. 19.2 Scatter plots of some methylation parameters extracted from the 18 successful experimental runs, and their correlations. (a) A scatter plot of the methylation rate vs the base methylation level. (b) A scatter plot of the (dimensionless) methylation onset parameter, b, vs the base methylation level. The linear model together with the p-value are marked on the plots.

Fitting the function $f(t)$ to the data yielded the numerical values of the parameters m (remethyation rate) and b (the dimensionless methylation onset). We will first focus on the remethylation rate. The parameter m characterizes the "steepness" of the slope of the methylation time-series in the regions where *de novo* methylation process takes place (and it does not capture other aspects of the methylation process). Figure 19.2(a) presents a scatter plot of the measured re-methylation rates versus the baseline methylation level for the 18 experiments. We performed a linear regression analysis of this scatter plot and determined that the methylation rate negatively correlates with the base methylation level, with the Pearson rank correlation of -0.499 and the Spearman rank correlation of -0.574 (for the sample size of n=18). This correlation is significant, with the p-value of 0.035 (or p =0.013 if using the Student's t-distribution with the Spearman rank correlation). This surprising result suggests that the rate at which the methylation level climbs up after 5-AZA treatment is the lowest for the cell lines with the highest base-level of methylation. This can be clearly seen in figure 19.3(a), where we plot the mean average methylation rate for the four cell lines against their mean base methylation level. In the figure, we group the four cell-lines investigated into two groups: lines HT29 and HCT116 comprise the non-CIMP group, while lines SW48 and RKO are classified as CIMP. This grouping is consistent with the characterization these cell lines received in the literature [Boland and Goel (2005)], and also corresponds to the higher base methylation levels for the CIMP cells. We can see that lines HT29 and HCT116 (which exhibit lower base methylation levels compared to their CIMP counterparts) are characterized by higher *de novo* methylation rates. Thus, CIMP cell lines show slower rates of re-methylation, while non-CIMP cell lines show faster rates of re-methylation.

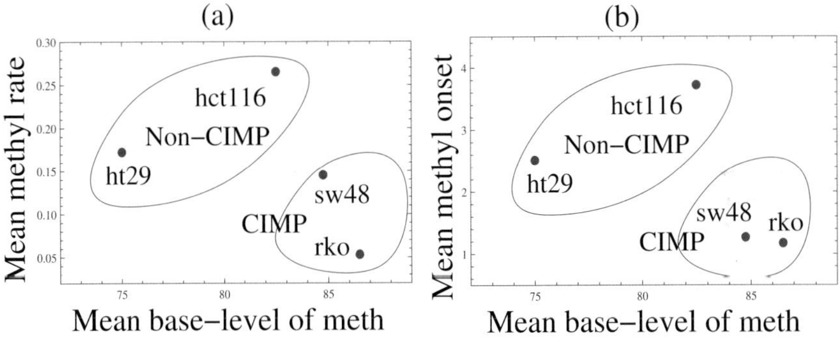

Fig. 19.3 Scatter plots of the mean characteristics presented in figure 19.2 for each cell line across the loci (a) The mean methylation rate vs the mean base methylation level. (b) The mean (dimensionless) methylation onset parameter vs the mean base methylation level. Each point corresponds to one cell-line, marked in the plots. Also, the CIMP status of each cell line is indicated.

To examine this phenomenon more closely, we performed further data analysis. While parameter m only provides information on how steeply *de novo* methylation curves climb up, it does not reflect the presence or the absence of a time-delay in the onset of the re-methylation process. To quantify these differences in a systematic way, we look at parameter b, which yields the dimensionless onset for remethylation. It turns out that there is a significant negative correlation between the base methylation level, and the onset parameter, b (the p-value is 0.038), see figures 19.2(b) and 19.3(b). There is also a very strong (with $p = 2 \times 10^{-6}$) positive correlation between the re-methylation rate, m, and the onset parameter, b (not shown). This suggests that the *de novo* methylation rate and the onset parameter (as specified by the function f) vary together, and a better description of the observed phenomena is provided by a function $\tilde{f}(t) = y_0 + y_1 \tanh(m(t - b'))$, where parameter $b' = b/m$ measures the delay time in days. The time-delay

b' does not show a significant correlation with the base methylation levels (not shown). The function $\tilde{f}(t)$ assumes explicitly that re-methylation time-series that climb faster tend to experience a longer delay between the 5-AZA treatment cessation and the onset of *de novo* methylation.

To summarize this analysis, we can say that the CIMP phenotypes, which are characterized by higher base methylation levels, tend to show a slower *de novo* methylation rate, but experience a steady climb in methylation levels which starts soon after the cessation of 5-AZA treatment. In contrast, non-CIMP cell lines with a lower base methylation level tend to exhibit a certain delay in re-gaining their methylation status, followed by a relatively rapid increase in methylation levels. Roughly speaking, non-CIMP phenotypes have a spurt of relatively rapid methylation increase following a relatively long delay. The methylator phenotypes start increasing their methylation levels relatively quickly and steadily, albeit slowly.

No other, statistically significant correlations were found in these data [Wodarz *et al.* (2013)]. The following section discusses a mathematical model that can help interpret the findings reviewed here.

19.3 Interpreting the results with the help of a mathematical model

A key finding from the above quantitative analysis was that CIMP cells tended to start the methylation process immediately upon cessation of 5-AZA treatment, but did so relatively slowly, while non-CIMP cells showed a relatively fast burst of re-methylation but only after a certain time delay following treatment cessation. The existence of a time delay before a phase of accelerated re-methylation could indicate the existence of a negative feedback loop in the regulation of the *de novo* methylation process. The basic idea is as follows. If methylation levels in the genome are around a certain homeostatic setpoint, *de novo* methylation ceases to occur due to negative feedback, and only maintenance methylation takes place. On the other hand, if the methylation levels are significantly reduced, e.g. following 5-AZA treatment, release of negative feedback induces appropriate DNA methyltransferase (MTase) activity. The process of activation typically is not instantaneous but requires the interactions among several factors, leading to a delay [Chen *et al.* (2005); Denis *et al.* (2011); McAdams and

Arkin (1997)]. MTase activation leads to a burst of *de novo* methylation, which is shut down again through negative feedback as methylation levels in the genome recover. On the other hand, it can be hypothesized that in CIMP cells this feedback regulatory mechanism is corrupted and the appropriate MTases are constantly active at a relatively low level, leading to slow but continuous *de novo* methylation. According to this scenario, re-methylation of CIMP cells would commence without a delay following 5-AZA treatment, and would proceed with a relatively slow rate because of the continuous activity of MTase. The exact MTase responsible for CpG island *de novo* methylation in cancer cells is debated. In the context of non-cancerous cells, it is thought that DNMT1 contributes mainly to maintenance methylation, while *de novo* methylation activity is ascribed to DNMT3a and DNMT3b [Hermann *et al.* (2004); Pradhan *et al.* (1999)]. Work in human cancer cell lines, however, demonstrated that DNMT1 can exhibit *de novo* methylation activity for CpG islands [Jair *et al.* (2006)], and DNMT1 has been shown to be up-regulated in different tumor types [Kanai *et al.* (2001); Lee *et al.* (1996); Saito *et al.* (2003)].

To investigate whether this hypothesis is consistent with the observed experimental patterns, we construct a mathematical model of *de novo* methylation in the context of negative feedback regulation of MTases that are responsible for *de novo* methylation. The model takes into account the following variables. The methylation level of the loci in question is denoted by x. The methylation level of loci that drives the negative feedback is denoted by w. These remain hypothetical loci for now. The molecular processes that regulate *de novo* methylation are not well understood [Denis *et al.* (2011); Turker (1999)], although feedback mechanisms have been implicated in the regulation of DNMT1 [Slack *et al.* (1999)]. It has been suggested that the methylation status of regulatory elements of DNMT1 determines the activity of this MTase, which could result in negative feedback. This element has been shown to be highly methylated in somatic tissues, and unmethylated in a mouse adrenal carcinoma cell line [Slack *et al.* (1999)], consistent with the notions explored here. For simplicity, we will refer to such regulatory elements as "feedback sensors", without assuming a particular mechanism that underlies feedback. Our model generally assumes a "sensor" that reduces and shuts down MTase activity if methylation levels in the genome rise towards some homeostatic level. The model can be adjusted as specific biological information becomes available. MTase activity that is required for *de novo* methylation is denoted by y_n. It is assumed that activation of MTase requires the interactions of different signaling components, which are denoted by y_i, where $i = 1, \ldots, n-1$. The model is given by the following set of ordinary differential equations, which describe the time-evolution of these variables.

$$\dot{x} = \lambda y_n \left(1 - \frac{x}{k_1}\right) - a_1 x, \tag{19.2}$$

$$\dot{w} = \gamma y_n \left(1 - \frac{w}{k_2}\right) - a_2 w, \tag{19.3}$$

$$\dot{y}_1 = \eta_{prod} - q y_1, \tag{19.4}$$

$$\dot{y}_i = q y_{i-1} - q y_i, \quad 1 < i < n, \tag{19.5}$$

$$\dot{y}_n = q y_{n-1} - b y_n. \tag{19.6}$$

The locus in question is methylated in the presence of active MTase (y_n) with a rate λ. It is assumed that k_1 methyl groups can be added. During 5-AZA treatment, de-methylation occurs with a rate a_1. Similarly, *de novo* methylation of feedback sensors occurs with a rate γ in the presence of MTase, y_n, and 5-AZA treatment causes de-methylation of these loci with a rate a_2. The maximum methylation level of feedback sensors is given by k_2. Activation of MTase, y_n, occurs via a signaling cascade, y_i. Regulation occurs in the first element of this signaling cascade, y_1, which is produced with a rate η_{prod} . Details of this production term depend on the nature of the cell line. For non-CIMP cells, we assume the presence of negative feedback. Thus, if the methylation levels of feedback sensors lie below a threshold, c, production occurs with a rate $\eta(c-w)$. If the methylation level of feedback sensors rises above this threshold, the rate of production is set to zero. On the other hand, for CIMP cells, it is assumed that production of y_1 occurs at a constant rate η. Once y_1 is produced, it induces MTase activity through interactions with elements of the signaling cascade, y_i. Finally, MTase activity decays with a rate b.

We will concentrate on the parameter regime where the MTase activity is relatively short-lived in the absence of the activation signals in the model. That is, the parameter b is sufficiently large. This ensures that when the activation signal is switched off, *de novo* methylation ceases without significant delay. Model properties will be described first for non-CIMP cells and then for CIMP cells under this assumption.

For non-CIMP cells, the methylation of feedback sensors, w, always rises towards a level given by the parameter c, after which the negative feedback kicks in and *de novo* methylation stops. Hence, the methylation of w remains constant at this level. On the other hand, the degree to which individual loci become methylated before *de novo* methylation is shut down by negative feedback depends on initial conditions, as explained below.

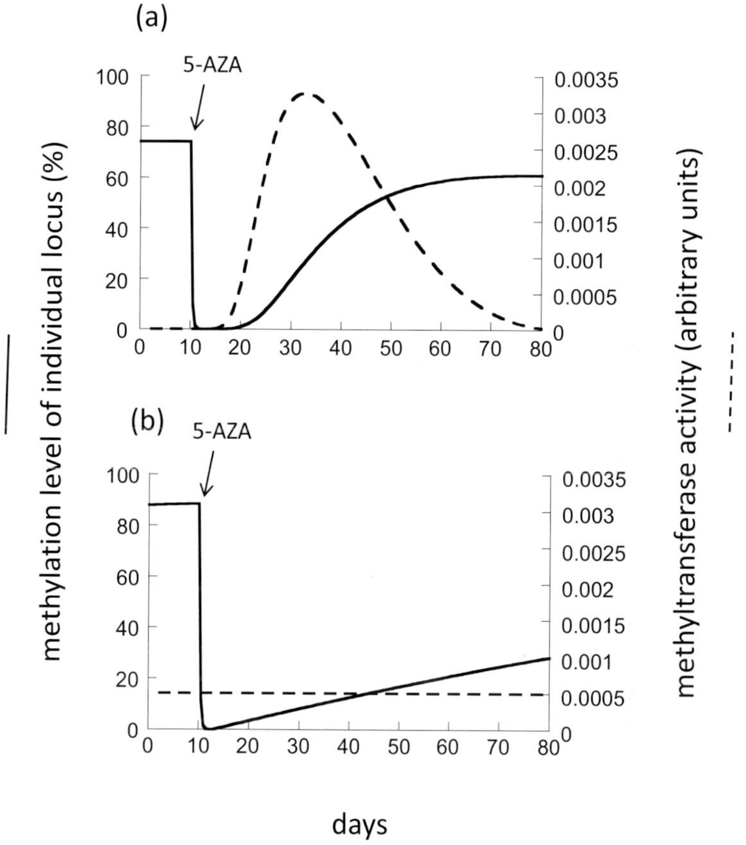

Fig. 19.4 Computer simulation of the re-methylation kinetics in (a) non-CIMP and (b) CIMP cells, according to the mathematical model described in the text. 5-AZA treatment is indicated by the arrow. The solid line represents the methylation level of an individual locus, x, while the dashed line represents *de novo* MTase activity, y_n. Parameters were chosen as follows. (a) $\lambda = 0.4$; $k_1 = 0.04$; $a_1 = 4$; $g = 80$; $k_2 = 100$; $a_2 = 0.4$ $\eta = 2$; $c = 10$; $q = 0.8$; $b = 4000$; (b) same, but $\eta = 2$.

Figure 19.4a has been generated by starting the computer simulation with a completely un-methylated cell and allowing the genome to become methylated. In this case, the individual methylation of locus x stabilizes around 75%. This stable state is shown in figure 19.4a before the start of

5-AZA treatment. Then, 5-AZA treatment is initiated in the simulation and maintained for 72 hours, after which treatment stops. Following the simulated 5-AZA treatment, both the site of interest, and the feedback sensors, become demethylated. However, the feedback sensors retain a higher level of methylation. This is compatible with the notion that different sites in the genome de-methylate at different rates in response to 5-AZA treatment. After cessation of 5-AZA treatment, after a certain time delay, active MTase levels rise and *de novo* methylation increases at a relatively fast rate, as seen in the experimental data. Over time, the methylation level stabilizes. In the simulation of figure 19.4a, it stabilizes at a lower level compared to the base methylation level, as observed in some of the experimental patterns described here (figure 19.1). This is a consequence of the assumption that the methylation level of feedback sensors, w, was reduced to a lesser degree than the methylation level of the locus under consideration, x. Therefore, upon re-methylation, feedback sensors reach their homeostatic set-point and shut down MTase activity before methylation of the locus under consideration has reached its pre-treatment base level. In general, the degree to which given loci become re-methylated following 5-AZA treatment depends on the exact state of the cell after treatment is complete. It can become less methylated, or it can achieve the same amount of methylation seen before treatment. Restoration of pre-treatment methylation levels is likely either if the amount of methylation of feedback sensors is reduced more during 5-AZA treatment, or if the loci in question become de-methylated to a lesser extend during treatment.

For CIMP cells, different dynamics are observed (figure 19.4b). Before 5-AZA treatment, the methylation level of the locus x is not stable but rises at a slow rate. This is the consequence of the assumption that MTase activity is constantly on at relatively low levels and that feedback regulation is corrupted. As before, the simulation assumes 5-AZA treatment for 72 hours. Re-methylation commences instantly and occurs at a relatively slow rate, as observed in the experimental data. Again, the reason is that MTase activity is constantly on. Thus, after de-methylation, it does not have to be activated and hence re-methylation starts immediately. Similarly, because feedback regulation is corrupted, low levels of methylation do not induce a sharp rise in the methylation rate.

In summary, the mathematical model reproduces the key phenomena found in the data: Following 5-AZA treatment, non-CIMP cells re-methylate with a faster rate following a certain time delay, while CIMP cells start re-methylation immediately, although at a slower rate. Moreover, in agreement with the data, the model predicts that in non-CIMP cells, individual loci can re-methylate to levels that are lower than those found before 5-AZA treatment.

19.4 *De novo* methylation kinetics in highly methylated cells

The slower re-methylation kinetics observed in CIMP cells comes as a surprise given the observed hyper-methylation of CpG islands in these cells. It also appears to be at odds with another study which investigated patterns of *de novo* methylation in CIMP and non-CIMP gastric cancer cell lines [Ushijima *et al.* (2005); Watanabe *et al.* (2006)]. In this study, however, cells were not de-methylated before measuring the kinetics. CIMP cells were characterized by a higher *de novo* methylation rate than non-CIMP cells. This study [Ushijima *et al.* (2005); Watanabe *et al.* (2006)] demonstrated reduced fidelity in replicating methylation patterns of CpG islands in CIMP gastric cell lines compared to non-CIMP lines, mostly caused by de novo methylation. The methylated status of CpG sites was more stably maintained than the unmethylated state. This could lead to the methylation of entire CpG islands in experiments that allowed clonal expansion of cells, which was not observed in non-CIMP cells.

The above described mathematical model, based on feedback-regulated activity of *de novo* methylation, can reconcile these data with the seemingly conflicting results obtained in de-methylated cells. If methylation levels are relatively low in non-CIMP cells, maximal MTase activity is attained in order to re-methylate the cell; *de novo* methylation stops in highly methylated cells once the degree of methylation of feedback sensors reaches a defined level. However, if we assume that the feedback mechanism is corrupted in CIMP cells, a low level of methylation activity occurs constantly. Hence, in de-methylated CIMP cells a slow rate of *de novo* methylation occurs without any time delay. However, this slow rate of *de novo* methylation continues with the same rate even when relatively high rates methylation levels have already been achieved. In CIMP cells, feedback does not shut down the *de novo* methylation process. Hence, in addition to the dynamics observed in de-methylated cells, this model can reproduce the results found in highly methylated cancer cells: *de novo* methylation continues unobstructedly in CIMP cells, while this is not the case in non-CIMP cells.

19.5 Importance of experimental verification

This theory provides a guide to investigate possible feedback senor mechanisms in CIMP and non-CIMP cells. The mechanisms that are responsible for regulating the activity of *de novo* methylation are not well understood. Data suggest a dynamic interplay between different posttranscriptional modifications [Denis *et al.* (2011)], and the occurrence of negative feedback has been suggested in the context of the DNA methyl transferase DNMT1 [Slack *et al.* (1999)]. However, more work needs to be done to investigate and document such effects more extensively in order to test the model. While this model presented here provides a nice framework to interpret experimental data, it has to be kept in mind that this is currently just a hypothesis that awaits experimental investigation, and that other mechanisms could also play a role at driving these dynamics. The overall effect of de-methylation on the gene expression profile of cells is likely to be highly complex and needs further examination. For example, there is indication that DNA methylation status alone cannot account for gene expression patterns, but that a coordination of DNA methylation and histone modifications can determine transcriptional status [Mossman and Scott (2011)].

19.6 Summary

The results described in this chapter clearly show that the concept of methylator phenotypes is a lot more complex than the concept of mutator phenotypes. A thorough understanding of the methylation kinetics, and the meaning of methylator phenotypes, will form an important basis for evaluating the relative role of genetic and epigenetic events for the emergence of cancerous phenotypes, such as cells with silenced tumor suppressor genes. This can in turn tell us whether one pathway is dominant over the other in carcinogenesis, or whether both mutations and CpG island methylation events contribute equally. This can have implications for treatment approaches. These explorations also show that computational models of cancer initiation and progression that are based on only genetic events (which are most models at the moment) might be too simplistic to capture the overall dynamics, and that more accurate predictions about the rates of cancer incidence and progression might require more complex models that take into account both genetic and epigenetic alterations. A good example of this is colon cancer. The introductory chapter (Chapter 2) reviewed the literature about colon cancer initiation and it was stated that the events leading to this disease are relatively well-understood, comprising a specific sequence of genetic events. At the same time, however, it is clear that the disease process is more complicated and heterogeneous, with some cases of colon cancer showing CIMP but not CIN, while others show CIN but not CIMP [Goel *et al.* (2007); Cheng *et al.* (2008)]. Hence, a more comprehensive framework is required to better understand the initiation and pathogenesis of specific cancers.

Problems

Problem 19.1. Research project. *Learn about the role of methylation in cancer initiation.*

Problem 19.2. Research project. *Learn about the CIMP phenotype and how it relates to CIN in colon cancer.*

Problem 19.3. Numerical project. *Learn about linear data fitting. (a) Create an artificial dataset by taking* $y_i = 1+2\eta_i i$, *with* $i = 1, \ldots, 20$, *where* η_i *is a random variable taken from a uniform distribution in* $[0, 1]$. *(b) Use any software to fit the function* $y = at + b$ *to this dataset. (c) Find the p-value of the fit. What does this value tell us?*

Problem 19.4. Numerical project. *Learn about nonlinear data fitting. (a) Create an artificial data set by taking* $y_i = y_0 + y_1 \tanh(mi - b\eta_i)$, *with* $i = 1, \ldots, 10$, *where* $y_0 = 1$, $y_1 = 10$, $m = 0.1$, $b = 0.1$, *and* η_i *is a random variable taken from a uniform distribution in* $[0, 1]$. *(b) Use any software to perform nonlinear fit of these data with function given by equation (19.1). (c) Repeat (b) several time with different random numbers, to see how much variation you get in the fitted values* y_0, y_1, m, *and* b.

Chapter 20

Telomeres and cancer protection

When modeling the process of carcinogenesis, we examined the probability that certain mutations are generated in cells, and assumed clonal expansion of transformed cells. Here, we add a layer of complexity to this situation by assuming that cells can undergo a limited number of cell divisions. This can greatly influence the potential to generate a progressing cancer, and is discussed in the current chapter.

Human somatic cells can undergo only a limited number of divisions in vitro [Hayflick and Moorhead (1961)]. This phenomenon known as replicative senescence or the Hayflick limit has long been attributed to the progressive shortening of telomeres with age, which occurs both in vivo and in vitro [Harley *et al.* (1990)]. Telomeres are specialized noncoding repetitive sequences of DNA that are highly conserved throughout evolution and are found at the end of eukaryotic chromosomes [Blackburn (1991); De Lange (2005)]. There are several processes, which are believed to contribute to telomere shortening during cell division; these include the incomplete replication of linear DNA molecules by DNA polymerases [Olovnikov (1973)], active degradation by an unknown exonuclease [Makarov *et al.* (1997)], and oxidative stress [von Zglinicki (2002)].

It has been suggested that replication limits in somatic cells evolved as a means to reduce the incidence of cancer in multicellular organisms. A transformed cell dividing without control must first evade the constrains imposed by the replication limit before it can establish a neoplasia of a significant size. The link between telomeres and cancer is supported by the fact that most colonies of transformed human cells initially proliferate but ultimately cease to divide and die [Counter *et al.* (1992); Chin *et al.* (2004)].

This extinction coincides with a phase termed telomere crisis, in which there is an abundance of cells with very short telomeres and widespread cell death (presumably due to chromosome instability) [Counter *et al.* (1992)]. Also, very significantly, between 85-90% of cancer cells express telomerase [Piatyszek *et al.* (1994)] (an enzyme that extends telomere length) allowing them to circumvent the limitations imposed by replicative limits.

The role of replication limits in the context of cancer biology has been seen as a mechanism to curtail the clonal expansion of cells. Conceptually if an oncogenic event causes uncontrolled proliferation of a cell and its progeny, replication limits place a cap on the maximum size of the cell colony and on the total number of divisions by transformed cells. According to the multi-hit theory of carcinogenesis, full progression toward malignancy requires the accumulation of several mutations in altered cells. Since mutations typically occur during cell division, a limit on the possible number of divisions reduces the probability of acquiring additional mutations. Hence, the lower the replication capacity (defined as the number of divisions left) of the originally transformed cell, the lower the chances of acquiring subsequent mutations that can lead to further cancer progression. This explains the goal of minimizing the average replication capacity of a dividing cell. We also note that a mutation that results in the activation of telomerase could allow cells to bypass the replicative limit [Piatyszek *et al.* (1994)], so the probability of escaping Hayflick's limit itself also depends on the replication capacity of the originally transformed cell.

In order to understand how replication limits protect against cancer, it is essential to understand how a tissue's architecture affects the replicative capacity of the cell population. Recently cell lineages have been viewed as the fundamental units of tissue development, maintenance, and regeneration [Reya *et al.* (2001); Shizuru *et al.* (2005); Frank (2007)]. At the starting points of lineages one finds stem cells, characterized by their ability to maintain their own numbers through self-replication [Reya *et al.* (2001)]. Stem cells give rise to intermediate more differentiated progenitor cells, which are often capable of at least some degree of self-replication [Shizuru *et al.* (2005)]. The end products of lineages are the fully differentiated mostly non-dividing cells associated with mature tissue functions.

We explore how different architectural characteristics of a cell lineage –the number of intermediate cell compartments, the self-renewal capabilities of cells, and the rates of cell division– impact the replication capacity of a cell population. In any given system, there are many theoretically possible architectures that are able to produce a fixed physiologically required output of differentiated cells from a small stem cell pool. Yet, we find that these alternative architectures may produce radically different results with regard to the replicative potential of the cell population. We describe specific features that define an optimal tissue architecture that minimizes the expected replication capacity of dividing cells and thus the risk of cancer.

20.1 Lineages and replication limits

Cell lineages follow specific differentiation pathways. The turnover rate, degree of differentiation and distinct function of different cells within a lineage can often be associated with the expression of specific markers [Frank (2007); Passegué *et al.* (2005)]. These observations have led to the idea of cell compartments as a sequence of distinct and distinguishable differentiation steps. The organization of lineages into cell compartments is a widely proposed model that has been studied both biologically and mathematically in many tissues including the hematopoietic and neural systems, epidermis, esophagus and colon crypt (see e.g. [Passegué *et al.* (2005); Gage (2000); Potten and Booth (2002); Okumura *et al.* (2003); Marshman *et al.* (2002)]). The level of differentiation of cells can change upon cell division [Shizuru *et al.* (2005)]. Alternatively for certain tissues it has been proposed that cells can also change their differentiation level by moving away from the stem cell niche [Reya and Clevers (2005)] leading to a continuous differentiation process. Given our interest in replication limits and their intrinsic connection to cell division here we focus on differentiation occurring through cell division.

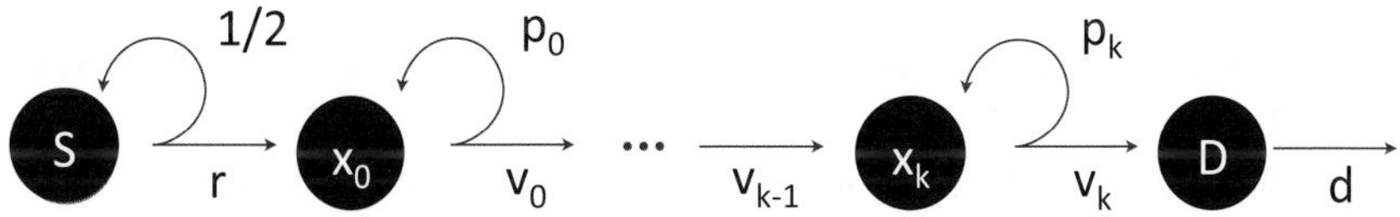

Fig. 20.1 Cell lineage model. Transit cells of type j divide at a rate v_j producing two j-type cells with probability p_j or two $(j+1)$- type cells with probability $1-p_j$. Stem cells S divide at a rate r. There are $k+1$ intermediate steps until cells become fully differentiated (D). Once they do they exit the cell cycle and die at a rate d.

We begin by introducing a variant of a widely proposed model of cell dynamics within a cell lineage [Johnston *et al.* (2007); Marciniak-Czochra *et al.* (2009a); Lander *et al.* (2009); Bocharov *et al.* (2011); Rodriguez-Brenes *et al.* (2011); Werner *et al.* (2011)]. In this model (figure 20.1) the starting point of the lineage are stem cells S. Downstream from the stem cell population, one finds a series of intermediate cell types often referred to in the literature as progenitor cells or transit amplifying cells (here named $X_0, \ldots, X_k$). If a stem cell divides, each daughter cell remains in the stem cell compartment with probability p_s or proceed to the X_0 compartment with probability $1 - p_s$. Similarly, if a cell in compartment j (hereafter called a j-type cell) divides, each daughter cell will remain in the j compartment with probability p_j or differentiate into a $(j + 1)$-type cell with probability $1 - p_j$. The end products of the cell lineage are fully differentiated cells D, which cannot divide any further and die at a certain rate d. The division rates are r for stem cells and v_j for a j-type cell.

Healthy tissue is highly regulated to ensure homeostasis with feedback loops playing a fundamental role in this regard [McPherron *et al.* (1997); Wu *et al.* (2003)]. Thus, if one wishes to examine certain dynamic properties, such as tissue regeneration after an injury, it is necessary to impose control mechanisms that guarantee the stability of the system. In practice this means that the division rates (r and v_j) and self-renewal probabilities (p_s and p_j) may be functions that take on different values depending on the state of the system [Johnston *et al.* (2007); Marciniak-Czochra *et al.* (2009a); Lander *et al.* (2009); Bocharov *et al.* (2011); Rodriguez-Brenes *et al.* (2011)]. In case of an injury the self-renewal probabilities and division rates might temporarily increase to ensure a faster recovery [Lander *et al.* (2009)], leading to an increased number of divisions per unit of time. Thus, repeated injury and repair might augment the risk of cancer by increasing the number of cell divisions. In particular, control must be imposed on the self-renewal probability of stem cells, because otherwise in a stochastic setting the fate of the lineage would inevitably result in extinction or uncontrolled growth. Here however, we are only concerned with properties at homeostasis. Hence, we are only interested in the values of r, p_s, v_j and p_j at equilibrium and the precise nature of the control mechanisms is irrelevant to our analysis.

Next we consider replication limits in the context of telomere biology. In vivo telomeres interact with a number of proteins, allowing them to be recognized by the cell as being in a protected or 'capped' state (for a Review see [De Lange (2005)]). As telomeres shorten they lose their ability to form these capped or protected structures and any further division is then halted. If a cell happens to exhaust its replication capacity but nevertheless attempts mitosis, then senescence or apoptosis would be triggered through the p53 or p16-RB pathways [Smogorzewska and De Lange (2002)]. Several models for cellular replication limits have been proposed. These models use a range of approaches from population dynamics [Marciniak-Czochra *et al.* (2009b); Glauche *et al.* (2011); Enderling *et al.* (2009b); Arkus (2005); Olofsson and Kimmel (1999)] to detailed considerations of the molecular mechanisms affecting telomere function [Rodriguez-Brenes and Peskin (2010); Proctor and Kirkwood (2002)]. Here, we discuss the following framework: associated with every cell there is a number that we call the replication capacity of the cell. When a non-stem cell divides, the replication capacity of the daughter cells will be one unit less than the replication capacity of the parent cell. First we consider the case where all stem cells have a fixed replication capacity ρ that does not change with time. This setting corresponds to a scenario in which stem cells express enough telomerase to maintain a stable telomere length. (Later we consider the case where the replication capacity of stem cells diminishes with time.) Here ρ plays the role of Hayflick's limit and is treated as a parameter. Experimental measurements typically set the value of ρ between 50–70 divisions [Huffman *et al.* (2000)]. Biologically ρ should be large enough to produce and replenish the necessary number of cells required to sustain tissue function during the lifespan of the organism.

We integrate the cell dynamics described above into a single stochastic agent-based model. In this agent-based formulation we track cells individually. Every cell has two attributes: a type determined by the compartment it belongs to, and a replication capacity. When an intermediate cell divides, the replication capacity of the daughter cells will be one unit less than that of the parent. The same thing occurs if a stem cell divides into two intermediate cells. The cell division events and the death of differentiated cells are decided probabilistically according to the death rate d and the division rates r and v_j. If a cell attempting division has exhausted its replication capacity, division is halted and the cell is removed from the cell population. If division occurs in compartment j the probability of self-renewal will be p_j. To decide the fate of a stem cell division we impose the following simple control mechanism. If the number of stem cells is larger than the equilibrium number S, then the stem cell division results in two intermediate daughter cells, otherwise division results in two daughter stem cells. At homeostasis equation (20.1) gives the analogous model in terms of ordinary differential equations.

$$\begin{cases} \dot{S} = 0 \\ \dot{X}_0 = (2p_0 - 1)v_0 X_0 + rS \\ \dot{X}_1 = (2p_1 - 1)v_1 X_1 + 2(1-p_0)v_0 X_0 \\ \quad\vdots \qquad\qquad\qquad \vdots \\ \dot{X}_k = (2p_k - 1)v_k X_k + 2(1-p_{k-1})v_{k-1}X_{k-1} \\ \dot{D} = 2(1-p_k)v_k X_k - dD. \end{cases} \tag{20.1}$$

The precise correspondence between the agent-based model and this analytical formulation is discussed later in the chapter.

20.2 Model analysis

20.2.1 *Population turnover and replication capacity: analytical results*

From system (20.1) we find two expressions for the steady state number of cells in compartment j (which we will need later):

$$\hat{x}_j = \left(\frac{2(1-p_{j-1})v_{j-1}}{(1-2p_j)v_j}\right)\hat{x}_{j-1} \qquad \& \qquad \hat{x}_j = \frac{rS}{v_j}\cdot\frac{2^j}{1-2p_j}\cdot\prod_{i=0}^{j-1}\frac{1-p_i}{1-2p_i}.$$

In compartment j at any given time there are: $v_j x_j$ cells leaving the compartment; $2p_j v_j x_j$ new j-type cells created through symmetric divisions; and $2(1-p_{j-1})v_{j-1}x_{j-1}$ cells arriving from compartment $j-1$. If the system is at equilibrium, then the expected replication capacity of the cells coming into the compartment must be the same as the expected replication capacity of the cells leaving the compartment. Thus, if we call a_i the expected replication capacity of the i-compartment at equilibrium, we find that:

$$a_j \cdot \hat{x}_j = (a_j - 1)\cdot 2p_j v_j \hat{x}_j + (a_{j-1} - 1)\cdot 2(1-p_{j-1})v_{j-1}\hat{x}_{j-1}$$

and using the relation previously found between $\hat{x}_j$ and $\hat{x}_{j-1}$ we find:

$$a_j \cdot v_j \hat{x}_j = (a_j - 1)\cdot 2p_j v_j \hat{x}_j + (a_{j-1} - 1)\cdot(1-2p_j)v_j\hat{x}_j.$$

From this have:

$$a_j = -1 - \frac{2p_j}{1-2p_j} + a_{j-1} \quad \Rightarrow \quad a_j = \rho - (j+1) - \sum_{i=0}^{j} \frac{2p_i}{1-2p_i}.$$

Proposition 1. *If the equilibrium number of stem cells S in not fixed, a cell lineage that minimizes the average replication capacity of a dividing cell necessarily has $S = 1$.*

Proof. In this case the system is constrained by the equation $\sum v_j x_j + rS = dD$ where r, d and D are fixed. Clearly S cannot be smaller than 1. Suppose now that there is a cell lineage that minimizes the average replication capacity of a dividing cell with $S \geq 2$. This lineage is defined by a given stem cell division rate r and a set of parameters $\{p_j, v_j\}_{j=0,\ldots,k}$. Let us define an alternative architecture with one more intermediate cell compartment characterized by the same stem cell division rate and a set of parameters $\{\tilde{p}_j, \tilde{v}_j\}_{j=0,\ldots,k+1}$ that satisfy $\tilde{p}_j = p_{j-1}$ and $\tilde{v}_j = v_{j-1}$ for $j > 0$.

If we make $\tilde{p}_0 = 0$, $\tilde{v}_0 = 1$ and $\tilde{S} = S/2$ then $\tilde{x}_0 = S/2$ and $\tilde{x}_j = x_{j-1}$ for all $j > 0$. It follows that this new cell lineage also satisfies $\sum \tilde{v}_j \tilde{x}_j + r\tilde{S} = dD$. Furthermore, if we respectively call the average replication capacities of the jth compartments a_j and $\tilde{a}_j$, we find $\tilde{a}_0 = \rho - 1$ and $\tilde{a}_j = a_{j-1} - 1$ for $j > 0$. The variable a_j refers to a specific compartment (the jth compartment). We are also interested in the variable A, the expected replication capacity of a dividing cell in the entire population. We find: The expected replication capacity of a dividing cell $A = (\rho r S + \sum_0^k a_j v_j x_j)/dD$ for the original cell linage and $\tilde{A} = (\rho r S/2 + (\rho - 1) r S/2 + \sum_0^k (a_j - 1) v_j x_j)/dD$ for the new cell lineage. Clearly $\tilde{A} < A$ which is a contradiction. □

Proposition 2. *Let v, r, S, d, D and k be fixed and assume there is at most one compartment j of transit amplifying cells for which $p_j > 0$. Then the value of p_j, and the distribution of the replication capacity of the transit cell population at equilibrium are independent of j.*

Proof. Let $N = \sum x_j$ be the total steady state number of transit amplifying cells. Using the previously derived expression for x_j we find after simplifying:

$$N = \frac{rS}{v} \cdot \frac{-1 + 2p_j + (1 - p_j)2^{k+1}}{1 - 2p_j}$$

which implicitly defines p_j as a function of N and k independent of j.

We want to look at the distribution of the replication capacity of the entire cell population at equilibrium. To simplify the notation we assume $rS/v = 1$ (the case $rS/v \neq 1$ follows immediately from this). Let $x^{(a)}$ be the number of cells in the entire population that have replication capacity a at equilibrium, and $x_j^{(a)}$ the corresponding number of j type cells. Let us assume that $p_j = 0 \quad \forall j \neq s$. Then for $j = 0, \ldots, s-1$ we have:

$$x_j^{(a)} = \begin{cases} 2^j \text{ if } & a = \rho - (j+1) \\ 0 \text{ otherwise.} \end{cases}$$

For $j = s$ we have:

$$x_s^{\rho-(s+1)-r} = \begin{cases} 2^s(2p)^r \text{ if } & r \geq 0 \\ 0 & \text{otherwise.} \end{cases}$$

Moreover it can be shown that $j = 1, \ldots, k-s$:

$$x_{s+j}^{\rho-(k+1)-r} = 2^s(1-p)(2p)^{(k-s)+r-j}2^j.$$

First, we will show that $x^{\rho-(k+1)-r}$ is independent of s for $r > 0$. We have:

$$x^{\rho-(k+1)-r} = 2^s(2p)^{(k-s)+r} + 2^s(1-p)\sum_{j=1}^{k-s}(2p)^{(k-s)+r-j}2^j.$$

But then after simplifying we get:

$$x^{\rho-(k+1)-r} = 2^{k+r}p^r.$$

Now we want to look at the values of $x^{\rho-(i+1)}$ for $0 \leq i \leq k-1$. If $i < s$, then clearly $x^{(i)} = 2^i$. If $i \geq s$, then we can call $r = i - s$ and we find:

$$x^{\rho-(i+1)} = x^{\rho-(s+1)-r} = 2^s(2p)^r + 2^s(1-p)\sum_{j=1}^{r}(2p)^{r-j}2^j = 2^{r+s} = 2^i.$$

Thus we find that the distribution of the cell replication capacity is independent of the choice of the self-renewing compartment. □

Proposition 3. *Suppose that all the v_j are equal and consider v, r, S, d and D fixed. If at most one $p_j > 0$ we want to find the pair (p, k) that minimizes the entire replication capacity of the transit cell population at equilibrium. Under this condition the entire replication capacity of the transit cell population at equilibrium is minimized by choosing p as large as possible subject to the restriction $a_k \geq 1$.*

Proof. We proved that if at most one $p_i > 0$, then the entire replication capacity of the transit cell population is independent of the choice of i. Thus without loss of generality we assume $i = 0$. Let $\alpha = (1-p)/(1-2p)$, N be the steady state number of transit cells and k the number of compartments, then:

$$N = \sum_{j=0}^{k} x_j = \frac{rS}{v}\left(2^{k+1}\alpha - 1\right).$$

We also have $x_k = (rS/v)2^k\alpha$ from where it follows that $2x_k = N + rS/v$. On the other hand, $dD = 2vx_k$, and we find that N is completely determined by rS, dD and v:

$$N = \frac{dD - rS}{v}.$$

Now the entire replication capacity of the j-compartment at equilibrium is $a_j = \rho - (j+1) - 2(\alpha - 1)$ for all j. We want to minimize $A = \sum a_j x_j$. We have:

$$A = [\rho - 2(\alpha - 1)]\sum x_j - \sum (j+1)x_j.$$

The first term in the LHS of the previous equation equals $N[\rho - 2(\alpha - 1)]$. Given that $x_0 = (rS/v)(2\alpha - 1)$ and $x_j = (rS/v)\cdot 2^j\alpha$ for $j > 0$, we can decompose the second term (call it B) in the following way:

$$\begin{aligned} B &= \sum (j+1)x_j \\ &= (rS/v)(\alpha - 1) + (rS/v)\sum (j+1)2^j\alpha \\ &= (rS/v)(\alpha - 1) + (rS/v)[(k+1)2^{k+2}\alpha - (k+2)2^{k+1}\alpha + \alpha]. \end{aligned}$$

Now call $c = rS/v$ and $n = k+1$. Then using the fact that $2^n\alpha c = N + c$ we find that:

$$\begin{aligned} B &= 2\alpha c + (N+c)n - (N+2c) \\ \Rightarrow A &= \{N(\rho+2) + (N+2c)\} - \{(N+c)(2\alpha+n)\}. \end{aligned}$$

Hence, to minimize A we should maximize $2\alpha + n$. Given that $n\log(2) + \log(\alpha) = \log(N/c+1)$, if we write $f(\alpha) = 2\alpha - \log(\alpha)/\log(2)$ we find that $2\alpha + n$ equals

$$f(\alpha) + \log(N/c+1)/\log(2).$$

It is easy to prove that the $f : [1,\infty] \to \mathbf{R}$ is a decreasing function. Hence to minimize A we should make α as large as possible, which is equivalent to choosing p as large as possible given the restriction $a_k \geq 1$. □

Lemma 1. *For any pair* (N,k) *let* $\{y_j\}$ *be the sequence defined by* $y_0 = 1/(1-2p)$, $y_j = 2^j(1-p)/(1-2p)$ *for* $0 < j < k$ *and* $\sum_{j=0}^{k} y_j = N$, *and* b_j *be the entire replication capacity of the* j^{th} *compartment at equilibrium. Then for any other sequence* $\{x_j\}$ *with entire replication capacities* a_j *that satisfies* $N = \sum_{j=0}^{k} x_j$ *we have:*

$$1)\quad b_k \leq a_k \quad \textit{and} \quad 2)\quad \sum b_j y_j \leq \sum a_j x_j.$$

Proof. *Preliminaries*

First let us write $\alpha_i = 2(1-p_i)/(1-2p_i) \Rightarrow 1/(1-2p_i) = \alpha_i - 1$. For $j > 0$ we then have:

$$x_j = \frac{1}{1-2p_j}\prod_{i=0}^{j-1}\frac{2(1-p_i)}{1-2p_i} = (\alpha_j - 1)\prod_{i=0}^{j-1}\alpha_i.$$

Calling $\beta_j = \prod_{i=0}^{j}\alpha_i$ we have $x_j = \beta_j - \beta_{j-1}$ for $j > 0$. Thus we can write:

$$N = \sum_{j=0}^{k} x_j = (\alpha_0 - 1) + \sum_{j=1}^{k}[\beta_j - \beta_{j-1}] = \beta_k - 1.$$

Thus we have:

$$N + 1 = \prod_{i=0}^{k}\alpha_i.$$

Proof of 1)

Note that following the previous definitions $a_k = \rho - (k+1) - \sum_{j=0}^{k}(\alpha_j - 2) = \rho + (k+1) - \sum_{j=0}^{k}\alpha_j$. Thus the problem reduces to the maximization of $\sum \alpha_j$ subject to the conditions: a) $N + 1 = \prod_{i=0}^{k}\alpha_i$, b) $\alpha_i \geq 2$ and c) $a_k \geq 0$. Let us assume that $\{\alpha_i\}$ satisfy the conditions stated above. Assume that at least two of the α_i are greater than two. Without loss of generality let them be α_0 and α_1. We can write $N + 1 = P\alpha_0\alpha_1$ and $S = s + (\alpha_0 + \alpha_1)$. We want to maximize $\alpha_0 + \alpha_1$ subject to $A = (N+1)/P = \alpha_0\alpha_1$. Which means we want to maximize $\alpha_0 + A/\alpha_0$. It is easy to see that this function has a unique minimum at $\alpha_0 = \sqrt{A}$ thus the maxima occur at the endpoints of its domain which is $[2\,, A/2]$.

Proof of 2)

We will prove this part of the proposition using the principle of mathematical induction.

Base step

Let $k = 2$. Then, $N + 1 = \alpha_0\alpha_1$ and:

$$\begin{aligned} S &= (\alpha_0 - 1)(\rho - 1 - (\alpha_0 - 2)) \\ &\quad + (\alpha_1 - 1)\alpha_0(\rho - 2 - (\alpha_0 - 2) - (\alpha_1 - 2)) \\ \Rightarrow S &= 2 + 3N - (\alpha_0 + \alpha_1)(1 + N) + Nr. \end{aligned}$$

Given this last expression the problem for minimizing S reduces to the maximization of $\alpha_0 + \alpha_1$, subject to the conditions $N + 1 = \alpha_0\alpha_1$ and $a_1 \geq 0 \Leftrightarrow r + 2 - (\alpha_0 + \alpha_1) \geq 0$. From the symmetry of these equations it is easy to prove that the minimization occurs when either α_0 or α_1 equals 2, and therefore either p_0 or p_1 equals 0.

Induction step

Assume the proposition is true for $n = k - 1$. Let $(x_0, \ldots, x_k)$ be defined by $P = (p_0, \ldots, p_k)$ and assume that the sequence $\{x_j\}$ minimizes the replication capacity of the population subject to the condition $N = \sum x_j$.

Case A: There is a single $p_j \geq 0$ for $j \leq k$. Then there is nothing to prove.

Case B: There are at least two $p_j > 0$ for $j \leq k - 1$ (we will prove this leads to a contradiction).

Let us call $N_{k-1} = \sum_{j=0}^{k-1} x_j$. Make $\tilde{p}_j = 0$ for $j > 0$ and $\tilde{p}_0$ such that $N_{k-1} = \sum_{j=0}^{k-1} y_j$. Now by the induction hypothesis

$S_{k-1} = \sum_{j=0}^{k-1} a_j x_j > \tilde{S}_{k-1} = \sum_{j=0}^{k-1} b_j y_j$. Note that:

$$1 + N_k = \beta_k = 2(\tilde{p}_k - 1)y_k \qquad \text{and} \qquad 1 + N_k = 2(p_k - 1)x_k.$$

Given that $x_k = y_k$ it follows that $p_k = \tilde{p}_k$ and then we have:

$$S_k = S_{k-1} + a_k x_k = S_{k-1} + \left[a_{k-1} - \frac{2p_k}{1 - 2p_k}\right] y_k.$$

From part 1) we have $b_{k-1} \leq a_{k-1}$, and thus it follows that $S_k > \tilde{S}_k$, which means that $\{x_j\}$ does not minimize the entire replication capacity of the transit cell population, which is a contradiction.

Case C: There is one $p_j \neq 0$ for $j < k - 1$ and $p_k \neq 0$.

If we prove that $A = \sum a_j x_j$ is invariant under a permutation $p_i \leftrightarrow p_j$, then the situation reduces to Case B. It is sufficient to prove that A is invariant under $p_j \leftrightarrow p_{j+1}$. Note that:

$$x_j = (\alpha_j - 1)\beta_{j-1} \qquad \& \qquad a_j = \rho + (j + 1) - \alpha_j - \sum_{i=0}^{j-1} \alpha_i.$$

After the permutation $x'_j = (\alpha_{j+1} - 1)\beta_{j-1}$ and $x'_{j+1} = (\alpha_j - 1)\alpha_{j+1}\beta_{j-1}$. Then:

$$a'_j = \rho + (j+1) - \alpha_{j+1} - \sum_{i=0}^{j-1} \alpha_i \quad \& \quad a'_{j+1} = \rho + (j+2) - \alpha_j - \alpha_{j+1} - \sum_{i=0}^{j-1} \alpha_i.$$

From where $a_j x_j + a_{j+1} x_{j+1} = a'_j x'_j + a'_{j+1} x'_{j+1}$ and it follows that the permutation A is invariant under this permutation. □

Proposition 4. *Suppose that all the v_j are equal and consider v, r, S, d and D fixed. Then to minimize the entire replication capacity of the transit cell population at equilibrium, make at most one $p_j > 0$ and choose the pair (p_j, k) such that p_j is the largest possible subject to the restriction $a_k \geq 0$. Furthermore, the value of p_j, and the distribution of the replication capacity of the transit cell population at equilibrium are independent of j.*

Proof. Let k and $P = (p_0, \ldots, p_k)$ be such that the sequence $(x_0, \ldots, x_k)$ they define minimizes $A = \sum a_j x_j$ subject to the restrictions imposed by the choice of parameters r, S, v and D. Because of

Proposition 2, there is at most one j such that $p_j > 0$. Furthermore because of Lemma 1 the pair (k, p_j) is such that amongst all pairs (K, P_j) that fit the parameters, p_j is maximum subject to the restriction $a_k \geq 1$. The last statement of the proposition follows straight from Proposition 1. □

20.2.2 *Agent-based model*

An agent-based stochastic version of the model is implemented as an algorithm. At any given time t the system is described by a set of cells, each of which has two attributes: a real number representing its replication capacity and a type. If we call $A = rS + dD(t) + \sum v_i x_i(t)$, then the probability that the next reaction involves a j-type cell is $v_i x_i(t)/A(t)$, and the probabilities that it involves a differentiated cell or a stem cell are $Sr/A(t)$ and $dD(t)/A(t)$ respectively. Once a type of cell is selected, a random cell amongst all cells of this particular type is selected. Then we proceed in the following way:

(1) If division occurs in an intermediate cell, the two offspring of the parent cell will have a replication capacity one unit smaller than that of the parent cell. If division occurs in a j-type cell the probability of self-renewal is p_j. If the cell attempting division has a zero replication capacity division is halted and the cell is removed from the cell population.
(2) If division occurs in a stem cell and the current number of stem cells $S(t)$ is less than the equilibrium value $\hat{S}$, then we make $p_s = 1$ and if $S(t) \geq \hat{S}$ we make $p_s = 0$. This is a simple way of establishing control in the stem cell population. As we mentioned before, we are only interested in equilibrium properties of the intermediate compartments, so any control mechanism on the number of stem cells will suffice.
(3) If a differentiated cell is selected then the only possible event is cell death.
(4) The time when the next reaction occurs is exponentially distributed with mean equal to $1/A(t)$.

The difference between the ode model and the agent-based model lies with the fraction of cells at equilibrium that exhaust their replication capacity and nevertheless attempt cell division. In the ode model there is no built-in mechanism to prevent such cells from dividing. In the agent-based model division is halted and the cells are removed from the population. For an optimal architecture this fraction f is given by:

$$f = \frac{(2p)^{\rho-(k+1)}}{2(1-p)} - \frac{1-2p}{2^{k+1}(1-p)} \leq (2p)^{\rho-(k+1)}.$$

Typical experimental measurements for ρ yield a value of approximately $50 - 70$ divisions [Huffman *et al.* (2000)], and proposed models for the number of transit amplifying cell types in several tissues, including blood, neurons and hair [Shizuru *et al.* (2005)], set the number of intermediate cell compartments $(k+1)$ between one and four. These values for ρ and $k+1$ produce a very large exponent on the right hand side of the previous inequality suggesting a small value for f. In figure 20.4(a) we show results from the agent-based model that demonstrate very good agreement with the corresponding analytical model, suggesting that the latter adequately captures the essential dynamics of the system.

20.2.3 *Decrease in the replication capacity of stem cells*

First, assuming that the cell population is at equilibrium, we derive an approximation formula for $a_j(t)$, the expected replication capacity of the j-type population as a function of time. Under this condition we have $rS = x_0 v_0(1-2p_0)$, which means that:

$$\dot{a}_0 = (2p_0 - 1)v_0 a_0 - 2pv + v_0(1-2p)(\rho_0 - 1) + v_0(2p-1)\epsilon.$$

Hence we can write $\dot{a}_0 = Ka_0 + B + K\epsilon t$ where K and B are constants. From here we find:

$$a_0 = -\frac{B+\epsilon}{K} - \epsilon t + C_0 e^{kt}$$

for some constant C_0 and it follows that:

$$a_0 = \rho(t) - 1 - \frac{2p_0}{1-2p_0} + \mathrm{O}(e^{(2p_0-1)v_0}).$$

Let us call $K_j = (2p_j - 1)v_j$. We find after simplifying:

$$\dot{a}_1 = K_1 a_1 + \left\{\rho - 1 - \frac{2p_0}{1-2p_0} + \frac{\epsilon}{(1-2p_0)v_0}\right\} + K_1\epsilon t + \mathrm{O}(e^{K_0 t})$$

and we find:

$$a_1 = C_1 e^{K_1 t} + (\rho - 2) - \frac{2p_0}{1-2p_0} - \frac{2p_1}{1-2p_1} + \frac{\epsilon}{(1-2p_0)v_0} + \frac{\epsilon}{(1-2p_1)v_1}$$
$$- t\epsilon + e^{K_1 t} \int e^{-K_1 t} \mathrm{O}(e^{K_o t}) dt.$$

Note that $e^{K_1 t} \int e^{-K_1 t} \mathrm{O}(e^{K_o t}) dt = \mathrm{O}(e^{K_o t})$ and $C e^{K_1 t} + \mathrm{O}(e^{K_o t}) = \mathrm{O}(e^{\max\{K_o, K_1\}})$. From these considerations it follows that:

$$a_j(t) = \rho(t) - (j+1) - \sum_{i=0}^{j} \frac{2p_i}{1-2p_i} + \sum_{i=0}^{j} \frac{\epsilon}{(1-2p_i)v_i} + \mathrm{O}(e^{max\{K_i\}}).$$

Finally note that all the K_i are negative, hence the $\mathrm{O}(e^{max\{K_i\}})$ in the previous expression goes to zero exponentially fast. We may thus neglect this term and find a good approximation to $a_j(t)$.

Now we would like to address the optimality results previously derived, this time in the context of a replication capacity of stem cells that decreases with time. Consider the case where all the $v_j = v$ are equal. Once again we assume that the cell population is at equilibrium and thus the $a_j(t)$ are well approximated by the following formula:

$$a_j(t) = \rho(t) - (j+1) - \sum_{i=0}^{j} \frac{2p_i}{1-2p_i} + \sum_{i=0}^{j} \frac{\epsilon}{(1-2p_i)v_i}.$$

Let us write $\alpha_i = 2(1-p_i)/(1-2p_i)$ and call $F(j) = (j+1) - \sum_{i=0}^{j} \alpha_i$. If $\epsilon = 0$ then we find after simplifying that $a_j = \rho + F(j)$. If $\epsilon > 0$ similarly we find $a_j(t) = \rho(t) + (1 - \epsilon/v)F(j)$. Thus to minimize $\sum a_j(t)x_j$ we only need to focus on minimizing $(1-\epsilon/v)\sum F(j)x_j$. It follows that the choice of parameters that minimizes S when $\epsilon = 0$ also minimizes $S(t)$ when $\epsilon > 0$.

20.3 Tissue architecture and the development of cancer

We are interested in finding an optimal cell lineage architecture that protects against cancer by minimizing the replication capacity of dividing cells. Stated in this form, however, the problem is not sufficiently constrained. In particular the target number of differentiated cells D and their death rate

d depend on other biological considerations such as tissue function and organismic physiology that lie clearly outside the scope of the optimization problem. Thus we consider the outflow of differentiated cells (dD) as a fundamental fixed quantity of the system. With this constraint at hand we arrive at our first insight: A cell lineage architecture concerned only with minimizing the replication capacity of dividing cells would have a stem cell compartment consisting of a single cell. Note that a cell lineage that depends on a single stem cell would be extremely fragile; thus in vivo the equilibrium number of stem cells must depend on other factors (independent of replication limits) that deal with the robustness of the system. Hence in broader terms what this result suggests is that a tissue architecture concerned with reducing the risk of cancer should have a very small number of stem cells compared to the total number of cells in the lineage (in the colon epithelium for example there might be as little as four stem cells per crypt [Bach *et al.* (2000); Marshman *et al.* (2002)]). This result becomes intuitive when we look at the proliferative potential of stem cells in specific tissues. For example, colon stem cells are estimated to divide up to 5000 times during a human lifespan [Marshman *et al.* (2002)]. Thus the cumulative number of divisions in clones originating from a transformed colon stem cell is potentially enormous, suggesting that replication limits are not an effective mechanism to protect against the accumulation of mutations in stem cells.

If the stem cell compartment is small, then most of the cell divisions required for normal tissue function must be carried out by non-stem cells, where the much smaller replication limits can protect against the sequential accumulation of mutations. As pointed out above, the effectiveness of this protection will depend on the replication capacity of the originally transformed cell as it directly influences the likelihood of acquiring subsequent mutations and of escaping the Hayflik limit itself. Even though the proliferative potential of non-stem cells is limited because they are responsible for the overwhelming majority of cell divisions within a tissue, it can be shown that statistically the accumulation of mutations in non-stem cells is possible [Komarova and Wang (2004)]. Furthermore for certain types of cancers there is evidence that the initiating mutations originate in progenitors (for a review see [Visvader (2011)]). Recently progenitor cell populations have been identified as targets for tumor initiation in a number of leukemias [Guibal *et al.* (2009); Krivtsov *et al.* (2006); Goardon *et al.* (2011); Wojiski *et al.* (2009); Huntly *et al.* (2004); Cozzio *et al.*

(2003)] and several types of solid cancer, including glioblastoma, medulloblastoma, prostate cancer, basal cell carcinoma and basal-like breast cancer [Friedmann-Morvinski *et al.* (2012); Schüller *et al.* (2008); Goldstein *et al.* (2010); Youssef *et al.* (2010); Lim *et al.* (2009)]. To study how replication limits and tissue architecture protect against mutations originating outside the stem cell compartment, we treat the influx of stem cells (rS) and the outflow of differentiated cells (dD) as the fundamental fixed quantities of the system, and ask how the number of intermediate cell compartments ($k+1$), the self-renewal probabilities (p_j), and the cell division rates (v_j) affect the replication capacity of dividing cells.

We start by describing some of the fundamental features of the system. As explained above, the flux of cells between the stem cell and differentiated cell compartments ($dD - rS$) is fixed. At homeostasis $dD - rS$ equals the number of intermediate cell divisions per unit of time. Thus, if we call the equilibrium number of j-type cells x_j, we have the constraint:

$$\sum v_j x_j = dD - rS. \tag{20.2}$$

The steady state number of cells in compartment j is:

$$x_j = \frac{rS}{v_j(1-2p_j)} \prod_{i=0}^{j-1} \frac{2(1-p_i)}{1-2p_i}. \tag{20.3}$$

From this last equation it follows that increasing the self-renewal probability in an intermediate compartment increases the compartments size and the number of divisions per unit of time in that compartment ($v_j x_j$). Hence, given the constraint found in equation (20.2), an increase in the self-renewal probability in one of the compartments must be offset by a change in some other variable of the system. Figure 20.2 illustrates this situation with two alternative architectures. The same target number of divisions may be reached by a lineage with smaller self-renewal probabilities and a larger number of compartments or by a lineage with larger self-renewal probabilities and fewer compartments.

An increase in the division rate in a compartment produces a decrease in the compartment's size, equation (20.3). If we multiply the expression for x_j in equation (20.3) by v_j, we find that at equilibrium the number of divisions per unit of time is independent of the division rate. Both of these phenomena are demonstrated in figure 20.3. Here, an increase in the division rate

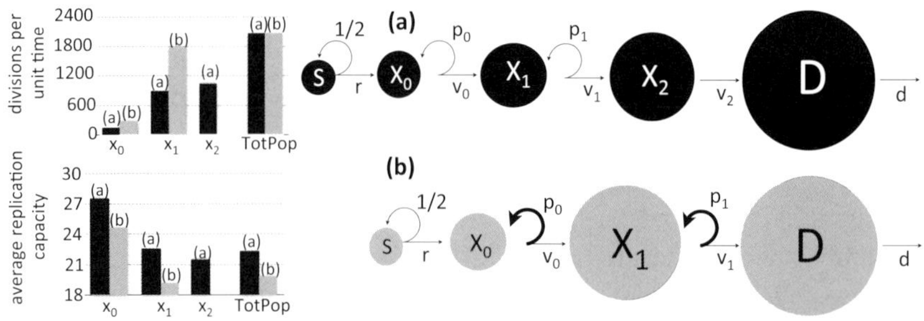

Fig. 20.2 Architectures for the same target number of divisions in the intermediate cell population. Increasing the self renewal probability increases the size and the number of divisions per unit of time of a compartment. The same target number of divisions in the intermediate cell population can be reached by a lineage with smaller self-renewal probabilities and three intermediate compartments (a), or by a lineage with larger self-renewal probabilities and only two intermediate cell compartments (b). The size of circles is indicative of population size. Bar plots: The average replication capacity and the number of divisions in the compartments depend on the architecture of the cell lineage. (See text for discussion.) In both figures $rS = 50$, $dD = 2100$, and all $v_i = 1$. Parameters (a): ($p_0 = 0.3$, $p_1 = 0.4$, $p_2 = 0$). Parameters (b): ($p_0 = 0.41$, $p_1 = 0.41$).

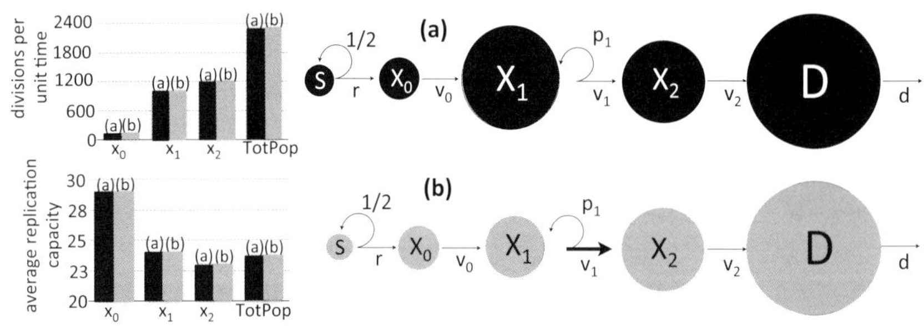

Fig. 20.3 An increase in the division rate of a compartment produces a decrease in the compartments size; however the number of divisions per unit of time and the average replication capacity in a compartment is independent of the division rate. Note that the number of divisions per unit of time increases with each compartment. (See text for discussion.) In both figures $rS = 100$, $dD = 2400$, $p_0 = 0$, $p_1 = 0.4$, $p_2 = 0$, $v_0 = 1$, $v_2 = 1.5$ and $k = 2$. Parameters (a): ($v_1 = 1$). Parameters (b): ($v_1 = 2$).

in one of the compartments results in a reduction of the population size; the number of divisions per unit of time however, does not change. There is also another feature of the system that is apparent from figure 20.3. The relative sizes of the compartments are not necessarily determined by their positions within the lineage; however, the number of divisions occurring in the compartments is. Thus, a more differentiated compartment produces at least the same number of divisions than any of its predecessors. Indeed, it is easy to see from equation (20.3) that $v_{j-1}x_{j-1} \leq v_j x_j$.

Let us call a_j the expected replication capacity of the j-compartment at equilibrium, which can be intuitively defined as the average number of divisions left for a typical cell in the compartment when the tissue is at homeostasis. There are two important things to remark: one, a_j decreases with differentiation (histograms in figures 20.2 and 20.3) and two, the architecture of a lineage affects the distribution of the replication capacity of the entire population (figures 20.2 and 20.3). From the point of view of replication limits, the optimal architecture to protect against cancer is one that minimizes the expected replication capacity of a dividing cell. Note that we emphasize the fact that we are interested in the properties of dividing cells and not just cells in general. Given that mutations typically occur during cell division, we need to take into account that compartments with a fast division rate carry an increased risk of producing a transformed cell. Hence, the quantity that we seek to minimize is the expected replication capacity of a dividing cell, which in mathematical terms equals $(\sum a_j v_j x_j)/(rS - dD)$.

Our next result examines the effect of position within the cell lineage on the replicative potential. If there is only one intermediate cell compartment with self-renewal capabilities, then the distribution of the replication capacity of dividing cells is independent of where this compartment lies within the order of the cell lineage (Proposition 1 in Section 20.2.1). Figure 20.4(a) exemplifies this behavior in a system with three transit amplifying cell compartments. Here only one compartment has a non-zero self-renewal probability and the distribution of the replication capacity of dividing cells does not change when the self-renewal compartment is alternatively chosen to be either the zeroth or second compartment.

Next we study what happens if we distribute the self-renewal potential amongst several compartments. If the number of compartments is fixed, then the average replication capacity of dividing cells is minimized when there is no more than one self-renewing compartment (Proposition 2 in Section 20.2.1). This is illustrated in figure 20.4(b) where we consider a system with two intermediate cell compartments and plot the average replication

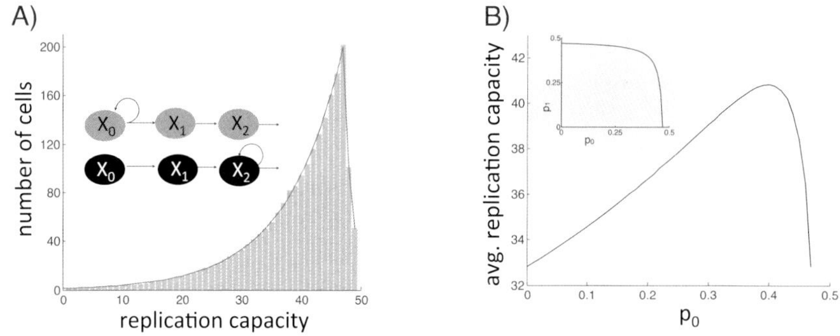

Fig. 20.4 (*A*) Distribution of the replication capacity of dividing cells ($k = 2$, $r = 1$, $v_j = 1$, $S = 50$, $\rho = 50$, p_0 or $p_2 = 0.45$). Here self-renewal only occurs in one transit amplifying cell compartment. Bar plots are produced using the agent-based model and allowing for self-renewal only in compartment 0. The continuous line is produced using the analytical model allowing for self-renewal only in compartment 2. The distribution does not change when the non-zero self-renewal probability is switched from compartment 0 to compartment 2. (Stem cell and differentiated cell compartments are not depicted in the inset.) (*B*) Replication capacity of the entire cell population at equilibrium as a function of p_0 ($k = 1$, $v = 1$, $dD - rS = 1716$, $\rho = 50$). Given that $dD - rS$ is fixed, the value of $p_0 = 0$ determines the value of p_1 (see inset). Note that the minimum occurs when there is self-renewal in only one of the compartments (i.e., p_0 or p_1 is equal to zero).

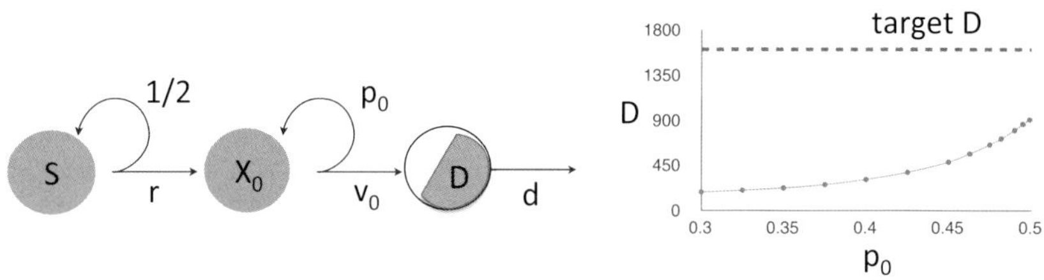

Fig. 20.5 It is not always possible to reach a specified target number of divisions with only one intermediate compartment. Here the number of differentiated cells produced by a system with a single intermediate compartment is incapable of producing the target number of differentiated cells (indicated by a half-filled compartment D). In this case $rS = 50$, $D = 1600$, $d = 1$, and $\rho = 20$. Results were obtained using the agent-based model. (See text for discussion.)

capacity for different values of the self-renewal probability of the zeroth compartment. In this instance, where there are only two compartments, the self-renewal probability of one of them completely determines the self-renewal probability of the other (see inset). From this figure we note that the average replication capacity is minimized when only one of the compartments has a positive probability of self-renewal.

Given a fixed target of intermediate cell divisions ($dD - rS$), there is an upper limit to the number of cell compartments. Indeed, if there are $k + 1$ intermediate compartments, the equilibrium number of cell divisions per unit of time is always greater than or equal to $rS(2^{k+1} - 1)$. From this, it is clear that we cannot choose k arbitrarily large. There may also be a lower limit to the number of compartments. First, having only one intermediate cell compartment may lead to too many cells exhausting their replication capacity, making it impossible for the compartment to reach the target number of divisions. For example, in figure 20.5 simulations using the agent-based model show that for a given set of values $dD - rS$ and ρ, it is impossible to produce the target number of divisions with only one intermediate cell compartment. Hence a target flux of cells $dD - rS$ and a given maximum replication capacity ρ might preclude certain tissue architectures. Second, it is important to note that every fork in the differentiation pathway of cells adds a new compartment to a cell lineage. Thus, there may be a minimal theoretical number of intermediate cell compartments when different types of mature cells arise from the same kind of stem cell (such is the case of the hematopoietic system, discussed in the next section).

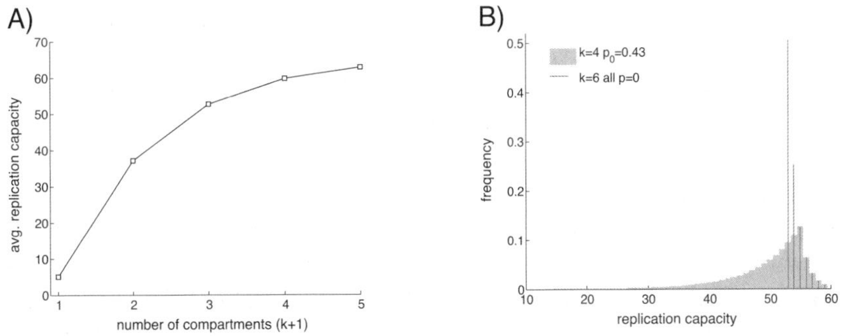

Fig. 20.6 (A) Average replication capacity as a function of the number of transit amplifying cell compartments (k+1). Here only one compartment has self-renewal capabilities ($v_j = 1$, $dD - rS = 6500$, $\rho = 70$). The average replication capacity increases with (k+1). (B) Frequency of the replication capacity of dividing cells. In both instances the number of intermediate cell divisions is the same. In both cases $v_j = 1$ for all j, $\rho = 60$ and $rS = 50$. Lines: $k = 6$ and all $p_j = 0$. Bars: $k = 4$, $p_0 = 0.43$ and all other $p_j = 0$.

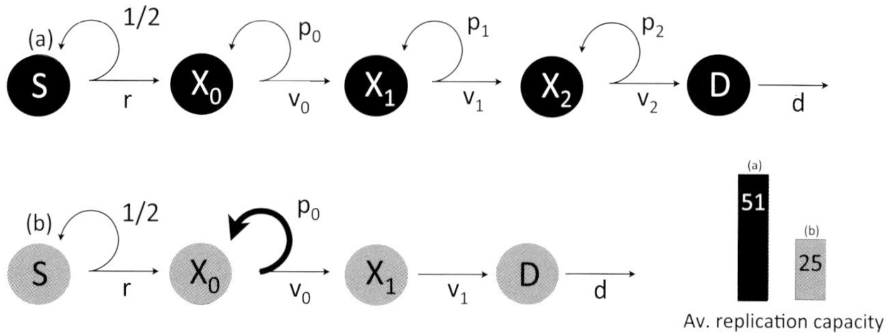

Fig. 20.7 Two alternative architectures for the same target number of intermediate cell divisions (3450). In the cell lineage depicted in (a) ($k = 2$, $p_0 = p_1 = p_2 = 0.341$) the resulting average replication capacity of dividing cells is 51. An optimal cell lineage depicted in (b) ($k = 1$, $p_0 = 0.485$, $p_1 = 0$) minimizes the average replication capacity of dividing cells by minimizing the number of compartments and allowing self-renewal in only one of them. In both cell lineages $rS = 50$, $\rho = 60$ and all $v = 1$. The average replication capacity of dividing cells is minimized by a tissue architecture in which at most one intermediate cell type has self-renewal capabilities and the number of compartments is kept as small as possible.

Finally to arrive at our core result we combine the previous observations with the relation between the self-renewal probabilities and the number of intermediate cell compartments. We find that the average replication capacity of dividing cells is minimized by a tissue architecture in which at most one cell compartment has self-renewal capabilities and the number of compartments is kept as small as possible. Moreover the replication capacity of the cell population is independent of the position that the self-renewing compartment holds in the hierarchy of the cell lineage (Proposition 3 in Section 20.2.1). Figure 20.6(a) demonstrates these results when there is only one self-renewing compartment. As the number of intermediate cell compartments increases, so does the average replication capacity. Also note that not only the average, but also the entire distribution of the replication capacity is deeply affected by the number of compartments and self-renewal probabilities (figure 20.6(b)). Figure 20.7 further highlights these results. Here, two alternative architectures for the same target of intermediate cell divisions are presented. The optimal cell lineage that both minimizes the number of intermediate cell compartments and has only one self-renewing compartment has a significantly lower average replication capacity than that of the alternative architecture.

Next we turn our attention to the division rates. If there is one intermediate compartment with a slower division rate than all the rest, then it would be optimal as a cancer preventing strategy if it were the first (zeroth) compartment. In this case the most 'dangerous cells' (i.e., those with the largest replication capacity) would be dividing slower. Indeed, it is reasonable to assume that if a cell starts behaving erratically and breaks away from tissue regulation, it would present a greater threat if it originally comes from a compartment that has a fast division rate. Assuming that the first compartment has the slowest division rate, it would then make sense as a cancer prevention strategy to have this same compartment be the one with self-renewal capabilities, as this would increase the number of cells with a slow division rate.

Hence, an optimal tissue architecture to protect against cancer is one where the less differentiated cells have a larger rate of self-renewal and a slower rate of cell division. These types of cell dynamics have been repeatedly observed in cell lineages, suggesting that they may have evolved to decrease the risk of cancer. It is important to note however, that there are other biological issues at play (which are not considered here) that may affect the choice of the cell division rates, such as the speed of tissue regeneration after an injury.

Finally, we look at the question of whether adult stem cells have a truly unlimited replication capacity. While it is widely acknowledged that adult stem cells have a greater replication capacity than more differentiated cell types, experimental evidence suggest that some adult stem cells experience a diminishment of their replicative potential during the lifespan of the host [Flores *et al.* (2006, 2008)]. To address this possibility we consider a cell lineage model in which the replication capacity of stem cells decreases with time and explore whether our previous results hold in this scenario. More precisely let $\rho(t)$ be the time dependent average replication capacity of the stem cell population. We assume a decrease in the replication capacity of the stem cell population that is linear with time. In mathematical terms: $\rho(t) = \rho_0 - \epsilon t$. Similarly let us call $a_j(t)$ the time dependent expected replication capacity of the transit cells in the j-compartment. If the cell population $(x_0, \ldots, x_k)$ is at equilibrium, we have:

$$\begin{cases} \dot{\rho} &= -\epsilon \\ x_0\dot{a}_0 &= (a_0-1)2p_0v_0x_0 \ + (\rho(t)-1)rS - a_0v_0x_0 \\ x_1\dot{a}_1 &= (a_1-1)2p_1v_1x_1 \ + (a_0-1)2(1-p_0)v_0x_1 - a_1v_1x_1 \\ \vdots & \qquad\qquad \vdots \\ x_k\dot{a}_k &= (a_k-1)2p_kv_kx_k + (a_{k-1}-1)2(1-p_{k-1})v_{k-1}x_k - a_kv_kx_k. \end{cases} \tag{20.4}$$

To analyze this system of ordinary differential equations (see Sections 20.2.2, 20.2.3) we develop an approximation formula and compare our results with the corresponding implementation of the agent-based model. We find that the central result regarding the optimal architecture to minimize the expected replication capacity of a dividing cell holds when the replicative capacity of stem cells decreases with time. This is demonstrated in figure 20.8: here we compare the distributions of the replication capacity of two cell lineages with the same target number of divisions (one with an optimal and one with a sub-optimal architecture). Each distribution is presented at two different times. In both instances the replication capacity of stem cells decreases at the same rate.

20.4 Theory and observed tissue architecture

Several of the features that characterize an optimal tissue architecture are found in various cell lineages. Consider the hematopoietic system. At the starting point of this lineage there are stem cells which are classified into two categories: long-term repopulating stem cells and short-term repopulating stem cells. There are three intermediate cell types: Multipotent progenitor cells, common progenitors, and precursor cells. Out of the intermediate cells there is self-renewal only in multipotent progenitors, which in the cell lineage appear immediately downstream from the stem cell population. The end products of the lineages are fully mature differentiated cells that perform tissue function [Shizuru *et al.* (2005); Lobo *et al.* (2007)]. Thus, in the hematopoietic system it appears that self-renewal occurs only in the first least differentiated intermediate cell compartment, which is one of the features that we found reduces the replication capacity of the non-stem cell population (although there is some recent ex vivo evidence of self-renewal downstream of progenitors [England *et al.* (2011)]. With regard to the number of intermediate cell compartments, we note that there

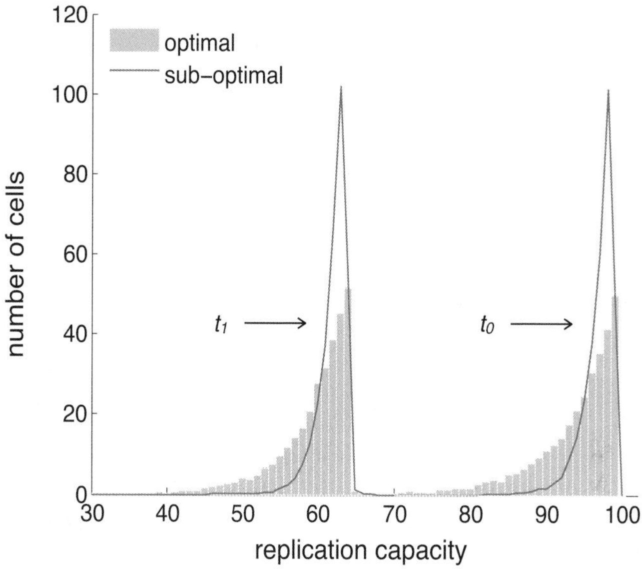

Fig. 20.8 Distribution of the replication capacity of dividing cells for the case when the stem cell replication capacity diminishes with time. Results from two alternative cell lineage architectures for the same target number of intermediate cell divisions are presented. The distributions are shown at two different times $t0 = 0$ and $t1 = 3500$ (units of time equal the mean cell division time of stem cells). In both cell lineages all $v_j = 1$, $r = 1$, $S = 50$ and $\epsilon = 0.02$. In the optimal architecture (bars) $k = 0$ and $p_0 = 0.42$; in the sub-optimal architecture (lines) $k = 1$, $p_0 = 0.31$ and $p_1 = 0.31$.

is no definite agreement on the number of stages of differentiation; however there appears to be at least two forks in the differentiation pathway. Multipotent progenitors give rise to two different types of common progenitors: Common lymphoid progenitors and common myeloid progenitors. These common progenitors further subdivide into two types of precursors, each of which ultimately gives rise to the mature lymphoid and myeloid cells that make up blood. The division rates follow an optimal pattern with division rates increasing with each more differentiated intermediate compartment [Passegué *et al.* (2005)]. Also the number of stem cells is small compared to the total number of blood cells [Shizuru *et al.* (2005)].

Neural tissues also exhibit characteristics of an optimal tissue architecture. In the process of adult neurogenesis, multipotent neural stem cells give rise to intermediate neuron progenitors that exhibit some degree of self-renewal. The neuron progenitors in turn give rise to cells that exit the cell cycle and differentiate into neurons [Lander *et al.* (2009); Sanai *et al.* (2005)]. A similar organization has been observed in the production of glia cells. Neural stem cells give rise to intermediate glia progenitors, which exhibit some degree of self-renewal and produce the different types of mature glia cells [Sanai *et al.* (2005)].

In most tissues, however, there is some uncertainty about the precise hierarchical structure of the cell lineage. For example, in adult neurogenesis the number of intermediate cell compartments is alternatively reported as one or two [Gage (2000); Lander *et al.* (2009)]. Part of this uncertainty is explained by a lack of a clear standard to distinguish between stem cells and progenitor cells. In many tissues there is also uncertainty about the self-renewal capabilities of intermediate cells. It is often unclear whether an experimentally observed transit-amplifying behavior is produced by a cell program that allows for a fixed number of divisions in progenitor cells, or by some degree of self-renewal. A cell program that calls for a fixed number of divisions would be represented in our framework as a lineage with numerous intermediate compartments and no self-renewal. By contrast through a self-renewal mechanism the cell's decision to differentiate would be independent of the number of previous divisions and instead be determined by the current state of the cell's microenvironment. As we have discussed here, the two mechanisms may be able to produce the same results in terms of population sizes. Yet, our model demonstrates that the precise architecture of a cell lineage has dramatic implications for the replication capacity of a cell population, and thus the risk of cancer. Experiments should be devised to characterize not only the transit-amplifying behavior of intermediate cells but to determine which mechanisms different systems use. Finally, we note that when interpreting the model's results in the context of a specific biological system, it is important that the biological description of a "cell compartment" agrees with the one presented here. In particular within the model's framework, a common surface marker cannot be used to define a cell compartment if it is expressed by a heterogeneous group of cells with inherently different self-renewal capabilities.

20.5 Summary

In this chapter we have demonstrated that a lineage's architecture can significantly impact the goal of reducing the replicative potential of cells. These findings underscore the importance of fully understanding a lineage's architecture as well as the precise mechanisms used to accomplish transit-amplifying behavior. The fact that at least some of the features that characterize an optimal architecture are present in various tissues suggests that they might have evolved to minimize cancer risk. This however does not mean that tissues must follow all aspects that define an optimal architecture. What we have described here is only one of possibly many evolutionary forces that shape a tissue's architecture. There could be other forces unrelated to reducing the risk of cancer, which also play a role in ultimately determining the architecture of a specific tissue. A better understanding of how a tissue's architecture and replicative limits impact the likelihood of cancer can provide insights into cancer biology that may lead to new targets of therapy.

Problems

Problem 20.1. Research project. *Learn more about replicative senescence, the Hayflick limit and telomeres.*

Problem 20.2. *What is the long-term behavior of equation (20.1)? Find the stable fixed point.*

Problem 20.3. *Show that equation (20.2) holds at steady state.*

Problem 20.4. Numerical project. *Consider equation (20.3) which defines steady-state sizes of compartments. How do compartment sizes depend on parameters p_i and v_i? Explore different numerical values for these quantities and observe the patterns of tissue architecture that you notice.*

Chapter 21

Gene therapy and oncolytic virus therapy

Oncolytic viruses replicate selectively in tumor cells and have been explored as a targeted treatment approach against cancers [Bell (2007); Bell *et al.* (2003); Moon Crompton and Kirn (2007); Davis and Fang (2005); Kaplan (2005); Kelly and Russell (2007); Kirn and McCormick (1996); McCormick (2003, 2005); O'Shea (2005); Parato *et al.* (2005); Post *et al.* (2005); Roberts *et al.* (2006); Vähä-Koskela *et al.* (2007); Wong *et al.* (2010)]. In principle an oncolytic virus will spread though the tumor cell population and lyse the infected cells, leading to eradication or control of the tumor. Because of the selectivity of such viruses for cancer cells rather than normal human cells, side effects should also be less pronounced than those associated with traditional treatments, such as chemotherapy or ionizing radiation. Oncolytic virus therapy has been explored in the context of several different virus species. While some non-human viruses display natural selectivity for cancer cells in humans [Koppers-Lalic and Hoeben (2011)], modern approaches use genetically engineered viruses to achieve tumor selectivity. The first engineered virus generated in the 1990s was a herpes simplex virus-1 [Martuza *et al.* (1991)]. Engineered adenoviruses have been of major interest in recent clinical trials, especially in the context of head and neck cancer [Wong *et al.* (2010)]. Indeed the adenovirus H101 (Shanghai Sunway Biotech, Shanghai, China) was approved in China for the treatment of head and neck cancer in combination with chemotherapy [Garber (2006)]. A variety of other virus types has also been explored [Eager and Nemunaitis (2011)]. However despite initial promising results and observations in the laboratory and clinic, oncolytic viruses have so far failed to demonstrate sustained and reliable treatment success [Wong *et al.* (2010)].

Besides experimental research, mathematical and computational modeling has increasingly become a tool to study the dynamics of oncolytic viruses. Mathematical models can help us understand the emerging properties of cancer-virus interactions, to interpret experimental results, and to design new experiments. The first mathematical models of oncolytic virus therapy considered ordinary differential equations that described the basic interactions between a replicating virus and a growing population of tumor cells, and also immune responses [Wodarz (2001b, 2003)]. Further work extended this type of approach in a number of ways, describing different scenarios and applying models to specific virus-tumor systems [Bajzer *et al.* (2008); Biesecker *et al.* (2010); Dingli *et al.* (2006, 2009); Friedman *et al.* (2006); Karev *et al.* (2006); Komarova and Wodarz (2010); Novozhilov *et al.* (2006); Wein *et al.* (2003); Wodarz (2004, 2009); Wodarz and Komarova (2009); Bagheri *et al.* (2011); Zurakowski and Wodarz (2007)]. One of the assumptions that is implicit in such modeling approaches is that cells and viruses mix perfectly with each other (mass action). While this might hold true in the context of some in vitro experiments, and while this might be a reasonable approximation of the dynamics occurring in some non-solid tumors, the majority of tumors have intricate spatial structures where cells and viruses do not mix well, but where interactions are limited to local neighborhoods. Hence, to gain a better understanding about the dynamics of oncolytic viruses, spatially explicit models are required. Some spatial modeling studies have been performed and have given rise to interesting results [Wein *et al.* (2003); Mok *et al.* (2009); Paiva *et al.* (2009); Reis *et al.* (2010)]. They commonly include, in addition to basic spatial dynamics, one or more additional assumptions that introduce further complexity.

This chapter summarizes our own modeling approaches to studying the dynamics of oncolytic viruses. The aim is to highlight the usefulness and limitations of different modeling approaches in the quest to obtain an accurate model that can be applied to specific treatments and provide predictive power. The chapter starts by discussing specific ordinary differential equation (ODE) models, continues to describe more general, axiomatic modeling approaches, and finally examines a spatially explicit model describing virus dynamics in relatively simple in vitro settings.

21.1 A basic ordinary differential equation model

This section introduces a simple mathematical model describing the development of a growing tumor and an oncolytic virus population over time. The model is based on [Wodarz (2001b)] and includes three variables: the growing, uninfected, cancer cells, x, infected cancer cells, y, and virus particles, v. It is given by the following set of differential equations which describe the development of these populations over time [Wodarz (2004)]:

$$\begin{aligned}
\dot{x} &= rx\left(1-\frac{x+y}{\omega}\right)-dx-\beta xv,\\
\dot{y} &= \beta xv-(d+a)y, \qquad (21.1)\\
\dot{v} &= ky-uv.
\end{aligned}$$

The cancer cells grow at a rate r, and this growth is density dependent, limited by a maximum size ω. In biological terms, this means that the cancer cells divide and that this results in exponential growth at small tumor cell densities, but that growth is slowed down as the tumor reaches larger sizes and runs out of space, nutrients, and other resources required for growth. Cancer cells die at a rate d. Therefore, the average life-span of the cancer cells is given by $1/d$. The cancer cells become infected by the virus proportional to βxv. The rate constant, β, describes the efficacy of this process, including the rate at which virus particles find uninfected cells, the rate of virus entry, and the rate of successful infection. The term βxv multiplied by a time interval gives the number of infection events occurring during the time interval (if it is assumed to be small). Infected cancer cells also have a death rate. The death rate of infected cells is a composite of the natural death rate, d, and the virus-inflicted death rate, a. Therefore, the average life-time of an infected cell is $1/(d+a)$. Infected cells produce virus at a rate k. The total amount of virus particles produced from one infected cell, or the "burst size", is hence given by $k/(a+d)$. Finally, the virus decays at a rate uv. Thus, the average life-span of a virus particle is given by $1/u$. In the context of replicating versus non-replicating viruses, we can make the following distinction. A non-replicating virus is characterized by $k=0$ (no virus production by the infected cell), while a replicating virus is characterized by $k>0$.

In this model, the tumor expands if its growth rate is greater than its death rate, i.e., if $r>d$. In the absence of treatment, the tumor will eventually grow to its maximum size given by $x^{(0)}=\omega(r-d)/r$. Treatment in the model corresponds to the introduction of virus into the system. The virus has the potential to spread if $\beta x^{(0)}>d+a$, that is if the replication rate of the virus is fast relative to the death rate of the infected cancer cells. The virus infection then takes the system to a new equilibrium outcome which is given by the following expressions. $x^{(1)}=(d+a)/\beta'; y^{(1)}=[\beta'\omega(r-d)-r(d+a)/[\beta'(r+\beta'\omega)]$; $v^{(1)}=ky^{(1)}/u$; where β' summarizes

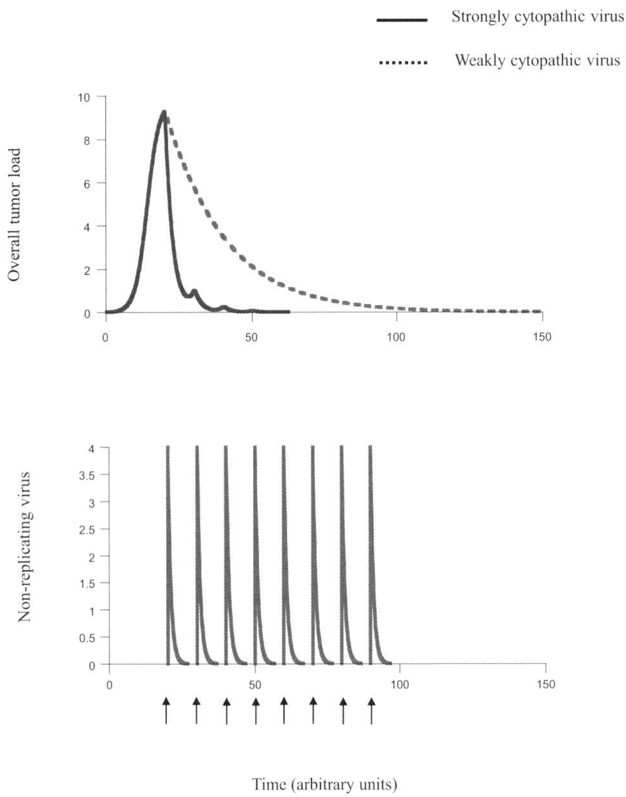

Fig. 21.1 Simulation of therapy using a non-replicating virus. The virus is administered repeatedly, as indicated by the arrows. The development of tumor load over time is shown assuming both a strongly cytopathic virus and a weakly cytopathic virus. The strongly cytopathic virus results in more efficient eradication of the tumor. Parameters were chosen as follows. $r = 0.5$, $\omega = 10$, $\beta' = 1.5$, $d = 0.01$, $k = 0$, $\delta = 1$. For the strongly cytopahtic virus, $\alpha = 0.4$. For the weakly cytopathic virus, $\alpha = 0.04$.

the overall replication rate of the virus and is given by $\beta' = \beta k/u$. The size of the overall tumor cell population during virus therapy is given by the sum of uninfected and infected tumor cells, $x^{(1)} + y^{(1)}$. The aim of therapy should be to reduce this population to low levels. If it has been reduced below a threshold in the model, the tumor population can be assumed to be extinct (number of tumor cells below one). In the following sections, model properties will be analyzed. Figures are shown to demonstrate specific aspects qualitatively, i.e., parameters were chosen for illustrative purposes and unites are arbitrary.

21.1.1 *Non-replicating viruses*

Assume the virus is not replicating ($k = 0$). If not all cancer cells become infected after the first administration, it has to be given repeatedly in order to ensure continued presence of the virus and hence remission of

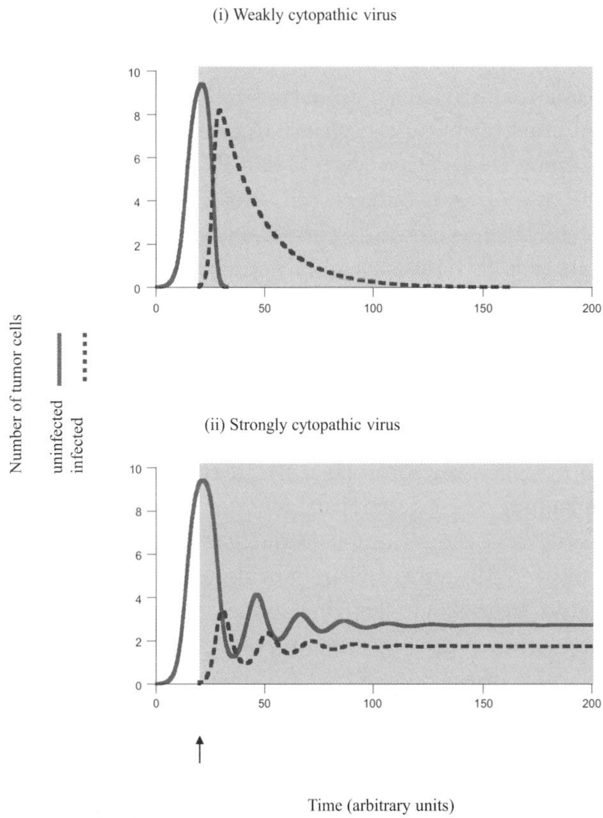

Fig. 21.2 Simulation of therapy using a replicating virus. The virus is administered once, as indicated by the arrow. Shading indicates the phase of the dynamics following administration of the virus. (i) Use of a weakly cytopathic virus results in sustained cancer remission. (ii) Use of a more cytopathic virus results in long term persistence of the cancer and the virus. Parameters were chosen as follows: $r = 0.5$, $\omega = 10$, $\beta' = 1.5$, $d = 0.01$, $k = 0.1$, $\delta = 1$. For (i) $\alpha = 0.04$. For (ii) $\alpha = 0.4$.

the cancer (figure 21.1). The goal is to eradicate the population of uninfected tumor cells. Since infected cancer cells do not divide, they do not pose a threat and will decay to extinction. The dynamics of the uninfected cancer cell population over time during treatment can be approximated by $x_t = x^{(0)} exp[(r - d - \beta v)t]$. This assumes that the tumor is still growing and has not yet reached levels close to maximum tumor size ($x \ll \omega$). For the uninfected cancer cells to decline, the level of virus has to be kept above a threshold during therapy, given by $v > (r - d)/\beta$. Thus, a high infectivity of the virus, and a low growth rate of the tumor facilitate tumor reduction. If this condition is fulfilled, the half-life of the uninfected tumor cell population is given by $t_{1/2} = \ln(1/2)/(r - d - \beta v)$. The time to eradication of the uninfected tumor cells is hence given by $t = \frac{\ln(x_0)}{\ln(2)} \frac{\ln(1/2)}{r-d-\beta v}$. After this time threshold, administration of the virus can be stopped, and the population of infected tumor cells decays with a half-life of $t_{1/2} = \ln(2)/a$. Hence, the faster the rate of virus-induced cell death, a, the faster the population of infected cells declines towards extinction (figure 21.1). To summarize, the best strategy is to (i) use a cytopathic virus, (ii) use a virus with high infectivity, and (iii) reduce the cancer growth rate which might be achieved by certain chemo- or radio-therapeutic regimes [You *et al.* (2000)].

21.1.2 *Replicating viruses*

Now assume a replicating virus ($k > 0$). If the patient is injected with a very high inoculum dose of the virus, most or all cancer cells immediately become infected and the situation is the same as for the non-replicating virus. If, however, the initial virus inoculum is not that large and does not immediately overwhelm the cancer, then the dynamics between virus replication and tumor growth will play out, resulting in an equilibrium outcome. The size of the tumor load (infected + uninfected cells) at equilibrium determines the level of success. If equilibrium tumor load in the model is very low, this corresponds to eradication in practical terms. Higher equilibrium tumor loads in the model corresponds to persistence of the tumor in the presence of the virus. Equilibrium tumor load is given by $x^{(1)} + y^{(1)} = \frac{\omega(a+r)}{r+\beta'\omega}$. An important result is that a lower virus-induced death rate of infected cells (small a) results in lower equilibrium tumor load (figure 21.2). If the rate of virus-induced tumor cell killing is too high, the outcome is persistence of the tumor in the face of ongoing viral replication (figure 21.2). The reason is as follows. Low viral cytopathicity increases virus load. Higher virus load results in more infection and in a greater decline of the uninfected cancer cells. Higher viral cytopathicity results in lower virus load. Low virus load results in less infection and in less reduction of the uninfected tumor cells. In addition, equilibrium tumor load is reduced by other parameters, most notably by a high replication rate of the virus (high value of β'), and a slow growth rate of the tumor (low value of r). As expected, the replication rate of the virus (value of β') has to lie above a threshold for tumor load to become low enough that tumor eradication is feasible in a stochastic setting. If the replication rate of the

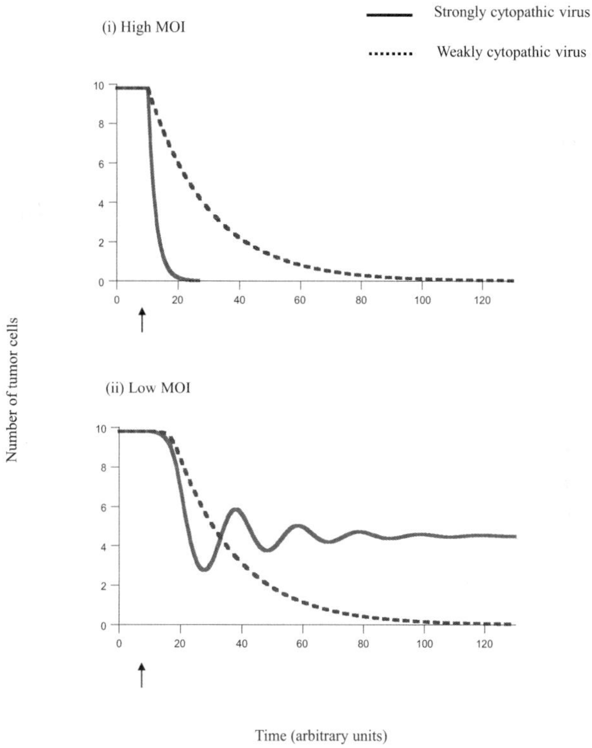

Fig. 21.3 Simulation showing the evaluation of potential replicating viruses in culture. A weakly and a strongly cytopathic virus are compared when the culture is inoculated (see arrow) with (i) a high MOI and (ii) a low MOI. While under high MOI infection the more cytopathic virus yields better results, the opposite is true for low MOI infection, which reflects virus spread conditions in vivo. Parameters were chosen as follows. $r = 0.5$, $\omega = 10$, $\beta' = 1.5$, $d = 0.01$, $k = 0.1$, $\delta = 1$. For the strongly cytopahtic virus, $\alpha = 0.4$. For the weakly cytopathic virus, $\alpha = 0.04$. Virus inoculum was $v = 10$ for high MOI and $v = 0.01$ for low MOI.

virus is too low, the virus does not spread sufficiently through the population of tumor cells. In summary, if the virus is replicating, the best strategy is to (i) use a weakly cytopathic virus, (ii) use a fast replicating virus, and (iii) reduce the growth rate of the tumor by alternative therapeutic means.

21.1.2.1 *Evaluation of replicating viruses in culture*

The mathematical model has given rise to an important difference in treatment strategy depending on whether the virus replicates or not. If the virus does not replicate, a high degree of cytopathicity is beneficial because it speeds up elimination of all tumor cells. On the other hand, if the virus replicates, success is promoted by using a weakly cytopathic virus. A high rate of virus-induced cell death is detrimental and leads to the persistence of both tumor and virus. These findings also have important implications for the methods used to evaluate potential viruses in culture. If the virus does not replicate, a high multiplicity of infection (MOI) has to be used. The virus with the strongest degree of tumor cell killing will remove the cancer cells fastest. Such a virus will also work best in the physiological situation, since the aim is to overwhelm the tumor by repeatedly injecting the agent, resulting in fast killing of as many cancer cells as possible (figure 21.1). On the other hand, if the virus replicates, a low MOI is required to evaluate the virus. The reason is that *in vivo*, the replicating virus has to spread through the cancer cell population, and this has to be mimicked in culture. Using a high MOI can lead to misleading evaluations. These notions are illustrated in figure 21.3 with computer simulations. This figure depicts the dynamics in culture for strongly and weakly cytopathic viruses, using different MOIs. Figure 21.3(i) shows the dynamics for a high MOI. In this simulation, the strongly cytopathic virus results in quick elimination of the tumor cells, while the weakly cytopathic virus is much less effective. Thus, if viruses are evaluated using a high MOI, the virus with the strongest degree of tumor cell killing receives the highest grades. Importantly, this is the virus which is predicted to be least efficient at reducing tumor load *in vivo*. The situation is different when viruses are evaluated in culture using a low MOI (figure 21.3(ii)). The less cytopathic virus results in elimination of tumor cells in culture, while the more cytopathic virus fails to eliminate tumor cells in culture. Therefore, the less cytopathic virus gets the better marks, and this is also the virus which is predicted to be more efficient at reducing tumor load *in vivo*.

21.2 Different mathematical formulations and the robustness of results

The above described model investigated the dynamics of oncolytic viruses in the context of specific mathematical formulations that are of uncertain biological nature. For example, consider the term that describes the infection of cells, given by βxv. It assumes perfect mixing of cells and viruses, which is hardly realistic in vivo, where most tumors exhibit intricate spatial structures. The larger the number of tumor cells, the faster the growth rate of the virus population. Similarly, the larger the number of viruses available, the more cells become infected, ignoring possible saturation effects. The "biologically correct" term to describe infection is currently not known. Other processes assumed in the model are also described by arbitrary mathematical expressions, such as the term describing cell growth, given by the logistic equation. While this is assumed in many models of tumor growth, the laws underlying tumor growth are likely to be more complex, and this could impact the properties of the virus dynamics model. To address this issue, we avoided concentrating on a particular model, but took a more general approach. Processes such as infection and cell growth were formulated in terms of general functions and results were thus independent from biologically uncertain and arbitrary mathematical formulations [Wodarz and Komarova (2009)]. Through specific restrictions about biological assumptions, we analyzed a class of mathematical models that aim to describe viral spread through a tumor in different settings. Details of the axiomatic analysis are given in [Komarova and Wodarz (2010)].

We found that based on the infection term, we can divide models into two categories with fundamentally different behavior [Wodarz and Komarova (2009)]. In one group, virus growth is relatively fast, which could be characteristic of a relatively well mixed system. In these models, there is a clear viral replication rate threshold beyond which the number of cancer cells drops to levels of the order of one or less, corresponding to extinction in practical terms. Under this parameter region, this is the only outcome in this class of model. In the other category, virus growth is relatively slow, which could be characteristic of more intricate tumor cell arrangements. In this class of model, the larger the number of infected cells, the smaller the proportion of cells that can pass on the virus. In this scenario, virus therapy is more difficult. Even in the parameter regions where the dynamics can converge to a tumor control or eradication outcome, there can be the possibility that the cancer can outrun the virus if the number of cancer cells lies above a threshold at the start of virus therapy. This is because of the existence of the saddle node equilibrium which ensures dependence of the outcome on initial conditions. This might be problematic in clinical settings, because tumors typically become detectable at relatively large sizes and there is only a relatively small window between the size at which the tumor becomes detectable (about 10^{10} cells) and the size at which it can induce mortality (around 10^{13} cells).

Another important finding of this study is that the basic results regarding the outcome of oncolytic virus therapy do not depend on the particular tumor growth terms used in the model [Wodarz and Komarova (2009)]. The exact kinetics of tumor growth are still poorly understood and a source of uncertainty [Rodriguez-Brenes *et al.* (2011)]. We examined straight exponential growth, as well as a number of more realistic options, including saturated but continued growth at high numbers of cancer cells, as well as cessation of growth as the number of tumor cells approaches an upper limit [Komarova and Wodarz (2010)]. While there are minor differences, the properties of the tumor control equilibrium are largely independent from the exact way in which tumor growth is modeled.

21.3 A spatially explicit model of oncolytic virus dynamics

Tumors are characterized by intricate spatial structures that cannot be captured by ordinary differential equations. Other modeling methods have to be called upon in order to capture this complexity. On the other hand, the spatial complexity of tumors is enormous and not well understood, which can undermine the predictive power of such spatial models. As a first step to tackle this problem, we considered a relatively simple in vitro setting, in which target cells are arranged in a two-dimensional monolayer and in which viruses can only spread from the infected source cell to the nearest neighboring target cells [Wodarz *et al.* (2012)]. This relatively simple experimental setting can be described by an agent-based model, and model predictions can be tested relatively easily by experiments. In the model, each cell is represented as an "agent" occupying a certain position on a grid, and interacting with other cells according to some (probabilistic) rules. Our modeling approach is spatial, that is, it takes into account the spatial distribution of the uninfected and infected cells. The model, based on [Satō *et al.* (1994)], describes target cell-virus dynamics on a two-dimensional grid that contains $N \times N$ spots [Wodarz *et al.* (2012)]. Each spot is either occupied by a cell (infected or uninfected), or it is empty. We model the development of the populations in discrete time. Given the state of the system at time t, a set of rules is applied to each spot, and this gives rise to the state of the system at time $t+1$. At each time step, the grid is randomly sampled N^2 times. If the chosen spot is occupied by an uninfected cell, it can die with a probability D, leaving the spot empty. Alternatively, the cell can reproduce with a probability R, and a destination spot is randomly chosen for the offspring from the set of eight nearest neighboring spots. If the destination spot is empty, the offspring is placed there, otherwise, no reproduction occurs. If the chosen spot contains an infected cell, it can die with a probability A, or attempt to transmit the virus with a probability B. A destination spot is chosen randomly from the eight nearest neighbors, and infection only proceeds when a susceptible cell is present. Infected cells are assumed not to reproduce. Typical oncolytic viruses lock the cell in the S-phase for replication, thus preventing further divisions [Bagheri *et al.* (2011)].

21.3.1 *Initial virus growth patterns*

In this section, we explore the initial virus growth patterns. As starting conditions we assume that the grid is filled with uninfected cells and that a relatively small square of infected cells (30×30 spots) is placed in the middle of the grid (which overall contains 300×300 spots). The emerging growth pattern depends on parameters that influence the rate of virus spread, in particular the probability for an infected cell to die, A, and the probability for an infected cell to transmit the virus, B. The patterns that we observe are presented in figure 21.4.

In figures 21.4(a) and (b), the infected cell population expands as a ring or wave that leaves no cell behind in its core. The two pictures differ in the death probability of infected cells. In figure 21.4(a), the probability for infected cells to die is relatively low such that during the time frame of the simulation a hollow ring has not yet formed and the infected cell population expands as a relatively solid mass. In figure 21.4(b), the death probability of infected cells is higher such that during the time frame of the simulation a hollow ring has formed. The total number of cells is proportional to $\frac{e^{-At}-1+At}{A^2}$ [Wodarz *et al.* (2012)], such that for short time-scales (or smaller death rates) the growth is quadratic in time, and for longer times scales (or larger death rates) it is linear in time. This is exactly what is observed. Figure 21.4(a), characterized by smaller values of A, shows a growth law of the infected cell population that is close to quadratic. In figure 21.4(b), where the death rate is larger, the infected cell population grows linearly once the hollow ring is present. Note that these two scenarios are identical in principle because in figure 21.4(a), the formation of the hollow ring requires more time (and a larger grid). The higher the death rate of infected cells,

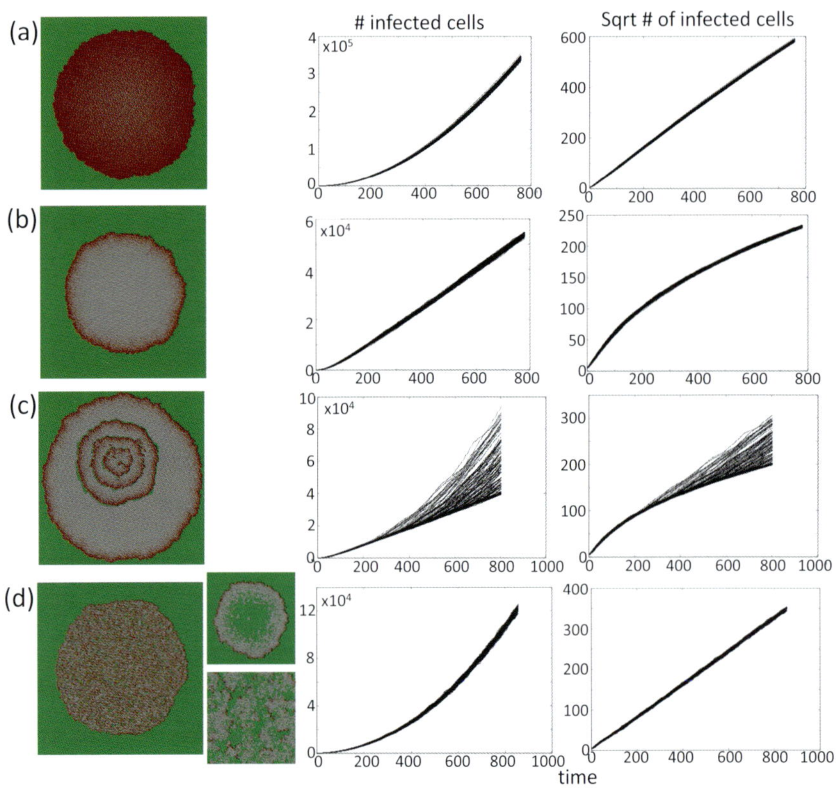

Fig. 21.4 Initial virus growth dynamics in the agent-based model. Green indicates uninfected cells, red infected cells, and grey empty spots. The middle graph shows the number of infected cells over time. The different lines represent 100 different instances of the simulation with the same parameter combination. The right graph shows the square root of the number of infected cells over time, again showing lines for 100 different runs. Parameter values were chosen as follows. (a) $R = 0.5$, $D = 0$, $B = 0.6$, $A = 0.601$. (b) $R = 0.5$, $D = 0$, $B = 0.6$, $A = 0.62$. (c) $R = 0.5$, $D = 0$; $B = 0.6$, $A = 0.628$. (d) $R = 0.5$, $D = 0$, $B = 0.6$, $A = 0.7$. The small graphs in (d) are characterized by $R = 0.04$, leading to fewer target cells in the area of infection and thus to slower viral spread.

the faster the ring is formed, and the faster the growth law changes from square to linear. Lowering the rate of virus spread (decreasing the value of B and increasing the value of A) gives rise to patterns of a different nature.

In figure 21.4(c), uninfected cells are left behind in the core of the expanding ring. When they grow and become infected by virus, a coupled expanding ring of uninfected and infected cells forms. This can occur repeatedly, giving rise to concentric rings. The persistence of cells in the core of the ring is probabilistic in nature, and that is reflected in the growth laws that are observed in multiple runs of the simulation. In cases where uninfected cells are not left behind inside the ring, the infected cell population grows linearly. When concentric rings do occur, the growth becomes quadratic.

Finally, no expanding ring structure is formed in figure 21.4(d) because the viral spread kinetics are even slower. Instead, the area of virus growth is characterized by a mix of infected and uninfected cells that expands over time. In this case, quadratic growth of infected cells is observed. Note that if the viral spread kinetics are in the lower end of this spectrum, it is possible to observe a variation of this pattern, shown in the inset of figure 21.4(d): While the spreading infection leaves uninfected cells behind, the viral spread kinetics are too low to maintain significant numbers of infected cells throughout this area. Most of the infected cells will be at the outer edge of the infection due to a higher density of target cells. In this case, a relatively thin, ring-like structure can be formed, with a large area of uninfected cells remaining in its core. This pattern, however, is temporary. With time, one of two scenarios can be observed. A mixed pattern can be generated, characterized by a large number of uninfected cells and a low number of infected cells, because the virus eventually spreads to the remaining susceptible cells. Alternatively, there is a chance that the virus population goes extinct due to the slow rate of spread. Long term outcomes are discussed further below.

21.3.2 *Growth patterns and the extinction of cells*

Here, we explore the long term dynamics, investigating how the above described patterns play out and correlate with the overall outcome if both the uninfected and infected cell population can expand in space. We seek to define conditions under which the virus can eliminate the target cell population in this system. All simulations are started with a small number of infected cells placed in a compact vicinity into a larger space filled with uninfected cells, which is in turn embedded into an even larger "empty" space (for the exact initial conditions for particular cases, see appropriate figure legends). In contrast to the simulations reported above, here we go beyond the initial virus growth stage, and focus on time-scales where the population of target cells experiences significant changes (grows in size in the absence of infection). The outcomes of this system include extinction of the target cells and thus the virus; extinction of the virus and persistence of the target cells; coexistence of virus and target cells. The dependency of these outcomes on the parameters is shown in figure 21.5, which is the result of at least 10^4 instances of the simulation, where the $\log_{10}$ of all the parameters was varied between -4 and 4. Figures 21.6 and 21.7 show corresponding spatial and temporal patterns. We examine the outcomes first in a relatively small 30×30 grid, and subsequently in a larger, 300×300 grid.

Small grid. In the 30×30 grid, the following outcomes are found (figure 21.5(a), and figures 21.6 & 21.7).

Two types of target cell extinction can be observed, both associated with the initial "hollow ring" structure. According to pattern **A**, ***virus-***

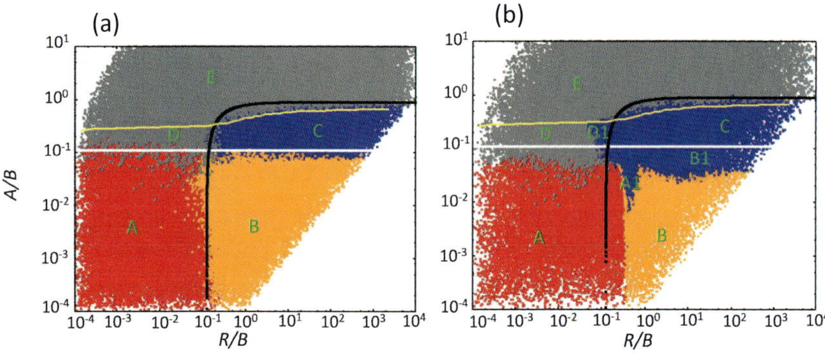

Fig. 21.5 Dependence of outcomes on parameters in the agent-based model for (a) a 30×30 grid, and (b) a 300×300 grid. Blue means coexistence of virus and cells. Red and orange indicate extinction of the cells and thus the virus. Red is used if extinction occurs before the boundary of the system has been reached, while orange is used if extinction occurs after cells have reached the boundary of the system. Grey indicates extinction of the virus while cells persist. Above the white line and below the black line, the "local" equilibrium number of uninfected and infected cells, respectively, is greater than one. Below the yellow line, the virus can successfully invade its target cell population. The capital letters indicate different spatial patterns that are described in the text and in figure 21.6. In these simulations, the probability for an uninfected cell to die was $D = 0$.

mediated target cell extinction, both the target cell and the virus populations spread outward in space as a wave, but the virus wave overtakes the target cell wave, leading to extinction of both populations. Pattern **B**, ***boundary-mediated extinction***, represents weaker viruses compared to case A. In pattern B, the virus wave catches up with the target cell wave, leaves no uninfected cells behind in its wake, but fails to eliminate the target cell wave. Instead, the two waves travel together with the same velocity until the boundary in reached. The target cell population can escape the virus only by spreading outward. Once the boundary is reached, this is not possible anymore, explaining the extinction. Note that although real

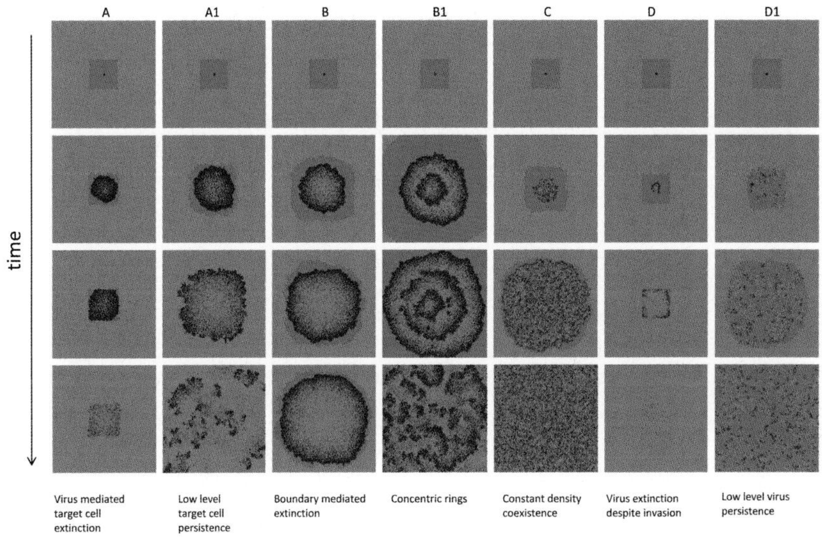

Fig. 21.6 Seven spatial patterns observed in the agent-based model. For each pattern, four snapshots in time are shown. Green indicates uninfected cells, red infected cells, and grey empty patches. See corresponding capital letters in figure 21.5, showing in which parameter regions the individual patterns are observed. The time series that are associated with the individual patterns are shown in figure 21.7. See text for details. The simulations were run on a 300×300 grid. The simulations were started by placing a small number of infected cells (5×5 cells) into a larger space filled with infected cells (13×13 cells). Parameters were chosen as follows: (A) $R = 0.013$, $D = 0$, $B = 0.14$, $A = 0.003$, (A1) $R = 0.15$, $D = 0$, $B = 0.32$, $A = 0.007$, (B) $R = 0.014$, $D = 0$, $B = 0.015$, $A = 0.00056$, (B1) $R = 0.04$, $D = 0$, $B = 0.032$, $A = 0.0016$, (C) $R = 0.014$, $D = 0$, $B = 0.032$, $A = 0.008$, (D) $R = 0.0002$, $D = 0$, $B = 0.019$, $A = 0.0032$, (D1) $R = 0.069$, $D = 0$, $B = 0.64$, $A = 0.18$.

tumors are capable of breaking out of homeostatic control and spreading beyond the "carrying capacity" of their environment, boundary-mediated extinction can still take place. Genetic transformations associated with waves of clonal expansion or induction angiogenesis generally happen on longer time-scales. Therefore, it is realistic to assume the existence of some geometric constraints (even temporary). Pattern B represents the situation where extinction is a consequence of such spatial constraints.

Another type of outcome is the coexistence of infected and uninfected cells, which is shown in pattern **C**, ***constant density coexistence***. As the virus population spreads out in space, it leaves behind uninfected cells with a high probability, leading to the disperse pattern of initial expansion and the absence of any clear traveling waves. Instead, the expanding virus population leaves behind a mix of both populations, which eventually is

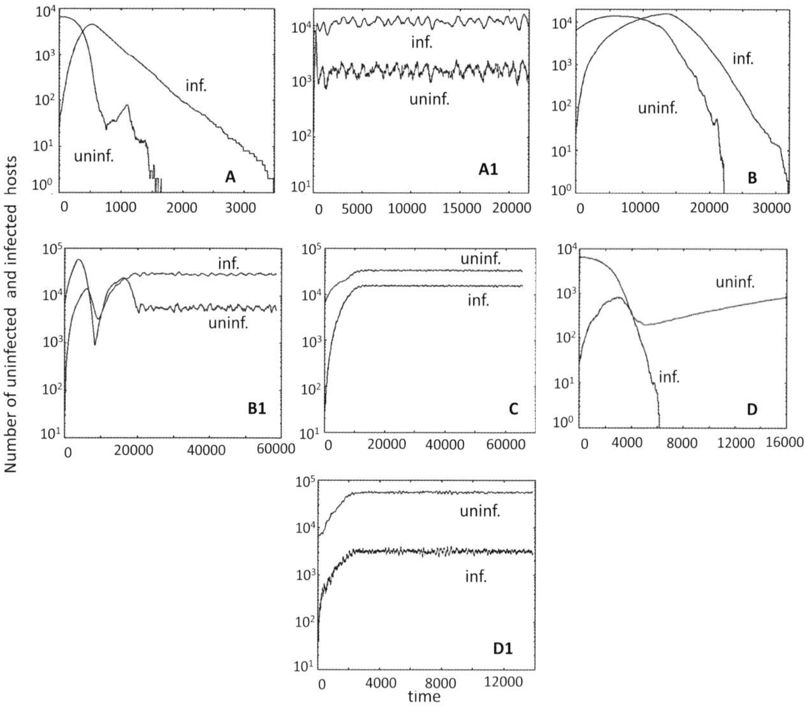

Fig. 21.7 Typical time series corresponding to the spatial patterns presented in figure 21.6, based on a single run of the spatial agent based model, assuming a 300 × 300 grid. See text for details. Parameter values and initial conditions are given in figure 21.6.

found across the whole space and is characterized by an equilibrium density that is determined by the parameters of the system, while the populations settle around a stochastic steady state.

Finally, there are two types of virus extinction patterns. Pattern **D**, ***virus extinction despite invasion***, represents a virus extinction regime where the virus can initially invade the target cell population, but does not persist in the long term. The virus reduces the target cell population, and subsequently goes extinct. This leaves the uninfected cell population to grow unopposed. The stronger the virus (lower value of A/B), the less likely this is observed in this regime, because the uninfected cell population is more likely to be driven extinct before the virus population hits extinction. The second extinction pattern **E**, ***lack of invasion***, is observed when the virus population cannot invade the target cell population and goes extinct (spatial and temporal pattern not shown).

Although the spatial stochastic predator-prey system studied here exhibits a variety of patterns, its dynamics can be understood by studying the local interactions of the agents. The idea of a "characteristic scale" has been proposed in the literature in the context of different predator-prey models [De Roos *et al.* (1991)] where the system's behavior was found most predictable on an intermediate scale defined by the agents' motility and interactions. In [Pascual *et al.* (2001)], it was shown that in a class of systems exhibiting oscillatory dynamics, the functional forms governing the local predator-prey interactions at those characteristic scales are the same as the ones describing a perfectly mixed, mass-action system, but contain different parameters. This allowed the authors to approximate the long-term dynamics of the spatial system at large scales with a temporal predator-prey model describing local interactions. Here we build on this idea, and show that the global outcomes of the spatially-distributed system can be predicted by utilizing the laws of local dynamics.

We start from the well-known system of ordinary differential equations that can be derived for our agent-based model if no spatial restrictions were in place, and reproduction and infection events were driven by laws of mass-action:

$$\frac{dS}{dt} = RS\left(1 - \tfrac{S+I}{K}\right) - \tfrac{BSI}{K}, \qquad (21.2)$$
$$\frac{dI}{dt} = \tfrac{BSI}{K} - AI,$$

where the number of uninfected cells is denoted by S, and the number of infected cells by I. In these equations, K has the meaning of carrying capacity. This well-known modified Lotka-Volterra system [Anderson *et al.* (1992); Nowak and May (2000)] is characterized by two equilibria: (i) The uninfected population persists at carrying capacity, while the virus population is extinct, i.e., $S^{(0)} = K$, $I^{(0)} = 0$, (ii) Alternatively, the virus establishes a successful infection, such that $S^{(1)} = AK/\mathrm{B}, I^{(1)} = RK(\mathrm{B} - A)/\mathrm{B}(R+\mathrm{B})$. The latter equilibrium is stable if the basic reproductive ratio of the virus is greater than one, which is equivalent to the inequality A<B. The approach to the coexistence equilibrium can be either monotonic, or can involve damped oscillations.

While these properties of the virus-cell system are well-known, it is usually thought that the ordinary differential equations can only be applied to a well-mixed system, and fail to describe a spatially-distributed system of cells. Contrary to this, figure 21.5 demonstrates that, if interpreted correctly, the above system can explain a lot of the patterns that arise in the spatial agent-based model. Let us think of the carrying capacity coefficient, K, as the size of the "local neighborhood" where cell-to-cell interactions happen in a spatial model. In our case, this neighborhood consists of K=9 cells (a cell plus its eight nearest neighbors, the relevant characteristic scale of our spatial model). Model (21.2) with the modified parameter K is capable of informing us of the local equilibrium density of the infected and uninfected cells, which in turn is correlated with the expected long term behavior of the spatial system.

In equations (21.2), the number of uninfected cells at equilibrium $S^{(1)}$ is proportional to K. In order for this equilibrium to be biologically meaningful, this value must be greater than one cell. The equation $S^{(1)} = 1$ defines the white line in figure 21.5. Similarly, the number of infected cells in local neighborhoods must be greater than one, which yields the black line, $I^{(1)} = 1$. We can see that the coexistence region in figure 21.5(a) (regime C) corresponds to the parameters for which both equilibrium values are larger than one; it is enclosed by the lines $S^{(1)} = 1$ and , $I^{(1)} = 1$ obtained directly from the cancer-virus equations. The white line $S^{(1)} = 1$ outlines the lower boundary of the coexistence region, while the black line $I^{(1)} = 1$ defines the upper boundary. (A more precise definition of the upper bound of the coexistence region is given by the yellow line in figure 21.5(a), below which the virus is strong enough to invade the cell population.)

Thus, in the spatial system, target cell extinction is observed if the local equilibrium number of uninfected cells is less than one (regions A & B, figure 21.5(a), below white line). Extinction of only the virus is observed either following initial invasion if its local equilibrium is less than one (region D, figure 21.5(a), area encased by white, black and yellow lines) or if invasion is impossible (region E, figure 21.5(a), above yellow line). The finding that equilibrium properties of simple ODE models that describe the dynamics in a small local neighborhood can predict the outcome of the spatial system has important practical implications. Note however that this method is unable to explain all the details of the diagram in figure 21.5. In particular, the proximity of the black ($I^{(1)} = 1$) line to the boundary between regions A and B is purely coincidental. The equilibrium analysis predicts extinction in regions A and B, but cannot distinguish between virus-mediated extinction (A) and boundary-mediated extinction.

Large grid. In a larger, 300×300, grid (figure 21.5(b), and figures 21.6 & 21.7), the basic patterns found in a small grid are still in place, but additional complexity is observed. In the parameter space where target cell extinction happens in the smaller grid, regions of coexistence can occur. In pattern **A1**, the expanding virus wave proceeds initially as a "hollow ring" structure, catches up with the target cell wave, leaves no uninfected cells in its wake, but only partially breaks the target cell wave. The virus is not efficient enough to eliminate the target cell wave, as observed in pattern A, but still strong enough not to leave it intact, as observed in pattern B. The partially broken wave structure allows the uninfected cells to escape not only outward, but in all directions. Hence, local extinction combined with continuous target cell movement away from the virus leads to persisting moving fronts, which can go extinct and give rise to new fronts over time. Thus, more extensive population fluctuations are observed in the long run (figure 21.7). This is the well-known regime of global persistence despite local extinction which is an important basis for the argument that space promotes coexistence [Hassell (2000)]. The levels at which the uninfected cell population persists, however, are relatively low (figure 21.7). A sufficiently large grid size is required to observe this behavior, such that enough space is available for the moving target cell fronts to persist. We refer to pattern A1 as ***low-level target cell persistence***. Region **B1** shows a different reason for target cell persistence at low levels, a pattern we call ***concentric rings***, which corresponds to the concentric ring pattern of initial virus spread described earlier. When the virus wave expands, the probability to leave behind uninfected cells is proportional to the local equilibrium number of uninfected cells. In the region where this equilibrium number is just slightly below one, this does not occur often enough to be observed on a small grid. On a larger grid, however, it can be observed. These infrequent events lead to renewed target cell growth, followed by virus growth, and a new wave structure is formed. This can lead to the occurrence of concentric expanding rings. With time, stochasticity breaks the ring structure, leading to traveling fronts that eventually go extinct, but occasionally leave behind uninfected cells to form new fronts, thus persisting in the long term. Consequently, populations show more extensive fluctuations around characteristic steady state values (figure 21.7). For lower values of A/B, the local equilibrium number of uninfected cells becomes too low for this to be observed in the grid size under consideration. Finally, in region **D1**, ***low-level virus persistence***, global persistence of the virus despite local extinction is observed, leading to relatively strong population fluctuations (figure 21.7). While the virus invades the target cell population, it converges to its local equilibrium value that is less than one. However, movement through space before extinction occurs allows coexistence if the grid is sufficiently large. For lower values of R/B, the local equilibrium number of infected cells is too low to observe this outcome even in the context of the larger grid.

These simulations show that increasing the grid size allows more complex outcomes to occur and increases the parameter region in which the cell populations persist. The additional patterns that emerge in larger grids are variations of those found in the smaller grid and involve non-equilibrium persistence, where extinction occurs locally, but movement through space allows cells to temporarily avoid extinction. These dynamics are well documented in the ecological literature [Hassell (2000)]. Besides allowing cells to move through space, a larger grid size also increases the chances that certain rare events can occur. For example, boundary-mediated extinction (pattern B, figure 21.5) is less likely to occur in large grids. The larger the grid the higher the probability that uninfected cells are left in the core of the ring before the uninfected cell population has moved to the boundary and is eliminated by the virus. All these non-equilibrium persistence outcomes in larger grids, however, are characterized by persistence of the cells at very low levels, which can be considered controlled persistence and does not involve uncontrolled cellular growth. Therefore, the outcome can still be predicted by the "local mass action equilibrium values" discussed above: if the local equilibrium of uninfected cells, predicted by the ODEs, is less than one, we can expect either extinction or controlled persistence. If the local equilibrium of uninfected cells is greater than one, we can expect to see uncontrolled cellular growth. The lower the local equilibrium of uninfected cells the less likely controlled persistence occurs and the more likely extinction is observed. However, this could not be demonstrated systematically for larger grids due to the extensive computational costs involved.

The long term outcomes shown in figure 21.6 are obviously related to the initial growth patterns described in figure 21.4. Patterns A, A1 and B in figure 21.6 arise out of the hollow-ring structure. Pattern B1 in figure 21.6 emerges from the concentric ring structure. Patterns C, D, and D1 are consequences of the disperse growth pattern / filled ring structure, which for faster viral spread rates typically leads to coexistence of the virus and cell populations, while extinction of the virus population can be observed for smaller virus spread rates.

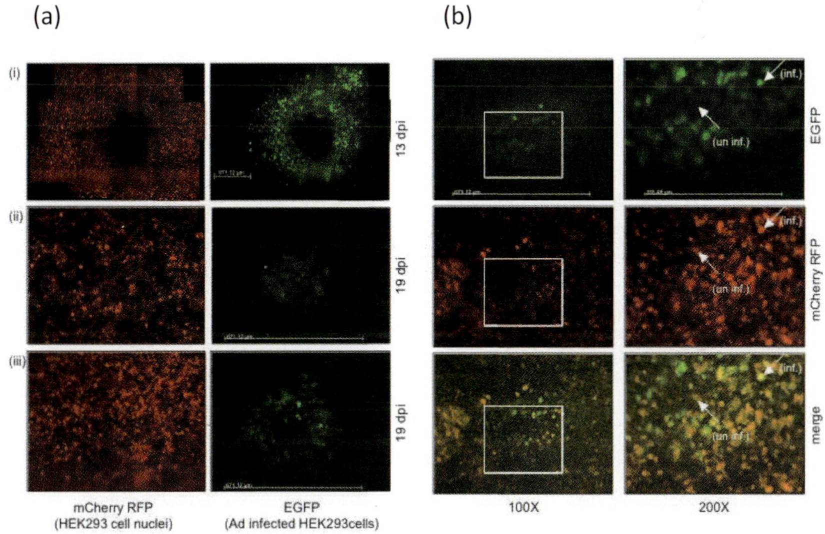

Fig. 21.8 Experimentally observed spatial patterns of virus infection, using the adenovirus AdEGFPuci on HEK293-mCherry cells [Wodarz *et al.* (2012)]. (a) Three patterns were observed: a hollow ring (top), a disperse pattern (middle) and a ring filled with uninfected cells that is eventually predicted to develop into a disperse pattern (bottom). Infected cells are shown in green on the right, and all cells (uninfected + infected) in red on the left. (b) Closer examination of the disperse, or limited, growth pattern, magnified to different degrees (100x and 200x). The top panels show infected cells in green, the middle panels show all cells in red. Because the mCherry red fluorescent protein cannot distinguish between infected and uninfected cells, the red and the green images are merged in the bottom panel, illustrating infected vs. uninfected cells. The arrows in the right panels point to an infected cell (inf.) and an uninfected cell (un inf.) within the center of the virus infected region of the cells.

21.4 Experimentally observed patterns of virus spread

In order to experimentally examine spatial virus spread in a setting corresponding to the assumptions of our model, a recombinant adenovirus type-5 (Ad5) was constructed that expresses enhanced jellyfish green fluorescent protein (EGFP), AdEGFPuci, and grows on human 293 embryonic kidney epithelial (293) cells [Hofacre *et al.* (2012)]. The experiment was set up such that cells are arranged in a two-dimensional layer, and virus spread is most likely to occur to neighboring cells. An agar overlay prevents long-range spread of the virus away from infected cells in the culture medium. This set-up allows to quantify not only the number of infected cells over time, but also the spatial patterns of infected cells that are formed as the virus population expands [Wodarz *et al.* (2012)]. In addition, fluorescent markers were used to visualize the spatial distribution of all cells (infected and uninfected) by generating HEK293-H2BmCherry cells, that stably express the core nuclear histone protein H2B fused to mCherry (a highly photostable, monomeric red fluorescent protein (RFP)) [Shaner *et al.* (2004)]. Thus, using HEK293-H2BmCherry cells allows visualization of all the cell nuclei (i.e., intact cells) in any particular culture. The culture was infected at a very low multiplicity of infection (MOI), such that any area of infection resulted from a single "founder" infected cell. Each culture contained several

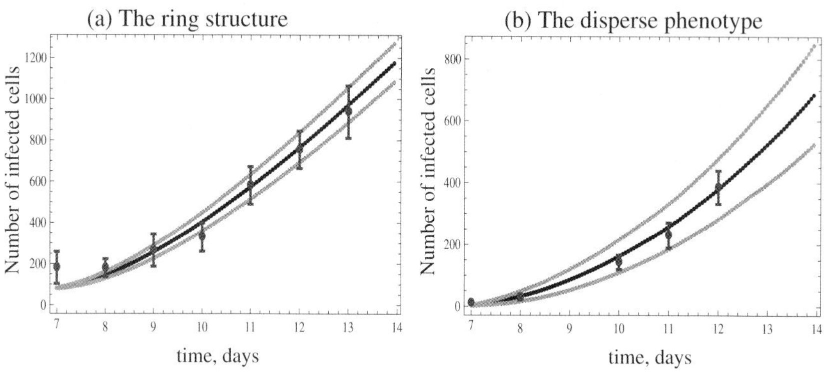

Fig. 21.9 Two different experimentally observed time series of adenovirus infection, fitted with the model. The black middle line through the data represents the time series predicted by the agent-based model. The upper and lower lines show the standard deviations. (a) A ring structure (figure 21.8(a)). (b) A disperse pattern (figure 21.8(b)). The best-fitting values for parameters R, B, and A are given as follows: (a) $R = 0.18$, $B = 0.26$, $A = 1.85 \cdot 10^{-2}$. (b) $R = 0.19$, $B = 0.52$, $A = 0.12$. The parameter D was kept constant at $D = 0$.

such founder cells that were sufficiently separated from each other, to allow tracking multiple growth foci across the dish. The earliest stages of virus growth starting from a single founder infected cell were described in [Hofacre *et al.* (2012)]. In [Wodarz *et al.* (2012)], we followed the growth of such spreading infections and characterized the consequent growth patterns. We observed three basic patterns of virus spread, which correspond to the patterns predicted by the model. The experimentally observed patterns are shown in figure 21.8(a) and described as follows. (i) In the first pattern, the virus infection spreads rapidly outwards as a ring, leaving no cells behind in the core of the ring (figure 21.8(a), pattern (i)). This classic plaque pattern is observed in virus growth experiments, and it corresponds to the hollow ring structure predicted by mathematical modeling above. In the second and third patters there is viral spread, but it is limited. (ii) In the second case, a disperse growth pattern is observed, where the virus population expands as a mixed cluster of infected and uninfected cells (figure 21.8(a), pattern (ii)). Finally, the virus population expands as a thinner ring, but

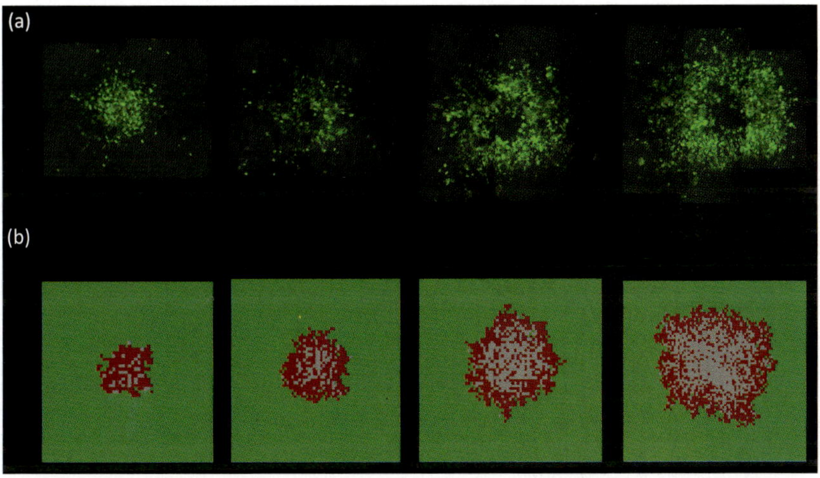

Fig. 21.10 Observed (a) and predicted (b) spatial pattern of adenovirus (AdEGFPuci) growth for the experiment that exhibits a ring structure (time series given in figure 3a). The predicted spatial pattern is the result of an individual run of the agent-based model with the parameter combination obtained from the model fitting procedure. Snapshots in time are shown, representing days 7, 9, 11, and 13 post infection. (a) The area of green fluorescence is shown, expressed by the infected cells, thus documenting the spatial spread of the virus through the population target cells arranged in a two-dimensional setting. (b) In the computer simulation, green indicates infected cells, red infected cells, and grey empty spots.

in contrast to the first case, uninfected cells are left behind in the core of the ring (figure 21.8(a), pattern (iii)). A limited growth pattern is magnified in figure 21.8(b), in which uninfected cells are visible within the center of the virus infected population. In the top right panel of figure 21.8(b), an AdEGFPuci infected (fluorescent) cell is indicated (arrow marked with "inf."), whereas an uninfected cell in the center of the spreading infection does not fluoresce green (arrow, un inf.). The same cells are indicated in the middle right panel of figure 21.8(b), showing red fluorescence. In the bottom left panel of figure 21.8(b), images of the top and middle panels are merged; infected cell (arrow, "inf.") fluoresces yellow, while the uninfected cell, (arrow, "un inf.") remains red. As mentioned, the area over which the infection spread, remained limited in patterns (ii) and (iii) and persisted throughout the infection (through 19 dpi). In contrast, in pattern (i), the ring of infected cells continued to spread outward as long as there was space; cell clearing in the center of the plaque was apparent at 13 dpi, as shown in figure 21.8(a). Similar patterns of spreading infection were also seen in

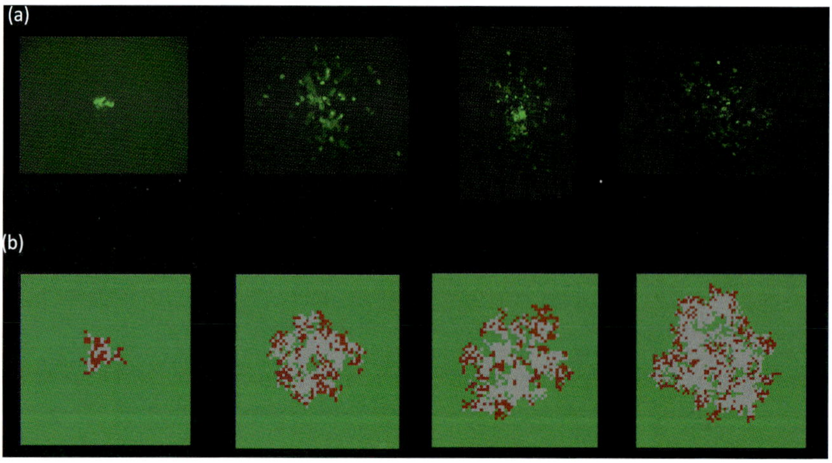

Fig. 21.11 Observed (a) and predicted (b) spatial pattern of adenovirus (AdEGFPuci) growth for the experiment that exhibits a disperse growth pattern (time series given in figure 3b). The predicted spatial pattern is the result of an individual run of the agent-based model with the parameter combination obtained from the model fitting procedure. Snapshots in time are shown, representing days 7, 10, 11, and 12 post infection. (a) The area of green fluorescence is shown, expressed by the infected cells, thus documenting the spatial spread of the virus through the population target cells arranged in a two dimensional setting. (b) In the computer simulation, green indicates infected cells, red infected cells, and grey empty spots.

Ad293 cells, a HEK293 cell derivative optimized for adenovirus plaque assays. Overall, among 436 scored growth foci, the hollow ring structure was found in 45%, and the limited patterns in 55% of cases.

In order to go beyond the qualitative comparison of model and data, we fit the model to two sets of experimental data, one showing an expanding hollow ring, and the other the disperse growth pattern, see figure 21.9. The number of cells was experimentally determined by measuring the fluorescent area of the infected cell population, divided by the fluorescent area of individual infected cells, using Photoshop. Each graph in figure 21.9 is based on a single experimental run. The area however, was measured independently four times, giving rise to the plotted error bars. The black middle line through the data represents the time series predicted by the agent-based model, using a parameter combination that was obtained by a least squares fitting procedure. Since the model is stochastic, the predicted time series represents the average over 1000 instances of the simulation. The upper and lower lines show the standard deviations added to and subtracted from the average. The experiment fitted in figure 21.9(a) shows virus growth characterized by the formation of a ring structure. Consequently there is a relatively short phase of quadratic growth, followed by a transition to linear growth. The experiment fitted in 21.9(b) shows disperse virus growth characterized by quadratic growth throughout time. The corresponding observed and predicted spatial patterns are shown in figures 21.10 and 21.11. The types of spatial patterns that emerged matched the observed ones qualitatively (figures 21.10 and 21.11). Note that although this procedure found best fitting parameter values, their biological meaning remains questionable, since different parameter combinations can give rise to similarly good fits.

While the experimentally observed spatial patterns correspond to predicted ones, the experiments give rise to further observations that are not seen in the model and that are likely due to additional biological process that are at work in this *in vitro* system and that are not part of the model. The most puzzling observation was that identical experimental conditions, using the same virus-target cell system, gave rise to different patterns of virus growth. This indicates the existence of so far unidentified factors that influence virus spread in this *in vitro* system. It is possible that initial events, stochastic in nature, might determine the remaining fate of the virus population. One hypothesis is that infection of cells triggers the production of anti-viral factors such as interferon, by the infected cell, which could induce an anti-viral state in neighboring cells.

21.5 Conclusions

This review summarized various modeling approaches that we have employed to study the dynamics of oncolytic viruses. Ordinary differential equations have been heavily employed in the field of virus dynamics in general, and have given rise to many biologically useful results [Nowak and May (2000)]. The same applies to oncolytic virus dynamics, especially with respect to some very fundamental concepts and predictions. At the same time, this modeling approach has weaknesses that limit the predictive power of such models in more advanced settings where precise treatment regimes need to be worked out in the context of specific viruses and cancers. One of the problems with the ODE approach is that equations contain arbitrary mathematical expressions that could be equally well described by alternative terms that change the properties of the model. In most cases, the "biologically correct" formulation is not known. Thus, a general, axiomatic modeling approach can be useful, where analysis does not focus on specific models, but on classes of models characterized by biological constraints. Another important limitation of the ODE approach is that spatial aspects of tumors cannot be accounted for explicitly. To do so, alternative modeling techniques need to be utilized, and we have described a spatially explicit agent-based model in this respect. In all modeling approaches, it is important to realize that the biological complexity of the system is enormous and largely not well understood, which again limits the predictive power of models. A reasonable strategy described here is to first focus on relatively simple settings, such as in vitro systems, in which most assumptions are relatively well understood. This allows the model to be tested and validated by specific experiments [Wodarz *et al.* (2012)]. Once the simplest setting can be described adequately with a model, additional biological complexities can be incorporated one step at a time. The ultimate goal of this work is to generate a biologically realistic model for the treatment of tumors in vivo, which has solid predictive power and can be used to optimize treatment strategies.

Problems

Problem 21.1. *Take the basic ordinary differential equation model for oncolytic virus therapy and modify it to include simultaneous treatment with a chemotherapeutic agent that either slows down the division rate of tumor* *cells or increases the death rate of tumor cells (both uninfected and infected). Does it make sense to combine oncolytic virus therapy with chemotherapies? Explore with a combination of analytical techniques and numerical simulations.*

Problem 21.2. Research project. *Find out about different types of oncolytic viruses, what types of clinical trials have been performed, and what outcomes have been observed in these trials.*

Problem 21.3. Numerical project. *In the basic ODE model of oncolytic virus replication, take the parameter sets explored in figure 21.2 and numerically explore the dynamics as the carrying capacity of the tumor cell population, ω, is increased to higher values. What happens in the extreme case of unbounded, exponential virus growth? What do you conclude about the model structure? How do the dynamics depend on slight modifications of the model, such as the assumption of an infection term that saturates in the number of uninfected cells?*

Chapter 22

Immune responses, tumor growth, and therapy

The immune system defends organisms from intruders such as pathogens which would otherwise kill them. It does so by specifically recognizing proteins derived from the pathogens (for example, viruses, bacteria, or parasites). Through complicated mechanisms which will be discussed briefly later on, the immune system knows that these proteins are foreign and that they are not derived from the organism that it is supposed to protect. What about cancer? As discussed throughout this book, carcinogenesis involves the accumulation of multiple mutations and in general often exhibits genetic instability. This means that many mutated proteins are produced which are different from the organism's own proteins and should thus appear foreign. In principle, these should be visible to the immune system which could potentially remove tumor cells and prevent the development of cancer.

A role of the immune system in the fight against cancer was first suggested in 1909 by Paul Ehrlich [Ehrlich (1909)]. It was not, however, until the 1950s, when the idea was pursued more vigorously and the immune surveillance hypothesis was formulated by Burnet [Burnet (1957)]. It stated that while cancers continuously arise, they are eliminated by specific immune responses. The successful establishment of cancer was thought to come about by the occasional escape of cancer cells from the immune responses.

Following a lot of enthusiastic research, clinical and experimental data cast doubt on the immune surveillance hypothesis [Dunn *et al.* (2002)], and indeed this concept has been subject to much controversy over the years and decades. Around the early 1990s, the immune surveillance hypothesis was largely ignored and immune-mediated extinction of tumor cells was not considered relevant [Vesely *et al.* (2011)]. This notion, however, started to change by the mid 1990s when additional research established more clearly the role of immune responses (both innate and adaptive) for protecting the organism against cancer. This was solidified significantly by work performed over the last decade, which demonstrated that the immune system has the ability to protect mice from the growth of different types of tumors [Dighe *et al.* (1994); Kaplan *et al.* (1998); Shankaran *et al.* (2001); Smyth *et al.* (2000b,a); Street *et al.* (2002); Girardi *et al.* (2001)]. A compelling piece of evidence for the occurrence of immune surveillance are paraneoplastic diseases, which are neurological disorders that arise from immune responses directed against cancers [Darnell (1996); Albert *et al.* (1998a)]. Interestingly, it was shown that tumors that emerged in the absence of a functional immune system were more immunogenic (i.e., generated more efficient anti-tumor responses) than tumors that arose in immunocompetent organisms. This means that besides protecting the organism, the presence of the immune system can also shape the nature of an emerging tumor, i.e. select for cells that are less susceptible to immune-mediated attack. This has been called immunoediting [Vesely *et al.* (2011)].

In general, a picture has emerged according to which the interactions between the immune system and newly generated tumor cells can lead to three qualitatively different outcomes [Vesely *et al.* (2011)]. (i) The immune system can eliminate the founding tumor cell population, thus preventing the development of disease. (ii) The dynamics between the founding tumor cells and immune responses can converge towards an equilibrium at which the tumor cell population remains present at sub-clinical levels, controlled by the immune response. (iii) The tumor can escape the immune responses, thus giving rise to clinical disease.

The tumor-fighting ability of immune responses has also been explored from a therapeutic perspective [Palucka and Banchereau (2012)]. In this respect, the induction of adaptive immunity, especially T cells, seems crucial. Different approaches have been taken. On the one hand, immune cells from a patient can be isolated, stimulated in vitro, and re-infused into the patient. On the other hand, tumor proteins (antigens), together with immuno-stimulatory factors (adjuvant) can be injected into patients with the aim to stimulate tumor-specific immune responses. In these respects, dendritic cells are crucial for the induction of tumor specific immunity. The overall goal of many immunotherapeutic approaches is to establish a solid immune response against the tumor that generates "immunological memory", i.e., the persistence of the immune cells at elevated levels following tumor control/clearance, conveying long term protection against relapse of the disease. While these approaches are still under development, promising outcomes have been observed in clinical trials in the context of several cancers [Palucka and Banchereau (2012)].

This chapter has two parts. The first part considers a mathematical model describing basic interactions between a growing tumor cell population and a tumor-specific immune response. This model will be used to examine the states of clearance, controlled persistence, and uncontrolled tumor growth. The second part of the chapter considers applications in the context of therapies, especially combining the use of small molecule inhibitors with immuno-therapeutic approaches. Before diving into these topics, however we will briefly review some basic facts about immune responses, which forms the necessary background.

22.1 Some facts about immune responses

This section will briefly review some basic immunological principles which form the basis for the rest of the chapter. More extensive descriptions of the immune system can be found in any standard immunology textbook, for example [Janeway *et al.* (2005)]. We can distinguish between two basic types of immune responses. *Innate immune responses* provide a first line of defense. They do not recognize foreign proteins specifically. They provide environments which generally inhibit the spread of intruders. While they may be important to limit the initial growth of a pathogen, they are usually not sufficient to resolve diseases. On the other hand, *adaptive immune responses* can specifically recognize foreign proteins, and they tend to be crucial for resolving diseases. We concentrate on this type of response here, although it is important to note that a complex interplay occurs between the innate and adaptive branches of the immune system, which enables the adaptive response to effectively carry out its functions. Adaptive immunity includes antibody and T cell responses. The T cell responses can in turn be subdivided into helper T cells and killer T cells. Helper T cells largely have regulatory functions, contributing to the activation of the effector responses that directly fight infectious agents: antibodies and killer T cells. Roughly speaking, antibodies recognize proteins outside the cells, such as free virus particles, extracellular bacteria or parasites, although they can also attach to proteins found on the cell surface. Killer T cells recognize foreign proteins which are displayed on cells. For example, viruses replicate inside cells. During this process, the cell captures some viral proteins and displays them on the cell surface. When the killer T cells recognize the foreign proteins on the cell surface, they release substances which kill. Mutated cancer proteins are displayed on the surface of cancer cells. Therefore, killer T cells are an important branch of the immune system in the fight against cancer. The rest of this chapter will discuss only the role of killer T cell responses. The scientific term is *cytotoxic T lymphocyte*, abbreviated as *CTL*. They can also be referred to as CD8+ T cells because they are characterized by the expression of the CD8 molecule on the cell surface.

The CTL are able to recognize the cell which displays a foreign protein in the following way. When proteins inside the cell are captured for display on the cell surface, they are presented in conjunction with so–called *major histocompatibility complex* (MHC) molecules. The MHC genes are highly variable, and different MHC genotypes present different proteins. This accounts for the variability between different people in immune responses against the same pathogen. The CTL carries the so–called T cell receptor or TCR. The TCR recognizes the protein-MHC complexes. This triggers the release of specific molecules such as perforin or FAS, which induce apoptosis in the cell that displays the foreign proteins. In immunology, the foreign protein which is recognized by the immune cells is also referred to as *antigen*. It is important to note that all proteins of a cell are processed and displayed in this way, not just the ones which are supposed to be recognized by the immune system. However, self-reactive immune cells are normally deleted, leaving only immune cells that are specific to invaders or to cells that bear altered proteins, such as products of mutated genes in cancer cells.

In order to induce CTL killing activity, however, they first need to be activated. Before CTL that are specific to a certain antigen have been exposed to that antigen, they exists at relatively low numbers and are not able to perform effector function, e.g. to kill target cells. They are said to be in a naive state. Contact with the antigen activates the CTL and leads to a phase of clonal expansion or proliferation, which significantly increases the size of this specific CTL population. This process also leads to the generation of effector cells, which are able to kill their target cells. Following this effector phase, a portion of the CTL die, while another portion differentiates into memory CTL that can persist at elevated levels in the absence of their specific antigen for relatively long periods of time. They have the capacity to protect against a secondary occurrence of the disease.

Activation and proliferation of CTL is not induced by contact with the target cells. Instead, it is induced by contact with so-called professional antigen-presenting cells or APCs, most notably dendritic cells. Dendritic cells internalize free antigens and display them on their cell surface. Contact between an antigen-presenting dendritic cell and a CTL leads to the activation and proliferation of the CTL. These interactions also involve the T helper cells or CD4+ T cells, which do not have significant effector activity, but are crucial in the generation of CTL and antibody responses. Because CTL are stimulated not directly by the target cells, but indirectly via dendritic cells, this process of stimulation is also called cross-stimulation or cross-priming [Heath and Carbone (2001); Carbone *et al.* (1998)]. These concepts are explained in more detail schematically in figure 22.1.

In the context of tumor-specific immune responses, an important question is how the tumor cell population can escape CTL-mediated elimination. Various processes have been proposed [Rabinovich *et al.* (2007)] It is possible that the tumor cells evade CTL-mediated activity. In this case, the CTL response is present but ineffective. Other, more direct modes of immune impairment are also possible, leading to the disappearance of the appropriate CTL response, a state that can be called tolerance towards the tumor. Understanding the dynamical interactions between a CTL response that fights tumor cells and a tumor cell population that impairs the CTL response is central to elucidating the conditions that lead to tumor elimination, controlled tumor persistence, and uncontrolled tumor growth. In the following section, this will be investigated in the context of a mathematical model.

22.2 The model

Here, we construct a mathematical model that describes the dynamics between a growing tumor cell population and a tumor-specific CTL response [Wodarz and Jansen (2003)]. This model is related to other models that describe the relationship between tumors and the immune system, e.g.

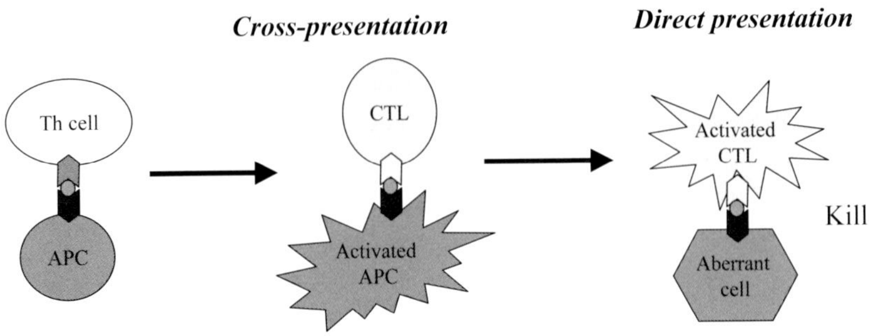

Fig. 22.1 Schematic representation of the concept of cross-priming, which is central to this chapter. So-called "antigen presenting cells" can take up antigen (proteins derived from pathogens or cells) and display them on their surface. Before the APCs can function, they need to be activated. This is achieved by so-called helper T cells (Th) which can recognize the antigen on the APC. The activated APC can subsequently interact with CTL. CTL can also specifically recognize the antigen on the APC. This interaction activates the CTL which can then turn into effector cells and kill the troubled target cells which display the antigen. These target cells are different from APCs and can be for example virus-infected cells or tumor cells. This process is called cross-priming because the CTL do not get activated directly by the troubled cells which need to be killed, but indirectly by the APCs which can take up and display the antigen.

[De Pillis and Radunskaya (2001); DePillis *et al.* (2013); Ledzewicz *et al.* (2011)]. It takes into account the cross-priming pathway described above and further assumes that while the CTL fight the tumor cells, the tumor can fight back by impairing the CTL response. The model contains four variables: cells directly displaying antigen such as tumor cells, T (which we will also refer to as "target cells"); non-activated APCs which do not present the antigen, A; loaded and activated APCs which have taken up antigen and display it, A^*; CTL, C. The model is given by the following system of differential equations which describe the development of these populations over time,

$$
\begin{aligned}
\dot{T} &= rT\left(1-\frac{T}{k}\right) - dT - \gamma TC,\\
\dot{A} &= \lambda - \delta_1 A - \alpha AT,\\
\dot{A}^* &= \alpha AT - \delta_2 A^*,\\
\dot{C} &= \frac{\eta A^* C}{\epsilon C + 1} - qTC - \mu C.
\end{aligned}
\tag{22.1}
$$

The tumor cells grow at a density dependent rate $rT(1-T/k)$. The parameter k denotes the maximum size the tumor can achieve, limited for example by spatial constraints. The cells die at a rate dT, and are in addition killed by CTL at a rate γTC. APCs, A, are produced at a constant rate λ and die at a rate $\delta_1 A$. They take up antigen and become activated at a rate αAT. The parameter α summarizes several processes: the rate at which antigen is released from the cells, T, and the rate at which this antigen is taken up by APCs and processed for display and cross-presentation. Loaded APCs, A^*, are lost at a rate $\delta_2 A^*$. This corresponds either to death of the loaded APC, or to loss of the antigen-MHC complexes on the APC. Upon cross-presentation, CTL expand at a rate $\eta A^* C/(\epsilon C + 1)$. The saturation term, $\epsilon C + 1$, has been included to prevent excessive expansion of CTL in the presence of strong cross-stimulation, which can lead to unstable dynamics. The activated and expanding population of CTL can kill the infected cells upon direct presentation. In addition, it is assumed that direct presentation can result in removal of CTL at a rate qTC. This can be brought about, for example, by antigen-induced cell death, or over-differentiation into effectors followed by death. Finally, CTL die at a rate μC.

Thus a central assumption of the model is that cross-presentation can induce CTL expansion, while direct presentation does not have that effect; instead it can result in the decline of the CTL population. This assumption implies that the magnitude of cross-presentation relative to direct presentation could be a decisive factor which determines the outcome of a CTL response: activation or tolerance. In the model, the ratio of cross-presentation to direct presentation is given by cA^*/qT.

We assume that $r > a$. That is, the rate of increase of the target cells, T, is greater than their death rate. This ensures that this population of cells can grow and remain present. If this is fulfilled, the system can converge to a number of different equilibria (figure 22.2). The expressions for the equilibria will not be written out here since most of them involve complicated expressions.

(1) The CTL response fails to expand, i.e., $C = 0$. The population of target cells grows to a high equilibrium level, unchecked by the CTL. The populations of unloaded and loaded APCs, A and A^*, also equilibrate.
(2) The CTL response expands, i.e., C > 0. In this case, the system can converge to one of two different outcomes. (*a*) The number of CTL is low and the number of target cells is high. This outcome is qualitatively

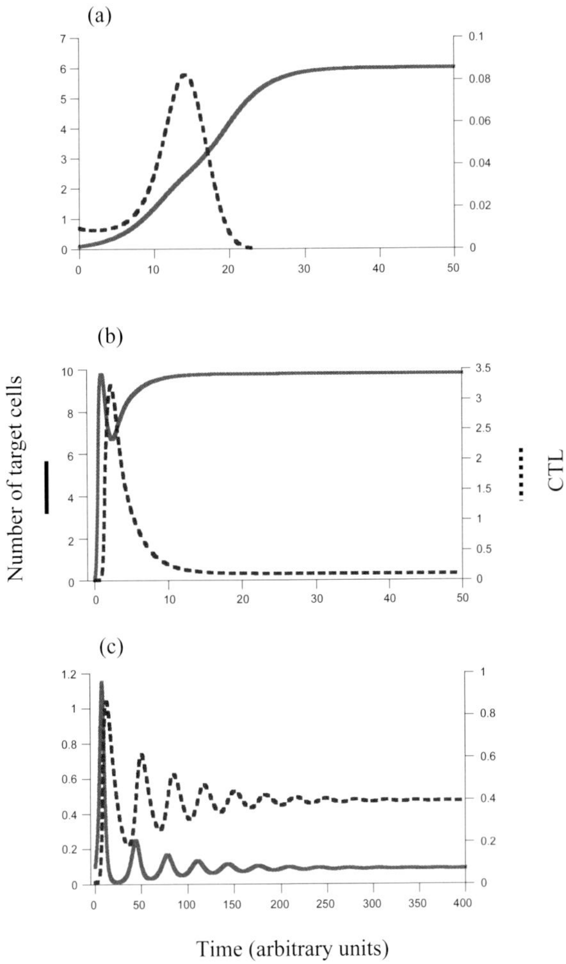

Fig. 22.2 Different outcomes of the model shown as time series. (a) Tolerance; CTL go extinct. (b) Tolerance outcome where CTL do not go extinct but are maintained at very low levels. (c) Immunity outcome. Parameters were chosen as follows: $r = 0.5$, $k = 10$, $d = 0.1$, $\gamma = 1$, $\lambda = 1$, $\delta_1 = 0.1$, $\delta_2 = 1.5$, $\eta = 2$, $\epsilon = 1$, $q = 0.5$, $\mu = 0.1$. $\alpha = 0$ for (a) and $\alpha = 0.2$ for (c). For (b) $\alpha = 0.05$, $r = 10$, $\eta = 10$.

similar to (1), because the CTL population does not fully expand, and the population of target remains high. (*b*) The number of CTL is high and the number of target cells is low. This can be considered the immune control equilibrium. If the population of target cells is reduced to very low levels, this can be considered equivalent to extinction (number of cells below one).

These equilibria therefore fall into two basic categories: (*a*) Tolerance; this is described by two equilibria. Either the immune response goes extinct, or it exists at low and ineffective levels. However, the latter equilibrium only occurs in a restrictive and biologically unrealistic parameter regime [Wodarz and Jansen (2003)] and will not be considered further here. For further details regarding this equilibrium, see [Wodarz and Jansen (2003)]. Therefore, when quoting the tolerance equilibrium, we from now on refer to the outcome where the CTL response goes extinct. (*b*) Reactivity; this is described by only one equilibrium. The immune response expands to higher levels and exerts significant levels of effector activity. The following sections will examine which outcomes are achieved under which circumstances.

22.3 Properties of equilibria and parameter dependencies

The two most important parameters in the present context are the rate of antigen uptake by APCs, α, and the growth rate of the target cells, r. This is because variation in these parameters can significantly influence the ratio of cross-presentation to direct presentation which is the subject of investigation. Hence, in the following sections we will examine the behavior of the model in dependence of these two parameters.

The rate of antigen uptake by APCs. The rate of antigen uptake by APCs comprises two processes: (i) the degree to which the antigen is made available for uptake; this can be determined for example by the amount of antigen released from the target cell, or the amount of apoptosis going on [Albert *et al.* (1998b)]. (ii) The rate at which the APCs take up the available antigen and process it for presentation. As the rate of antigen uptake by APCs, α, decreases, the ratio of cross-presentation to direct presentation decreases (figure 22.3(a)). When the value of α is high, the outcome is immunity. If the value of α is decreased and crosses a threshold, we enter a region of bistability (figure 22.3(a)): both the immunity and the tolerance equilibria are stable. Which outcome is achieved depends on the initial conditions. If the value of α is further decreased and crosses another threshold, the immune control equilibrium loses stability. The only stable outcome is tolerance (figure 22.3(a)).

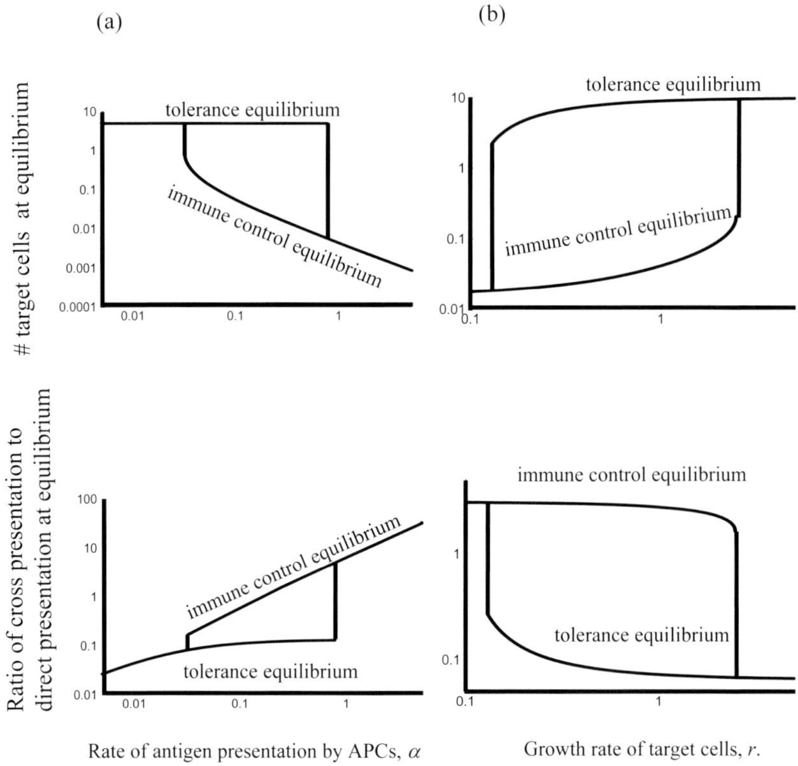

Fig. 22.3 Bifurcation diagram showing the outcome of the model as a function of (a) the rate of antigen presentation by APCs, α, and (b) the growth rate of target cells, r. Tumor cell load and the ratio of cross-presentation to direct presentation at equilibrium are shown. Parameters were chosen as follows: $r = 0.5$, $k = 10$, $d = 0.1$, $\gamma = 1$, $\lambda = 1$, $\delta_1 = 0.1$, $\alpha = 0.5$, $\delta_2 = 1.5$, $\eta = 2$, $\epsilon = 1$, $q = 0.5$, $\mu = 0.1$.

In the region of bistability, the dependence on initial conditions is as follows. Convergence to the immune control equilibrium is promoted by low initial numbers of target cells, high initial numbers of presenting APCs, and high initial numbers of CTL. This is because under these initial conditions, the dynamics start out with a high ratio of cross-presentation to direct presentation and this promotes the expansion of the CTL. On the other hand, if the initial number of target cells is high and the initial number of presenting APCs and CTL is low, then the initial ratio of cross-presentation to direct presentation is low and this promotes tolerance. There are some slight variations to this general picture. As they do not alter the basic results, however, the reader is referred to [Wodarz and Jansen (2001)] for details.

In summary, as the rate of antigen uptake by APCs is decreased, the ratio of cross-presentation to direct presentation decreases, and this shifts the dynamics of the CTL response in the direction of tolerance. This can include a parameter region in which both the tolerance and the immunity outcome are stable, depending on the initial conditions. If the CTL responsiveness to cross-presentation is very strong, tolerance becomes an unlikely outcome.

The growth rate of target cells. An increase in the growth rate of target cells, r, results in a decrease in the ratio of cross-presentation to direct presentation in the model. Hence an increase in the growth rate of target cells shifts the dynamics of the CTL from a responsive state towards tolerance. The dependence of the dynamics on the parameter r is shown in figure 22.3(b). The growth rate of target cells needs to lie above a threshold to enable the CTL to potentially react. This is because for very low values of r, the number of target cells is very low, not sufficient to trigger immunity. If the growth rate of target cells is sufficiently high to potentially induce immunity, we observe the following behavior (figure 22.3(b)). If the value of r lies below a threshold, the only stable outcome is immunity. If the value of r is increased and crosses a threshold, we enter a region of bistability. That is, both the immunity and the tolerance outcomes are possible, depending on the initial conditions. The dependence on the initial conditions is the same as explained in the last section. If the value of r is further increased and crosses another threshold, the immunity equilibrium loses stability and the only possible outcome is tolerance. Again, there are some slight variations to this general picture. As they do not alter the basic results, however, the reader is referred to [Wodarz and Jansen (2001)] for details.

In summary, an increase in the growth rate of target cells has a similar effect as a decrease in the rate of antigen uptake by APCs: the ratio of cross-presentation to direct presentation becomes smaller, and the outcome of the dynamics is driven from immunity towards tolerance. Again, this includes a parameter region where both the immunity and tolerance outcomes are stable and where the outcome depends on the initial conditions. The higher the overall responsiveness of the CTL to cross-stimulation, the less likely it is that a high growth rate of target cells can induce tolerance.

22.4 Immunity versus tolerance

The models have investigated the topic of CTL regulation from a dynamical point of view. We showed that the immune system can switch between two states: tolerance and activation. Which state is reached need not depend on the presence or absence of signals, but on the relative magnitude of cross-presentation to direct presentation. This shows that regulation can be accomplished without signals but in response to a continuously varying parameter. This relies on the assumption that cross-presentation stimulates immunity, while direct presentation can lead to removal of CTL. Such effects have been documented to occur, e.g. through antigen-induced cells death [Baumann *et al.* (2002); Budd (2001); Hildeman *et al.* (2002)]. This mechanism can account for various phenomena of tolerance vs reactivity, which is outlined as follows.

For self antigen displayed on cells of the body, the ratio of cross-presentation to direct presentation is normally low. This is because these cells do not die at a high enough rate or release the antigen at a high enough rate for the amount of cross presentation to be strong. On the other hand, large amounts of this antigen can be available on the surface of the cells expressing them (direct presentation). In terms of our model, this situation can best be described by a low value of α. Hence, in our model, CTL responses are not predicted to become established against self antigens. Instead, the outcome is tolerance. In addition, the initial conditions favor tolerance in this scenario. When immune cells with specificity for self are created and try to react, the number of these immune cells is very low and the number of target cells (tissue) is relatively high. This promotes failure of the CTL response to expand and to become established. On the other hand, with infectious agents, free antigen is abundantly available for cross-presentation. For example, virus particles are released from infected cells, ready to be taken up by APCs for cross-presentation. Therefore, the immune responses react and become fully established. In contrast, tumors may fail to induce CTL responses because tumor antigens are largely displayed on the surface of the tumor cells, but relatively little tumor antigen is made available for uptake by dendritic cells and hence for cross-presentation.

22.5 Cancer initiation

According to the model analysis presented so far, when tumor cells are generated, they can have different fates, depending on the parameters that determine the interactions between the tumor and the immune system. These are summarized in figure 22.4. If the growth rate of the tumor cells is relatively low or if the rate of cross-presentation of tumor antigen is relatively high, e.g. due to relatively large amounts of cell death and apoptosis, then a strong CTL response can develop in the model that reduces the tumor cell population to very low levels that practically correspond to extinction. If the growth rate of the tumor cells is faster and less cross-presentation occurs due to reduced cell death, the CTL response becomes weaker and this can lead to controlled persistence of the tumor. Finally if the growth rate of the tumor is sufficiently fast, or the amount of cross presentation is sufficiently reduced due to loss of cell death, then a CTL response is not established and the tumor can grow unopposed, causing disease. These are the types of tumors that are expected to be seen in patients, selected by the inability of the immune system to become established and control the tumor cell population. Hence, this corresponds to the concept of immunoediting. The following section will take these considerations further from an evolutionary perspective.

22.6 Tumor dormancy, evolution, and progression

Here, we investigate in more detail the scenario where the growth rate of the tumor is intermediate, and both the tolerance and the CTL control outcomes are possible, depending on the initial conditions. Assume the CTL control equilibrium is attained because the initial tumor size is small at the time at which the disease is initiated. The number of tumor cells is kept at low levels, but the tumor is unlikely to be cleared because in this bistable parameter region the ratio of cross-presentation to direct presentation is already reduced. If the tumor persists at low levels, the cells can keep evolving over time. They can evolve, through selection and accumulation of mutations, either towards a higher growth rate, r, or towards a reduced rate of apoptosis which leads to reduced levels of antigen uptake by dendritic cells, α. Both cases result in similar evolutionary dynamics. This is illustrated in figure 22.5 assuming that the cancer cells evolve towards faster growth rates (higher values of r).

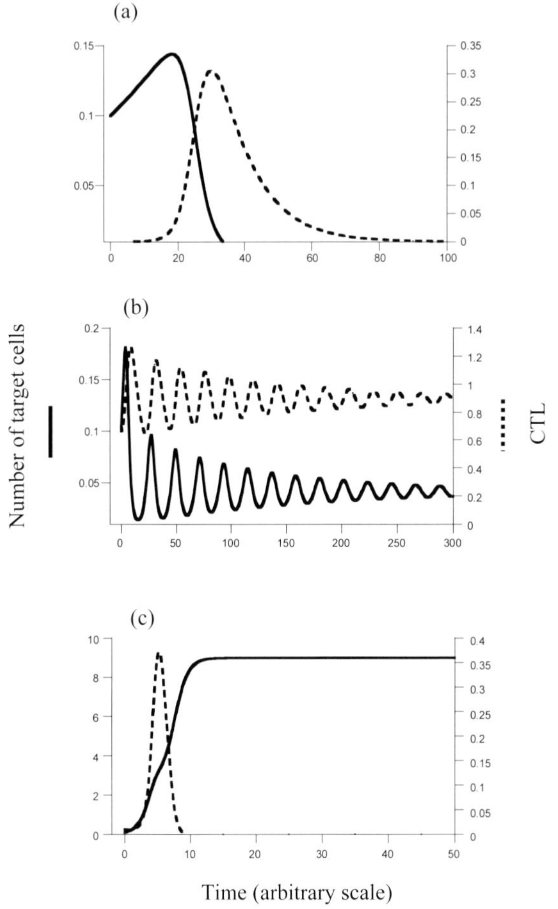

Fig. 22.4 Time series plots showing the different possible outcomes when a tumor is generated. (a) Reduction of the tumor cell population to very low levels, which in practical terms corresponds to clearance. (b) Immune control of the tumor at higher levels, practically corresponding to controlled persistence. (c) Tolerance and uncontrolled growth. Parameters were chosen as follows: $k = 10$, $d = 0.1$, $\gamma = 1$, $\lambda = 1$, $\delta_1 = 0.1$, $\alpha = 0.5$, $\delta_2 = 1.5$, $\eta = 2$, $\epsilon = 1$, $q = 0.5$, $\mu = 0.1$. (a) $r = 0.13$. (b, c) $r = 1$. The difference between graphs (b) and (c) lies in the initial number of CTL, z.

An increase in the growth rate of tumor cells does not lead to a significant increase in tumor load. At the same time, it results in an increase in the number of tumor-specific CTL. The reason is that a faster growth rate of tumor cells stimulates more CTL which counter this growth and keep

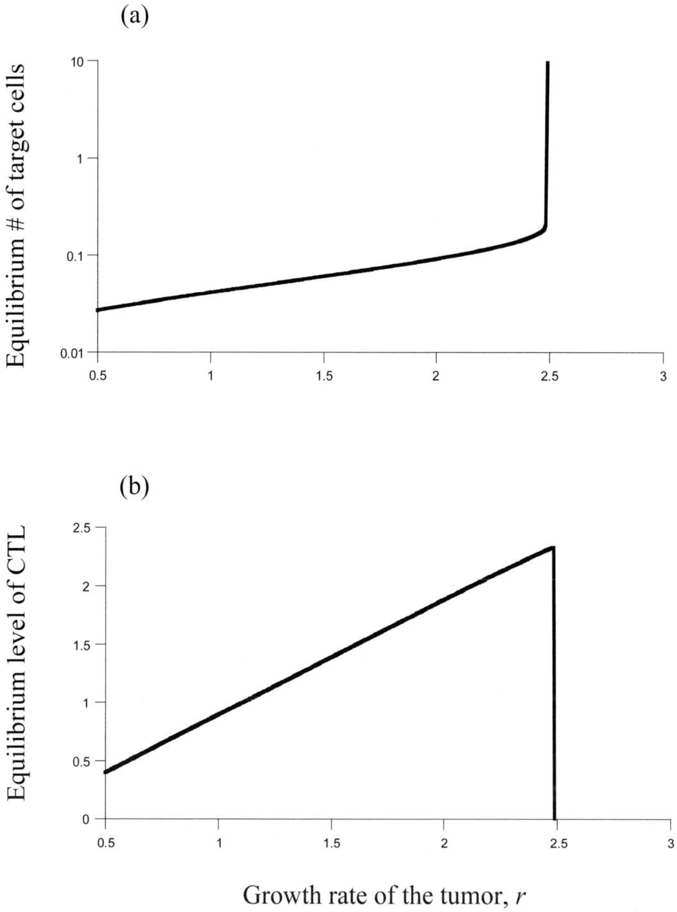

Fig. 22.5 Equilibrium tumor load (a) and the number of tumor specific CTL (b) as a function of the growth rate of tumor cells, r. This graph can be interpreted to show the effect of tumor evolution towards faster growth rates over time. As evolution increase the value of r over time, the tumor population and the CTL attain a new equilibrium. Parameters were chosen as follows: $r = 0.5$, $k = 10$, $d = 0.1$, $\gamma = 1$, $\lambda = 1$, $\delta_1 = 0.1$, $\alpha = 0.5$, $\delta_2 = 1.5$, $\eta = 2$, $\epsilon = 1$, $q = 0.5$, $\mu = 0.1$.

the number of tumor cells at low levels. When the growth rate of the tumor cells evolves beyond a threshold, the equilibrium describing CTL-mediated control of the cancer becomes unstable. Consequently, the CTL response collapses and the tumor can grow to high levels.

The dynamics of tumor growth and progression can include a phase called "dormancy". During this phase the tumor size remains steady at a low level over a prolonged period of time before breaking out of dormancy and progressing further. Several mechanisms could account for this phenomenon. The limitation of blood supply, or inhibition of angiogenesis, can prevent a tumor from growing above a certain size threshold [Folkman (2006)]. When angiogenic tumor cell lines evolve, the cancer can progress further. Other mechanisms that have been suggested to account for dormancy are immune mediated, although a precise nature of this regulation remains elusive [Uhr and Marches (2001); Page and Uhr (2005)]. As shown in this section, the model presented here can account for a dormancy phase in tumor progression. If the overall growth rate of the cancer cells evolves beyond this threshold, dormancy is broken: the CTL response collapses and the tumor progresses.

22.7 Immunotherapy against cancers

Assuming that the CTL response has failed and the cancer can grow unchecked, we investigate how immunotherapy can be used to restore CTL mediated control or to eradicate the tumor. In the context of the model, the aim of immunotherapy should be to increase the ratio of cross-presentation to direct presentation. The most straightforward way to do this is dendritic cell vaccination. In the model, this corresponds to an increase in the number of activated and presenting dendritic cells, A^*. We have to distinguish between two scenarios: (i) The tumor cells have evolved sufficiently so that the CTL control equilibrium is not stable anymore, and the only stable outcome is tolerance. (ii) The tumor has evolved and progressed less; the equilibrium describing CTL mediated control is still stable.

First we consider the situation where the tumor has progressed far enough so that the CTL control equilibrium is not stable anymore. Upon dendritic cell vaccination, tolerance is temporarily broken (figure 22.6). That is, the CTL expand and reduce the tumor cell population. This CTL expansion is, however, not sustained and tumor growth relapses (figure 22.6). The reason is as follows. Upon dendritic cell vaccination, the ratio of cross-presentation to direct presentation is increased sufficiently, enabling the CTL to expand. However, this boost in the level of cross-presentation subsequently declines, allowing the tumor to get the upper hand and regrow. The model suggests, however, that the tumor can be eradicated if the

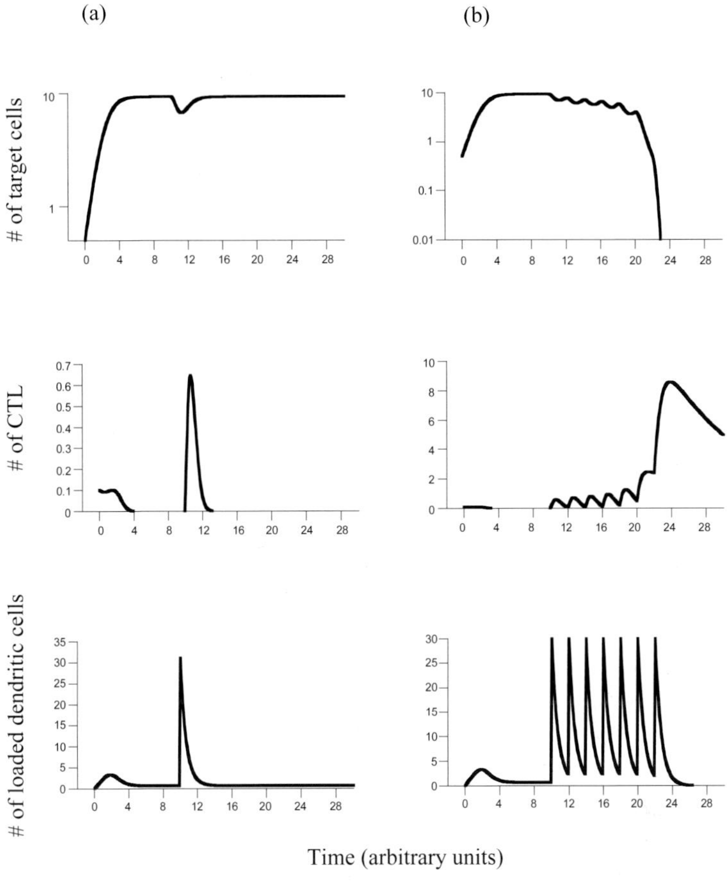

Fig. 22.6 Effect of dendritic cell vaccination on tumor dynamics assuming that the growth rate of the tumor has evolved to high values, where only the tolerance outcome is stable. (a) A single vaccination event induces a temporary reduction in tumor load, followed by a relapse. (b) Repeated vaccination events can drive the tumor load below a threshold which corresponds to extinction in practical terms. Parameters were chosen as follows: $r = 1.5$, $k = 10$, $d = 0.1$, $\gamma = 1$, $\lambda = 1$, $\delta_1 = 0.01$, $\alpha = 0.5$, $\delta_2 = 1.5$, $\eta = 0.5$, $\epsilon = 1$, $q = 0.5$, $\mu = 0.1$.

level of cross-presentation is continuously maintained at high levels. This can be achieved by repeated vaccination events (figure 22.6). The next vaccination event has to occur before the level of cross-presentation has significantly declined. This will drive tumor load below a threshold level which practically corresponds to extinction (figure 22.6).

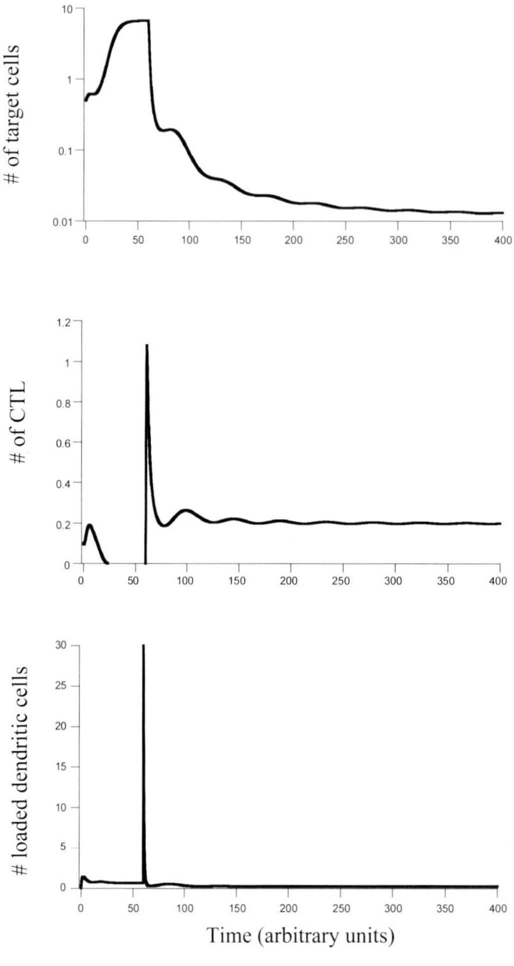

Fig. 22.7 Effect of dendritic cell vaccination on tumor dynamics assuming that the growth rate has not yet progressed beyond a threshold, so that we are in the bistable parameter region of the model. A single vaccination event can induce immunity which can control the tumor at low levels. Parameters were chosen as follows: $r = 0.3; k = 10; d = 0.1; \gamma = 1; \lambda = 1; \delta_1 = 0.01; \alpha = 0.5; \delta_2 = 1.5; \eta = 0.5; \epsilon = 1; q = 0.5; \mu = 0.1$.

Next, we consider the more benign scenario in which the tumor has not progressed that far and the CTL control equilibrium is still stable. Now a single vaccination event can shift the dynamics from the tolerance outcome to the CTL control outcome (figure 22.7). The reason is that an elevation in the number of presenting dendritic cells shifts the system into a space where the trajectories lead to CTL control and not to tolerance. This is likely to be achieved if the size of the tumor is not very large. The larger the size of the tumor, the stronger the vaccination has to be (higher A^*) in order to achieve success. If the tumor size is very large, then an elevated level of dendritic cells cannot shift the ratio of cross-presentation to direct presentation sufficiently to induce sustained immunity. A combination of vaccination and chemotherapy can, however, result in success. This is because chemotherapy reduces the size of the tumor and also induces death of tumor cells. Both factors contribute to a higher ratio of cross-presentation to direct presentation and to induction of immunity. Once a sustained CTL response has been induced, tumor cells are kept at low levels. However, the cancer is unlikely to be eradicated. Consequently, it can evolve over time. Thus, induction of CTL mediated control in the model is likely to result in a temporary phase of tumor dormancy. This phase is again broken after the overall growth rate of the tumor has evolved beyond the threshold at which the CTL control outcome becomes unstable.

These considerations result in the following suggestions. Dendritic cell vaccination should be administered repeatedly until the last tumor cell has been eradicated. If the tumor has already progressed relatively far, this is the only way to prevent immediate relapse of the cancer. If the tumor is less progressed, temporary tumor dormancy can be achieved by a single vaccination event. Tumor persistence and evolution will, however, break this dormancy phase, resulting in renewed cancer growth after a certain period of time. Thus, in this case, repeated vaccination is also advisable in order to keep the level of cross-presentation above a threshold and to avoid tumor persistence. In all cases, the model suggests that a combination of immunotherapy with conventional therapy is beneficial because conventional therapy can reduce the growth rate of the tumor. If conventional therapy increases the chances of developing immunological control of the tumor, conventional therapy would have to be applied only temporarily which would have significant clinical benefits.

22.8 Case study: immune responses and the treatment for chronic myeloid leukemia

So far, we have examined general mathematical models describing the interactions among tumor cells and CTL, and also explored general aspects of immunotherapy. In the second part of this chapter, we will discuss possible roles of immune responses in the context of a specific cancer, i.e., chronic myeloid leukemia (CML). In this cancer, tumor specific CTL have been found and could play an important role determining the outcome of treatment with small molecule inhibitors [Kim *et al.* (2008)]. In particular, they might contribute to the dynamics of tumor cell decline during therapy, which is explored in this chapter.

An introduction to CML biology and therapy has been given in Chapter 16. Average responses to treatment with imatinib display a biphasic decline of BCR-ABL transcript levels. Initially a first phase of decline is observed, followed by a second, slower phase of decline. However, there is considerable heterogeneity in individual responses among different patients, and in the rates of tumor cell decline. This can be seen in a variety of treatment responses published int he context of a German cohort study [Roeder *et al.* (2006)]. Consistent with the average picture, in many patients, the dynamics appear to begin with a relatively fast phase of exponential decline. In some patients, this decline continues for the duration of the study. In most patients, however, this fast phase of decline is followed by a slower phase of exponential decline. Finally, in many patients, BCR-ABL transcript numbers resurge subsequently. Examples of such patterns are seen in figure 22.8. In most patients, there are three important slopes if the dynamics of BCR-ABL numbers are documented on a log-scale: (1) The slope of the fast decline phase, (2) the slope of the slower decline phase, and (3) the slope of the eventual rise. The value of these slopes was quantified by fitting a three-phase exponential growth/decline model to the data, using non-linear least squares regression (figure 22.9(a)). A significant positive correlation was found between the slopes of the fast and the slow phases of decline. A significant negative correlation was found between the slopes of the slow phase of decline and the re-growth. Finally, no significant correlation was found between the slope of the fast phase of decline and the re-growth.

The negative correlation between the slopes of the slow phase of decline and the rise implies that both must probably be influenced by the presence of drug resistant mutants [Shah *et al.* (2004); Druker (2004)]. The rise of drug resistant mutants can slow down the decline of BCR-ABL transcript numbers, and is responsible for the eventual rise of BCR-ABL transcripts. The fact that the slopes of the slow and fast decline phases are positively correlated means that these two phases share a common mechanism of cell death that varies over time in strength. One such factor could be the

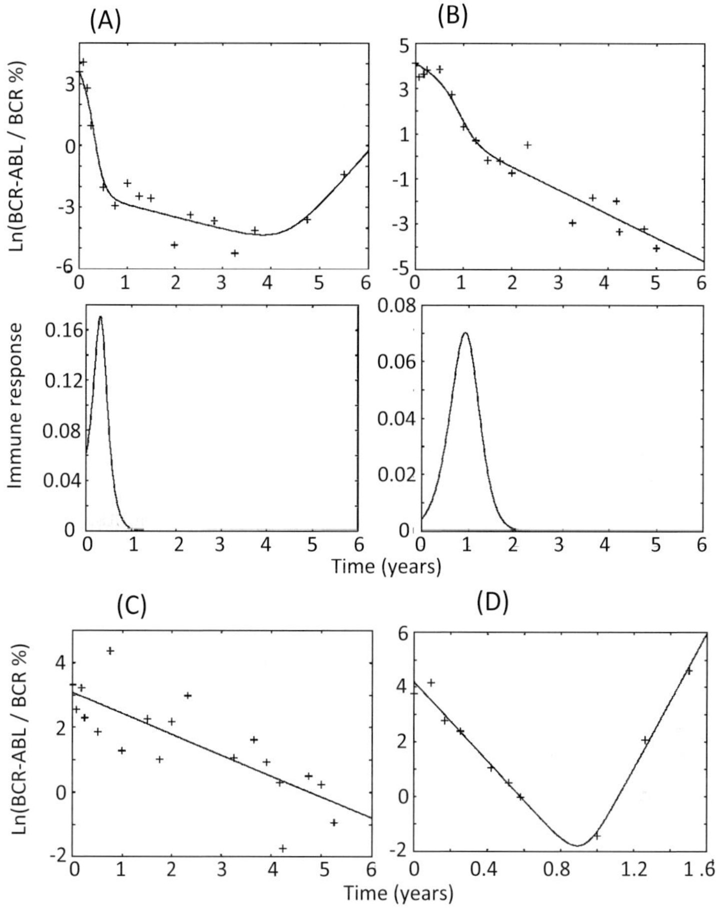

Fig. 22.8 Examples of individual treatment response data from the German cohort of the IRIS study, taken from [Roeder *et al.* (2006)]. The dots are the actual clinical data. The line is the model fit to the data, obtained by non-linear least squares regression. In (A) and (B) immune responses rise temporarily during therapy, contributing to the overall treatment dynamics. In (C) and (D), the model predicts that immune responses did not rise during therapy. Drug resistant mutants play a role in (A) and (D), but not in (B) and (C). Model parameter values for the model fits can be found in [Wodarz (2010)].

immune system, which has been suggested to play a role in CML therapy before [Kim *et al.* (2008); Chen *et al.* (2008); Wang *et al.* (2005)]. In particular, T cell responses (both CD4+ and CD8+) have been implicated.

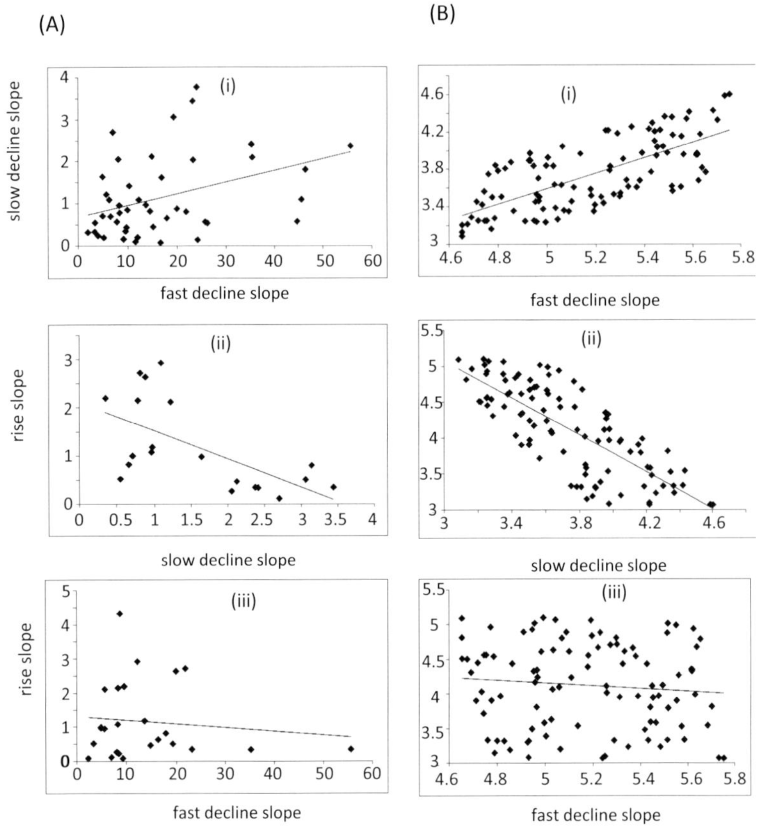

Fig. 22.9 Correlations between the slopes of the fast decline, slow decline, and the re-growth of BCR-ABL transcript numbers. (A) Correlations found in the clinical data from the German cohort of the IRIS study [Roeder *et al.* (2006)]. For (i) and (ii), the correlation is significant ($p = 0.009$ and $p = 0.003$, respectively), but not for (iii), $p = 0.59$. (B) Corresponding correlations as predicted by computer simulations. Simulations were run 100 times, varying randomly both the rate of specific immune proliferation, c, and the growth rate of the resistant cancer cells, r. For (i) and (ii), the correlations are significant ($p < 0.0001$ in each case), while the correlation is not significant for (iii), $p = 0.32$. $\delta = 1$, $p = 3$, $b = 0.5$, $q = 0.01$, $\epsilon = 1$, $\eta = 1$ (units are year^{-1}). $x_0 = 100$, $y_0 = 0.001$, $z_0 = 0.5$.

22.9 Role of immunity and resistance in driving treatment dynamics

Data indicate that the level of immune responses against CML is low before treatment, rises as treatment is administered, and declines again as the BCR-ABL transcript numbers fall to low levels [Kim *et al.* (2008); Chen *et al.* (2008); Wang *et al.* (2005)]. These are similar immune response dynamics as observed in human immunodeficiency virus (HIV) and hepatitis C virus (HCV) infected patients that receive anti-viral drug therapy [Barnes *et al.* (2002); Kalams *et al.* (1999); Ogg *et al.* (1999)], and can be explained in the same way. While the cancer cells impair immunity, treatment reduces the amount of impairment, leading to a rise of immunity. As the number of cancer cells declines, withdrawal of antigenic stimulation causes a drop in immune responses. Such impairment dynamics have been explored extensively with mathematical models in the context of viral infections [Komarova *et al.* (2003a); Wodarz (2001a); Wodarz and Nowak (1999)], and subsequently also in the context of CML [Kim *et al.* (2008)]. The following will use a basic mathematical model that captures these assumptions. Model fitting to individual patient data demonstrates consistency with clinical observations, and the model further predicts the correlations found in the data.

A mathematical model is adapted that was previously published in the context of viral infections [Komarova *et al.* (2003a)] and that captures the necessary assumptions. The model contains two variables: a growing virus population that has the ability to impair the immune system, and a specific immune response. In the present context, we are interested in CML and thus consider a population of growing CML cells that have the potential to impair immunity, and an immune response that reacts against CML antigens. The equations are the same and are given by [Wodarz (2010)]:

$$\dot{x} = Lx(1 - x/k) - dx - pxz, \tag{22.2}$$

$$\dot{z} = cxz/[(z + \eta)(x + \epsilon)] - qxz - bz. \tag{22.3}$$

The CML population is denoted by x, while the immune cell population is denoted by z. The CML population grows logistically. That is, at low numbers of cells, growth is exponential while growth slows down as the number of cells increases. The tumor cells die with a rate d and become removed by the immune response with a rate p. The immune response expands upon antigenic stimulation by tumor cells with a rate c. Immune expansion saturates if the number of tumor cells and the number of immune cells are high. The tumor cells impair immunity with a rate q. Finally, immune cells decay in the absence of antigenic stimulation with a rate b. The mathematical model is part of a general class of models that share common properties [Komarova *et al.* (2003a)]. Results obtained from the model are therefore not dependent on the particular mathematical formulation used, but are robust. The equations that describe the immune response make the very basic assumptions that cancer cells can both stimulate the specific immune cells to proliferate, and impair the response, similar to the assumptions made in the first part of this chapter. Because of the general nature of the model, this can be applied to any branch of the adaptive immune system including the T cell responses that are thought to play a role during CML therapy [Kim *et al.* (2008); Chen *et al.* (2008); Wang *et al.* (2005)]. The basic model is characterized by the following outcomes, assuming that the degree of stimulation of immune cells is strong enough to drive immune expansion: (i) An immune response is successfully established and the system converges to a stable equilibrium in which the cancer is controlled at relatively low levels. (ii) The immune response goes extinct, and the cancer cell population converges to an equilibrium characterized by a large number of tumor cells. If the degree of immune impairment is relatively low, only the cancer control outcome is stable. If the degree of immune impairment lies above a threshold, both outcomes are stable. To which outcome the system converges depends on the initial conditions, with a high initial number of immune cells and a low initial number of tumor cells promoting the cancer control outcome. Although this model is considerably simpler and less descriptive of specific biological events that drive the development of immune responses than the CTL-APC model described in the first part of this chapter, the properties are very similar.

For the current purposes, this general model was modified to include two sub-populations of cancer cells: drug sensitive and drug resistant cells. Denoting drug sensitive tumor cells by x and drug resistant tumor cells by y, the model is given by the following ordinary differential equations.

$$\dot{x} = -\delta x - pxz, \tag{22.4}$$

$$\dot{y} = ry - pyz, \tag{22.5}$$

$$\dot{z} = c(x+y)z/[(z+\eta)(x+y+\epsilon] - q(x+y)z - bz, \tag{22.6}$$

where we denoted

$$L - D - H \equiv -\delta < 0, \quad L - D \equiv r > 0.$$

Because drug-sensitive cells, x, are susceptible to therapy, this population of cells is assumed to decline during treatment with a rate δ. As before, immune responses contribute to cell death with a rate p. On the other hand, drug-resistant cells, y, are not susceptible to therapy and are thus assumed to expand exponentially during therapy with a rate r. Because the initial growth phase of the resistant CML cells is interesting in the current context, we ignore growth saturation at high numbers of cells for simplicity. As with drug-sensitive cells, drug-resistant cells are removed by immune responses with a rate p. The equation for immune responses is the same as in the simple model described previously. However, both antigenic stimulation and immune impairment are now proportional to the number of drug-sensitive and drug-resistant cells, $x + y$.

The model can give rise to treatment dynamics that can describe diverse clinical data well, shown by non-linear least squares fits of the model to selected patient data (figure 22.8). Figure 22.8(a) shows a response that involves a faster, followed by a slower phase of decline, eventually leading to a resurgence of the cancer. Immunity rises as the number of CML cells starts to decline during treatment because of reduced immune impairment. This accounts for the fast phase of CML decline. Immunity subsequently falls due to lack of antigenic stimulation. This, together with the rise of a resistant mutant, accounts for the slower phase of decline. The rise of the resistant mutant eventually leads to resurgence of the cancer. Figure 22.8(b) shows the same kind of profile without eventual resurgence of the cancer. In this case, a resistant mutant is not present. Figure 22.8(c) shows a single phase exponential decline of CML cells, while figure 22.8(d) shows an exponential decline followed by re-growth of the cancer. In these cases, immunity does not expand during the treatment dynamics according to the model, correlating with a significantly slower decline rate of the cancer. In figure 22.8(c), no resistant mutant exists, while in figure 22.8(d), a resistant mutant grows during therapy.

To examine predicted correlations between the slopes, the simulation was run many times, randomly varying the growth rate of drug-resistant CML cells and the rate of immune expansion against CML. Consistent with the experimental data, the model predicts a significant positive correlation between the slopes of the fast and slow phase of decline, a significant negative correlation between the slopes of the slow phase of decline and the re-growth, as well as a lack of a significant correlation between the slopes of the fast phase of decline and the re-growth (figure 22.9(b)). The reason is as follows. The immune response largely drives the fast phase of decline. Part of the reason for the slower decline is a fall of immune responses as the number of CML cells drops to low levels. However, the strength of the declining immunity is still proportional to the strength of immunity at its peak, hence the positive correlation. Another reason for the slowing decline is the rise of drug-resistant mutants. In addition, growth of resistant mutants completely determines the eventual rise of CML cells, hence the

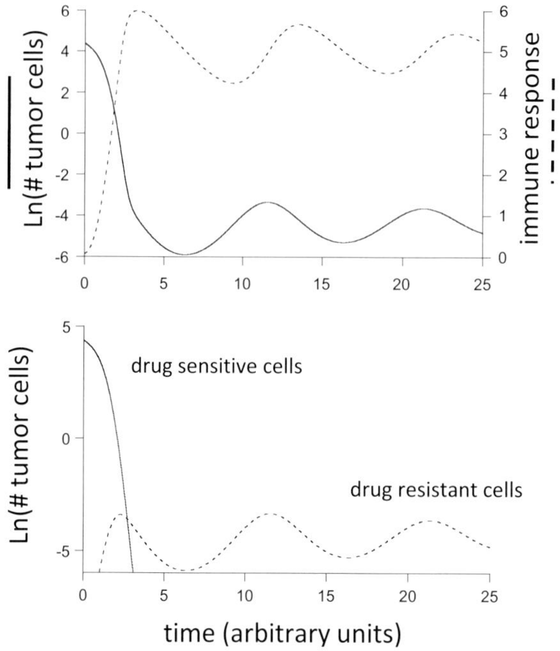

Fig. 22.10 Computer simulation showing the possibility that during therapy, the immune response is maintained at higher levels rather than dropping to insignificant levels, leading to long-term control of CML. The lower panel shows the CML dynamics separately for the populations of drug sensitive and drug resistant cells. Parameters were chosen as follows: $r = 5$, $c = 5$, $\delta = 0.5$, $p = 1$, $b = 0.1$, $q = 0.01$, $\epsilon = 0.1$, $\eta = 1$, $x_0 = 80$, $y_0 = 2.7 \cdot 10^{-5}$, $z_0 = 0.06$. The units of the axes are arbitrary, as the parameter set was chosen for illustrative purposes and is not based on measured parameters that are currently unknown.

negative correlation. Since the fast phase of decline is mostly determined by immunity while the initial rise of resistant mutants is mostly determined by their replication rate, there is no correlation between these slopes.

22.10 Possible role of immune stimulation for long-term remission

If immune responses that arise during treatment are sustained, they could in principle suppress the population of drug-resistant cancer cells. Interestingly, it has been found that a set of patients who discontinued imatinib treatment after maintaining a complete molecular response for at least two years remained in remission without relapse [Ross *et al.* (2013); Deininger (2011); Mahon *et al.* (2010)]. It can be hypothesized that in these cases, immunity became fully established during therapy rather than rising only temporarily. This possibility is demonstrated by computer simulation in figure 22.10. Establishment of sustained immune responses during therapy of immunosuppressive diseases has been found to occur in some experimental HCV and SIV therapy regimes [Barnes *et al.* (2002); Lifson *et al.* (2001)].

22.11 Summary

This chapter explored mathematical models that describe the dynamical interactions between a tumor cell population and a tumor-specific immune response. The first model took into account the basic dynamics of CTL activation and proliferation by dendritic cells via cross-priming, as well as the ability of the tumor to fight back and impair the immune response against itself. This model gave rise to all the documented outcomes of these interactions: clearance, controlled persistence at a stable equilibrium, and uncontrolled growth of tumor cells that can escape immunity (immunoediting). The model showed how tumor evolution can shift the balance in this battle away from controlled persistence towards the breakout of a tumor cell population that can significantly impair the response and thus achieve uncontrolled growth. The model was further used to explore implications for immunotherapeutic approaches. Subsequently, this type of model was applied to analyze the importance of immune responses in determining the treatment dynamics of CML during imatinib therapy. Data documenting the kinetics of tumor cell decline during imatinib therapy were analyzed in the context of the model, and this suggested that the different phases of the decline are in part determined by the immune cell dynamics that have been observed to occur during imatinib therapy of the disease. Based on these notions, the possibility was explored that specific drug treatment regimes could boost anti-tumor immunity and lead to efficient long-term immune control of the tumor after discontinuation of drug therapy.

Problems

Problem 22.1. *In equations (22.2)-(22.3), explore how the region of bistability depends on the parameter q.*

Problem 22.2. Research project. *Find out more about the concept of immunoediting in cancer. How does it relate to the concept of antigenic escape in viral infections such as human immunodeficiency virus (HIV)?*

Chapter 23

Towards higher complexities: social interactions

This book covered a lot of material that showed how mathematical models can be used to study the dynamics of cancer. Both, the basic dynamics, and evolutionary dynamics were covered. In the sections on basic dynamics, we considered single species growth models, competition dynamics in various settings, and spatial models of tumor growth in the context of deterministic models. In the evolutionary dynamics section, we investigated how cells accumulate different types of mutations that are required for uncontrolled growth, using stochastic models and considering a variety of scenarios including well-mixed cell populations, spatially structured cell populations, and hierarchically structured cell populations. Among advanced topics, we discussed further, biologically interesting topics that have been investigated with the types of mathematical methods described before. While these topics cover many fundamental aspects that are required to study the dynamics of cancer initiation and progression, it is obvious that there are a variety of relevant topics that go beyond this basic introduction, but that are nevertheless biologically important.

23.1 Microenvironment

A very important such topic is the effect of the microenvironemnt on cancer initiation and progression. A large number of studies support the notion that the microenvironment plays an important role in tumorigenesis, both promoting and suppressing the disease through various mechanisms, e.g. [Rubin (2003); Bhowmick *et al.* (2004); Kuperwasser *et al.* (2004); Bissell *et al.* (2002); Matrisian *et al.* (2001); Whiteside (2008); Bissell and Hines (2011); Polyak *et al.* (2009)]. The effect on the microenvironment on the evolutionary dynamics of cancers has been demonstrated mathematically using spatial, multi-scale models [Anderson *et al.* (2006); Anderson (2005)]. In particular, it has been shown that the type of the microenvironment can shape the genetic composition and characteristics of the emerging tumor. Thus, a harsh microenvironment (e.g. hypoxia or a heterogeneous extracellular matrix) was shown to lead to the emergence of tumors that are dominated by one or a few aggressive clones and are invasive, displaying fingering margins. On the other hand, mild microenvironmental conditions have been shown to lead to the absence of clonal dominance and to the coexistence of more and less aggressive tumor cell types. In addition, such tumors were predicted to display smooth, non-invasive margins. Thus, the microenvironment can not only promote or prevent the establishments of tumors, but can also determine their characteristics. This is similar to the "immunoediting" concept discussed in Chapter 22 where tumor-specific immune responses not only suppress tumor growth, but can shape what types of tumors do eventually grow when they break out of control. In general, the interactions between tumor cells and their microenvironment give rise to many evolutionary questions, and mathematical models can provide important insights [Gatenby and Gillies (2008); Gerisch and Chaplain (2008); Lowengrub *et al.* (2010); Gatenby *et al.* (2007); Martin *et al.* (2010); Gatenby and Vincent (2003b); Gillies and Gatenby (2007); Alarcón *et al.* (2004); Smallbone *et al.* (2005); Kim *et al.* (2011)].

23.2 Cooperation and division of labor

The concept of cooperation [Nowak and Highfield (2011)] is often brought up when studying the interactions between tumors and their microenvironment. The tumor cells and the cells of the microenvironment are said to conspire in order to enable uncontrolled tumor growth [Hsu *et al.* (2002); Lathia *et al.* (2011)]. However, cooperative interactions are also possible among different tumor cells. Throughout this book, we concentrated on the sequential accumulation of mutations in cells. An alternative model of evolution can involve the accumulation of different mutations not sequentially in one cell, but parallel in different cells. Assume that tumor progression requires the presence of three gene products, call them A, B, and C. A cell can evolve to accumulate all three mutations, thus achieving the malignant phenotype. On the other hand, the three gene products can arise independently in different cells, and they can share the gene product as a public good. In this setting, the malignant phenotype arises as an emergent property from a collection of cooperating cells rather than from an individual cell. Such evolution through division of labor / cooperation has been first suggested to be important by [Axelrod *et al.* (2006)], and indeed evidence for such cooperative interactions in cancers is mounting in the literature. It has been suggested that metastatic cells cooperate to prepare the microenvironment at a distant site by clustered migration and the release of various signaling factors [Bidard *et al.* (2008)]. Deisboeck and Couzin argued in favor of collective, swarm-like behavior of tumor cells that can confer advantageous properties to the cancer [Deisboeck and Couzin (2009)]. Angiogenesis and the formation of new blood supply for the tumor is an obvious situation in which cooperation can occur. If one cell induces the formation of new blood supply, this benefits other tumor cells that are located in the vicinity. A variety of possible cooperative interactions in cancer are summarized in [Sprouffske and Maley (2010)], although in most cases conclusive evidence for cooperation is still outstanding.

Mathematical models compared the rate at which multi-hit mutants emerge in the context of sequential evolution and in the context of a division of labor scenario where different gene products are acquired in different cells and shared as a public good [Komarova *et al.* (2012)]. It was found that multi-hit mutants emerged significantly faster in the context of division of labor than in the sequential evolution scenario, which means that such interactions could play a vial role in carcinogenesis. Moreover, the model indicated the occurrence of complex evolutionary processes in division of labor scenarios that might be crucially important for the accelerated emergence of malignant multi-hit cells [Komarova *et al.* (2012)]. If there is cooperation, cheaters typically emerge that have the potential to destroy the cooperative interactions [Nowak (2006)]. If division of labor occurs in the tumor cell population, and if cheaters can arise, it has been shown mathematically, that cells can quickly arise that simultaneously harbor all mutations (i.e. A, B and C in a single cell) [Komarova *et al.* (2012)]. In other words, cooperative interactions and the presence of cheaters can lead to the emergence of multi-hit mutants at a rate that is significantly faster than in the absence of cooperative interactions [Komarova *et al.* (2012)]. If such dynamics indeed occur in cancer, this mechanism could be an important driving force for the emergence of the disease during the natural life-span on the organism. These explorations showed that the evolutionary processes that might occur in cancer can be significantly more complex than the relatively simple somatic evolutionary processes outlined in this book.

23.3 Conclusion

The issues highlighted in this chapter only scratch the surface of the complexities that are at work in the initiation and progression of cancer. Mathematical models have been very useful for understanding the process of carcinogenesis in a variety of settings, and some of this material has been reviewed in this book. It will be important to solidify the modeling efforts in relatively simple settings because this framework will form the basis for the construction of more complex models that will be able to capture the intricacies of carcinogenesis in more realistic ways as more biological information and relevant data become available. While an enormous amount of data is continuously becoming available in the field of cancer research, progress in our understanding might be facilitated by a more concerted effort, spanning different disciplines. Not only should mathematical models be strongly based on experimental data, but the generation of the data themselves should be planned in the light of the guidance of predictions generated by mathematical models. Such tight interactions provide specific data and information that is required to push the field ahead. This interdisciplinary approach will help us gain better insights into the highly complex dynamics of cancer.

Bibliography

Abdelrahim, M., Konduri, S., Basha, R., Philip, P. A., Baker, C. H. *et al.* (2010). Angiogenesis: an update and potential drug approaches (review). *International Journal of Oncology* **36**, 1, p. 5.

Adams, G. and Scadden, D. (2007). A niche opportunity for stem cell therapeutics, *Gene Therapy* **15**, 2, pp. 96–99.

Afenya, E. K. and Calderón, C. P. (2000). Diverse ideas on the growth kinetics of disseminated cancer cells, *Bulletin of Mathematical Biology* **62**, 3, pp. 527–542.

Aguirre-Ghiso, J. A. (2007). Models, mechanisms and clinical evidence for cancer dormancy, *Nature Reviews Cancer* **7**, 11, pp. 834–846.

Ahuja, N., Mohan, A. L., Li, Q., Stolker, J. M., Herman, J. G., Hamilton, S. R., Baylin, S. B. and Issa, J. P. (1997). Association between CpG island methylation and microsatellite instability in colorectal cancer, *Cancer Research* **57**, 16, pp. 3370–3374.

Al-Hajj, M., Wicha, M. S., Benito-Hernandez, A., Morrison, S. J. and Clarke, M. F. (2003). Prospective identification of tumorigenic breast cancer cells, *Proceedings of the National Academy of Sciences U S A* **100**, 7, pp. 3983–8, doi:10.1073/pnas.0530291100.

Alarcón, T., Byrne, H. and Maini, P. (2004). A mathematical model of the effects of hypoxia on the cell-cycle of normal and cancer cells, *Journal of Theoretical Biology* **229**, 3, pp. 395–411.

Alarcón, T., Byrne, H. M. and Maini, P. K. (2003). A cellular automaton model for tumour growth in inhomogeneous environment, *Journal of Theoretical Biology* **225**, 2, pp. 257–274.

Alarcón, T., Byrne, H. M. and Maini, P. K. (2005). A multiple scale model for tumor growth, *Multiscale Modeling & Simulation* **3**, 2, pp. 440–475.

Albert, M. L., Darnell, J. C., Bender, A., Francisco, L. M., Bhardwaj, N. and Darnell, R. B. (1998a). Tumor-specific killer cells in paraneoplastic cerebellar degeneration, *Nature Medicine* **4**, 11, pp. 1321–1324.

Albert, M. L., Sauter, B. and Bhardwaj, N. (1998b). Dendritic cells acquire antigen from apoptotic cells and induce class i-restricted CTLs, *Nature* **392**, 6671, pp. 86–89.

Albertson, D. G., Collins, C., McCormick, F. and Gray, J. W. (2003). Chromosome aberrations in solid tumors, *Nature Genetics* **34**, 4, pp. 369–376.

Albini, A. and Sporn, M. B. (2007). The tumour microenvironment as a target for chemoprevention, *Nature Reviews Cancer* **7**, 2, pp. 139–147.

Alvarez-Buylla, A. and Lim, D. (2004). For the long run: maintaining germinal niches in the adult brain, *Neuron* **41**, 5, pp. 683–686.

An, X., Tiwari, A., Sun, Y., Ding, P., Ashby, C. and Chen, Z. (2010). BCR-ABL tyrosine kinase inhibitors in the treatment of Philadelphia chromosome positive chronic myeloid leukemia: a review, *Leukemia Research* **34**, 10, pp. 1255–1268.

Andea, A. A., Wallis, T., Newman, L. A., Bouwman, D., Dey, J. and Visscher, D. W. (2002). Pathologic analysis of tumor size and lymph node status in multifocal/multicentric breast carcinoma, *Cancer* **94**, 5, pp. 1383–1390.

Anderson, A. and Chaplain, M. (1998a). Continuous and discrete mathematical models of tumor-induced angiogenesis, *Bulletin of Mathematical Biology* **60**, 5, pp. 857–899.

Anderson, A. and Chaplain, M. A. (1998b). A mathematical model for capillary network formation in the absence of endothelial cell proliferation, *Applied Mathematics Letters* **11**, 3, pp. 109–114.

Anderson, A. and Quaranta, V. (2008). Integrative mathematical oncology, *Nature Reviews Cancer* **8**, 3, pp. 227–234.

Anderson, A. R. (2005). A hybrid mathematical model of solid tumour invasion: the importance of cell adhesion, *Mathematical Medicine and Biology* **22**, 2, pp. 163–186.

Anderson, A. R. (2007). A hybrid multiscale model of solid tumour growth and invasion: Evolution and the microenvironment, in *Single-cell-based Models in Biology and Medicine* (Springer), pp. 3–28.

Anderson, A. R., Chaplain, M. A. and McDougall, S. (2012). A hybrid discrete-continuum model of tumour induced angiogenesis, in *Modeling Tumor Vasculature* (Springer), pp. 105–133.

Anderson, A. R., Weaver, A. M., Cummings, P. T. and Quaranta, V. (2006). Tumor morphology and phenotypic evolution driven by selective pressure from the microenvironment, *Cell* **127**, 5, pp. 905–915.

Anderson, R. M., May, R. M. and Anderson, B. (1992). *Infectious Diseases of Humans: Dynamics and Control*, Vol. 28 (Wiley Online Library).

Andreoiu, M. and Cheng, L. (2010). Multifocal prostate cancer: biologic, prognostic, and therapeutic implications, *Human Pathology* **41**, 6, pp. 781–793.

Antonescu, C. R., Elahi, A., Healey, J. H., Brennan, M. F., Lui, M. Y., Lewis, J., Jhanwar, S. C., Woodruff, J. M. and Ladanyi, M. (2000). Monoclonality of multifocal myxoid liposarcoma: confirmation by analysis of TLS-CHOP or EWS-CHOP rearrangements, *Clinical Cancer Research* **6**, 7, pp. 2788–2793.

Arai, F. and Suda, T. (2007). Maintenance of quiescent hematopoietic stem cells in the osteoblastic niche, *Annals of the New York Academy of Sciences* **1106**, 1, pp. 41–53.

Araujo, R. and McElwain, D. (2004). A history of the study of solid tumor growth: The contribution of mathematical modeling, *Bulletin of Mathematical Biology* **66**, 5, pp. 1039–1091.

Arkus, N. (2005). A mathematical model of cellular apoptosis and senescence through the dynamics of telomere loss, *Journal of Theoretical Biology* **235**, 1, pp. 13–32, doi:10.1016/j.jtbi.2004.12.016.

Armitage, P. and Doll, R. (1954). The age distribution of cancer and a multi-stage theory of carcinogenesis, *British Journal of Cancer* **8**, 1, p. 1.

Arrowsmith, D. K. and Place, C. (1990). *An Introduction to Dynamical Systems* (Cambridge University Press).

Artandi, S. E. and DePinho, R. A. (2000). A critical role for telomeres in suppressing and facilitating carcinogenesis, *Current Opinion in Genetics and Develpment* **10**, 1, pp. 39–46.

Ashkenazi, R., Gentry, S. N. and Jackson, T. L. (2008). Pathways to tumorigenesis – modeling mutation acquisition in stem cells and their progeny, *Neoplasia (New York, NY)* **10**, 11, p. 1170.

Axelrod, D. E., Baggerly, K. A. and Kimmel, M. (1994). Gene amplification by unequal sister chromatid exchange: probabilistic modeling and analysis of drug resistance data, *Journal of Theoretical Biology* **168**, 2, pp. 151–159.

Axelrod, R., Axelrod, D. E. and Pienta, K. J. (2006). Evolution of cooperation among tumor cells, *Proceedings of the National Academy of Sciences* **103**, 36, pp. 13474–13479.

Bach, S. P., Renehan, A. G. and Potten, C. S. (2000). Stem cells: the intestinal stem cell as a paradigm, *Carcinogenesis* **21**, 3, pp. 469–76.

Bagheri, N., Shiina, M., Lauffenburger, D. A. and Korn, W. M. (2011). A dynamical systems model for combinatorial cancer therapy enhances oncolytic adenovirus efficacy by MEK-inhibition, *PLoS Computational Biology* **7**, 2, p. e1001085.

Bajzer, Ž., Carr, T., Josić, K., Russell, S. J. and Dingli, D. (2008). Modeling of cancer virotherapy with recombinant measles viruses, *Journal of Theoretical Biology* **252**, 1, pp. 109–122.

Bardeesy, N. and DePinho, R. A. (2002). Pancreatic cancer biology and genetics, *Nature Reviews Cancer* **2**, 12, pp. 897–909.

Bardelli, A., Cahill, D. P., Lederer, G., Speicher, M. R., Kinzler, K. W., Vogelstein, B. and Lengauer, C. (2001). Carcinogen-specific induction of genetic instability, *Proceedings of the National Academy of Sciences* **98**, 10, pp. 5770–5775.

Barnes, E., Harcourt, G., Brown, D., Lucas, M., Phillips, R., Dusheiko, G. and Klenerman, P. (2002). The dynamics of T-lymphocyte responses during combination therapy for chronic hepatitis C virus infection, *Hepatology* **36**, 3, pp. 743–754.

Barton, N. and Rouhani, S. (1987). The frequency of shifts between alternative equilibria, *Journal of Theoretical Biology* **125**, 4, pp. 397–418.

Bartoszyński, R., Edler, L., Hanin, L., Kopp-Schneider, A., Pavlova, L., Tsodikov, A., Zorin, A. and Yakovlev, A. Y. (2001). Modeling cancer detection: tumor size as a source of information on unobservable stages of carcinogenesis, *Mathematical Biosciences* **171**, 2, pp. 113–142.

Baumann, K. (2010). Stem cells: Dividing with symmetry, *Nature Reviews Molecular Cell Biology* **11**, 11, p. 752.

Baumann, S., Krueger, A., Kirchhoff, S. and Krammer, P. H. (2002). Regulation of T cell apoptosis during the immune response, *Current Molecular Medicine* **2**, 3, pp. 257–272.

Baylin, S. B. (2005). DNA methylation and gene silencing in cancer, *Nature Clinical Practice Oncology* **2**, pp. S4–S11.

Baylin, S. B. and Herman, J. G. (2000). DNA hypermethylation in tumorigenesis: epigenetics joins genetics, *Trends in Genetics* **16**, 4, pp. 168–174.

Beerenwinkel, N., Antal, T., Dingli, D., Traulsen, A., Kinzler, K. W., Velculescu, V. E., Vogelstein, B. and Nowak, M. A. (2007). Genetic progression and the waiting time to cancer, *PLoS Computational Biology* **3**, 11, p. e225.

Begon, M., Townsend, C. R. and Harper, J. L. (2009). *Ecology: From Individuals to Ecosystems* (Wiley-Blackwell).

Bell, J. C. (2007). Oncolytic viruses: What's next? *Current Cancer Drug Targets* **7**, 2, pp. 127–131.

Bell, J. C., Lichty, B. and Stojdl, D. (2003). Getting oncolytic virus therapies off the ground, *Cancer Cell* **4**, 1, pp. 7–11.

Bellomo, N., Bellouquid, A. and Delitala, M. (2004). Mathematical topics in the modeling complex multicellular systems and tumor immune cells competition, *Mathematical Modeling and Methods in Applied Sciences* **14**, pp. 1683–1733.

Bellomo, N., Li, N. and Maini, P. (2008). On the foundations of cancer modelling: selected topics, speculations, and perspectives, *Mathematical Models and Methods in Applied Sciences* **18**, 04, pp. 593–646.

Bellomo, N. and Maini, P. (2007). Challenging mathematical problems in cancer modelling, *Mathematical Models and Methods in Applied Sciences* **17**, supp01, pp. 1641–1645.

Bellomo, N. and Preziosi, L. (2000). Modeling and mathematical problems related to tumor evolution and its interaction with the immune system, *Mathematical and Computational Modeling* **22**, 3-4, pp. 413–452.

Benz, C. C. and Yau, C. (2008). Ageing, oxidative stress and cancer: paradigms in parallax, *Nature Reviews Cancer* **8**, 11, pp. 875–879.

Bhowmick, N. A., Neilson, E. G. and Moses, H. L. (2004). Stromal fibroblasts in cancer initiation and progression, *Nature* **432**, 7015, pp. 332–337.

Bidard, F.-C., Pierga, J.-Y., Vincent-Salomon, A. and Poupon, M.-F. (2008). A "class action" against the microenvironment: do cancer cells cooperate in metastasis? *Cancer and Metastasis Reviews* **27**, 1, pp. 5–10.

Biesecker, M., Kimn, J.-H., Lu, H., Dingli, D. and Bajzer, Ž. (2010). Optimization of virotherapy for cancer, *Bulletin of Mathematical Biology* **72**, 2, pp. 469–489.

Binetruy, B., Heasley, L., Bost, F., Caron, L. and Aouadi, M. (2007). Concise review: regulation of embryonic stem cell lineage commitment by mitogen-activated protein kinases, *Stem Cells* **25**, pp. 1090–1095.

Bissell, M. J. and Hines, W. C. (2011). Why don't we get more cancer? A proposed role of the microenvironment in restraining cancer progression, *Nature Medicine* **17**, 3, pp. 320–329.

Bissell, M. J., Radisky, D. C., Rizki, A., Weaver, V. M. and Petersen, O. W. (2002). The organizing principle: microenvironmental influences in the normal and malignant breast, *Differentiation* **70**, 9-10, pp. 537–546.

Blackburn, E. (1991). Structure and function of telomeres, *Nature* **350**, 6319, pp. 569–573.

Blagosklonny, M. V. (2002). P53: an ubiquitous target of anticancer drugs, *International Journal of Cancer* **98**, 2, pp. 161–166.

Bocharov, G., Quiel, J., Luzyanina, T., Alon, H., Chiglintsev, E., Chereshnev, V., Meier-Schellersheim, M., Paul, W. E. and Grossman, Z. (2011). Feedback regulation of proliferation vs. differentiation rates explains the dependence of CD4 t-cell expansion on precursor number, *Proceedings of the National Academy of Sciences U S A* **108**, 8, pp. 3318–23, doi: 10.1073/pnas.1019706108.

Boland, C. R. and Goel, A. (2005). Somatic evolution of cancer cells, in *Seminars in Cancer Biology*, Vol. 15 (Elsevier), pp. 436–450.

Boland, C. R. and Goel, A. (2010). Microsatellite instability in colorectal cancer, *Gastroenterology* **138**, 6, pp. 2073–2087.

Boland, C. R., Komarova, N. L. and Goel, A. (2009). Chromosomal instability and cancer: not just one CINgle mechanism, *Gut* **58**, 2, pp. 163–164.

Boland, C. R., Sato, J., Saito, K., Carethers, J. M., Marra, G., Laghi, L., Chauhan, D. *et al.* (1998). Genetic instability and chromosomal aberrations in colorectal cancer: a review of the current models. *Cancer Detection and Prevention* **22**, 5, p. 377.

Bolin, R., Robinson, W., Sutherland, J. and Hamman, R. (1982). Busulfan versus hydroxyurea in long-term therapy of chronic myelogenous leukemia, *Cancer* **50**, 9, pp. 1683–1686.

Bonnet, D. and Dick, J. E. (1997). Human acute myeloid leukemia is organized as a hierarchy that originates from a primitive hematopoietic cell, *Nature Medicine* **3**, 7, pp. 730–7.

Boots, M. and Sasaki, A. (2002). Parasite-driven extinction in spatially explicit host-parasite systems, *The American Naturalist* **159**, 6, pp. 706–713.

Bosniak, M. (1995). Observation of small incidentally detected renal masses. *Seminars in Urologic Oncology* **13**, 4, p. 267.

Boyages, J., Jayasinghe, U. W. and Coombs, N. (2010). Multifocal breast cancer and survival: each focus does matter particularly for larger tumours, *European Journal of Cancer* **46**, 11, pp. 1990–1996.

Bozic, I., Allen, B. and Nowak, M. A. (2012). Dynamics of targeted cancer therapy, *Trends in Molecular Medicine* **18**, 6, pp. 311–316.

Bozic, I., Reiter, J. G., Allen, B., Antal, T., Chatterjee, K., Shah, P., Moon, Y. S., Yaqubie, A., Kelly, N., Le, D. T. *et al.* (2013). Evolutionary dynamics of cancer in response to targeted combination therapy, *eLife* **2**.

Bramson, M., Durrett, R. and Swindle, G. (1989). Statistical mechanics of crabgrass, *The Annals of Probability*, pp. 444–481.

Branford, S., Rudzki, Z., Walsh, S., Parkinson, I., Grigg, A., Szer, J., Taylor, K., Herrmann, R., Seymour, J., Arthur, C. *et al.* (2003). Detection of BCR-ABL mutations in patients with CML treated with imatinib is virtually always accompanied by clinical resistance, and mutations in the ATP phosphate-binding loop (P-loop) are associated with a poor prognosis, *Blood* **102**, 1, pp. 276–283.

Breivik, J. (2001). Don't stop for repairs in a war zone: Darwinian evolution unites genes and environment in cancer development, *Proceedings of the National Academy of Sciences* **98**, 10, pp. 5379–5381.

Breivik, J. and Gaudernack, G. (1999a). Carcinogenesis and natural selection: a new perspective to the genetics and epigenetics of colorectal cancer, *Advances in Cancer Research* **76**, pp. 187–212.

Breivik, J. and Gaudernack, G. (1999b). Genomic instability, DNA methylation, and natural selection in colorectal carcinogenesis, in *Seminars in Cancer Biology*, Vol. 9 (Elsevier), pp. 245–254.

Brú, A., Albertos, S., Subiza, J. L., García-Asenjo, J. L. and Brú, I. (2003). The universal dynamics of tumor growth, *Biophysical Journal* **85**, 5, p. 2948.

Brú, A., Pastor, J. M., Fernaud, I., Brú, I., Melle, S. and Berenguer, C. (1998). Super-rough dynamics on tumor growth, *Physical Review Letters* **81**, 18, pp. 4008–4011.

Bryson, A. E. and Ho, Y. C. (1969). *Applied Optimal Control* (Ginn and Company, Waltham MA).

Budd, R. C. (2001). Activation-induced cell death, *Current Opinion in Immunology* **13**, 3, pp. 356–362.

Burley, S. (2006). Application of fasttm fragment-based lead discovery and structure-guided design to discovery of small molecule inhibitors of BCR-ABL tyrosine kinase active against the T315I imatinib-resistant mutant, *Proceedings of the American Association for Cancer Research* **2006**, 1, p. 1139.

Burnet, M. (1957). Cancer – biological approach: I. The processes of control. II. The significance of somatic mutation, *British Medical Journal* **1**, 5022, p. 779.

Burrell, R. A., McClelland, S. E., Endesfelder, D., Groth, P., Weller, M.-C., Shaikh, N., Domingo, E., Kanu, N., Dewhurst, S. M., Gronroos, E. *et al.* (2013). Replication stress links structural and numerical cancer chromosomal instability, *Nature* **494**, 7438, pp. 492–496.

Burton, A. C. *et al.* (1966). Rate of growth of solid tumours as a problem of diffusion. *Growth* **30**, 2, p. 157.

Butel, J. S. (2000). Viral carcinogenesis: revelation of molecular mechanisms and etiology of human disease, *Carcinogenesis* **21**, 3, pp. 405–426.

Byrne, H., Alarcon, T., Owen, M., Webb, S. and Maini, P. (2006). Modelling aspects of cancer dynamics: a review, *Philosophical Transactions of the Royal Society A: Mathematical, Physical and Engineering Sciences* **364**, 1843, pp. 1563–1578.

Byrne, H. and Chaplain, M. (1996). Growth of necrotic tumors in the presence and absence of inhibitors, *Mathematical Biosciences* **135**, 2, pp. 187–216.

Byrne, H. and Chaplain, M. A. (1995). Growth of nonnecrotic tumors in the presence and absence of inhibitors, *Mathematical Biosciences* **130**, 2, pp. 151–181.

Byrne, H. and Preziosi, L. (2003). Modelling solid tumour growth using the theory of mixtures, *Mathematical Medicine and Biology* **20**, 4, pp. 341–366.

Cahill, D. P., Kinzler, K. W., Vogelstein, B. and Lengauer, C. (1999). Genetic instability and Darwinian selection in tumours, *Trends in Cell Biology* **9**, 12, pp. M57–M60.

Cairns, J. (2002). Somatic stem cells and the kinetics of mutagenesis and carcinogenesis, *Proceedings of the National Academy of Sciences* **99**, 16, pp. 10567–10570.

Calabrese, P., Tavaré, S. and Shibata, D. (2004). Pretumor progression: clonal evolution of human stem cell populations, *The American Journal of Pathology* **164**, 4, pp. 1337–1346.

Campbell, F., Williams, G., Appleton, M., Dixon, M., Harris, M. and Williams, E. (1996). Post-irradiation somatic mutation and clonal stabilisation time in the human colon. *Gut* **39**, 4, pp. 569–573.

Campisi, J. and Vijg, J. (2009). Does damage to DNA and other macromolecules play a role in aging? If so, how? *The Journals of Gerontology Series A: Biological Sciences and Medical Sciences* **64**, 2, pp. 175–178.

Campisi, J. and Warner, H. R. (2001). Aging in mitotic and post-mitotic cells, *Advances in Cell Aging and Gerontology* **4**, pp. 1–16.

Carbone, F. R., Kurts, C., Bennett, S. R., Miller, J. F. and Heath, W. R. (1998). Cross-presentation: a general mechanism for CTL immunity and tolerance, *Immunology Today* **19**, 8, pp. 368–373.

Carter, A. J. and Wagner, G. P. (2002). Evolution of functionally conserved enhancers can be accelerated in large populations: a population–genetic model, *Proceedings of the Royal Society of London. Series B: Biological Sciences* **269**, 1494, pp. 953–960.

Carter, T., Wodicka, L., Shah, N., Velasco, A., Fabian, M., Treiber, D., Milanov, Z., Atteridge, C., Biggs III, W., Edeen, P. *et al.* (2005). Inhibition of drug-resistant mutants of ABL, KIT, and EGF receptor kinases, *Proceedings of the National Academy of Sciences USA* **102**, 31, pp. 11011–11016.

Chaplain, M. (1996). Avascular growth, angiogenesis and vascular growth in solid tumors, *Mathematical and Computational Modeling* **23**, 6, pp. 47–87.

Chaplain, M. and Lolas, G. (2005). Spatio-temporal heterogeneity arising in a mathematical model of cancer invasion of tissue, *Mathematical Modeling and Methods in Applied Sciences* **15**, pp. 1685–1734.

Chaplain, M., McDougall, S. and Anderson, A. (2006). Mathematical modeling of tumor-induced angiogenesis, *Annual Reviews of Biomedical Engineering* **8**, pp. 233–257.

Chen, C. I., Maecker, H. T. and Lee, P. P. (2008). Development and dynamics of robust T-cell responses to CML under imatinib treatment, *Blood* **111**, 11, pp. 5342–5349.

Chen, G. K., Duran, G. E., Mangili, A., Beketic-Oreskovic, L. and Sikic, B. I. (2000). MDR 1 activation is the predominant resistance mechanism selected by vinblastine in MES-SA cells, *British Journal of Cancer* **83**, 7, pp. 892–898.

Chen, J., Li, Y., Yu, T.-S., McKay, R. M., Burns, D. K., Kernie, S. G. and

Parada, L. F. (2012). A restricted cell population propagates glioblastoma growth after chemotherapy, *Nature* **488**, 7412, pp. 522–526.

Chen, Z.-X., Mann, J. R., Hsieh, C.-L., Riggs, A. D. and Chédin, F. (2005). Physical and functional interactions between the human DNMT3L protein and members of the de novo methyltransferase family, *Journal of Cellular Biochemistry* **95**, 5, pp. 902–917.

Cheng, Y.-W., Pincas, H., Bacolod, M. D., Schemmann, G., Giardina, S. F., Huang, J., Barral, S., Idrees, K., Khan, S. A., Zeng, Z. *et al.* (2008). CpG island methylator phenotype associates with low-degree chromosomal abnormalities in colorectal cancer, *Clinical Cancer Research* **14**, 19, pp. 6005–6013.

Cheshier, S. H., Morrison, S. J., Liao, X. and Weissman, I. L. (1999). In vivo proliferation and cell cycle kinetics of long-term self-renewing hematopoietic stem cells, *Proceedings of the National Academy of Sciences* **96**, 6, pp. 3120–3125.

Chin, K., de Solorzano, C. O., Knowles, D., Jones, A., Chou, W., Rodriguez, E. G., Kuo, W.-L., Ljung, B.-M., Chew, K., Myambo, K., Miranda, M., Krig, S., Garbe, J., Stampfer, M., Yaswen, P., Gray, J. W. and Lockett, S. J. (2004). In situ analyses of genome instability in breast cancer, *Nature Genetics* **36**, 9, pp. 984–8, doi:10.1038/ng1409.

Choe, S. C., Zhao, G., Zhao, Z., Rosenblatt, J. D., Cho, H.-M., Shin, S.-U. and Johnson, N. F. (2011). Model for in vivo progression of tumors based on co-evolving cell population and vasculature, *Scientific Reports* **1**.

Clark, S. J., Melki, J. *et al.* (2002). DNA methylation and gene silencing in cancer: which is the guilty party? *Oncogene* **21**, 35, pp. 5380–5387.

Clayton, E., Doupé, D. P., Klein, A. M., Winton, D. J., Simons, B. D. and Jones, P. H. (2007). A single type of progenitor cell maintains normal epidermis, *Nature* **446**, 7132, pp. 185–189.

Cleaver, J. E. (2005). Cancer in xeroderma pigmentosum and related disorders of DNA repair, *Nature Reviews Cancer* **5**, 7, pp. 564–573.

Clevers, H. (2011). The cancer stem cell: premises, promises and challenges, *Nature Medicine* **17**, 3, pp. 313–9, doi:10.1038/nm.2304.

Cojocaru, L. and Agur, Z. (1992). A theoretical analysis of interval drug dosing for cell-cycle-phase-specific drugs, *Mathematical Biosciences* **109**, 1, pp. 85–97.

Coldman, A. J. and Goldie, J. H. (1985). Role of mathematical modeling in protocol formulation in cancer chemotherapy, *Cancer Treatment Reports* **69**, 10, pp. 1041–1048.

Coldman, A. J. and Murray, J. M. (2000). Optimal control for a stochastic model of cancer chemotherapy, *Mathematical Biosciences* **168**, 2, pp. 187–200.

Collins, V. P. (1956). Observation on growth rates of human tumors, *American Journal of Roentgenology* **76**, pp. 988–1000.

Counter, C. M., Avilion, A. A., LeFeuvre, C. E., Stewart, N. G., Greider, C. W., Harley, C. B. and Bacchetti, S. (1992). Telomere shortening associated with chromosome instability is arrested in immortal cells which express telomerase activity, *EMBO Journal* **11**, 5, pp. 1921–9.

Cozzio, A., Passegué, E., Ayton, P. M., Karsunky, H., Cleary, M. L. and Weissman, I. L. (2003). Similar MLL-associated leukemias arising from self-renewing stem cells and short-lived myeloid progenitors, *Genes & Development* **17**, 24, pp. 3029–35, doi:10.1101/gad.1143403.

Cristini, V. and Lowengrub, J. (2010). *Multiscale Modeling of Cancer: an Integrated Experimental and Mathematical Modeling Approach* (Cambridge University Press).

Cunningham, J. M., Christensen, E. R., Tester, D. J., Kim, C. Y., Roche, P. C., Burgart, L. J. and Thibodeau, S. N. (1998). Hypermethylation of the hMLH1 promoter in colon cancer with microsatellite instability, *Cancer Research* **58**, 15, pp. 3455–3460.

Dalerba, P., Cho, R. W. and Clarke, M. F. (2007). Cancer stem cells: models and concepts, *Annual Reviews of Medicine* **58**, pp. 267–284.

Daluiski, A., Engstrand, T., Bahamonde, M. E., Gamer, L. W., Agius, E., Stevenson, S. L., Cox, K., Rosen, V. and Lyons, K. M. (2001). Bone morphogenetic protein-3 is a negative regulator of bone density, *Nature Genetics* **27**, 1, pp. 84–8, doi:10.1038/83810.

Darnell, R. B. (1996). Onconeural antigens and the paraneoplastic neurologic disorders: at the intersection of cancer, immunity, and the brain, *Proceedings of the National Academy of Sciences* **93**, 10, pp. 4529–4536.

Davis, J. and Fang, B. (2005). Oncolytic virotherapy for cancer treatment: challenges and solutions, *The Journal of Gene Medicine* **7**, 11, pp. 1380–1389.

Day, R. S. (1986). Treatment sequencing, asymmetry, and uncertainty: protocol strategies for combination chemotherapy, *Cancer Research* **46**, pp. 3876–3885.

De Graaf, C., Kauppi, M., Baldwin, T., Hyland, C., Metcalf, D., Willson, T., Carpinelli, M., Smyth, G., Alexander, W. and Hilton, D. (2010). Regulation of hematopoietic stem cells by their mature progeny, *Proceedings of the National Academy of Sciences* **107**, 50, pp. 21689–21694.

De Lange, T. (2005). Shelterin: the protein complex that shapes and safeguards human telomeres, *Genes & Development* **19**, 18, pp. 2100–2110, doi:10.1101/gad.1346005.

De Matteis, G., Graudenzi, A. and Antoniotti, M. (2013). A review of spatial computational models for multi-cellular systems, with regard to intestinal crypts and colorectal cancer development, *Journal of Mathematical Biology* **66**, pp. 1409–62.

De Pillis, L., Gu, W., Fister, K., Head, T., Maples, K., Murugan, A., Neal, T. and Yoshida, K. (2007). Chemotherapy for tumors: An analysis of the dynamics and a study of quadratic and linear optimal controls, *Mathematical Biosciences* **209**, pp. 292–315.

De Pillis, L., Mallet, D. and Radunskaya, A. (2006). Spatial tumor-immune modeling, *Computational and Mathematical Methods in Medicine* **7**, 2-3, pp. 159–176.

De Pillis, L. G. and Radunskaya, A. (2001). A mathematical tumor model with immune resistance and drug therapy: an optimal control approach, *Computational and Mathematical Methods in Medicine* **3**, 2, pp. 79–100.

De Pillis, L. G. and Radunskaya, A. (2003). The dynamics of an optimally controlled tumor model: A case study, *Mathematical and Computer Modelling* **37**, 11, pp. 1221–1244.

De Roos, A. M., Mccauley, E. and Wilson, W. G. (1991). Mobility versus density-limited predator–prey dynamics on different spatial scales, *Proceedings of the Royal Society of London. Series B: Biological Sciences* **246**, 1316, pp. 117–122.

de Vries, A., Flores, E. R., Miranda, B., Hsieh, H.-M., van Oostrom, C. T. M., Sage, J. and Jacks, T. (2002). Targeted point mutations of p53 lead to dominant-negative inhibition of wild-type p53 function, *Proceedings of the National Academy of Sciences* **99**, 5, pp. 2948–2953.

Dehay, C. and Kennedy, H. (2007). Cell-cycle control and cortical development, *Nature Reviews Neuroscience* **8**, 6, pp. 438–450.

Deininger, M. (2011). Hematology: Curing CML with imatinib – a dream come true? *Nature Reviews Clinical Oncology* **8**, 3, pp. 127–128.

Deininger, M., Buchdunger, E. and Druker, B. J. (2005). The development of imatinib as a therapeutic agent for chronic myeloid leukemia, *Blood* **105**, 7, pp. 2640–2653.

Deininger, M., Goldman, J. and Melo, J. (2000). The molecular biology of chronic myeloid leukemia, *Blood* **96**, 10, pp. 3343–3356.

Deininger, M. W. and Druker, B. J. (2003). Specific targeted therapy of chronic myelogenous leukemia with imatinib, *Pharmacological Reviews* **55**, 3, pp. 401–23.

Deisboeck, T., Berens, M., Kansal, A., Torquato, S., Stemmer-Rachamimov, A. and Chiocca, E. (2008a). Pattern of self-organization in tumour systems: complex growth dynamics in a novel brain tumour spheroid model, *Cell Proliferation* **34**, 2, pp. 115–134.

Deisboeck, T., Zhang, L., Yoon, J. and Costa, J. (2008b). In silico cancer modeling: is it ready for prime time? *Nature Clinical Practice Oncology* **6**, 1, pp. 34–42.

Deisboeck, T. S. and Couzin, I. D. (2009). Collective behavior in cancer cell populations, *Bioessays* **31**, 2, pp. 190–197.

Deisboeck, T. S. and Stamatakos, G. S. (2011). *Multiscale Cancer Modeling* (CRC Press).

Denis, H., Ndlovu, M. N. and Fuks, F. (2011). Regulation of mammalian DNA methyltransferases: a route to new mechanisms, *EMBO Reports* **12**, 7, pp. 647–656.

DePillis, L., Gallegos, A. and Radunskaya, A. (2013). A model of dendritic cell therapy for melanoma, *Frontiers in Oncology* **3**, p. 56.

Derynck, R., Akhurst, R. J. and Balmain, A. (2001). TGF-beta signaling in tumor suppression and cancer progression, *Nature Genetics* **29**, 2, pp. 117–29, doi: 10.1038/ng1001-117.

Dewanji, A., Luebeck, E. and Moolgavkar, S. (2005). A generalized Luria–Delbrück model, *Mathematical Biosciences* **197**, 2, pp. 140–152.

Diaz Jr, L. A., Williams, R. T., Wu, J., Kinde, I., Hecht, J. R., Berlin, J., Allen, B., Bozic, I., Reiter, J. G., Nowak, M. A. *et al.* (2012). The molecular evolution of acquired resistance to targeted EGFR blockade in colorectal cancers, *Nature* **486**, 7404, pp. 537–540.

Dighe, A. S., Richards, E., Old, L. J. and Schreiber, R. D. (1994). Enhanced in vivo growth and resistance to rejection of tumor cells expressing dominant negative ifnγ receptors, *Immunity* **1**, 6, pp. 447–456.

Dingli, D., Cascino, M. D., Josić, K., Russell, S. J. and Bajzer, Ž. (2006). Mathematical modeling of cancer radiovirotherapy, *Mathematical Biosciences* **199**, 1, pp. 55–78.

Dingli, D., Offord, C., Myers, R., Peng, K., Carr, T., Josic, K., Russell, S. and Bajzer, Z. (2009). Dynamics of multiple myeloma tumor therapy with a recombinant measles virus, *Cancer Gene Therapy* **16**, 12, pp. 873–882.

Dingli, D. and Pacheco, J. (2011). Stochastic dynamics and the evolution of mutations in stem cells, *BMC Biology* **9**, 1, p. 41.

Dingli, D. and Pacheco, J. M. (2010). Modeling the architecture and dynamics of hematopoiesis, *Wiley Interdisciplinary Reviews: Systems Biology and Medicine* **2**, 2, pp. 235–244.

Dingli, D., Traulsen, A., Lenaerts, T. and Pacheco, J. M. (2010). Evolutionary dynamics of chronic myeloid leukemia, *Genes & Cancer* **1**, 4, pp. 309–315.

Dingli, D., Traulsen, A. and Pacheco, J. M. (2007). Stochastic dynamics of hematopoietic tumor stem cells, *Cell Cycle* **6**, 4, pp. 461–466.

Dingli, D., Traulsen, A. and Pacheco, J. M. (2008). Chronic myeloid leukemia: origin, development, response to therapy, and relapse, *Clinical Leukemia* **2**, 2, pp. 133–139.

Discher, D., Mooney, D. and Zandstra, P. (2009). Growth factors, matrices, and forces combine and control stem cells, *Science* **324**, 5935, pp. 1673–1677.

Doupé, D. P., Klein, A. M., Simons, B. D. and Jones, P. H. (2010). The ordered architecture of murine ear epidermis is maintained by progenitor cells with random fate, *Developmental Cell* **18**, 2, pp. 317–323.

Drasdo, D. and Höhme, S. (2005). A single-cell-based model of tumor growth in vitro: monolayers and spheroids, *Physical Biology* **2**, 3, pp. 133–147.

Driessens, G., Beck, B., Caauwe, A., Simons, B. D. and Blanpain, C. (2012). Defining the mode of tumour growth by clonal analysis, *Nature* **488**, p. 527.

Druker, B. J. (2004). Imatinib as a paradigm of targeted therapies, *Advances in Cancer Research* **91**, pp. 1–30.

Ducrot, A., Le Foll, F., Magal, P., Murakawa, H., Pasquier, J. and Webb, G. F. (2011). An in vitro cell population dynamics model incorporating cell size, quiescence, and contact inhibition, *Mathematical Models and Methods in Applied Sciences* **21**, supp01, pp. 871–892.

Duncan, E., Goetz, C., Stein, S., Mayo, K., Skaggs, B., Ziegelbauer, K., Sawyers, C. and Baldwin, A. (2008). Iκb kinase β inhibition induces cell death in imatinib-resistant and T315I dasatinib-resistant BCR-ABL+ cells, *Molecular Cancer Therapeutics* **7**, 2, pp. 391–397.

Dunn, G. P., Bruce, A. T., Ikeda, H., Old, L. J. and Schreiber, R. D. (2002). Cancer immunoediting: from immunosurveillance to tumor escape, *Nature Immunology* **3**, 11, pp. 991–998.

Durrett, R., Foo, J. and Leder, K. (2012). Spatial moran models II. Tumor growth and progression.

Durrett, R. and Moseley, S. (2012). Spatial moran models I. Tunneling in the neutral case, to appear in *Annals Applied Probability.*

Durrett, R. and Schmidt, D. (2008). Waiting for two mutations: with applications to regulatory sequence evolution and the limits of Darwinian evolution, *Genetics* **180**, 3, pp. 1501–1509.

Dvorak, H. F., Weaver, V. M., Tlsty, T. D. and Bergers, G. (2011). Tumor microenvironment and progression, *Journal of Surgical Oncology* **103**, 6, pp. 468–474.

Dworaczek, H. and Xiao, W. (2007). Xeroderma pigmentosum: A glimpse into nucleotide excision/repair, genetic instability, and cancer, *Critical Reviews in Oncogenesis* **13**, 2, pp. 159–178.

Eager, R. and Nemunaitis, J. (2011). Clinical development directions in oncolytic viral therapy, *Cancer Gene Therapy* **18**, 5, pp. 305–317.

Egger, B., Gold, K. S. and Brand, A. H. (2010). Notch regulates the switch from symmetric to asymmetric neural stem cell division in the drosophila optic lobe, *Development* **137**, 18, pp. 2981–2987.

Egger, B., Gold, K. S. and Brand, A. H. (2011). Regulating the balance between symmetric and asymmetric stem cell division in the developing brain, *Fly* **5**, 3, pp. 237–241.

Ehrlich, M. *et al.* (2002). DNA methylation in cancer: too much, but also too little. *Oncogene* **21**, 35, p. 5400.

Ehrlich, P. (1909). *Ueber den jetzigen Stand der Karzinomforschung*, Vol. 5.

Eigen, M. (1971). Selforganization of matter and the evolution of biological macromolecules, *Naturwissenschaften* **58**, 10, pp. 465–523.

Eigen, M. and Schuster, P. (1979). *The Hypercycle: A Principle of Natural Self-Organization* (Springer, Berlin).

Elgjo, K. and Reichelt, K. L. (2004). Chalones: from aqueous extracts to oligopeptides, *Cell Cycle* **3**, 9, pp. 1208–11.

Enderling, H., Anderson, A., Chaplain, M., Rowe, G. *et al.* (2006). Visualisation of the numerical solution of partial differential equation systems in three space dimensions and its importance for mathematical models in biology, *Mathematical Biosciences and Engineering* **3**, 4, p. 571.

Enderling, H., Anderson, A. R. A., Chaplain, M. A. J., Beheshti, A., Hlatky, L. and Hahnfeldt, P. (2009a). Paradoxical dependencies of tumor dormancy and progression on basic cell kinetics, *Cancer Research* **69**, 22, pp. 8814–21, doi:10.1158/0008-5472.CAN-09-2115.

Enderling, H., Chaplain, M. A., Anderson, A. R. and Vaidya, J. S. (2007). A mathematical model of breast cancer development, local treatment and recurrence, *Journal of Theoretical Biology* **246**, 2, pp. 245–259.

Enderling, H., Hlatky, L. and Hahnfeldt, P. (2009b). Migration rules: tumours are conglomerates of self-metastases, *British Journal of Cancer* **100**, 12, pp. 1917–25, doi:10.1038/sj.bjc.6605071.

Engelhart, M., Lebiedz, D. and Sager, S. (2011). Optimal control for selected cancer chemotherapy ode models: A view on the potential of optimal schedules and choice of objective function, *Mathematical Biosciences* **229**, 1, pp. 123–134.

England, S. J., McGrath, K. E., Frame, J. M. and Palis, J. (2011). Immature erythroblasts with extensive ex vivo self-renewal capacity emerge from the early mammalian fetus, *Blood* **117**, 9, pp. 2708–17, doi:10.1182/blood-2010-07-299743.

Esteller, M. (2008). Epigenetics in cancer, *New England Journal of Medicine* **358**, 11, pp. 1148–1159.

Esteller, M. and Herman, J. G. (2002). Cancer as an epigenetic disease: DNA methylation and chromatin alterations in human tumours, *The Journal of Pathology* **196**, 1, pp. 1–7.

Evans, N. D., Errington, R. J., Shelley, M., Feeney, G. P., Chapman, M. J., Godfrey, K. R., Smith, P. J. and Chappell, M. J. (2004). A mathematical model for the in vitro kinetics of the anti-cancer agent topotecan, *Mathematical Biosciences* **189**, 2, pp. 185–217.

Fauth, C. and Speicher, M. (2001). Classifying by colors: Fish-based genome analysis, *Cytogenetic and Genome Research* **93**, 1-2, pp. 1–10.

Felsher, D. W. (2003). Cancer revoked: oncogenes as therapeutic targets, *Nature Reviews. Cancer* **3**, 5, pp. 375–80, doi:10.1038/nrc1070.

Filoche, M. and Schwartz, L. (2004). Cancer death statistics: analogy between epidemiology and critical systems in physics, *Medical Hypotheses* **62**, 5, pp. 704–709.

Flores, I., Benetti, R. and Blasco, M. A. (2006). Telomerase regulation and stem cell behaviour, *Current Opinion in Cell Biology* **18**, 3, pp. 254–60, doi:10.1016/j.ceb.2006.03.003.

Flores, I., Canela, A., Vera, E., Tejera, A., Cotsarelis, G. and Blasco, M. A. (2008). The longest telomeres: a general signature of adult stem cell compartments, *Genes & Development* **22**, 5, pp. 654–67, doi:10.1101/gad.451008.

Folkman, J. (1971). Tumor angiogenesis: therapeutic implications, *New England Journal of Medicine* **285**, pp. 1182–1186.

Folkman, J. (1995). Angiogenesis in cancer, vascular, rheumatoid and other disease, *Nature Medicine* **1**, 1, pp. 27–30.

Folkman, J. (2002). Role of angiogenesis in tumor growth and metastasis, in *Seminars in oncology*, Vol. 29 (Elsevier), pp. 15–18.

Folkman, J. (2006). Angiogenesis, *Annual Reviews of Medicine* **57**, pp. 1–18.

Folkman, J. and Kalluri, R. (2004). Cancer without disease, *Nature* **427**, 6977, pp. 787–787.

Folkman, J. *et al.* (1992). The role of angiogenesis in tumor growth. *Seminars in Cancer Biology* **3**, 2, p. 65.

Fox, E. J., Prindle, M. J. and Loeb, L. A. (2013). Do mutator mutations fuel tumorigenesis? *Cancer and Metastasis Reviews* **32**, pp. 1–9.

Frank, S. A. (2003). Somatic mosaicism and cancer: inference based on a conditional Luria–Delbrück distribution, *Journal of Theoretical Biology* **223**, 4, pp. 405–412.

Frank, S. A. (2004). Age-specific acceleration of cancer, *Current Biology* **14**, 3, pp. 242–246.

Frank, S. A. (2007). *Dynamics of Cancer: Incidence, Inheritance, and Evolution* (Princeton University Press).

Frank, S. A. and Nowak, M. A. (2003). Cell biology: Developmental predisposition to cancer, *Nature* **422**, 6931, p. 494.

Frank, S. A. and Nowak, M. A. (2004). Problems of somatic mutation and cancer, *Bioessays* **26**, 3, pp. 291–299.

Franks, S., Byrne, H., Mudhar, H., Underwood, J. and Lewis, C. (2003). Mathematical modelling of comedo ductal carcinoma in situ of the breast, *Mathematical Medicine and Biology* **20**, 3, pp. 277–308.

Freyer, J. and Sutherland, R. (1985). A reduction in the in situ rates of oxygen and glucose consumption of cells in EMT6/Ro spheroids during growth, *Journal of Cellular Physiology* **124**, 3, pp. 516–524.

Freyer, J. P. and Sutherland, R. M. (1986). Regulation of growth saturation and development of necrosis in EMT6/Ro multicellular spheroids by the glucose and oxygen supply, *Cancer Research* **46**, 7, pp. 3504–3512.

Friberg, S., Mattson, S. *et al.* (1997). On the growth rates of human malignant tumors: implications for medical decision making, *Journal of Surgical Oncology* **65**, 4, pp. 284–297.

Friedman, A., Tian, J. P., Fulci, G., Chiocca, E. A. and Wang, J. (2006). Glioma virotherapy: effects of innate immune suppression and increased viral replication capacity, *Cancer Research* **66**, 4, pp. 2314–2319.

Friedmann-Morvinski, D., Bushong, E. A., Ke, E., Soda, Y., Marumoto, T., Singer, O., Ellisman, M. H. and Verma, I. M. (2012). Dedifferentiation of neurons and astrocytes by oncogenes can induce gliomas in mice, *Science* **338**, 6110, pp. 1080–4, doi:10.1126/science.1226929.

Frigyesi, A., Gisselsson, D., Mitelman, F. and Höglund, M. (2003). Power law distribution of chromosome aberrations in cancer, *Cancer Research* **63**, 21, pp. 7094–7097.

Fuchs, E., Tumbar, T. and Guasch, G. (2004). Socializing with the neighbors: stem cells and their niche, *Cell* **116**, 6, pp. 769–778.

Gaffney, E. (2004). The application of mathematical modelling to aspects of adjuvant chemotherapy scheduling, *Journal of Mathematical Biology* **48**, 4, pp. 375–422.

Gaffney, E. A. (2005). The mathematical modelling of adjuvant chemotherapy scheduling: incorporating the effects of protocol rest phases and pharmacokinetics, *Bulletin of Mathematical Biology* **67**, pp. 563–611.

Gage, F. (2000). Mammalian neural stem cells, *Science* **287**, 5457, pp. 1433–1438.

Gambacorti-Passerini, C. B., Gunby, R. H., Piazza, R., Galietta, A., Rostagno, R. and Scapozza, L. (2003). Molecular mechanisms of resistance to imatinib in Philadelphia-chromosome-positive leukaemias, *The Lancet Oncology* **4**, 2, pp. 75–85.

Gan, Q., Yoshida, T., McDonald, O. G. and Owens, G. K. (2007). Concise review: epigenetic mechanisms contribute to pluripotency and cell lineage determination of embryonic stem cells, *Stem Cells* **25**, pp. 2–9.

Garber, K. (2006). China approves world's first oncolytic virus therapy for cancer treatment, *Journal of the National Cancer Institute* **98**, 5, pp. 298–300.

Gardner, S. N. and Fernandes, M. (2003). New tools for cancer chemotherapy: computational assistance for tailoring treatments, *Molecular Cancer Therapy* **2**, 10, pp. 1079–1084.

Gastl, G., Hermann, T., Steurer, M., Zmija, J., Gunsilius, E., Unger, C. and Kraft, A. (2009). Angiogenesis as a target for tumor treatment, *Oncology* **54**, 3, pp. 177–184.

Gatenby, R. and Maini, P. (2003). Mathematical oncology: cancer summed up, *Nature* **421**, 6921, pp. 321–321.

Gatenby, R., Smallbone, K., Maini, P., Rose, F., Averill, J., Nagle, R., Worrall, L. and Gillies, R. (2007). Cellular adaptations to hypoxia and acidosis during somatic evolution of breast cancer, *British Journal of Cancer* **97**, 5, pp. 646–653.

Gatenby, R. A. and Frieden, B. R. (2004). Information dynamics in carcinogenesis and tumor growth, *Mutation research/ Fundamental and Molecular Mechanisms of Mutagenesis* **568**, 2, pp. 259–273.

Gatenby, R. A. and Gawlinski, E. T. (2003). The glycolytic phenotype in carcinogenesis and tumor invasion insights through mathematical models, *Cancer Research* **63**, 14, pp. 3847–3854.

Gatenby, R. A. and Gillies, R. J. (2008). A microenvironmental model of carcinogenesis, *Nature Reviews Cancer* **8**, 1, pp. 56–61.

Gatenby, R. A. and Vincent, T. L. (2003a). Application of quantitative models from population biology and evolutionary game theory to tumor therapeutic strategies, *Molecular Cancer Therapeutics* **2**, 9, pp. 919–927.

Gatenby, R. A. and Vincent, T. L. (2003b). An evolutionary model of carcinogenesis, *Cancer Research* **63**, 19, pp. 6212–6220.

Gelfand, I. M. and Fomin, S. V. (1963). *Calculus of Variations* (Prentice Hall).

Gerisch, A. and Chaplain, M. (2008). Mathematical modelling of cancer cell invasion of tissue: local and non-local models and the effect of adhesion, *Journal of Theoretical Biology* **250**, 4, pp. 684–704.

Gerlee, P. and Anderson, A. (2007). An evolutionary hybrid cellular automaton model of solid tumour growth, *Journal of Theoretical Biology* **246**, 4, pp. 583–603.

Giles, F., Cortes, J., Jones, D., Bergstrom, D., Kantarjian, H. and Freedman, S. (2007). Mk-0457, a novel kinase inhibitor, is active in patients with chronic myeloid leukemia or acute lymphocytic leukemia with the T315I BCR-ABL mutation, *Blood* **109**, 2, pp. 500–502.

Gillies, R. J. and Gatenby, R. A. (2007). Hypoxia and adaptive landscapes in the evolution of carcinogenesis, *Cancer and Metastasis Reviews* **26**, 2, pp. 311–317.

Girardi, M., Oppenheim, D. E., Steele, C. R., Lewis, J. M., Glusac, E., Filler, R., Hobby, P., Sutton, B., Tigelaar, R. E. and Hayday, A. C. (2001). Regulation of cutaneous malignancy by gamma delta T cells, *Science Signaling* **294**, 5542, p. 605.

Glauche, I., Thielecke, L. and Roeder, I. (2011). Cellular aging leads to functional heterogeneity of hematopoietic stem cells: a modeling perspective, *Aging Cell* **10**, 3, pp. 457–65, doi:10.1111/j.1474-9726.2011.00692.x.

Goardon, N., Marchi, E., Atzberger, A., Quek, L., Schuh, A., Soneji, S., Woll, P., Mead, A., Alford, K. A., Rout, R., Chaudhury, S., Gilkes, A., Knapper, S., Beldjord, K., Begum, S., Rose, S., Geddes, N., Griffiths, M., Standen, G., Sternberg, A., Cavenagh, J., Hunter, H., Bowen, D., Killick, S., Robinson, L., Price, A., Macintyre, E., Virgo, P., Burnett, A., Craddock, C., Enver, T., Jacobsen, S. E. W., Porcher, C. and Vyas, P. (2011). Coexistence of LMPP-like and GMP-like leukemia stem cells in acute myeloid leukemia, *Cancer Cell* **19**, 1, pp. 138–52, doi:10.1016/j.ccr.2010.12.012.

Gockenbach, M. S. (2010). *Partial Differential Equations: Analytical and Numerical Methods* (SIAM).

Goel, A. and Boland, C. R. (2010). Recent insights into the pathogenesis of colorectal cancer, *Current Opinion in Gastroenterology* **26**, 1, p. 47.

Goel, A. and Boland, C. R. (2012). Epigenetics of colorectal cancer, *Gastroenterology* **143**, pp. 1442–60.

Goel, A., Nagasaka, T., Arnold, C. N., Inoue, T., Hamilton, C., Niedzwiecki, D., Compton, C., Mayer, R. J., Goldberg, R., Bertagnolli, M. M. *et al.* (2007). The CpG island methylator phenotype and chromosomal instability are inversely correlated in sporadic colorectal cancer, *Gastroenterology* **132**, 1, pp. 127–138.

Gokhale, C., Iwasa, Y., Nowak, M. A. and Traulsen, A. (2009). The pace of evolution across fitness valleys, *Journal of Theoretical Biology* **259**, 3, p. 613.

Goldie, J. H. and Coldman, A. J. (1979). A mathematic model for relating the drug sensitivity of tumors to their spontaneous mutation rate, *Cancer Treatment Reports* **63**, 11-12, pp. 1727–1733.

Goldie, J. H. and Coldman, A. J. (1983a). A model for resistance of tumor cells to cancer chemotherapeutic agents, *Mathematical Biosciences* **65**, pp. 291–307.

Goldie, J. H. and Coldman, A. J. (1983b). Quantitative model for multiple levels of drug resistance in clinical tumors, *Cancer Treatment Reports* **67**, 10, pp. 923–931.

Goldie, J. H. and Coldman, A. J. (1985). A model for tumor response to chemotherapy: an integration of the stem cell and somatic mutation hypotheses, *Cancer investigation* **3**, 6, pp. 553–564.

Goldie, J. H. and Coldman, A. J. (1998). *Drug Resistance in Cancer: Mechanisms and Models* (Cambridge University Press).

Goldstein, A. S., Huang, J., Guo, C., Garraway, I. P. and Witte, O. N. (2010). Identification of a cell of origin for human prostate cancer, *Science* **329**, 5991, pp. 568–571.

Gonzalez-Suarez, E., Samper, E., Flores, J. M. and Blasco, M. A. (2000). Telomerase-deficient mice with short telomeres are resistant to skin tumorigenesis, *Nature Genetics* **26**, 1, pp. 114–117.

Gorre, M. E., Mohammed, M., Ellwood, K., Hsu, N., Paquette, R., Rao, P. N. and Sawyers, C. L. (2001). Clinical resistance to STI-571 cancer therapy caused by BCR-ABL gene mutation or amplification, *Science* **293**, 5531, pp. 876–880.

Graham, S. M., Jørgensen, H. G., Allan, E., Pearson, C., Alcorn, M. J., Richmond, L. and Holyoake, T. L. (2002). Primitive, quiescent, Philadelphia-positive stem cells from patients with chronic myeloid leukemia are insensitive to STI571 in vitro, *Blood* **99**, 1, pp. 319–325.

Greaves, M. (2002). *Cancer: the Evolutionary Legacy* (Oxford University Press).

Greaves, M. (2007). Darwinian medicine: a case for cancer, *Nature Reviews Cancer* **7**, 3, pp. 213–221.

Group, C. M. L. T. C. (1997). Interferon alfa versus chemotherapy for chronic myeloid leukemia: a meta-analysis of seven randomized trials, *Journal of National Cancer Institute* **89**, pp. 1616–1620.

Guibal, F. C., Alberich-Jorda, M., Hirai, H., Ebralidze, A., Levantini, E., Di Ruscio, A., Zhang, P., Santana-Lemos, B. A., Neuberg, D., Wagers, A. J. *et al.* (2009). Identification of a myeloid committed progenitor as the cancer-initiating cell in acute promyelocytic leukemia, *Blood* **114**, 27, pp. 5415–5425.

Guilak, F., Cohen, D., Estes, B., Gimble, J., Liedtke, W. and Chen, C. (2009). Control of stem cell fate by physical interactions with the extracellular matrix, *Cell Stem Cell* **5**, 1, pp. 17–26.

Guiot, C., Degiorgis, P. G., Delsanto, P. P., Gabriele, P. and Deisboeck, T. S. (2003). Does tumor growth follow a "universal law"? *Journal of Theoretical Biology* **225**, 2, pp. 147–151.

Gupta, P. B., Chaffer, C. L. and Weinberg, R. A. (2009). Cancer stem cells: mirage or reality? *Nature Medicine* **15**, 9, pp. 1010–1012.

Gyllenberg, M. and Webb, G. F. (1991). Quiescence in structured population dynamics, *Lecture Notes in Pure and Applied Mathematics, Dekker* **131**, pp. 45–62.

Hafner, C., Knuechel, R., Stoehr, R. and Hartmann, A. (2002). Clonality of multifocal urothelial carcinomas: 10 years of molecular genetic studies, *International Journal of Cancer* **101**, 1, pp. 1–6.

Hahnfeldt, P., Panigrahy, D., Folkman, J. and Hlatky, L. (1999). Tumor development under angiogenic signaling a dynamical theory of tumor growth, treatment response, and postvascular dormancy, *Cancer Research* **59**, 19, pp. 4770–4775.

Hanin, L. (2002). Identification problem for stochastic models with application to carcinogenesis, cancer detection and radiation biology, *Discrete Dynamics in Nature and Society* **7**, 3, pp. 177–189.

Harley, C. B., Futcher, A. B. and Greider, C. W. (1990). Telomeres shorten during ageing of human fibroblasts, *Nature* **345**, 6274, pp. 458–60, doi: 10.1038/345458a0.

Harnevo, L. E. and Agur, Z. (1991). The dynamics of gene amplification described as a multitype compartmental model and as a branching process, *Mathematical Biosciences* **103**, 1, pp. 115–138.

Harnevo, L. E. and Agur, Z. (1993). Use of mathematical models for understanding the dynamics of gene amplification, *Mutation Research* **292**, 1, pp. 17–24.

Harrington, E., Bebbington, D., Moore, J., Rasmussen, R., Ajose-Adeogun, A., Nakayama, T., Graham, J., Demur, C., Hercend, T., Diu-Hercend, A. *et al.* (2004). VX-680, a potent and selective small-molecule inhibitor of the Aurora kinases, suppresses tumor growth in vivo, *Nature Medicine* **10**, 3, pp. 262–267.

Hart, D., Shochat, E. and Agur, Z. (1998). The growth law of primary breast cancer as inferred from mammography screening trials data. *British Journal of Cancer* **78**, 3, p. 382.

Hartmann, A., Rösner, U., Schlake, G., Dietmaier, W., Zaak, D., Hofstaedter, F. and Knuechel, R. (2000). Clonality and genetic divergence in multifocal low-grade superficial urothelial carcinoma as determined by chromosome 9 and p53 deletion analysis, *Laboratory Investigation* **80**, 5, pp. 709–718.

Hassell, M. P. (2000). *The Spatial and Temporal Dynamics of Host-Parasitoid Interactions* (Oxford University Press).

Hasty, P., Campisi, J., Hoeijmakers, J., van Steeg, H. and Vijg, J. (2003). Aging and genome maintenance: lessons from the mouse? *Science* **299**, 5611, pp. 1355–1359.

Hayflick, L. and Moorhead, P. (1961). Serial cultivation of human diploid cell strains, *Experimental Cell Research* **25**, 3, pp. 585–621.

Heath, W. R. and Carbone, F. R. (2001). Cross-presentation in viral immunity and self-tolerance, *Nature Reviews Immunology* **1**, 2, pp. 126–134.

Heisterkamp, N., Stam, K., Groffen, J., de Klein, A. and Grosveld, G. (1985). Structural organization of the bcr gene and its role in the Ph' translocation, *Nature* **315**, pp. 758–761.

Hermann, A., Gowher, H. and Jeltsch, A. (2004). Biochemistry and biology of mammalian DNA methyltransferases, *Cellular and Molecular Life Sciences CMLS* **61**, 19-20, pp. 2571–2587.

Hildeman, D. A., Zhu, Y., Mitchell, T. C., Kappler, J. and Marrack, P. (2002). Molecular mechanisms of activated T cell death in vivo, *Current Opinion in Immunology* **14**, 3, pp. 354–359.

Hingorani, S. R., Petricoin III, E. F., Maitra, A., Rajapakse, V., King, C., Jacobetz, M. A., Ross, S., Conrads, T. P., Veenstra, T. D., Hitt, B. A. *et al.* (2003). Preinvasive and invasive ductal pancreatic cancer and its early detection in the mouse, *Cancer Cell* **4**, 6, pp. 437–450.

Hinow, P., Gerlee, P., McCawley, L., Quaranta, V., Ciobanu, M., Wang, S., Graham, J., Ayati, B., Claridge, J., Swanson, K. *et al.* (2009). A spatial model of tumor-host interaction: application of chemotherapy, *Mathematical Biosciences and Engineering* **6**, 3, p. 521.

Hinow, P., Rogers, C., Barbieri, C., Pietenpol, J., Kenworthy, A. and DiBenedetto, E. (2006). The DNA binding activity of p53 displays reaction-diffusion kinetics, *Biophysical Journal* **91**, 1, pp. 330–342.

Hirth, F. (2011). Stem cells and asymmetric cell division, in *Regenerative Medicine* (Springer), pp. 103–123.

Ho, A. D. (2005). Kinetics and symmetry of divisions of hematopoietic stem cells, *Experimental Hematology* **33**, 1, pp. 1–8.

Hoeijmakers, J. H. (2001). Genome maintenance mechanisms for preventing cancer, *Nature* **411**, 6835, pp. 366–374.

Hofacre, A., Wodarz, D., Komarova, N. L. and Fan, H. (2012). Early infection and spread of a conditionally replicating adenovirus under conditions of plaque formation, *Virology* **423**, 1, pp. 89–96.

Höglund, M., Gisselsson, D., Hansen, G. B., Säll, T. and Mitelman, F. (2002a). Multivariate analysis of chromosomal imbalances in breast cancer delineates cytogenetic pathways and reveals complex relationships among imbalances, *Cancer Research* **62**, 9, pp. 2675–2680.

Höglund, M., Gisselsson, D., Hansen, G. B., Säll, T., Mitelman, F. and Nilbert, M. (2002b). Dissecting karyotypic patterns in colorectal tumors two distinct but overlapping pathways in the adenoma-carcinoma transition, *Cancer Research* **62**, 20, pp. 5939–5946.

Höglund, M., Säll, T., Heim, S., Mitelman, F., Mandahl, N. and Fadl-Elmula, I. (2001). Identification of cytogenetic subgroups and karyotypic pathways in transitional cell carcinoma, *Cancer Research* **61**, 22, pp. 8241–8246.

Holland, E. C. (2000). Glioblastoma multiforme: the terminator, *Proceedings of the National Academy of Sciences* **97**, 12, pp. 6242–6244.

Holyoake, T., Jiang, X., Eaves, C. and Eaves, A. (1999). Isolation of a highly quiescent subpopulation of primitive leukemic cells in chronic myeloid leukemia, *Blood* **94**, 6, pp. 2056–2064.

Holyoake, T. L., Jiang, X., Jorgensen, H. G., Graham, S., Alcorn, M. J., Laird, C., Eaves, A. C. and Eaves, C. J. (2001). Primitive quiescent leukemic cells from patients with chronic myeloid leukemia spontaneously initiate factor-independent growth in vitro in association with up-regulation of expression of interleukin-3, *Blood* **97**, 3, pp. 720–728.

Horn, M., Loeffler, M. and Roeder, I. (2008). Mathematical modeling of genesis and treatment of chronic myeloid leukemia, *Cells Tissues Organs* **188**, 1-2, pp. 236–247.

Hornsby, C., Page, K. M. and Tomlinson, I. P. (2007). What can we learn from the population incidence of cancer? Armitage and Doll revisited, *The Lancet Oncology* **8**, 11, pp. 1030–1038.

Howlader, N., Noone, A., Krapcho, M., Neyman, N., Aminou, R., Altekruse, S., Kosary, C., Ruhl, J., Tatalovich, Z., Cho, H. *et al.* (2012). SEER cancer statistics review, 1975–2009 (vintage 2009 populations), *Bethesda, MD: National Cancer Institute* .

Hsieh, J. (2012). Orchestrating transcriptional control of adult neurogenesis, *Genes & Development* **26**, 10, pp. 1010–1021.

Hsu, M.-Y., Meier, F. and Herlyn, M. (2002). Melanoma development and progression: a conspiracy between tumor and host, *Differentiation* **70**, 9-10, pp. 522–536.

Huang, J., Papadopoulos, N., McKinley, A. J., Farrington, S. M., Curtis, L. J., Wyllie, A. H., Zheng, S., Willson, J. K., Markowitz, S. D., Morin, P., Kinzler, K. W., Vogelstein, B. and Dunlop, M. G. (1996). APC mutations in colorectal tumors with mismatch repair deficiency, *Proceedings of the National Academy of Sciences U S A* **93**, 17, pp. 9049–9054.

Huebner, R. J. and Todaro, G. J. (1969). Oncogenes of RNA tumor viruses as determinants of cancer, *Proceedings of the National Academy of Sciences* **64**, 3, pp. 1087–1094.

Huffman, K. E., Levene, S. D., Tesmer, V. M., Shay, J. W. and Wright, W. E. (2000). Telomere shortening is proportional to the size of the G-rich telomeric 3'-overhang, *Journal of Biological Chemistry* **275**, 26, pp. 19719–22, doi:10.1074/jbc.M002843200.

Huntly, B. J. P., Shigematsu, H., Deguchi, K., Lee, B. H., Mizuno, S., Duclos, N., Rowan, R., Amaral, S., Curley, D., Williams, I. R., Akashi, K. and Gilliland, D. G. (2004). MOZ-TIF2, but not BCR-ABL, confers properties of leukemic stem cells to committed murine hematopoietic progenitors, *Cancer Cell* **6**, 6, pp. 587–96, doi:10.1016/j.ccr.2004.10.015.

Huttner, W. B. and Kosodo, Y. (2005). Symmetric versus asymmetric cell division during neurogenesis in the developing vertebrate central nervous system, *Current Opinion in Cell Biology* **17**, 6, pp. 648–657.

Iacobuzio-Donahue, C. A. (2009). Epigenetic changes in cancer, *Annual Review of Pathological Mechanical Disease* **4**, pp. 229–249.

Issa, J.-P. (2004). CpG island methylator phenotype in cancer, *Nature Reviews Cancer* **4**, 12, pp. 988–993.

Iwasa, Y. and Levin, S. A. (1995). The timing of life history events, *Journal of Theoretical Biology* **172**, pp. 33–42.

Iwasa, Y., Michor, F. and Nowak, M. A. (2004a). Evolutionary dynamics of invasion and escape, *Journal of Theoretical Biology* **226**, 2, pp. 205–214.

Iwasa, Y., Michor, F. and Nowak, M. A. (2004b). Stochastic tunnels in evolutionary dynamics, *Genetics* **166**, 3, pp. 1571–1579.

Jäättelä, M. (2004). Multiple cell death pathways as regulators of tumour initiation and progression, *Oncogene* **23**, 16, pp. 2746–2756.

Jackson, A. L., Chen, R. and Loeb, L. A. (1998). Induction of microsatellite instability by oxidative DNA damage, *Proceedings of the National Academy of Sciences* **95**, 21, pp. 12468–12473.

Jackson, T. L. (2012). *Modeling Tumor Vasculature: Molecular, Cellular, and Tissue Level Aspects and Implications* (Springer).

Jackson, T. L. and Byrne, H. M. (2000). A mathematical model to study the effects of drug resistance and vasculature on the response of solid tumors to chemotherapy, *Mathematical Biosciences* **164**, 1, pp. 17–38.

Jaffrézou, J.-P., Chen, G., Durán, G. E., Kühl, J.-S. and Sikic, B. I. (1994). Mutation rates and mechanisms of resistance to etoposide determined from flucattion analysis, *Journal of the National Cancer Institute* **86**, 15, pp. 1152–1158.

Jair, K.-W., Bachman, K. E., Suzuki, H., Ting, A. H., Rhee, I., Yen, R.-W. C., Baylin, S. B. and Schuebel, K. E. (2006). De novo CpG island methylation in human cancer cells, *Cancer Research* **66**, 2, pp. 682–692.

Janeway, C. A., Travers, P., Walport, M. and Shlomchik, M. J. (2005). *Immunobiology: The immune system in health and disease* (Garland Science, New York).

Janssen, A. and Medema, R. H. (2013). Cancer: Stress mixes chromosomes, *Nature* **494**, 7438, pp. 439–441.

Jefferson, K. K. (2004). What drives bacteria to produce a biofilm? *FEMS microbiology letters* **236**, 2, pp. 163–173.

Johnston, M. D., Edwards, C. M., Bodmer, W. F., Maini, P. K. and Chapman, S. J. (2007). Mathematical modeling of cell population dynamics in the colonic crypt and in colorectal cancer, *Proceedings of the National Academy of Sciences U S A* **104**, 10, pp. 4008–13, doi:10.1073/pnas.0611179104.

Johnston, M. D., Maini, P. K., Jonathan Chapman, S., Edwards, C. M. and Bodmer, W. F. (2010). On the proportion of cancer stem cells in a tumour, *Journal of Theoretical Biology* **266**, 4, pp. 708–711.

Jones, P. A. and Baylin, S. B. (2002). The fundamental role of epigenetic events in cancer, *Nature Reviews. Genetics* **3**, 6, pp. 415–428.

Jordan, C. T., Guzman, M. L. and Noble, M. (2006). Cancer stem cells, *New England Journal of Medicine* **355**, 12, pp. 1253–1261.

Junker, K., Schlichter, A., Hindermann, W. and Schubert, J. (1999). Genetic characterization of multifocal tumor growth in renal cell carcinoma, *Kidney International* **56**, 4, pp. 1291–1294.

Junker, K., Schlichter, A., Junker, U., Knöfel, B., Kosmehl, H., Schubert, J. and Claussen, U. (1997). Cytogenetic, histopathologic, and immunologic studies of multifocal renal cell carcinoma, *Cancer* **79**, 5, pp. 975–981.

Junker, K., Thrum, K., Schlichter, A., Müller, G., Hindermann, W. and Schubert, J. (2002). Clonal origin of multifocal renal cell carcinoma as determined by microsatellite analysis, *The Journal of Urology* **168**, 6, pp. 2632–2636.

Kalams, S. A., Buchbinder, S., Rosenberg, E., Billingsley, J., Colbert, D., Jones, N., Shea, A., Trocha, A. and Walker, B. (1999). Association between virus-specific cytotoxic T-lymphocyte and helper responses in human immunodeficiency virus type 1 infection, *Journal of Virology* **73**, 8, pp. 6715–6720.

Kanai, Y., Ushijima, S., Kondo, Y., Nakanishi, Y. and Hirohashi, S. (2001). DNA methyltransferase expression and DNA methylation of CPG islands and peri-centromeric satellite regions in human colorectal and stomach cancers, *International Journal of Cancer* **91**, 2, pp. 205–212.

Kane, M. F., Loda, M., Gaida, G. M., Lipman, J., Mishra, R., Goldman, H., Jessup, J. M. and Kolodner, R. (1997). Methylation of the hMLH1 promoter correlates with lack of expression of hMLH1 in sporadic colon tumors and mismatch repair-defective human tumor cell lines, *Cancer Research* **57**, 5, pp. 808–811.

Kansal, A., Torquato, S., Harsh IV, G., Chiocca, E. and Deisboeck, T. (2000). Simulated brain tumor growth dynamics using a three-dimensional cellular automaton, *Journal of Theoretical Biology* **203**, 4, pp. 367–382.

Kantarjian, H., Baccarani, M., Jabbour, E., Saglio, G. and Cortes, J. (2011). Second-generation tyrosine kinase inhibitors: the future of frontline CML therapy, *Clinical Cancer Research* **17**, 7, pp. 1674–1683.

Kantarjian, H., Talpaz, M., Giles, F., O'Brien, S., Cortes, J. *et al.* (2006). New insights into the pathophysiology of chronic myeloid leukemia and imatinib resistance. *Annals of Internal Medicine* **145**, 12, p. 913.

Kaplan, D. H., Shankaran, V., Dighe, A. S., Stockert, E., Aguet, M., Old, L. J. and Schreiber, R. D. (1998). Demonstration of an interferon γ-dependent tumor surveillance system in immunocompetent mice, *Proceedings of the National Academy of Sciences* **95**, 13, pp. 7556–7561.

Kaplan, J. M. (2005). Adenovirus-based cancer gene therapy, *Current Gene Therapy* **5**, 6, pp. 595–605.

Karev, G. P., Novozhilov, A. S., Koonin, E. V. *et al.* (2006). Mathematical modeling of tumor therapy with oncolytic viruses: effects of parametric heterogeneity on cell dynamics, *Biology Direct* **1**, 30, p. 19.

Karlin, S. and Taylor, H. E. (1975). *A First Course in Stochastic Processes* (Academic Press).

Katouli, A. and Komarova, N. (2010). Optimizing combination therapies with existing and future CML drugs, *PLoS One* **5**, 8, p. e12300.

Katouli, A. A. and Komarova, N. L. (2011). The worst drug rule revisited: mathematical modeling of cyclic cancer treatments, *Bulletin of Mathematical Biology* **73**, 3, pp. 549–584.

Keller, J. J., Offerhaus, G. J. A., Drillenburg, P., Caspers, E., Musler, A., Ristimäki, A. and Giardiello, F. M. (2001). Molecular analysis of sulindac-resistant adenomas in familial adenomatous polyposis, *Clinical Cancer Research* **7**, 12, pp. 4000–4007.

Kelly, E. and Russell, S. J. (2007). History of oncolytic viruses: genesis to genetic engineering, *Molecular Therapy* **15**, 4, pp. 651–659.

Kemp, C. J., Donehower, L. A., Bradley, A. and Balmain, A. (1993). Reduction of p53 gene dosage does not increase initiation or promotion but enhances malignant progression of chemically induced skin tumors, *Cell* **74**, 5, pp. 813–822.

Kendal, W. S. and Frost, P. (1988). Pitfalls and practice of Luria-Delbrück fluctuation analysis: a review, *Cancer Research* **48**, 5, pp. 1060–1065.

Kern, S. E. and Shibata, D. (2007). The fuzzy math of solid tumor stem cells: a perspective, *Cancer Research* **67**, 19, pp. 8985–8988.

Kim, K.-M., Calabrese, P., Tavaré, S. and Shibata, D. (2004). Enhanced stem cell survival in familial adenomatous polyposis, *The American Journal of Pathology* **164**, 4, pp. 1369–1377.

Kim, P. S., Lee, P. P. and Levy, D. (2008). Dynamics and potential impact of the immune response to chronic myelogenous leukemia, *PLoS Computational Biology* **4**, 6, p. e1000095.

Kim, Y., Stolarska, M. A. and Othmer, H. G. (2011). The role of the microenvironment in tumor growth and invasion, *Progress in Biophysics and Molecular Biology* **106**, 2, pp. 353–379.

Kimmel, M. (2010). Evolution and cancer: a mathematical biology approach, *Biology Direct* **5**, p. 29.

Kimmel, M. and Axelrod, D. E. (1990). Mathematical models of gene amplification with applications to cellular drug resistance and tumorigenicity, *Genetics* **125**, 3, pp. 633–644.

Kimmel, M. and Stivers, D. N. (1994). Time-continuous branching walk models of unstable gene amplification, *Bulletin of Mathematical Biology* **56**, 2, pp. 337–357.

Kimmel, M., Swierniak, A. and Polanski, A. (1998). Infinite-dimensional model of evolution of drug resistance of cancer cells, *Journal of Mathematical Systems Estimation and Control* **8**, pp. 1–16.

Kimura, M. (1985). The role of compensatory neutral mutations in molecular evolution, *Journal of Genetics* **64**, 1, pp. 7–19.

Kimura, M. (1995). *Population Genetics, Molecular Evolution, and the Neutral Theory: Selected Papers* (University of Chicago Press).

Kinzler, K. and Vogelstein, B. (2002). *The Genetic Basis of Human Cancer* (McGraw-Hill).

Kirn, D. H. and McCormick, F. (1996). Replicating viruses as selective cancer therapeutics, *Molecular Medicine Today* **2**, 12, pp. 519–527.

Kirschner, D., Lenhart, S. and Serbin, S. (1997). Optimal control of the chemotherapy of HIV, *Journal of Mathematical Biology* **35**, 7, pp. 775–792.

Klein, A. M., Nakagawa, T., Ichikawa, R., Yoshida, S. and Simons, B. D. (2010). Mouse germ line stem cells undergo rapid and stochastic turnover, *Cell Stem Cell* **7**, 2, pp. 214–224.

Klein, A. M. and Simons, B. D. (2011). Universal patterns of stem cell fate in cycling adult tissues, *Development* **138**, 15, pp. 3103–3111.

Klein, G. and Révész, L. (1953). Quantitative studies on the multiplication of neoplastic cells in vivo. I. Growth curves of the Ehrlich and MClM ascites tumors, *Journal of the National Cancer Institute* **14**, 2, pp. 229–277.

Knighton, D., Ausprunk, D., Tapper, D. and Folkman, J. (1977). Avascular and vascular phases of tumour growth in the chick embryo, *British Journal of Cancer* **35**, 3, p. 347.

Knoblich, J. A. (2008). Mechanisms of asymmetric stem cell division, *Cell* **132**, 4, pp. 583–597.

Knudson, A. G. (1971). Mutation and cancer: statistical study of retinoblastoma, *Proceedings of the National Academy of Sciences* **68**, 4, pp. 820–823.

Kolodner, R. D., Cleveland, D. W. and Putnam, C. D. (2011). Aneuploidy drives a mutator phenotype in cancer, *Science* **333**, 6045, pp. 942–943.

Komarova, N. (2006a). Loss- and gain-of-function mutations in cancer: mass-action, spatial and hierarchical models, *Journal of Statistical Physics* DOI 10.1007/s10955-006-9238-0.

Komarova, N. (2006b). Stochastic modeling of drug resistance in cancer, *Journal of Theoretical Biology* **239**, 3, pp. 351–366.

Komarova, N., Wu, L. and Baldi, P. (2007). The fixed-size Luria–Delbrück model with a nonzero death rate, *Mathematical Biosciences* **210**, 1, pp. 253–290.

Komarova, N. L. (2004). Does cancer solve an optimization problem? *Cell Cycle* **3**, 7, pp. 838–842.

Komarova, N. L. (2006c). Spatial stochastic models for cancer initiation and progression, *Bulletin of Mathematical Biology* **68**, 7, pp. 1573–1599.

Komarova, N. L. (2007). Stochastic modeling of loss-and gain-of-function mutations in cancer, *Mathematical Models and Methods in Applied Sciences* **17**, supp01, pp. 1647–1673.

Komarova, N. L., Barnes, E., Klenerman, P. and Wodarz, D. (2003a). Boosting immunity by antiviral drug therapy: a simple relationship among timing, efficacy, and success, *Proceedings of the National Academy of Sciences* **100**, 4, pp. 1855–1860.

Komarova, N. L. and Boland, C. R. (2013). Cancer: Calculated treatment, *Nature* **499**, 7458, pp. 291–292.

Komarova, N. L. and Mironov, V. (2005). On the role of endothelial progenitor cells in tumor neovascularization, *Journal of Theoretical Biology* **235**, 3, pp. 338–349.

Komarova, N. L., Sadovsky, A. V. and Wan, F. Y. (2008a). Selective pressures for and against genetic instability in cancer: a time-dependent problem, *Journal of The Royal Society Interface* **5**, 18, pp. 105–121.

Komarova, N. L., Sengupta, A. and Nowak, M. A. (2003b). Mutation–selection networks of cancer initiation: tumor suppressor genes and chromosomal instability, *Journal of Theoretical Biology* **223**, 4, pp. 433–450.

Komarova, N. L., Urwin, E. and Wodarz, D. (2012). Accelerated crossing of fitness valleys through division of labor and cheating in asexual populations, *Scientific Reports* **2**, p. 917.

Komarova, N. L. and Wang, L. (2004). Initiation of colorectal cancer: where do the two hits hit? *Cell Cycle* **3**, 12, pp. 1558–1565.

Komarova, N. L. and Wodarz, D. (2003). Evolutionary dynamics of mutator phenotypes in cancer implications for chemotherapy, *Cancer Research* **63**, 20, pp. 6635–6642.

Komarova, N. L. and Wodarz, D. (2004). The optimal rate of chromosome loss for the inactivation of tumor suppressor genes in cancer, *Proceedings of the National Academy of Sciences USA* **101**, 18, pp. 7017–7021.

Komarova, N. L. and Wodarz, D. (2005). Drug resistance in cancer: principles of emergence and prevention, *Proceedings of the National Academy of Sciences U S A* **102**, 27, pp. 9714–9.

Komarova, N. L. and Wodarz, D. (2007). Effect of cellular quiescence on the success of targeted CML therapy, *PLoS One* **2**, 10, p. e990.

Komarova, N. L. and Wodarz, D. (2009). Combination therapies against chronic myeloid leukemia: short-term versus long-term strategies, *Cancer Research* **69**, 11, pp. 4904–4910.

Komarova, N. L. and Wodarz, D. (2010). ODE models for oncolytic virus dynamics, *Journal of Theoretical Biology* **263**, 4, pp. 530–543.

Komarova, N. L., Wodarz, D., Boland, C. R. and Goel, A. (2008b). Genetic and epigenetic pathways to colon cancer: Relating experimental evidence with modeling, in *Selected Topics in Cancer Modeling* (Springer), pp. 1–30.

Koppers-Lalic, D. and Hoeben, R. C. (2011). Non-human viruses developed as therapeutic agent for use in humans, *Reviews in Medical Virology* **21**, 4, pp. 227–239.

Krivtsov, A. V., Twomey, D., Feng, Z., Stubbs, M. C., Wang, Y., Faber, J., Levine, J. E., Wang, J., Hahn, W. C., Gilliland, D. G., Golub, T. R. and Armstrong, S. A. (2006). Transformation from committed progenitor to leukaemia stem cell initiated by MLL-aF9, *Nature* **442**, 7104, pp. 818–22, doi:10.1038/nature04980.

Kroemer, G. and Martin, S. J. (2005). Caspase-independent cell death, *Nature Medicine* **11**, 7, pp. 725–730.

Kuperwasser, C., Chavarria, T., Wu, M., Magrane, G., Gray, J. W., Carey, L., Richardson, A. and Weinberg, R. A. (2004). Reconstruction of functionally normal and malignant human breast tissues in mice, *Proceedings of the National Academy of Sciences USA* **101**, 14, pp. 4966–4971.

Kupryjanczyk, J., Thor, A. D., Beauchamp, R., Poremba, C., Scully, R. E., Yandell, D. W. *et al.* (1996). Ovarian, peritoneal, and endometrial serous carcinoma: clonal origin of multifocal disease, *Modern Pathology* **9**, 3, p. 166.

Laird, A. K. (1964). Dynamics of tumour growth, *British Journal of Cancer* **18**, 3, p. 490.

Laird, P. W. and Jaenisch, R. (1996). The role of DNA methylation in cancer genetics and epigenetics, *Annual Review of Genetics* **30**, 1, pp. 441–464.

Lamlum, H., Ilyas, M., Rowan, A., Clark, S., Johnson, V., Bell, J., Frayling, I., Efstathiou, J., Pack, K., Payne, S. *et al.* (1999). The type of somatic mutation at APC in familial adenomatous polyposis is determined by the site of the germline mutation: a new facet to Knudson's 'two-hit' hypothesis, *Nature Medicine* **5**, 9, pp. 1071–1075.

Lander, A. D., Gokoffski, K. K., Wan, F. Y., Nie, Q. and Calof, A. L. (2009). Cell lineages and the logic of proliferative control, *PLoS Biology* **7**, 1, p. e1000015.

Lathia, J. D., Heddleston, J. M., Venere, M. and Rich, J. N. (2011). Deadly teamwork: neural cancer stem cells and the tumor microenvironment, *Cell Stem Cell* **8**, 5, pp. 482–485.

Lavado, A., Lagutin, O., Chow, L., Baker, S. and Oliver, G. (2010). Prox1 is required for granule cell maturation and intermediate progenitor maintenance during brain neurogenesis, *PLoS Biology* **8**, 8, p. e1000460.

Leach, S. D. (2004). Mouse models of pancreatic cancer: the fur is finally flying! *Cancer Cell* **5**, 1, pp. 7–11.

Ledzewicz, U., Naghnaeian, M. and Schättler, H. (2011). Dynamics of tumor-immune interaction under treatment as an optimal control problem, *Dynamical Systems* , pp. 971–980.

Ledzewicz, U., Naghnaeian, M. and Schättler, H. (2012). Optimal response to chemotherapy for a mathematical model of tumor–immune dynamics, *Journal of Mathematical Biology* **64**, 3, pp. 557–577.

Ledzewicz, U. and Schättler, H. (2002). Analysis of a cell-cycle specific model for cancer chemotherapy, *Journal of Biological Systems* **10**, 03, pp. 183–206.

Ledzewicz, U. and Schättler, H. (2007). Optimal controls for a model with pharmacokinetics maximizing bone marrow in cancer chemotherapy, *Mathematical Biosciences* **206**, 2, pp. 320–342.

Ledzewicz, U. and Schättler, H. (2008). Optimal and suboptimal protocols for a class of mathematical models of tumor anti-angiogenesis, *Journal of Theoretical Biology* **252**, 2, pp. 295–312.

Lee, J., Son, M. J., Woolard, K., Donin, N. M., Li, A., Cheng, C. H., Kotliarova, S., Kotliarov, Y., Walling, J., Ahn, S., Kim, M., Totonchy, M., Cusack,

T., Ene, C., Ma, H., Su, Q., Zenklusen, J. C., Zhang, W., Maric, D. and Fine, H. A. (2008). Epigenetic-mediated dysfunction of the bone morphogenetic protein pathway inhibits differentiation of glioblastoma-initiating cells, *Cancer Cell* **13**, 1, pp. 69–80, doi:10.1016/j.ccr.2007.12.005.

Lee, P. J., Washer, L. L., Law, D. J., Boland, C. R., Horon, I. L. and Feinberg, A. P. (1996). Limited up-regulation of DNA methyltransferase in human colon cancer reflecting increased cell proliferation, *Proceedings of the National Academy of Sciences* **93**, 19, pp. 10366–10370.

Leitner, A., Hochhaus, A., Müller, M. *et al.* (2011). Current treatment concepts of CML. *Current Cancer Drug Targets* **11**, 1, p. 31.

Lenaerts, T., Pacheco, J. M., Traulsen, A. and Dingli, D. (2010). Tyrosine kinase inhibitor therapy can cure chronic myeloid leukemia without hitting leukemic stem cells, *Haematologica* **95**, 6, pp. 900–907.

Lengauer, C., Kinzler, K. and Vogelstein, B. (1997). Genetic instability in colorectal cancers, *Nature* **386**, 6625, pp. 623–627.

Lengauer, C., Kinzler, K. W. and Vogelstein, B. (1998). Genetic instabilities in human cancers, *Nature* **396**, 6712, pp. 643–649.

Lenhart, S. and Workman, J. T. (2007). *Optimal Control Applied to Biological Models* (Chapman and Hall/CRC).

Li, L. and Clevers, H. (2010). Coexistence of quiescent and active adult stem cells in mammals, *Science* **327**, 5965, pp. 542–545.

Lieberman, E., Hauert, C. and Nowak, M. (2005). Evolutionary dynamics on graphs, *Nature* **433**, pp. 312–316.

Lien, W., Klezovitch, O., Fernandez, T., Delrow, J. and Vasioukhin, V. (2006). α E-catenin controls cerebral cortical size by regulating the hedgehog signaling pathway, *Science's STKE* **311**, 5767, p. 1609.

Lifson, J. D., Rossio, J. L., Piatak, M., Parks, T., Li, L., Kiser, R., Coalter, V., Fisher, B., Flynn, B. M., Czajak, S. *et al.* (2001). Role of CD8+ lymphocytes in control of simian immunodeficiency virus infection and resistance to rechallenge after transient early antiretroviral treatment, *Journal of Virology* **75**, 21, pp. 10187–10199.

Liggett, T. M. (1999). *Stochastic Interacting Systems: Contact, Voter and Exclusion Processes*, Vol. 324 (Springer).

Lim, C.-S. and Campisi, J. (2001). Recent advances in cellular senescence, cancer and aging, *Biotechnology and Bioprocess Engineering* **6**, 4, pp. 231–236.

Lim, D. A., Tramontin, A. D., Trevejo, J. M., Herrera, D. G., García-Verdugo, J. M. and Alvarez-Buylla, A. (2000). Noggin antagonizes BMP signaling to create a niche for adult neurogenesis, *Neuron* **28**, 3, pp. 713–26.

Lim, E., Vaillant, F., Wu, D., Forrest, N. C., Pal, B., Hart, A. H., Asselin-Labat, M.-L., Gyorki, D. E., Ward, T., Partanen, A., Feleppa, F., Huschtscha, L. I., Thorne, H. J., kConFab, Fox, S. B., Yan, M., French, J. D., Brown, M. A., Smyth, G. K., Visvader, J. E. and Lindeman, G. J. (2009). Aberrant luminal progenitors as the candidate target population for basal tumor development in BRCA1 mutation carriers, *Nature Medicine* **15**, 8, pp. 907–13, doi:10.1038/nm.2000.

Little, M. and Wright, E. (2003). A stochastic carcinogenesis model incorporating genomic instability fitted to colon cancer data, *Mathematical Biosciences* **183**, 2, pp. 111–134.

Little, M. P. and Li, G. (2006). Stochastic modelling of colon cancer: is there a role for genomic instability? *Carcinogenesis* **28**, 2, pp. 479–487.

Liu, M., Pleasure, S., Collins, A., Noebels, J., Naya, F., Tsai, M. and Lowenstein, D. (2000). Loss of beta2/neurod leads to malformation of the dentate gyrus and epilepsy, *Proceedings of the National Academy of Sciences* **97**, 2, pp. 865–870.

Lobo, N. A., Shimono, Y., Qian, D. and Clarke, M. F. (2007). The biology of cancer stem cells, *Annual Review of Cell and Developmental Biology* **23**, pp. 675–699, doi:10.1146/annurev.cellbio.22.010305.104154.

Lodish, H., Berk, A., Zipursky, S. L., Matsudaira, P., Baltimore, D. and Darnell, J. (2000). *Molecular Cell Biology* (WH Freeman).

Loeb, L. A. (1991). Mutator phenotype may be required for multistage carcinogenesis, *Cancer Research* **51**, 12.

Loeb, L. A. (2001). A mutator phenotype in cancer, *Cancer Research* **61**, 8, pp. 3230–3239.

Loeb, L. A. (2011). Human cancers express mutator phenotypes: origin, consequences and targeting, *Nature Reviews Cancer* **11**, 6, pp. 450–457.

Loeb, L. A., Loeb, K. R. and Anderson, J. P. (2003). Multiple mutations and cancer, *Proceedings of the National Academy of Sciences* **100**, 3, pp. 776–781.

Loeb, L. A., Springgate, C. F. and Battula, N. (1974). Errors in DNA replication as a basis of malignant changes, *Cancer Research* **34**, 9, pp. 2311–2321.

Loeffler, M. and Roeder, I. (2002). Tissue stem cells: definition, plasticity, heterogeneity, self-organization and models–a conceptual approach, *Cells Tissues Organs* **171**, 1, pp. 8–26.

Lopez-Garcia, C., Klein, A. M., Simons, B. D. and Winton, D. J. (2010). Intestinal stem cell replacement follows a pattern of neutral drift, *Science* **330**, 6005, pp. 822–825.

Louhelainen, J., Wijkström, H. and Hemminki, K. (2000). Allelic losses demonstrate monoclonality of multifocal bladder tumors, *International Journal of Cancer* **87**, 4, pp. 522–527.

Lowengrub, J. S., Frieboes, H. B., Jin, F., Chuang, Y., Li, X., Macklin, P., Wise, S. and Cristini, V. (2010). Nonlinear modelling of cancer: bridging the gap between cells and tumours, *Nonlinearity* **23**, 1, p. R1.

Luebeck, E. G. and Moolgavkar, S. H. (2002). Multistage carcinogenesis and the incidence of colorectal cancer, *Proceedings of the National Academy of Sciences* **99**, 23, pp. 15095–15100.

Luo, G., Santoro, I. M., McDaniel, L. D., Nishijima, I., Mills, M., Youssoufian, H., Vogel, H., Schultz, R. A. and Bradley, A. (2000). Cancer predisposition caused by elevated mitotic recombination in Bloom mice, *Nature Genetics* **26**, 4, pp. 424–429.

Luria, S. E. and Delbrück, M. (1943). Mutations of bacteria from virus sensitivity to virus resistance, *Genetics* **28**, 6, p. 491.

MacKey, M. C. (2001). Cell kinetic status of haematopoietic stem cells, *Cell Proliferation* **34**, pp. 71–83.

Macklin, P., McDougall, S., Anderson, A., Chaplain, M., Cristini, V. and Lowengrub, J. (2009). Multiscale modelling and nonlinear simulation of vascular tumour growth, *Journal of Mathematical Biology* **58**, 4, pp. 765–798.

Mahon, F.-X., Réa, D., Guilhot, J., Guilhot, F., Huguet, F., Nicolini, F., Legros, L., Charbonnier, A., Guerci, A., Varet, B. *et al.* (2010). Discontinuation of imatinib in patients with chronic myeloid leukaemia who have maintained complete molecular remission for at least 2 years: the prospective, multicentre Stop Imatinib (STIM) trial, *The Lancet Oncology* **11**, 11, pp. 1029–1035.

Makarov, V., Hirose, Y. and Langmore, J. (1997). Long G tails at both ends of human chromosomes suggest a C strand degradation mechanism for telomere shortening, *Cell* **88**, 5, pp. 657–666.

Maldarelli, F., Palmer, S., King, M. S., Wiegand, A., Polis, M. A., Mican, J., Kovacs, J. A., Davey, R. T., Rock-Kress, D., Dewar, R. *et al.* (2007). ART suppresses plasma HIV-1 RNA to a stable set point predicted by pretherapy viremia, *PLoS Pathogens* **3**, 4, p. e46.

Mandonnet, E., Delattre, J.-Y., Tanguy, M.-L., Swanson, K. R., Carpentier, A. F., Duffau, H., Cornu, P., Van Effenterre, R., Alvord, E. C. and Capelle, L. (2003). Continuous growth of mean tumor diameter in a subset of grade II gliomas, *Annals of Neurology* **53**, 4, pp. 524–528.

Mantzaris, N. V., Webb, S. and Othmer, H. G. (2004). Mathematical modeling of tumor-induced angiogenesis, *Journal of Mathematical Biology* **49**, 2, pp. 111–187.

Marciniak-Czochra, A. and Kimmel, M. (2007). Modelling of early lung cancer progression: Influence of growth factor production and cooperation between partially transformed cells, *Mathematical Models and Methods in Applied Sciences* **17**, supp01, pp. 1693–1719.

Marciniak-Czochra, A. and Kimmel, M. (2008). Reaction-difusion model of early carcinogenesis: The effects of influx of mutated cells, *Mathematical Modelling of Natural Phenomena* **3**, 07, pp. 90–114.

Marciniak-Czochra, A., Stiehl, T., Ho, A. D., Jäger, W. and Wagner, W. (2009a). Modeling of asymmetric cell division in hematopoietic stem cells–regulation of self-renewal is essential for efficient repopulation, *Stem Cells Development* **18**, 3, pp. 377–85, doi:10.1089/scd.2008.0143.

Marciniak-Czochra, A., Stiehl, T. and Wagner, W. (2009b). Modeling of replicative senescence in hematopoietic development, *Aging* **1**, 8, pp. 723–32.

Marsh, D. J. and Zori, R. T. (2002). Genetic insights into familial cancers–update and recent discoveries, *Cancer Letters* **181**, 2, pp. 125–164.

Marshman, E., Booth, C. and Potten, C. S. (2002). The intestinal epithelial stem cell, *Bioessays* **24**, 1, pp. 91–98.

Martin, N. K., Gaffney, E. A., Gatenby, R. A. and Maini, P. K. (2010). Tumour–stromal interactions in acid-mediated invasion: a mathematical model, *Journal of Theoretical Biology* **267**, 3, pp. 461–470.

Martuza, R. L., Malick, A., Markert, J. M., Ruffner, K. L. and Coen, D. M. (1991). Experimental therapy of human glioma by means of a genetically engineered virus mutant, *Science* **252**, 5007, pp. 854–856.

Marušić, M., Bajzer, Ž., Freyer, J. and Vuk-Pavlović, S. (1994). Analysis of growth of multicellular tumour spheroids by mathematical models, *Cell Proliferation* **27**, 2, pp. 73–94.

Mascré, G., Dekoninck, S., Drogat, B., Youssef, K. K., Brohée, S., Sotiropoulou, P. A., Simons, B. D. and Blanpain, C. (2012). Distinct contribution of stem and progenitor cells to epidermal maintenance, *Nature* **489**, 7415, pp. 257–262.

Massagué, J. (2000). TGF-β in cancer, *Cell* **103**, 2, pp. 295–309.

Massagué, J. (2001). G1 cell cycle control and cancer, *Nature* **432**, 7015, pp. 298–306.

Matrisian, L. M., Cunha, G. R. and Mohla, S. (2001). Epithelial-stromal interactions and tumor progression: meeting summary and future directions, *Cancer Research* **61**, 9, pp. 3844–3846.

Maynard Smith, J. (1982). *Evolution and the Theory of Games* (Cambridge University Press).

Mayneord, W. (1932). On a law of growth of Jensen's rat sarcoma, *The American Journal of Cancer* **16**, 4, pp. 841–846.

Mbeunkui, F. and Johann Jr, D. J. (2009). Cancer and the tumor microenvironment: a review of an essential relationship, *Cancer Chemotherapy and Pharmacology* **63**, 4, pp. 571–582.

McAdams, H. H. and Arkin, A. (1997). Stochastic mechanisms in gene expression, *Proceedings of the National Academy of Sciences* **94**, 3, pp. 814–819.

McCormick, F. (2001). New-age drug meets resistance, *Nature* **412**, 6844, pp. 281–282.

McCormick, F. (2003). Cancer specific viruses and the development of ONYX-015, *Cancer Biology & Therapy* **2**, pp. 0–3.

McCormick, F. (2005). Future prospects for oncolytic therapy, *Oncogene* **24**, 52, pp. 7817–7819.

McKinnell, R. G. (1998). *The Biological Basis of Cancer* (Cambridge University Press).

McPherron, A. C., Lawler, A. M. and Lee, S. J. (1997). Regulation of skeletal muscle mass in mice by a new TGF-beta superfamily member, *Nature* **387**, 6628, pp. 83–90, doi:10.1038/387083a0.

Melo, J. and Chuah, C. (2007). Resistance to imatinib mesylate in chronic myeloid leukaemia, *Cancer Letters* **249**, 2, pp. 121–132.

Melo, J. V. and Barnes, D. J. (2007). Chronic myeloid leukaemia as a model of disease evolution in human cancer, *Nature Reviews Cancer* **7**, 6, pp. 441–453.

Melo, J. V., Hughes, T. P. and Apperley, J. F. (2003). Chronic myeloid leukemia, *ASH Education Program Book* **2003**, 1, pp. 132–152.

Menigatti, M., Di Gregorio, C., Borghi, F., Sala, E., Scarselli, A., Pedroni, M., Foroni, M., Benatti, P., Roncucci, L., Ponz de Leon, M. *et al.* (2001). Methylation pattern of different regions of the MLH1 promoter and silencing of gene expression in hereditary and sporadic colorectal cancer, *Genes, Chromosomes and Cancer* **31**, 4, pp. 357–361.

Merlo, L. M., Pepper, J. W., Reid, B. J. and Maley, C. C. (2006). Cancer as an evolutionary and ecological process, *Nature Reviews Cancer* **6**, 12, pp. 924–935.

Meza, R., Jeon, J., Moolgavkar, S. H. and Luebeck, E. G. (2008). Age-specific incidence of cancer: Phases, transitions, and biological implications, *Proceedings of the National Academy of Sciences* **105**, 42, pp. 16284–16289.

Michor, F. (2008). Mathematical models of cancer stem cells, *Journal of Clinical Oncology* **26**, 17, pp. 2854–2861.

Michor, F., Hughes, T. P., Iwasa, Y., Branford, S., Shah, N. P., Sawyers, C. L. and Nowak, M. A. (2005a). Dynamics of chronic myeloid leukaemia, *Nature* **435**, 7046, pp. 1267–1270.

Michor, F., Hughes, T. P., Iwasa, Y., Branford, S., Shah, N. P., Sawyers, C. L. and Nowak, M. A. (2005b). Dynamics of chronic myeloid leukaemia, *Nature* **435**, 7046, pp. 1267–70.

Middleton, L. P., Vlastos, G., Mirza, N. Q., Eva Singletary, S. and Sahin, A. A. (2002). Multicentric mammary carcinoma, *Cancer* **94**, 7, pp. 1910–1916.

Milde, F., Bergdorf, M. and Koumoutsakos, P. (2008). A hybrid model for three-dimensional simulations of sprouting angiogenesis, *Biophysical Journal* **95**, 7, pp. 3146–3160.

Mitelman, F. (2000). Recurrent chromosome aberrations in cancer, *Mutation Research/Reviews in Mutation Research* **462**, 2, pp. 247–253.

Miyake, H., Nakamura, H., Hara, I., Gohji, K., Arakawa, S., Kamidono, S. and Saya, H. (1998). Multifocal renal cell carcinoma: evidence for a common clonal origin. *Clinical Cancer Research* **4**, 10, pp. 2491–2494.

Mok, W., Stylianopoulos, T., Boucher, Y. and Jain, R. K. (2009). Mathematical modeling of herpes simplex virus distribution in solid tumors: implications for cancer gene therapy, *Clinical Cancer Research* **15**, 7, pp. 2352–2360.

Moolgavkar, S. H. (1978). The multistage theory of carcinogenesis and the age distribution of cancer in man, *Journal of the National Cancer Institute* **61**, 1, pp. 49–52.

Moolgavkar, S. H., Day, N. E. and Stevens, R. G. (1980). Two-stage model for carcinogenesis: epidemiology of breast cancer in females, *Journal of the National Cancer Institute* **65**, 3, pp. 559–569.

Moolgavkar, S. H., Dewanji, A. and Venzon, D. J. (1988). A stochastic two-stage model for cancer risk assessment. I. The hazard function and the probability of tumor, *Risk Analysis* **8**, 3, pp. 383–392.

Moolgavkar, S. H. and Knudson, A. G. (1981). Mutation and cancer: a model for human carcinogenesis, *Journal of the National Cancer Institute* **66**, 6, pp. 1037–1052.

Moon Crompton, A. and Kirn, D. H. (2007). From ONYX-015 to armed vaccinia viruses: the education and evolution of oncolytic virus development, *Current Cancer Drug Targets* **7**, 2, pp. 133–139.

Moore, H. and Li, N. K. (2004). A mathematical model for chronic myelogenous leukemia (CML) and T cell interaction, *Journal of Theoretical Biology* **227**, 4, pp. 513–523.

Moore, K. A. and Lemischka, I. R. (2006). Stem cells and their niches, *Science* **311**, 5769, pp. 1880–1885.

Morrison, S. J. and Kimble, J. (2006). Asymmetric and symmetric stem-cell divisions in development and cancer, *Nature* **441**, 7097, pp. 1068–1074.

Morrison, S. J. and Spradling, A. C. (2008). Stem cells and niches: mechanisms that promote stem cell maintenance throughout life, *Cell* **132**, 4, pp. 598–611.

Mossman, D. and Scott, R. J. (2011). Long term transcriptional reactivation of epigenetically silenced genes in colorectal cancer cells requires DNA hypomethylation and histone acetylation, *PLoS One* **6**, 8, p. e23127.

Motzer, R. J., Bander, N. H. and Nanus, D. M. (1996). Renal-cell carcinoma, *New England Journal of Medicine* **335**, 12, pp. 865–875.

Murray, J. M. and Coldman, A. J. (2003). The effect of heterogeneity on optimal regimens in cancer chemotherapy, *Mathematical Biosciences* **185**, 1, pp. 73–87.

Nardi, V., Azam, M. and Daley, G. Q. (2004). Mechanisms and implications of imatinib resistance mutations in BCR-ABL, *Current Opinion in Hematology* **11**, 1, pp. 35–43.

Negrini, S., Gorgoulis, V. G. and Halazonetis, T. D. (2010). Genomic instability – an evolving hallmark of cancer, *Nature Reviews Molecular Cell Biology* **11**, 3, pp. 220–228.

Nguyen, L. V., Vanner, R., Dirks, P. and Eaves, C. J. (2012). Cancer stem cells: an evolving concept, *Nature Reviews Cancer* **12**, 2, pp. 133–143.

Nicolas, P., Kim, K.-M., Shibata, D. and Tavaré, S. (2007). The stem cell population of the human colon crypt: analysis via methylation patterns, *PLoS Computational Biology* **3**, 3, p. e28.

Noctor, S. C., Martínez-Cerdeño, V., Ivic, L. and Kriegstein, A. R. (2004). Cortical neurons arise in symmetric and asymmetric division zones and migrate through specific phases, *Nature Neuroscience* **7**, 2, pp. 136–144.

Noguchi, S., Aihara, T., Koyama, H., Motomura, K., Inaji, H. and Imaoka, S. (1994). Discrimination between multicentric and multifocal carcinomas of the breast through clonal analysis, *Cancer* **74**, 3, pp. 872–877.

Norton, L. (1988). A Gompertzian model of human breast cancer growth, *Cancer Research* **48**, 24 Part 1, pp. 7067–7071.

Norton, L. and Massagué, J. (2006). Is cancer a disease of self-seeding? *Nature Medicine* **12**, 8, pp. 875–878.

Nosho, K., Irahara, N., Shima, K., Kure, S., Kirkner, G. J., Schernhammer, E. S., Hazra, A., Hunter, D. J., Quackenbush, J., Spiegelman, D. *et al.* (2008). Comprehensive biostatistical analysis of CpG island methylator phenotype in colorectal cancer using a large population-based sample, *PLoS One* **3**, 11, p. e3698.

Novozhilov, A. S., Berezovskaya, F. S., Koonin, E. V., Karev, G. P. *et al.* (2006). Mathematical modeling of tumor therapy with oncolytic viruses: Regimes with complete tumor elimination within the framework of deterministic models, *Biology Direct* **1**, 6, p. 18.

Nowak, M. and Highfield, R. (2011). *SuperCooperators: Altruism, Evolution, and Why We Need Each Other to Succeed* (Simon and Schuster).
Nowak, M. and May, R. M. (2000). *Virus Dynamics: Mathematical Principles of Immunology and Virology* (Oxford University Press).
Nowak, M. A. (2006). Five rules for the evolution of cooperation, *Science* **314**, 5805, pp. 1560–1563.
Nowak, M. A., Komarova, N. L., Sengupta, A., Jallepalli, P. V., Shih, I.-M., Vogelstein, B. and Lengauer, C. (2002). The role of chromosomal instability in tumor initiation, *Proceedings of the National Academy of Sciences* **99**, 25, pp. 16226–16231.
Nowak, M. A., Michor, F. and Iwasa, Y. (2003). The linear process of somatic evolution, *Proceedings of the National Academy of Sciences* **100**, 25, pp. 14966–14969.
Nowicki, M. O., Falinski, R., Koptyra, M., Slupianek, A., Stoklosa, T., Gloc, E., Nieborowska-Skorska, M., Blasiak, J. and Skorski, T. (2004). BCR/ABL oncogenic kinase promotes unfaithful repair of the reactive oxygen species–dependent DNA double-strand breaks, *Blood* **104**, 12, pp. 3746–3753.
Nusse, R. (2008). Wnt signaling and stem cell control, *Cell research* **18**, 5, pp. 523–527.
Nyberg, P., Salo, T., Kalluri, R. *et al.* (2008). Tumor microenvironment and angiogenesis. *Frontiers in Bioscience* **13**, p. 6537.
Oda, T., Miyao, N., Takahashi, A., Yanase, M., Masumori, N., Itoh, N., Tamakawa, M. and Tsukamoto, T. (2001). Growth rates of primary and metastatic lesions of renal cell carcinoma, *International Journal of Urology* **8**, 9, pp. 473–477.
Ogg, G., Jin, X., Bonhoeffer, S., Moss, P., Nowak, M., Monard, S., Segal, J., Cao, Y., Rowland-Jones, S., Hurley, A. *et al.* (1999). Decay kinetics of human immunodeficiency virus-specific effector cytotoxic T lymphocytes after combination antiretroviral therapy, *Journal of Virology* **73**, 1, pp. 797–800.
O'Hare, T., Corbin, A. and Druker, B. (2006). Targeted CML therapy: controlling drug resistance, seeking cure, *Current Opinion in Genetics & Development* **16**, 1, pp. 92–99.
O'Hare, T., Eide, C. and Deininger, M. (2007). Bcr-abl kinase domain mutations, drug resistance, and the road to a cure for chronic myeloid leukemia, *Blood* **110**, 7, pp. 2242–2249.
Okumura, T., Shimada, Y., Imamura, M. and Yasumoto, S. (2003). Neurotrophin receptor p75(NTR) characterizes human esophageal keratinocyte stem cells in vitro, *Oncogene* **22**, 26, pp. 4017–4026, doi:10.1038/sj.onc.1206525.
Olofsson, P. and Kimmel, M. (1999). Stochastic models of telomere shortening, *Mathematical Biosciences* **158**, 1, pp. 75–92.
Olovnikov, A. M. (1973). A theory of marginotomy. The incomplete copying of template margin in enzymic synthesis of polynucleotides and biological significance of the phenomenon, *Journal of Theoretical Biology* **41**, 1, pp. 181–90.

Ordóñez-Morán, P. and Huelsken, J. (2012). Lrig1: a new master regulator of epithelial stem cells, *The EMBO Journal*.

O'Reilly, M. S., Boehm, T., Shing, Y., Fukai, N., Vasios, G., Lane, W. S., Flynn, E., Birkhead, J. R., Olsen, B. R. and Folkman, J. (1997). Endostatin: an endogenous inhibitor of angiogenesis and tumor growth, *Cell* **88**, 2, pp. 277–285.

O'Reilly, M. S., Holmgren, L., Chen, C. and Folkman, J. (1996). Angiostatin induces and sustains dormancy of human primary tumors in mice, *Nature Medicine* **2**, 6, pp. 689–692.

Oren, M. (2003). Decision making by p53: life, death and cancer, *Cell Death & Differentiation* **10**, 4, pp. 431–442.

Orford, K. and Scadden, D. (2008). Deconstructing stem cell self-renewal: genetic insights into cell-cycle regulation, *Nature Reviews Genetics* **9**, 2, pp. 115–128.

Osborne, J., Walter, A., Kershaw, S., Mirams, G., Fletcher, A., Pathmanathan, P., Gavaghan, D., Jensen, O., Maini, P. and Byrne, H. (2010). A hybrid approach to multi-scale modelling of cancer, *Philosophical Transactions of the Royal Society A: Mathematical, Physical and Engineering Sciences* **368**, 1930, pp. 5013–5028.

O'Shea, C. C. (2005). Viruses–seeking and destroying the tumor program, *Oncogene* **24**, 52, pp. 7640–7655.

Ozono, S., Miyao, N., Igarashi, T., Marumo, K., Nakazawa, H., Fukuda, M., Tsushima, T., Tokuda, N., Kawamura, J., Murai, M. *et al.* (2004). Tumor doubling time of renal cell carcinoma measured by CT: collaboration of Japanese Society of Renal Cancer, *Japanese Journal of Clinical Oncology* **34**, 2, pp. 82–85.

Page, K. and Uhr, J. (2005). Mathematical models of cancer dormancy, *Leukemia & Lymphoma* **46**, 3, pp. 313–327.

Paiva, L. R., Binny, C., Ferreira, S. C. and Martins, M. L. (2009). A multiscale mathematical model for oncolytic virotherapy, *Cancer Research* **69**, 3, pp. 1205–1211.

Palucka, K. and Banchereau, J. (2012). Cancer immunotherapy via dendritic cells, *Nature Reviews Cancer* **12**, 4, pp. 265–277.

Parato, K. A., Senger, D., Forsyth, P. A. and Bell, J. C. (2005). Recent progress in the battle between oncolytic viruses and tumours, *Nature Reviews Cancer* **5**, 12, pp. 965–976.

Parzen, E. (1962). *Stochastic Processes* (Holden-Day, San Francisco).

Pascual, M., Mazzega, P. and Levin, S. A. (2001). Oscillatory dynamics and spatial scale: the role of noise and unresolved pattern, *Ecology* **82**, 8, pp. 2357–2369.

Passegué, E., Jamieson, C. H. M., Ailles, L. E. and Weissman, I. L. (2003). Normal and leukemic hematopoiesis: are leukemias a stem cell disorder or a reacquisition of stem cell characteristics? *Proceedings of the National Academy of Sciences U S A* **100 Suppl 1**, pp. 11842–9, doi:10.1073/pnas.2034201100.

Passegué, E., Wagers, A. J., Giuriato, S., Anderson, W. C. and Weissman, I. L.

(2005). Global analysis of proliferation and cell cycle gene expression in the regulation of hematopoietic stem and progenitor cell fates, *Journal of Experimental Medicine* **202**, 11, pp. 1599–611, doi:10.1084/jem.20050967.

Paul, J. (1984). Oncogenes, *The Journal of Pathology* **143**, 1, pp. 1–10.

Pelayo, R., Miyazaki, K., Huang, J., Garrett, K. P., Osmond, D. G. and Kincade, P. W. (2006). Cell cycle quiescence of early lymphoid progenitors in adult bone marrow, *Stem Cells* **24**, 12, pp. 2703–2713.

Petersen, O. W. and Polyak, K. (2010). Stem cells in the human breast, *Cold Spring Harbor Perspectives In Biology* **2**, 5, doi:10.1101/cshperspect.a003160.

Piatyszek, M., Prowse, K., Harley, C., West, M., Ho, P., Coviello, G., Wright, W., Weinrich, S. and Shay, J. (1994). Specific association of human telomerase activity with immortal cells and cancer, *Science* **266**, 5193, pp. 2011–2015.

Piccirillo, S. G. M., Reynolds, B. A., Zanetti, N., Lamorte, G., Binda, E., Broggi, G., Brem, H., Olivi, A., Dimeco, F. and Vescovi, A. L. (2006). Bone morphogenetic proteins inhibit the tumorigenic potential of human brain tumour-initiating cells, *Nature* **444**, 7120, pp. 761–5, doi:10.1038/nature05349.

Pierotti, M., Sozzi, G. and Croce, C. (2003). Mechanisms of oncogene activation, in D. W. Kufe, R. E. Pollock, R. R. Weichselbaum, R. C. Bast, T. S. Gansler, J. F. Holland and E. Frei (eds.), *Holland-Frei Cancer Medicine, 6th Edition* (BC Decker).

Pinilla-Ibarz, J. and Flinn, I. (2012). The expanding options for front-line treatment in patients with newly diagnosed CML, *Critical Reviews in Oncology/Hematology* **84**, pp. 287–99.

Pino, M. S. and Chung, D. C. (2010). The chromosomal instability pathway in colon cancer, *Gastroenterology* **138**, 6, pp. 2059–2072.

Polyak, K., Haviv, I. and Campbell, I. G. (2009). Co-evolution of tumor cells and their microenvironment, *Trends in Genetics* **25**, 1, pp. 30–38.

Pontryagin, L. S., Boltyanski, V. G., Gamkrelidze, R. V. and Mishchenko, E. F. (1962). *The Mathematical Theory of Optimal Control Processes* (Interscience Publishers, New York).

Popat, S., Hubner, R. and Houlston, R. (2005). Systematic review of microsatellite instability and colorectal cancer prognosis, *Journal of Clinical Oncology* **23**, 3, pp. 609–618.

Post, D. E., Shim, H., Toussaint-Smith, E. and Van Meir, E. G. (2005). Cancer scene investigation: how a cold virus became a tumor killer, *Future Oncology* **1**, 2, pp. 247–258.

Potten, C. and Booth, C. (2002). Keratinocyte stem cells: a commentary, *Journal of Investigative Dermatology* **119**, 4, pp. 888–899.

Potten, C. S., Kellett, M., Roberts, S. A., Rew, D. A. and Wilson, G. D. (1992). Measurement of in vivo proliferation in human colorectal mucosa using bromodeoxyuridine, *Gut* **33**, 1, pp. 71–8.

Pradhan, S., Bacolla, A., Wells, R. D. and Roberts, R. J. (1999). Recombinant human DNA (cytosine-5) methyltransferase I. Expression, purification, and comparison of de novo and maintenance methylation, *Journal of Biological Chemistry* **274**, 46, pp. 33002–33010.

Preziosi, L. (2003). *Cancer Modelling and Simulation* (CRC Press).

Proctor, C. J. and Kirkwood, T. B. L. (2002). Modelling telomere shortening and the role of oxidative stress, *Mechanisms of Ageing and Development* **123**, 4, pp. 351–63.

Rabinovich, G. A., Gabrilovich, D. and Sotomayor, E. M. (2007). Immunosuppressive strategies that are mediated by tumor cells, *Annual Review of Immunology* **25**, p. 267.

Rabinovitch, P. S., Dziadon, S., Brentnall, T. A., Emond, M. J., Crispin, D. A., Haggitt, R. C. and Bronner, M. P. (1999). Pancolonic chromosomal instability precedes dysplasia and cancer in ulcerative colitis, *Cancer Research* **59**, 20, pp. 5148–5153.

Rajagopalan, H. and Lengauer, C. (2004). Aneuploidy and cancer, *Nature* **432**, 7015, pp. 338–341.

Ramanujan, S., Koenig, G. C., Padera, T. P., Stoll, B. R. and Jain, R. K. (2000). Local imbalance of proangiogenic and antiangiogenic factors: a potential mechanism of focal necrosis and dormancy in tumors, *Cancer Research* **60**, 5, pp. 1442–1448.

Rathmell, W. K., Martz, C. A. and Rini, B. I. (2007). Renal cell carcinoma, *Current Opinion in Oncology* **19**, 3, pp. 234–240.

Reis, C. L., Pacheco, J. M., Ennis, M. K. and Dingli, D. (2010). In silico evolutionary dynamics of tumour virotherapy, *Integrative Biology* **2**, 1, pp. 41–45.

Rejniak, K. A. (2007). An immersed boundary framework for modelling the growth of individual cells: an application to the early tumour development, *Journal of Theoretical Biology* **247**, 1, pp. 186–204.

Rejniak, K. A. and Anderson, A. R. (2011). Hybrid models of tumor growth, *Wiley Interdisciplinary Reviews: Systems Biology and Medicine* **3**, 1, pp. 115–125.

Retsky, M., Swartzendruber, D., Wardwell, R. and Bame, P. (1990). Is Gompertzian or exponential kinetics a valid description of individual human cancer growth? *Medical Hypotheses* **33**, 2, pp. 95–106.

Reuss, R., Ludwig, J., Shirakashi, R., Ehrhart, F., Zimmermann, H., Schneider, S., Weber, M., Zimmermann, U., Schneider, H. and Sukhorukov, V. (2004). Intracellular delivery of carbohydrates into mammalian cells through swelling-activated pathways, *The Journal of Membrane Biology* **200**, 2, pp. 67–81.

Reya, T. and Clevers, H. (2005). Wnt signalling in stem cells and cancer, *Nature* **434**, 7035, pp. 843–850.

Reya, T., Morrison, S. J., Clarke, M. F. and Weissman, I. L. (2001). Stem cells, cancer, and cancer stem cells, *Nature* **414**, 6859, pp. 105–111.

Ribeiro, R. M., Bonhoeffer, S. and Nowak, M. A. (1998). The frequency of resistant mutant virus before antiviral therapy, *Aids* **12**, 5, pp. 461–465.

Rix, U., Hantschel, O., Dürnberger, G., Rix, L., Planyavsky, M., Fernbach, N., Kaupe, I., Bennett, K., Valent, P., Colinge, J. *et al.* (2007). Chemical proteomic profiles of the BCR-ABL inhibitors imatinib, nilotinib, and dasatinib reveal novel kinase and nonkinase targets, *Blood* **110**, 12, pp. 4055–4063.

Roberts, M. S., Lorence, R. M., Groene, W. S., Bamat, M. K. *et al.* (2006). Naturally oncolytic viruses. *Current Opinion in Molecular Therapeutics* **8**, 4, p. 314.

Robertson, K. D. *et al.* (2001). DNA methylation, methyltransferases, and cancer, *Oncogene* **20**, 24, pp. 3139–3155.

Robertson, T. B. (1923). *The Chemical Basis of Growth and Senescence* (JB Lippincott company).

Rodriguez-Brenes, I., Komarova, N. and Wodarz, D. (2011). Evolutionary dynamics of feedback escape and the development of stem-cell–driven cancers, *Proceedings of the National Academy of Sciences* **108**, 47, pp. 18983–18988.

Rodriguez-Brenes, I. A., Komarova, N. L. and Wodarz, D. (2013). Tumor growth dynamics: insights into evolutionary processes, *Trends in Ecology & Evolution* **28**, 10, pp. 597–604.

Rodriguez-Brenes, I. A. and Peskin, C. S. (2010). Quantitative theory of telomere length regulation and cellular senescence, *Proceedings of the National Academy of Sciences U S A* **107**, 12, pp. 5387–92, doi:10.1073/pnas.0914502107.

Roeder, I. and Glauche, I. (2008). Pathogenesis, treatment effects, and resistance dynamics in chronic myeloid leukemia-insights from mathematical model analyses, *Journal of Molecular Medicine* **86**, 1, pp. 17–27.

Roeder, I., Horn, M., Glauche, I., Hochhaus, A., Mueller, M. C. and Loeffler, M. (2006). Dynamic modeling of imatinib-treated chronic myeloid leukemia: functional insights and clinical implications, *Nature Medicine* **12**, 10, pp. 1181–1184.

Rosenberg, L. (2000). Pancreatic cancer, *Drugs* **59**, 5, pp. 1071–1089.

Rosenthal, A. N., Ryan, A., Hopster, D. and Jacobs, I. J. (2002). Molecular evidence of a common clonal origin and subsequent divergent clonal evolution in vulval intraepithelial neoplasia, vulval squamous cell carcinoma and lymph node metastases, *International Journal of Cancer* **99**, 4, pp. 549–554.

Ross, D., Branford, S., Seymour, J., Schwarer, A., Arthur, C., Yeung, D., Dang, P., Goyne, J., Slader, C., Filshie, R. *et al.* (2013). Safety and efficacy of imatinib cessation for CML patients with stable undetectable minimal residual disease: results from the TWISTER Study, *Blood* **122**, pp. 515–522.

Rozenblum, E., Schutte, M., Goggins, M., Hahn, S. A., Panzer, S., Zahurak, M., Goodman, S. N., Sohn, T. A., Hruban, R. H., Yeo, C. J. and Kern, S. E. (1997). Tumor-suppressive pathways in pancreatic carcinoma, *Cancer Research* **57**, 9, pp. 1731–4.

Rubin, H. (2003). Microenvironmental regulation of the initiated cell, *Advances in Cancer Research* **90**, pp. 1–62.

Rudolph, K. L., Millard, M., Bosenberg, M. W. and DePinho, R. A. (2001). Telomere dysfunction and evolution of intestinal carcinoma in mice and humans, *Nature Genetics* **28**, 2, pp. 155–159.

Ruijter, E. T. G., Miller, G. J., van de Kaa, C. A., van Bokhoven, A., Bussemakers, M. J., Debruyne, F. M., Ruiter, D. J. and Schalken, J. A. (1999).

Molecular analysis of multifocal prostate cancer lesions, *The Journal of Pathology* **188**, 3, pp. 271–277.

Saha, S., Ji, L., de Pablo, J. and Palecek, S. (2006). Inhibition of human embryonic stem cell differentiation by mechanical strain, *Journal of Cellular Physiology* **206**, 1, pp. 126–137.

Saha, S., Ji, L., De Pablo, J. and Palecek, S. (2008). TGF [beta]/activin/nodal pathway in inhibition of human embryonic stem cell differentiation by mechanical strain, *Biophysical Journal* **94**, 10, pp. 4123–4133.

Saito, Y., Kanai, Y., Nakagawa, T., Sakamoto, M., Saito, H., Ishii, H. and Hirohashi, S. (2003). Increased protein expression of DNA methyltransferase (DNMT) 1 is significantly correlated with the malignant potential and poor prognosis of human hepatocellular carcinomas, *International Journal of Cancer* **105**, 4, pp. 527–532.

Salomoni, P. and Calegari, F. (2010). Cell cycle control of mammalian neural stem cells: putting a speed limit on G1, *Trends in Cell Biology* **20**, 5, pp. 233–243.

Samper, E., Flores, J. M. and Blasco, M. A. (2001). Restoration of telomerase activity rescues chromosomal instability and premature aging in Terc-/- mice with short telomeres, *EMBO Reports* **2**, 9, pp. 800–807.

Sanai, N., Alvarez-Buylla, A. and Berger, M. S. (2005). Neural stem cells and the origin of gliomas, *New England Journal of Medicine* **353**, 8, pp. 811–22, doi:10.1056/NEJMra043666.

Satō, K., Matsuda, H. and Sasaki, A. (1994). Pathogen invasion and host extinction in lattice structured populations, *Journal of Mathematical Biology* **32**, 3, pp. 251–268.

Schabel Jr, F., Skipper, H., Trader, M., Laster Jr, W., Griswold Jr, D. and Corbett, T. (1983). Establishment of cross-resistance profiles for new agents, *Cancer Treatment Reports* **67**, 10, pp. 905–922.

Schabel Jr, F., Trader, M., Laster Jr, W., Corbett, T. and Griswold Jr, D. (1979). cis-Dichlorodiammineplatinum (II): combination chemotherapy and cross-resistance studies with tumors of mice, *Cancer Treatment Reports* **63**, 9-10, p. 1459.

Schepers, A. G., Snippert, H. J., Stange, D. E., van den Born, M., van Es, J. H., van de Wetering, M. and Clevers, H. (2012). Lineage tracing reveals Lgr5+ stem cell activity in mouse intestinal adenomas, *Science Signaling* **337**, 6095, p. 730.

Schrek, R. (1935). A quantitative study of the growth of the Walker rat tumor and the Flexner-Jobling rat carcinoma, *The American Journal of Cancer* **24**, 4, pp. 807–822.

Schüller, U., Heine, V. M., Mao, J., Kho, A. T., Dillon, A. K., Han, Y.-G., Huillard, E., Sun, T., Ligon, A. H., Qian, Y., Ma, Q., Alvarez-Buylla, A., McMahon, A. P., Rowitch, D. H. and Ligon, K. L. (2008). Acquisition of granule neuron precursor identity is a critical determinant of progenitor cell competence to form Shh-induced medulloblastoma, *Cancer Cell* **14**, 2, pp. 123–34, doi:10.1016/j.ccr.2008.07.005.

Segditsas, S. and Tomlinson, I. (2006). Colorectal cancer and genetic alterations in the wnt pathway, *Oncogene* **25**, 57, pp. 7531–7537.

Sen, B., Xie, Z., Case, N., Ma, M., Rubin, C. and Rubin, J. (2008). Mechanical strain inhibits adipogenesis in mesenchymal stem cells by stimulating a durable β-catenin signal, *Endocrinology* **149**, 12, pp. 6065–6075.

Sen, S. (2000). Aneuploidy and cancer, *Current Opinion in Oncology* **12**, 1, pp. 82–88.

Seoane, J., Le, H.-V. and Massagué, J. (2002). Myc suppression of the p21Cip1 Cdk inhibitor influences the outcome of the p53 response to DNA damage, *Nature* **419**, 6908, pp. 729–734.

Shachaf, C. M. and Felsher, D. W. (2005). Tumor dormancy and MYC inactivation: pushing cancer to the brink of normalcy, *Cancer Research* **65**, 11, pp. 4471–4, doi:10.1158/0008-5472.CAN-05-1172.

Shachaf, C. M., Kopelman, A. M., Arvanitis, C., Karlsson, A., Beer, S., Mandl, S., Bachmann, M. H., Borowsky, A. D., Ruebner, B., Cardiff, R. D., Yang, Q., Bishop, J. M., Contag, C. H. and Felsher, D. W. (2004). MYC inactivation uncovers pluripotent differentiation and tumour dormancy in hepatocellular cancer, *Nature* **431**, 7012, pp. 1112–7, doi:10.1038/nature03043.

Shackney, S. E. (1970). A computer model for tumor growth and chemotherapy, and its application to L1210 leukemia treated with cytosine arabinoside (NSC-63878), *Cancer Chemotherapy Reports* **54**, 6, p. 399.

Shah, N., Skaggs, B., Branford, S., Hughes, T., Nicoll, J., Paquette, R., Sawyers, C. *et al.* (2007). Sequential ABL kinase inhibitor therapy selects for compound drug-resistant BCR-ABL mutations with altered oncogenic potency, *Journal of Clinical Investigation* **117**, 9, p. 256an2010bc2.

Shah, N. P., Tran, C., Lee, F. Y., Chen, P., Norris, D. and Sawyers, C. L. (2004). Overriding imatinib resistance with a novel ABL kinase inhibitor, *Science Signaling* **305**, 5682, p. 399.

Shaner, N. C., Campbell, R. E., Steinbach, P. A., Giepmans, B. N. G., Palmer, A. E. and Tsien, R. Y. (2004). Improved monomeric red, orange and yellow fluorescent proteins derived from Discosoma sp red fluorescent protein, *Nature Biotechnology* **22**, 12, pp. 1567–1572.

Shankaran, V., Ikeda, H., Bruce, A. T., White, J. M., Swanson, P. E., Old, L. J. and Schreiber, R. D. (2001). IFNγ and lymphocytes prevent primary tumour development and shape tumour immunogenicity, *Nature* **410**, 6832, pp. 1107–1111.

Shannon, K. M. (2002). Resistance in the land of molecular cancer therapeutics, *Cancer Cell* **2**, 2, pp. 99–102.

Sharma, S., Kelly, T. K. and Jones, P. A. (2010). Epigenetics in cancer, *Carcinogenesis* **31**, 1, pp. 27–36.

Shen, Q., Goderie, S. K., Jin, L., Karanth, N., Sun, Y., Abramova, N., Vincent, P., Pumiglia, K. and Temple, S. (2004). Endothelial cells stimulate self-renewal and expand neurogenesis of neural stem cells, *Science* **304**, 5675, pp. 1338–1340.

Shih, I.-M., Zhou, W., Goodman, S. N., Lengauer, C., Kinzler, K. W. and Vogelstein, B. (2001). Evidence that genetic instability occurs at an early stage of colorectal tumorigenesis, *Cancer Research* **61**, 3, pp. 818–822.

Shizuru, J. A., Negrin, R. S. and Weissman, I. L. (2005). Hematopoietic stem and progenitor cells: clinical and preclinical regeneration of the hematolymphoid system, *Annual Review of Medicine* **56**, pp. 509–38, doi: 10.1146/annurev.med.54.101601.152334.

Shochat, E., Hart, D. and Agur, Z. (1999). Using computer simulations for evaluating the efficacy of breast cancer chemotherapy protocols, *Mathematical Models and Methods in Applied Sciences* **9**, 04, pp. 599–615.

Sieber, O. M., Heinimann, K. and Tomlinson, I. P. (2003). Genomic instability: the engine of tumorigenesis? *Nature Reviews Cancer* **3**, 9, pp. 701–708.

Simeoni, M., Magni, P., Cammia, C., De Nicolao, G., Croci, V., Pesenti, E., Germani, M., Poggesi, I. and Rocchetti, M. (2004). Predictive pharmacokinetic-pharmacodynamic modeling of tumor growth kinetics in xenograft models after administration of anticancer agents, *Cancer Research* **64**, 3, pp. 1094–1101.

Simmons, C., Matlis, S., Thornton, A., Chen, S., Wang, C. and Mooney, D. (2003). Cyclic strain enhances matrix mineralization by adult human mesenchymal stem cells via the extracellular signal-regulated kinase (ERK1/2) signaling pathway, *Journal of Biomechanics* **36**, 8, pp. 1087–1096.

Simon, R., Eltze, E., Schäfer, K.-L., Bürger, H., Semjonow, A., Hertle, L., Dockhorn-Dworniczak, B., Terpe, H.-J. and Böcker, W. (2001). Cytogenetic analysis of multifocal bladder cancer supports a monoclonal origin and intraepithelial spread of tumor cells, *Cancer Research* **61**, 1, pp. 355–362.

Simon, R. and Norton, L. (2006). The Norton–Simon hypothesis: designing more effective and less toxic chemotherapeutic regimens, *Nature Clinical Practice Oncology* **3**, 8, pp. 406–407.

Simons, B. D. and Clevers, H. (2011a). Stem cell self-renewal in intestinal crypt, *Experimental Cell Research* **317**, 19, pp. 2719–2724.

Simons, B. D. and Clevers, H. (2011b). Strategies for homeostatic stem cell self-renewal in adult tissues, *Cell* **145**, 6, pp. 851–862.

Simpson-Herren, L., Lloyd, H. H. *et al.* (1970). Kinetic parameters and growth curves for experimental tumor systems. *Cancer Chemotherapy Reports. Part 1* **54**, 3, p. 143.

Skehan, P. (1986). On the normality of growth dynamics of neoplasms in vivo: a data base analysis. *Growth* **50**, 4, p. 496.

Skipper, H. E., Schabel Jr, F. and Wilcox, W. S. (1964). Experimental evaluation of potential anticancer agents. XIII. On the criteria and kinetics associated with "curability" of experimental leukemia. *Cancer Chemotherapy Reports. Part 1* **35**, p. 1.

Slack, A., Cervoni, N., Pinard, M. and Szyf, M. (1999). Feedback regulation of DNA methyltransferase gene expression by methylation, *European Journal of Biochemistry* **264**, 1, pp. 191–199.

Smallbone, K., Gavaghan, D. J., Gatenby, R. A. and Maini, P. K. (2005). The role of acidity in solid tumour growth and invasion, *Journal of Theoretical Biology* **235**, 4, pp. 476–484.

Smieja, J. and Swierniak, A. (2003). Different models of chemotherapy taking into account drug resistance stemming from gene amplification, *International Journal of Applied Mathematical and Computational Sciences* **13**, 3, pp. 297–305.

Smogorzewska, A. and De Lange, T. (2002). Different telomere damage signaling pathways in human and mouse cells, *EMBO Journal* **21**, 16, pp. 4338–48.

Smyth, M. J., Thia, K. Y., Street, S. E., Cretney, E., Trapani, J. A., Taniguchi, M., Kawano, T., Pelikan, S. B., Crowe, N. Y. and Godfrey, D. I. (2000a). Differential tumor surveillance by natural killer (NK) and NKT cells, *The Journal of Experimental Medicine* **191**, 4, pp. 661–668.

Smyth, M. J., Thia, K. Y., Street, S. E., MacGregor, D., Godfrey, D. I. and Trapani, J. A. (2000b). Perforin-mediated cytotoxicity is critical for surveillance of spontaneous lymphoma, *The Journal of Experimental Medicine* **192**, 5, pp. 755–760.

Snippert, H. J., van der Flier, L. G., Sato, T., van Es, J. H., van den Born, M., Kroon-Veenboer, C., Barker, N., Klein, A. M., van Rheenen, J., Simons, B. D. *et al.* (2010). Intestinal crypt homeostasis results from neutral competition between symmetrically dividing Lgr5 stem cells, *Cell* **143**, 1, pp. 134–144.

Sontag, E. D. (2004). Some new directions in control theory inspired by systems biology, *Systems Biology* **1**, 1, pp. 9–18.

Sorace, R. and Komarova, N. L. (2012). Accumulation of neutral mutations in growing cell colonies with competition, *Journal of Theoretical Biology* **314**, pp. 84–94.

Sotillo, R., Schvartzman, J.-M. and Benezra, R. (2009). Very CIN-ful: whole chromosome instability promotes tumor suppressor loss of heterozygosity, *Cancer Cell* **16**, 6, pp. 451–452.

Sottoriva, A., Vermeulen, L. and Tavaré, S. (2011). Modeling evolutionary dynamics of epigenetic mutations in hierarchically organized tumors, *PLoS Computational Biology* **7**, 5, p. e1001132.

Speer, J. F., Petrosky, V. E., Retsky, M. W. and Wardwell, R. H. (1984). A stochastic numerical model of breast cancer growth that simulates clinical data, *Cancer Research* **44**, 9, pp. 4124–4130.

Spencer, S. L., Berryman, M. J., García, J. A. and Abbott, D. (2004). An ordinary differential equation model for the multistep transformation to cancer, *Journal of Theoretical Biology* **231**, 4, pp. 515–524.

Spencer, S. L., Gerety, R. A., Pienta, K. J. and Forrest, S. (2006). Modeling somatic evolution in tumorigenesis, *PLoS Computational Biology* **2**, 8, p. e108.

Spiegel, A., Kalinkovich, A., Shivtiel, S., Kollet, O. and Lapidot, T. (2008). Stem cell regulation via dynamic interactions of the nervous and immune systems with the microenvironment, *Cell Stem Cell* **3**, 5, pp. 484–492.

Spradling, A., Drummond-Barbosa, D. and Kai, T. (2001). Stem cells find their niche, *Nature* **414**, 6859, pp. 98–104.

Spratt, J. A., Von Fournier, D., Spratt, J. S. and Weber, E. E. (1993a). Decelerating growth and human breast cancer, *Cancer* **71**, 6, pp. 2013–2019.

Spratt, J. A., Von Fournier, D., Spratt, J. S. and Weber, E. E. (1993b). Mammographic assessment of human breast cancer growth and duration, *Cancer* **71**, 6, pp. 2020–2026.

Spratt Jr, J. S. and Spratt, T. L. (1964). Rates of growth of pulmonary metastases and host survival, *Annals of Surgery* **159**, 2, p. 161.

Sprouffske, K. and Maley, C. C. (2010). Cooperation and cancer, in A. Thomas-Tikhonenko (ed.), *Cancer Genome and Tumor Microenvironment* (Springer), pp. 471–485.

Squartini, F. (1961). Strain differences in growth of mouse mammary tumors, *Journal of National Cancer Institute* **26**, pp. 813–28.

Steel, G. G. (1977). *Growth Kinetics of Tumours: Cell Population Kinetics in Relation to the Growth and Treatment of Cancer* (Clanesdon press).

Stehelin, D., Varmus, H. E., Bishop, J. M. and Vogt, P. K. (1976). DNA related to the transforming gene(s) of avian sarcoma viruses is present in normal avian DNA, *Nature* **260**, pp. 170–173.

Stolarska, M. A., Kim, Y. and Othmer, H. G. (2009). Multi-scale models of cell and tissue dynamics, *Philosophical Transactions of the Royal Society A: Mathematical, Physical and Engineering Sciences* **367**, 1902, pp. 3525–3553.

Strachan, T. and Read, A. (1996). *Human Molecular Genetics* (BIOS Scientific, New York).

Stratton, M. R., Campbell, P. J. and Futreal, P. A. (2009). The cancer genome, *Nature* **458**, 7239, pp. 719–724.

Street, S. E., Trapani, J. A., MacGregor, D. and Smyth, M. J. (2002). Suppression of lymphoma and epithelial malignancies effected by interferon γ, *The Journal of Experimental Medicine* **196**, 1, pp. 129–134.

Sugimura, T. and Ushijima, T. (2000). Genetic and epigenetic alterations in carcinogenesis, *Mutation Research/Reviews in Mutation Research* **462**, 2, pp. 235–246.

Swan, G. W. (1990). Role of optimal control theory in cancer chemotherapy, *Mathematical Biosciences* **101**, 2, pp. 237–284.

Swanson, K. R., Bridge, C., Murray, J. and Alvord Jr, E. C. (2003). Virtual and real brain tumors: using mathematical modeling to quantify glioma growth and invasion, *Journal of the Neurological Sciences* **216**, 1, pp. 1–10.

Swierniak, A., Kimmel, M. and Smieja, J. (2009). Mathematical modeling as a tool for planning anticancer therapy, *European Journal of Pharmacology* **625**, 1, pp. 108–121.

Swierniak, A., Ledzewicz, U. and Schattler, H. (2003). Optimal control for a class of compartmental models in cancer chemotherapy, *International Journal of Applied Mathematics and Computer Science* **13**, 3, pp. 357–368.

Swierniak, A. and Smieja, J. (2001). Cancer chemotherapy optimization under evolving drug resistance, *Nonlinear Analysis* **47**, pp. 375–386.

Tauchi, T. and Ohyashiki, K. (2004). Molecular mechanisms of resistance of leukemia to imatinib mesylate, *Leukemia Research* **28 Suppl 1**, pp. S39–45.

Teixeira, M. R., Pandis, N. and Heim, S. (2003). Multicentric mammary carcinoma, *Cancer* **97**, 3, pp. 715–717.

Thalhauser, C., Lowengrub, J., Stupack, D. and Komarova, N. (2010). Selection in spatial stochastic models of cancer: Migration as a key modulator of fitness, *Biology Direct* **5**, pp. 1–17.

Thiagalingam, S., Laken, S., Willson, J. K., Markowitz, S. D., Kinzler, K. W., Vogelstein, B. and Lengauer, C. (2001). Mechanisms underlying losses of heterozygosity in human colorectal cancers, *Proceedings of the National Academy of Sciences* **98**, 5, pp. 2698–2702.

Thompson, S. L., Bakhoum, S. F. and Compton, D. A. (2010). Mechanisms of chromosomal instability, *Current Biology* **20**, 6, pp. R285–R295.

Tipping, A. J., Mahon, F. X., Lagarde, V., Goldman, J. M. and Melo, J. V. (2001). Restoration of sensitivity to STI571 in STI571-resistant chronic myeloid leukemia cells, *Blood* **98**, 13, pp. 3864–3867.

Tischfield, J. A. and Shao, C. (2003). Somatic recombination redux, *Nature Genetics* **33**, 1, pp. 5–6.

Tlsty, T. D. (2001). Stromal cells can contribute oncogenic signals, in *Seminars in Cancer Biology*, Vol. 11 (Elsevier), pp. 97–104.

Tlsty, T. D. and Hein, P. W. (2001). Know thy neighbor: stromal cells can contribute oncogenic signals, *Current Opinion in Genetics & Development* **11**, 1, pp. 54–59.

Tlsty, T. D., Margolin, B. H. and Lum, K. (1989). Differences in the rates of gene amplification in nontumorigenic and tumorigenic cell lines as measured by Luria-Delbrück fluctuation analysis, *Proceedings of the National Academy of Sciences* **86**, 23, pp. 9441–9445.

Tomé, T. and de Carvalho, K. C. (2007). Stable oscillations of a predator–prey probabilistic cellular automaton: a mean-field approach, *Journal of Physics A: Mathematical and Theoretical* **40**, 43, p. 12901.

Tomlinson, I. P., Novelli, M. and Bodmer, W. (1996). The mutation rate and cancer, *Proceedings of the National Academy of Sciences* **93**, 25, pp. 14800–14803.

Tothova, Z. and Gilliland, D. G. (2007). FoxO transcription factors and stem cell homeostasis: insights from the hematopoietic system, *Cell Stem Cell* **1**, pp. 140–152.

Toyota, M., Ahuja, N., Ohe-Toyota, M., Herman, J. G., Baylin, S. B. and Issa, J.-P. J. (1999). CpG island methylator phenotype in colorectal cancer, *Proceedings of the National Academy of Sciences* **96**, 15, pp. 8681–8686.

Toyota, M. and Issa, J.-P. J. (1999). CpG island methylator phenotypes in aging and cancer, in *Seminars in Cancer Biology*, Vol. 9 (Elsevier), pp. 349–357.

Traulsen, A., Pacheco, J. M., Luzzatto, L. and Dingli, D. (2010). Somatic mutations and the hierarchy of hematopoiesis, *Bioessays* **32**, 11, pp. 1003–1008.

Tsuda, H. and Hirohashi, S. (1995). Identification of multiple breast cancers of multicentric origin by histological observations and distribution of allele loss on chromosome 16q, *Cancer Research* **55**, 15, pp. 3395–3398.

Turker, M. S. (1999). The establishment and maintenance of DNA methylation patterns in mouse somatic cells, in *Seminars in Cancer Biology*, Vol. 9 (Elsevier), pp. 329–337.

Tzeng, Y.-S., Li, H., Kang, Y.-L., Chen, W.-C., Cheng, W.-C. and Lai, D.-M. (2011). Loss of Cxcl12/sdf-1 in adult mice decreases the quiescent state of hematopoietic stem/progenitor cells and alters the pattern of hematopoietic regeneration after myelosuppression, *Blood* **117**, 2, pp. 429–39, doi:10.1182/blood-2010-01-266833.

Uhr, J. W. and Marches, R. (2001). Dormancy in a model of murine B cell lymphoma, in *Seminars in Cancer Biology*, Vol. 11 (Elsevier), pp. 277–283.

Ushijima, T., Watanabe, N., Shimizu, K., Miyamoto, K., Sugimura, T. and Kaneda, A. (2005). Decreased fidelity in replicating CpG methylation patterns in cancer cells, *Cancer Research* **65**, 1, pp. 11–17.

Vähä-Koskela, M. J., Heikkilä, J. E. and Hinkkanen, A. E. (2007). Oncolytic viruses in cancer therapy, *Cancer Letters* **254**, 2, pp. 178–216.

Van Dekken, H., Vissers, C., Tilanus, H., Tanke, H. and Rosenberg, C. (1999). Clonal analysis of a case of multifocal oesophageal (Barrett's) adenocarcinoma by comparative genomic hybridization, *The Journal of Pathology* **188**, 3, pp. 263–266.

Vermeulen, L., Sprick, M., Kemper, K., Stassi, G. and Medema, J. (2008). Cancer stem cells–old concepts, new insights, *Cell Death & Differentiation* **15**, 6, pp. 947–958.

Vesely, M. D., Kershaw, M. H., Schreiber, R. D. and Smyth, M. J. (2011). Natural innate and adaptive immunity to cancer, *Annual Review of Immunology* **29**, pp. 235–271.

Vilar, E. and Gruber, S. B. (2010). Microsatellite instability in colorectal cancer – the stable evidence, *Nature Reviews Clinical Oncology* **7**, 3, pp. 153–162.

Visvader, J. E. (2011). Cells of origin in cancer, *Nature* **469**, 7330, pp. 314–322.

Visvader, J. E. and Lindeman, G. J. (2008). Cancer stem cells in solid tumours: accumulating evidence and unresolved questions, *Nature Reviews Cancer* **8**, 10, pp. 755–768.

Vogelstein, B. and Kinzler, K. (2004). Cancer genes and the pathways they control, *Nature Medicine* **10**, 8, pp. 789–799, doi:10.1038/nm1087.

Vogelstein, B. and Kinzler, K. W. (2002). *The Genetic Basis of Human Cancer* (McGraw-Hill, New York).

Vogelstein, B., Lane, D. and Levine, A. J. (2000). Surfing the p53 network, *Nature* **408**, 6810, pp. 307–310.

Volpe, A., Panzarella, T., Rendon, R. A., Haider, M. A., Kondylis, F. I. and Jewett, M. A. (2004). The natural history of incidentally detected small renal masses, *Cancer* **100**, 4, pp. 738–745.

Volpe, G., Panuzzo, C., Ulisciani, S., Cilloni, D. *et al.* (2009). Imatinib resistance in CML. *Cancer Letters* **274**, 1, p. 1.

von Bertalanffy, L. (1960). Principles and theory of growth, *Fundamental Aspects of Normal and Malignant Growth* **493**, pp. 137–259.

von Zglinicki, T. (2002). Oxidative stress shortens telomeres, *Trends In Biochemical Sciences* **27**, 7, pp. 339–344.

Wajed, S. A., Laird, P. W. and DeMeester, T. R. (2001). DNA methylation: an alternative pathway to cancer, *Annals of Surgery* **234**, 1, p. 10.

Wan, F. Y., Sadovsky, A. V. and Komarova, N. L. (2010). Genetic instability in cancer: An optimal control problem, *Studies in Applied Mathematics* **125**, 1, pp. 1–38.

Wan, F. Y. M. (1995). *Introduction to the Calculus of Variations and Its Applications* (Chapman & Hall/CRC; 2nd edition).

Wang, H., Cheng, F., Cuenca, A., Horna, P., Zheng, Z., Bhalla, K. and Sotomayor, E. M. (2005). Imatinib mesylate (STI-571) enhances antigen-presenting cell function and overcomes tumor-induced CD4+ T-cell tolerance, *Blood* **105**, 3, pp. 1135–1143.

Wang, R., Chadalavada, K., Wilshire, J., Kowalik, U., Hovinga, K. E., Geber, A., Fligelman, B., Leversha, M., Brennan, C. and Tabar, V. (2010). Glioblastoma stem-like cells give rise to tumour endothelium, *Nature* **468**, 7325, pp. 829–33, doi:10.1038/nature09624.

Wang, X., Wang, M., MacLennan, G. T., Abdul-Karim, F. W., Eble, J. N., Jones, T. D., Olobatuyi, F., Eisenberg, R., Cummings, O. W., Zhang, S. *et al.* (2009). Evidence for common clonal origin of multifocal lung cancers, *Journal of the National Cancer Institute* **101**, 8, pp. 560–570.

Watanabe, N., Okochi-Takada, E., Yagi, Y., Furuta, J.-I. and Ushijima, T. (2006). Decreased fidelity in replicating DNA methylation patterns in cancer cells leads to dense methylation of a CpG island, in *DNA Methylation: Development, Genetic Disease and Cancer* (Springer), pp. 199–210.

Weedon-Fekjær, H., Lindqvist, B. H., Vatten, L. J., Aalen, O. O., Tretli, S. *et al.* (2008). Breast cancer tumor growth estimated through mammography screening data, *Breast Cancer Research* **10**, 3, p. R41.

Wein, L. M., Wu, J. T. and Kirn, D. H. (2003). Validation and analysis of a mathematical model of a replication-competent oncolytic virus for cancer treatment implications for virus design and delivery, *Cancer Research* **63**, 6, pp. 1317–1324.

Weinberg, R. A. (2007). *The Biology of Cancer* (Garland Science, New York) URL http://www.loc.gov/catdir/toc/ecip067/2006001825.html.

Weinreich, D. M. and Chao, L. (2005). Rapid evolutionary escape by large populations from local fitness peaks is likely in nature, *Evolution* **59**, 6, pp. 1175–1182.

Weis, S. M. and Cheresh, D. A. (2011). Tumor angiogenesis: molecular pathways and therapeutic targets, *Nature Medicine* **17**, 11, pp. 1359–1370.

Weisberg, E., Manley, P., Cowan-Jacob, S., Hochhaus, A. and Griffin, J. (2007). Second generation inhibitors of BCR-ABL for the treatment of imatinib-resistant chronic myeloid leukaemia, *Nature Reviews Cancer* **7**, 5, pp. 345–356.

Weissenbacher, T. M., Zschage, M., Janni, W., Jeschke, U., Dimpfl, T., Mayr, D., Rack, B., Schindlbeck, C., Friese, K. and Dian, D. (2010). Multicentric and multifocal versus unifocal breast cancer: is the tumor-node-metastasis classification justified? *Breast Cancer Research and Treatment* **122**, 1, pp. 27–34.

Weissman, D. B., Desai, M. M., Fisher, D. S. and Feldman, M. W. (2009). The rate at which asexual populations cross fitness valleys, *Theoretical Population Biology* **75**, 4, pp. 286–300.

Weissman, D. B., Feldman, M. W. and Fisher, D. S. (2010). The rate of fitness-valley crossing in sexual populations, *Genetics* **186**, 4, pp. 1389–1410.

Weissman, I. L. (2000). Stem cells: units of development, units of regeneration, and units in evolution, *Cell* **100**, 1, pp. 157–168.

Werner, B., Dingli, D., Lenaerts, T., Pacheco, J. M. and Traulsen, A. (2011). Dynamics of mutant cells in hierarchical organized tissues, *PLoS Computational Biology* **7**, 12, p. e1002290, doi:10.1371/journal.pcbi.1002290.

West, G. B., Brown, J. H. and Enquist, B. J. (2001). A general model for ontogenetic growth, *Nature* **413**, 6856, pp. 628–631.

Whiteside, T. (2008). The tumor microenvironment and its role in promoting tumor growth, *Oncogene* **27**, 45, pp. 5904–5912.

Whitlock, M. C., Phillips, P. C., Moore, F. B.-G. and Tonsor, S. J. (1995). Multiple fitness peaks and epistasis, *Annual Review of Ecology and Systematics* **26**, pp. 601–629.

Wijnhoven, S. W., Kool, H. J., van Teijlingen, C. M., van Zeeland, A. A. and Vrieling, H. (2001). Loss of heterozygosity in somatic cells of the mouse: An important step in cancer initiation? *Mutation Research/Fundamental and Molecular Mechanisms of Mutagenesis* **473**, 1, pp. 23–36.

Wilcox, W. S. (1966). The last surviving cancer cell: the chances of killing it. *Cancer Chemotherapy Reports. Part 1* **50**, 8, p. 541.

Wilkens, L., Bredt, M., Flemming, P., Klempnauer, J. and Heinrich Kreipe, H. (2000). Differentiation of multicentric origin from intra-organ metastatic spread of hepatocellular carcinomas by comparative genomic hybridization, *The Journal of Pathology* **192**, 1, pp. 43–51.

Winter, G., Rix, U., Carlson, S., Gleixner, K., Grebien, F., Gridling, M., Müller, A., Breitwieser, F., Bilban, M., Colinge, J. *et al.* (2012). Systems-pharmacology dissection of a drug synergy in imatinib-resistant CML, *Nature Chemical Biology* **8**, 11, pp. 905–912.

Wodarz, A. (2005a). Molecular control of cell polarity and asymmetric cell division in Drosophila neuroblasts, *Current Opinion in Cell Biology* **17**, 5, pp. 475–481.

Wodarz, A. and Huttner, W. B. (2003). Asymmetric cell division during neurogenesis in Drosophila and vertebrates, *Mechanisms of Development* **120**, 11, pp. 1297–1309.

Wodarz, D. (2001a). Helper-dependent vs. helper-independent CTL responses in HIV infection: implications for drug therapy and resistance, *Journal of Theoretical Biology* **213**, 3, pp. 447–459.

Wodarz, D. (2001b). Viruses as antitumor weapons defining conditions for tumor remission, *Cancer Research* **61**, 8, pp. 3501–3507.

Wodarz, D. (2003). Gene therapy for killing p53-negative cancer cells: use of replicating versus nonreplicating agents, *Human Gene Therapy* **14**, 2, pp. 153–159.

Wodarz, D. (2004). Computational approaches to study oncolytic virus therapy: insights and challenges, *Gene Therapy and Molecular Biology* **8**, pp. 137–146.

Wodarz, D. (2005b). Somatic evolution of cancer cells, in *Seminars in Cancer Biology*, Vol. 15 (Elsevier Ltd), pp. 421–422.

Wodarz, D. (2009). Use of oncolytic viruses for the eradication of drug-resistant cancer cells, *Journal of The Royal Society Interface* **6**, 31, pp. 179–186.

Wodarz, D. (2010). Heterogeneity in chronic myeloid leukaemia dynamics during imatinib treatment: role of immune responses, *Proceedings of the Royal Society B: Biological Sciences* **277**, 1689, pp. 1875–1880.

Wodarz, D., Boland, C. R., Goel, A. and Komarova, N. L. (2013). Methylation kinetics and CpG-island methylator phenotyope status in colorectal cancer cell lines, *Biology Direct* **8**, 1, p. 14.

Wodarz, D., Hofacre, A., Lau, J. W., Sun, Z., Fan, H. and Komarova, N. L. (2012). Complex spatial dynamics of oncolytic viruses in vitro: mathematical and experimental approaches, *PLoS Computational Biology* **8**, 6, p. e1002547.

Wodarz, D., Iwasa, Y. and Komarova, N. L. (2004). On the emergence of multifocal cancers, *Journal of Carcinogenesis* **3**, 1, p. 13.

Wodarz, D. and Jansen, V. A. (2001). The role of T cell help for anti-viral CTL responses, *Journal of Theoretical Biology* **211**, 4, pp. 419–432.

Wodarz, D. and Jansen, V. A. (2003). A dynamical perspective of CTL cross-priming and regulation: implications for cancer immunology, *Immunology Letters* **86**, 3, pp. 213–227.

Wodarz, D. and Komarova, N. L. (2005). *Computational Biology of Cancer: Lecture Notes and Mathematical Modeling* (World Scientific Publishing Company).

Wodarz, D. and Komarova, N. L. (2009). Towards predictive computational models of oncolytic virus therapy: basis for experimental validation and model selection, *PLoS One* **4**, 1, p. e4271.

Wodarz, D. and Krakauer, D. C. (2001). Genetic instability and the evolution of angiogenic tumor cell lines (review), *Oncology Reports* **8**, 6, pp. 1195–1202.

Wodarz, D. and Nowak, M. A. (1999). Specific therapy regimes could lead to long-term immunological control of HIV, *Proceedings of the National Academy of Sciences* **96**, 25, pp. 14464–14469.

Wojiski, S., Guibal, F. C., Kindler, T., Lee, B. H., Jesneck, J. L., Fabian, A., Tenen, D. G. and Gilliland, D. G. (2009). PML-RARalpha initiates leukemia by conferring properties of self-renewal to committed promyelocytic progenitors, *Leukemia* **23**, 8, pp. 1462–71, doi:10.1038/leu.2009.63.

Wong, H. H., Lemoine, N. R. and Wang, Y. (2010). Oncolytic viruses for cancer therapy: overcoming the obstacles, *Viruses* **2**, 1, pp. 78–106.

Woodford-Richens, K. L., Rowan, A. J., Gorman, P., Halford, S., Bicknell, D. C., Wasan, H. S., Roylance, R. R., Bodmer, W. F. and Tomlinson, I. P. (2001). Smad4 mutations in colorectal cancer probably occur before chromosomal instability, but after divergence of the microsatellite instability pathway, *Proceedings of the National Academy of Sciences U S A* **98**, 17, pp. 9719–23, doi:10.1073/pnas.171321498.

Wu, A., Oh, S., Wiesner, S. M., Ericson, K., Chen, L., Hall, W. A., Champoux, P. E., Low, W. C. and Ohlfest, J. R. (2008). Persistence of CD133+ cells in human and mouse glioma cell lines: detailed characterization of GL261 glioma cells with cancer stem cell-like properties, *Stem Cells Development* **17**, 1, pp. 173–84, doi:10.1089/scd.2007.0133.

Wu, H.-H., Ivkovic, S., Murray, R. C., Jaramillo, S., Lyons, K. M., Johnson, J. E. and Calof, A. L. (2003). Autoregulation of neurogenesis by GDF11, *Neuron* **37**, 2, pp. 197–207.

Yamasaki, K., Toriu, N., Hanakawa, Y., Shirakata, Y., Sayama, K., Takayanagi, A., Ohtsubo, M., Gamou, S., Shimizu, N., Fujii, M., Miyazono, K. and Hashimoto, K. (2003). Keratinocyte growth inhibition by high-dose epidermal growth factor is mediated by transforming growth factor beta autoinduction: a negative feedback mechanism for keratinocyte growth, *Journal of Investigative Dermatology* **120**, 6, pp. 1030–7, doi:10.1046/j.1523-1747.2003.12239.x.

Yatabe, Y., Tavaré, S. and Shibata, D. (2001). Investigating stem cells in human colon by using methylation patterns, *Proceedings of the National Academy of Sciences* **98**, 19, pp. 10839–10844.

Yin, T., Li, L. *et al.* (2006). The stem cell niches in bone, *Journal of Clinical Investigation* **116**, 5, p. 1195.

You, L., Yang, C.-T. and Jablons, D. M. (2000). ONYX-015 works synergistically with chemotherapy in lung cancer cell lines and primary cultures freshly made from lung cancer patients, *Cancer Research* **60**, 4, pp. 1009–1013.

Young, M., Shah, N., Chao, L., Seeliger, M., Milanov, Z., Biggs III, W., Treiber, D., Patel, H., Zarrinkar, P., Lockhart, D. *et al.* (2006). Structure of the kinase domain of an imatinib-resistant ABL mutant in complex with the Aurora kinase inhibitor VX-680, *Cancer Research* **66**, 2, pp. 1007–1014.

Youssef, K. K., Van Keymeulen, A., Lapouge, G., Beck, B., Michaux, C., Achouri, Y., Sotiropoulou, P. A. and Blanpain, C. (2010). Identification of the cell lineage at the origin of basal cell carcinoma, *Nature Cell Biology* **12**, 3, pp. 299–305, doi:10.1038/ncb2031.

Yurgelun, M. B., Goel, A., Hornick, J. L., Sen, A., Turgeon, D. K., Ruffin, M. T., Marcon, N. E., Baron, J. A., Bresalier, R. S., Syngal, S. *et al.* (2012). Microsatellite instability and DNA mismatch repair protein deficiency in lynch syndrome colorectal polyps, *Cancer Prevention Research* **5**, 4, pp. 574–582.

Zhang, Y. V., Cheong, J., Ciapurin, N., McDermitt, D. J. and Tumbar, T. (2009). Distinct self-renewal and differentiation phases in the niche of infrequently dividing hair follicle stem cells, *Cell Stem Cell* **5**, 3, pp. 267–278.

Zheng, Q. (1999). Progress of a half century in the study of the Luria–Delbrück distribution, *Mathematical Biosciences* **162**, 1, pp. 1–32.

Zhong, W. and Chia, W. (2008). Neurogenesis and asymmetric cell division, *Current Opinion in Neurobiology* **18**, 1, pp. 4–11.

Zurakowski, R. and Wodarz, D. (2007). Model-driven approaches for in vitro combination therapy using ONYX-015 replicating oncolytic adenovirus, *Journal of Theoretical Biology* **245**, 1, pp. 1–8.

Index